Network Analysis

and

Synthesis

By the same author

- Electronics Engineering

Network Analysis
and
Synthesis

Prof. O.N. Pandey

JSS Academy of Technical Education,

Noida

Ane Books Pvt. Ltd.

New Delhi ♦ Chennai

Network Analysis and Synthesis

Prof. O.N. Pandey

Pages: xiv+394=408 (PB)

Size: 6.25 x 9.50 inch

Paper Quality (GSM): 80

Published by

Ane Books Pvt. Ltd.

4821, Parwana Bhawan, 1st Floor, 24 Ansari Road, Darya Ganj,
New Delhi - 110 002, Tel.: +91(011) 23276843-44, Fax: +91(011) 23276863
e-mail: kapoor@anebooks.com, Website: www.anebooks.com

Branch Office:

- Avantika Niwas, 1st Floor, 19 Doraiswamy Road, T. Nagar,
  Chennai - 600 017, Tel.: +91(044) 28141554, 28141209
  e-mail: anebookschennai@gmail.com, rathinam@anebooks.com

ISBN : 978-93-8061-809-8

Printed at : R. K. Offset Process, New Delhi.

Dedicated to

Ranjana, Nutan, Ankur,

Nishith, Prachi, Aahan & Aarav

Preface

A study of electric networks based on a system theoretic approach is covered in the book. The excitation-response properties have been used and the major theme is concerned with how a linear network behaves as a signal processor. The description of a linear network by its system function in the frequency domain and its impulse response in the time domain is considered. The transition from network analysis to synthesis is accomplished with relative ease by use of the system function as a unifying link.

This has been written after going through a large number of references. The objective has been to present the matter in simple, straight forward and easy form without loosing any important information or detail.

I am very thankful to JSS Academy of Technical Education, Noida for encouragements leading to such technical contribution.

I appreciate the cooperation and help extended by my wife Mrs. Ranjana Pandey relatives and friends. I also appreciate the efforts and helps extended by Mrs. Amita Rana, Mr. Rahul Gupta and others.

Warmest thanks are also extended to the publishers Mr. Sunil Saxena and his team for the quick and nice printing and get up of the book.

Prof. O.N. Pandey

Contents

1

Fundamentals of Engineering

1.1 INTRODUCTION

Electrical Engineering is closely related to technologies of mechanics, heat, light and sound. For example, electrical motors are used for driving machines such as cranes, electric heaters are used a for making air or water warmer and electric lamps are used for medical and scientific purposes etc. At this stage, it is presumed that the reader has some knowledge of basic mechanics, heat, light and sound.

1.2 SI UNITS

The International System of Units (SI) is the world's most widely used system of measurement. It is based on the seven physical quantities *i.e.*, mass, length, time, electric current, absolute temperature, luminous intensity and amount of substance. All other units are derived units and are related to these basic units by definition. Table 1.1 gives the details of basic SI units.

Table 1.1 *Basic SI units*

Physical quantity	*Symbol*	*Unit*	*Abbreviated unit*
Mass	m	kilogram	kg
Length	l	metre	m
Time	t	second	s
Electric current	I	ampere	A
Temperature (absolute)	T	kelvin	K 0°C = 273 K Interval of 1°C = 1K
Temperature (celsius)	Q	degree celsius	°C
Luminous int ensity		candela	
Amount of substance	m	mole	mol

The capital letters are normally used to represent constant quantities whereas the lower case symbols are used to represent the varying quantity. Table 1.2 gives derived SI units.

Table 1.2 *Derived SI units*

Physical quantity	*Symbol*	*Unit*	*Abbreviation*	*Relationships*
Area	A	square metre	m^2	$A = l \times w$
Value	V	cubic metre	m^3	$V = l \times w \times h$
Velocity	v	metre per second	m/s	$v = \frac{dx}{dt}$
Acceleration	a	meter per second square	m/s^2	$a = \frac{du}{dt}$
Angular velocity	ω	radian per second	rad/s	$\omega = 2\pi f = \frac{2\pi N}{60}$
Force	F	newton	$N = kg \times m/s^2$	$F = ma$
Weight	w	force	9.81 newton	$w = 9.81$ m
Torque or turning moment	T(or M)	newton meter	Nm	$T = Fr$
Work or energy	W	Joule	J	$w = fl$ $= \frac{1}{2} mu^2 = mgh$
Power	P	watt	W	$P = fu = mw$
Rotational speed	N	Revolution per minute	r/min	
Electric energy	W	kilowatt hour	kWh	kWh = 3.6 mega J

1.3 ELECTRIC CHARGE

Electric charge is the excess of negative or positive electricity on a body or in space. If the excess is negative, the body is said to have a negative charge or *vice versa.* Unit of charge is coulomb (c).

1.4 ELECTRONS

An electron is an elementary particle charged with a small and constant quantity of negative electricity. The charge of an electron is 1.602×10^{-19} coulomb (c).

1.5 ELECTROMOTIVE FORCE AND POTENTIAL DIFFERENCE

Electromotive force (*e.m.f*) is the driving influence which causes a current to flow. The energy introduced into a circuit is transferred to the load unit by the transmission system. The energy transferred due to the passage of unit charge between two points in a circuit is known as the potential difference (p.d.). If all the energy is transferred to the load unit, the p.d. across the load unit is equal to the source e.m.f.

The p.d. or e.m.f is measured in volts. The p.d. is also referred as voltage drop or just voltage.

1.6 ELECTRIC CURRENT

Electrons move freely from one energy level to another in the medium (material) in which they exist. The movement of electrons is known as electric current flow. The current flows from a point of high energy level to a point of low level energy. These points of energy levels are known as potential. The point of high potential and low potential are also known as positive (+) and negative (–) potentials respectively.

Mathematically, current is defined as rate of electric charge flow *i.e.*,

$$I = \frac{dQ}{dt} \text{ coulomb/second or Ampere (A).}$$

1.7 CHARGE, CURRENT, VOLTAGE AND POWER RELATIONSHIP

$$Q = It$$

$$I = \frac{Q}{t}$$

$$V = \frac{P}{I}$$

$$P = VI = \frac{W}{t} \cdot \frac{t}{Q} = \frac{W}{Q}$$

where, Q = charge in C

I = current in A

t = time in second

V = p.d. in V

P = power

W = energy.

1.8 RESISTANCE

When a p.d. is applied between two points of a conductor, the flow of current is opposed by a property of the conductor. This property of the conductor is known as resistance of the conductor.

$$R = \frac{V}{I}$$

1.8.1 Resistor

The device which offers resistance in an electric circuit is known as resistor. These are various type of resistors as given in Table 1.3

Table 1.3 *Various type of Resistors and their symbols*

Representation	*Symbols*
Fixed resistor	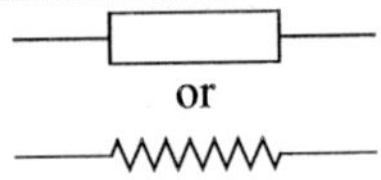
Variable resistor or rheostat	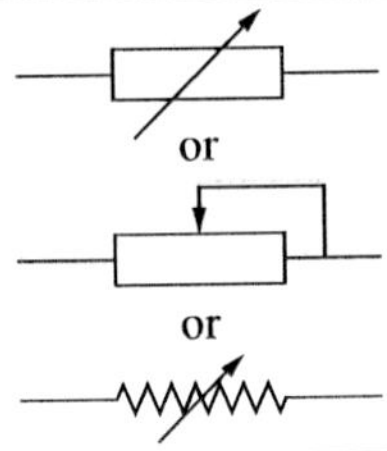
Potentiometer	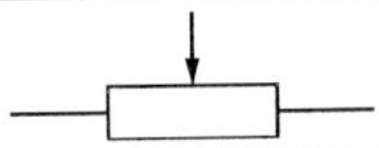

1.8.2 Resistor Coding

The resistors are made from carbon mouldings or metal-oxide film. These are very small in size, therefore, ratings if written on the body of the resistor is very difficult. Hence, in the case of carbon resistors, it is usual to identify the ratings by means of rings painted around the resistors as shown in Fig.1.1. One of the bands is always placed near one of the end of resistor which is taken as first band. The first, second and third bands indicate the resistance value by

Digit	Colour
0	Black
1	Brown
2	Red
3	Orange
4	Yellow
5	Green
6	Blue
7	Violet
8	Grey
9	White
Tolerance	**Colour**
5%	gold
10%	Silver
20%	No colour hand

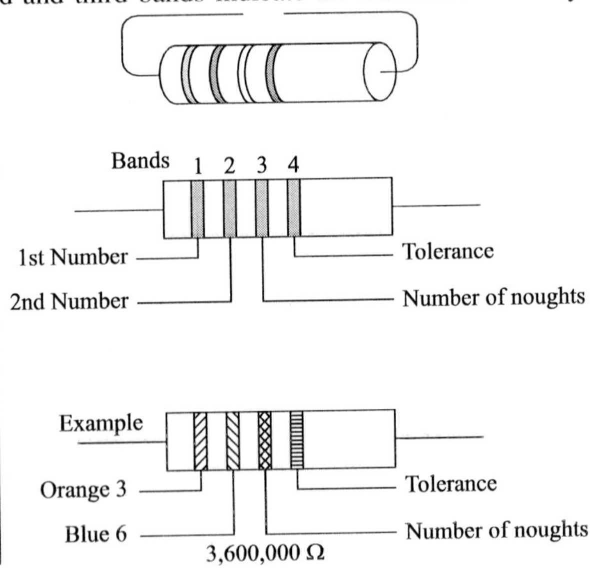

Fig. 1.1. *Colour coding of resistors.*

means of a colour code as given the figure. In order to understand the colour coding, consider the example given in the figure. In this case, the first two bands are orange and blue which give the number 36 will reference to the colour codes. The third band is green which means digit 5 third band value decides, the number of zero to be put after 36. Thus resistance value is 36,00000 or $36 \times 10^5\ \Omega$. The fourth band gives the tolerance of the resistor value. Thus, the resistor value is $36 \times 10^5\ \Omega \pm 10\%$. A list of readily available standard value resistors are given in Table 1.3. All are available with 5% tolerance while those in bold type are available with 10% tolerance.

Table 1.4 *Standard values of variable resistors.*

Ohms (Ω)					*Kilohms (kΩ)*		*Megaohms (MΩ)*	
0.10	1.0	10	100	1000	10	100	1.0	10.0
0.11	1.1	11	110	1100	11	110	1.1	11.0
0.12	1.2	12	120	1200	12	120	1.2	12.0
0.13	1.3	13	130	1300	13	130	1.3	13.0
0.15	1.5	15	150	1500	15	150	1.5	15.0
0.16	1.6	16	160	1600	16	160	1.6	16.0
0.18	1.8	18	180	1800	18	180	1.8	18.0
0.20	2.0	20	200	2000	20	200	2.0	20.0
0.22	2.2	22	220	2200	22	220	2.2	22.0
0.24	2.4	24	240	2400	24	240	2.4	
0.27	2.7	27	270	2700	27	270	2.7	
0.30	3.0	30	300	3000	30	300	3.0	
0.33	3.3	33	330	3300	33	330	3.3	
0.36	3.6	36	360	3600	36	360	3.6	
0.39	3.9	39	390	3900	39	390	3.9	
0.43	4.3	43	430	4300	43	430	4.3	
0.47	4.7	47	470	4700	47	470	4.7	
0.51	5.1	51	510	5100	51	510	5.1	
0.56	5.6	56	560	5600	56	560	5.6	
0.62	6.2	62	620	6200	62	620	6.2	
0.68	6.8	68	680	6800	68	680	6.8	
0.75	7.5	75	750	7500	75	750	7.5	
0.82	8.2	82	820	8200	82	820	8.2	
0.91	9.1	91	910	9100	91	910	9.1	

Some resistors such as metal-oxide resistors are so small that the colour coding is difficult to use. In such cases a letter code is used which is given as explained in Table 1.5.

Table 1.5 *Resistor values in letter codes.*

Resistance	*Meshing*
0.47 Ω	R47
4.7 Ω	4R7
47 Ω	47R
470 Ω	470R
4.7 kΩ	4K7
47 kΩ	47k
4.7 MΩ	4M7

1.9 CONDUCTORS AND INSULATORS

Non-metallic materials permit very little current to flow as compared to metallic materials. These can be characterised as the one which permit current flow and the other one which does not permit current flow. The materials which permit current flow are known as conductors and the materials which do not permit current flow are known as insulators. Table 1.6 gives some of the materials.

Table 1.6 *Conductors and insulators.*

Conductors	*Insulators*
	Glass
Copper	Rubber
Aluminium	Plastic
Silver	Air
Platinum	Varnish
Platinum	Paper
Bronze	Wood
Gold	Mica
	Ceramics

1.10 INDUCTANCE

It is the property of a device which creates capability of holding magnetic energy, the device which creates such capability is known as inductance. Inductance symbol is L and unit is Henry. Circuit symbol of an inductor is given in Fig. 1.2.

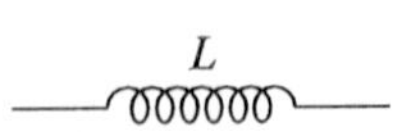

Fig. 1.2. *Circuit symbol of an inductor.*

The relationship between steady state current through an inductor and voltage across it is given by:

$$I = \frac{V}{L}$$

Energy stored in an inductor is given by:

$$W = \frac{1}{2} LI^2 \text{ Joules.}$$

1.11 CAPACITANCE

Capacitance is a property of a device which is capable of holding electric charge. Symbol of capacitance is C and unit is Farad. The device which stores electric charge is known as capacitor. Circuit symbols of a capacitor is shown in Fig .1.3.

(a) International symbol *(b) Symbols used in old diagrams*

Fig. 1.3. *Circuit symbol of a capacitor.*

The relationship between steady state voltage across a capacitor or charge on the plates of capacitor is given by:

$$Q = CV \text{ coulombs}$$

Energy stored in a capacitor is given by:

$$W = \frac{1}{2} CV^2 \text{ Joules.}$$

1.12 ELECTRICAL UNITS AND PREFIXES

Table 1.7 SI units of electrical parameters and Table 1.8 gives letter prefixes to the unit symbol.

Table 1.7 *SI units of electrical parameters*

Physical quantity	*Symbol*	*Unit*	*Abbreviated unit*
Electric charge	Q	coulomb	C
Electric potential	V	volt	V
Current	I	Ampere	A
Resistance	R	ohm	Ω
Capacitance	C	Farad	F
Inductance	L	Henry	H

Table 1.8 *Letter prefixes*

Multiplication factor	*Letter prefix*	*Multiplication factor*	*Letter prefix*
10^{18}	E (exa)	10^{-1}	d (deci)
10^{15}	P (peta)	10^{-2}	c (centi)
10^{12}	T (tera)	10^{-3}	m (milli)
10^{9}	G (giga)	10^{-6}	μ (micro)
10^{6}	M (mega)	10^{-9}	n (nano)
10^{3}	k (kilo)	10^{-12}	p (pico)
		10^{-15}	f (femto)
		10^{-18}	a (atto)

SUMMARY

1. **SI units:** International System of Units are kilogram (mass), metre (length), second (time), ampere (current), kelvin (temperature) and candela (luminous itensity and amount of substance (mol).
2. **Current:** It is the rate of electric charge flow in a circuit.
3. **Electric charge:** It is the excess of negative or positive electricity on a body or in space.
4. **Electron:** It is an elementary particle charged with a small and constant quantity of electricity.
5. **E.M.F./P.D:** Electromotive force (e.m.f) is the driving influence which causes a current to flow. The potential difference (p.d.) is the voltage which is same as e.m.f applied.
6. **Resistance:** It is a property of a device which opposes the current flow.
7. **Conductors and insulators:** The materials which permit current flow are known as conductors whereas materials which do not permit current flow are known as insulators.
8. **Inductance:** It is the property of a device which creates capability of holding magnetic energy.
9. **Capacitance:** It is the property of a device which creates capability of holding electric charge.
10. **Important formulae**

Electric charge, $Q = It$ coulombs

Voltage, $V = P/I = \frac{W}{Q} = IR$ volts

Power, $P = VI = I^2R = \frac{V^2}{R}$ watts.

$$I = \frac{V}{R} \quad \text{Ampere}$$

$$W = \frac{1}{2}LI^2 \quad \text{Joule}$$

$$Q = CV \quad \text{Coulomb.}$$

EXERCISES

1.1. What do you understand by SI units ?

1.2. What is an electron and its electric charge ?

1.3. What do you understand by e.m.f and p.d? Explain.

1.4. What is resistance ? Explain colour coding of a resistor.

1.5. Explain inductance and its relationship with the energy stored in an inductor.

1.6. What is capacitance and its relationship with the energy stored in a capacitor?

❑❑❑

Signals for Analysis, DC Circuits and AC Circuits

2.1 SIGNALS FOR ANALYSIS

Systems having properties of linearity and time invariance are known as LTI systems. LTI feature allows system to be modelled *i.e.*, expressed in mathematical equations and superposition theorem holds good. This helps in analysis of systems. The system analysis is done by studying output responses of the systems when various types of signals are input. These signals are functions of time, therefore, these are also known as functions. Various types of signals/functions are described here.

2.1.1 Continuous-Time and Discrete-Time Signals

A continuous-time signal is one that is defined for each and every instant of time within a given range, *see* Fig. 2.1

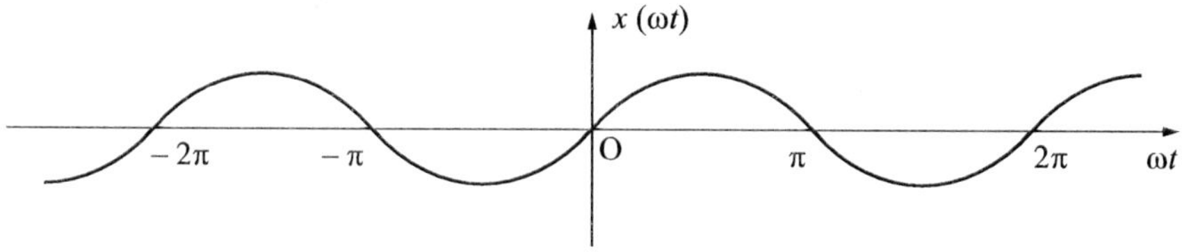

Fig. 2.1. *Continuous signal* $x(t) = A \sin \omega t$

A discrete-time signal is one that is defined at a regular interval of time, *see* Fig. 2.2.

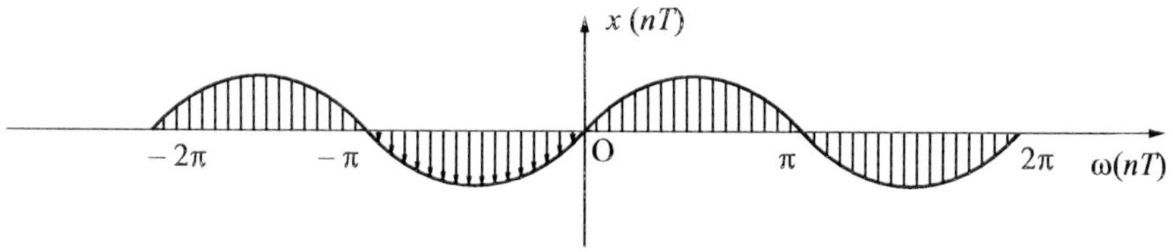

Fig. 2.2. *Discrete-time signal* $x(nT) = A \sin \left(\frac{2\pi n}{T}\right)$

2.1.2 Even and Odd Signals

A signal $x(t)$ is an even signal if it is identical to its time-reversed counter part *i.e.*,

$$x(-t) = x(t)$$

2.1.3 Periodic and Unperiodic Signals

A signal $x(t)$ is a periodic function if:

$$x(t + T) = x(t) \text{ for } - t < \infty$$

2.1.4 Singularity Signals/Functions

These are some standard signals which are easily realisable.

2.1.4.1 Step Function

A step function $f(t)$ (Fig. 2.3) is defined as :

$$f(t) = \begin{cases} 0 & \text{for } t < 0 \\ k & \text{for } t > 0 \end{cases}$$

Fig. 2.3. *A step function*

A delayed step function (Fig. 2.4) is defined as:

$$(t - t_0) = \begin{cases} 0 & \text{for } t < t_0 \\ k & \text{for } t > t_0 \end{cases}$$

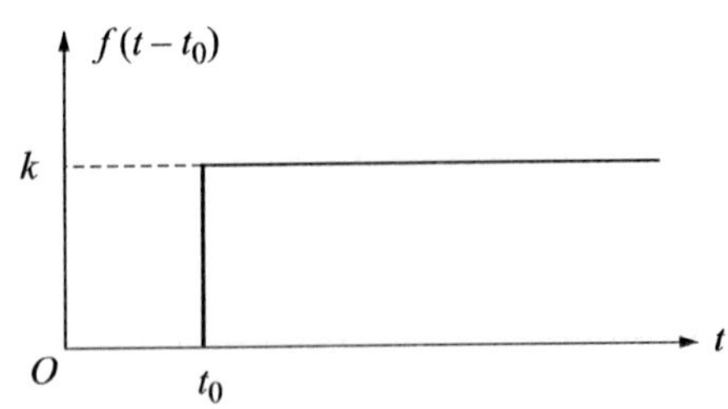

Fig. 2.4. *A delayed step function*

For a unit step function $k = 1$

2.1.4.2 Impulse Function

An impulse function (Fig. 2.5) is defined as :

$$s(t) = \begin{cases} 0 & \text{for } t \neq 0 \\ k & \text{for } t = 0 \end{cases}$$

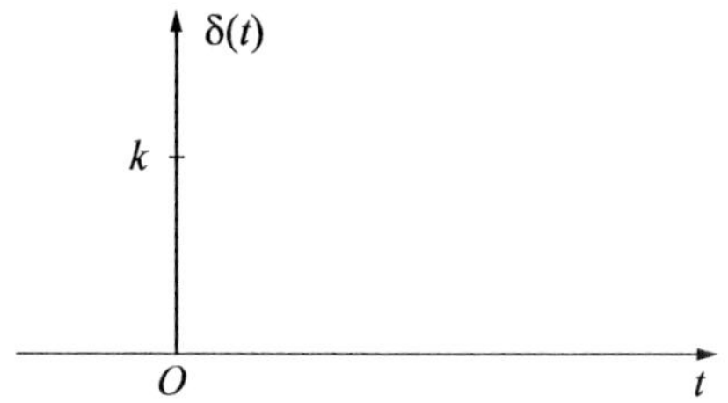

Fig. 2.5. *An impulse function*

Area of impulse function is *k*, which is obtained over an infinitesimal interval around $t = 0$. It $k = 1$, it is known as unit impulse function.

2.1.4.3 Ramp Function

Ramp function $f(t)$ (Fig. 2.6) is defined as:

$$f(t) = \begin{cases} 0 & \text{for } t < 0 \\ kt & \text{for } t \geq 0 \end{cases}$$

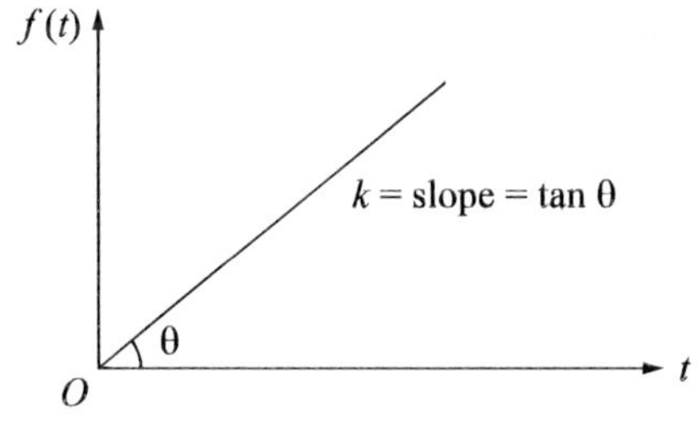

Fig. 2.6. *A ramp function*

For unit ramp function, $k = 1$

2.1.5 Exponential Function

Exponential function $f(t)$ (Fig. 2.7) is defined as:

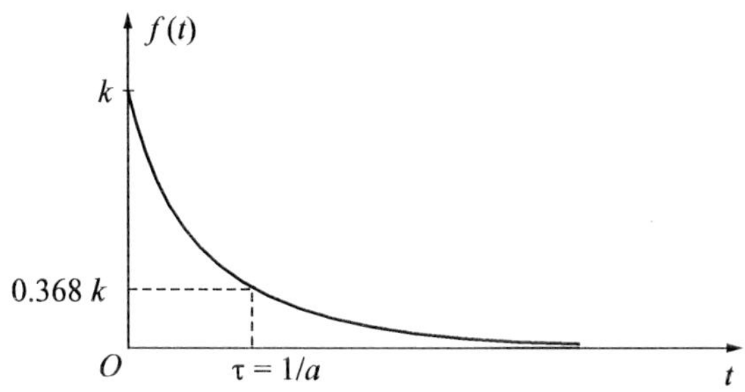

Fig. 2.7. *An exponential function*

$$f(t) = \begin{cases} 0 & \text{for } t \le 0 \\ ke^{-at} & \text{for } t \ge 0 \end{cases}$$

where, k and a are real constants.

2.1.6 Sinusoidal Function

Sinusoidal function $f(t)$ (Fig. (2.8) is defined as:

$$f(t) = \begin{cases} 0 & \text{for } t < 0 \\ A_m \sin \omega t & \text{for } t \ge 0 \end{cases}$$

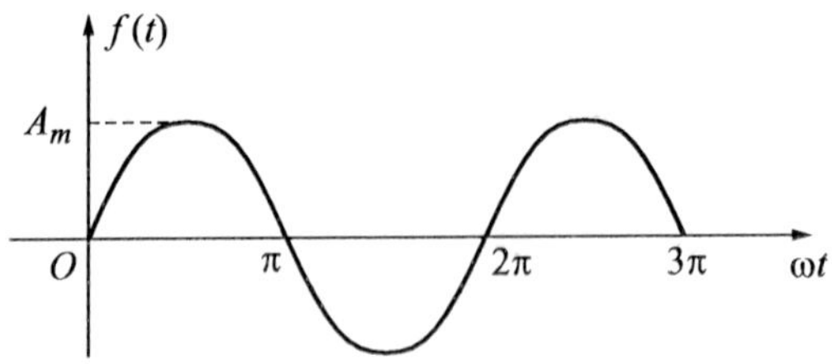

Fig. 2.8. *A sinusoidal function*

2.1.7 Gate Function

Unit gate function $f(t)$ (Fig. 2.9) is defined as:

$$f(t) = u(t - a) - u(t - b)$$

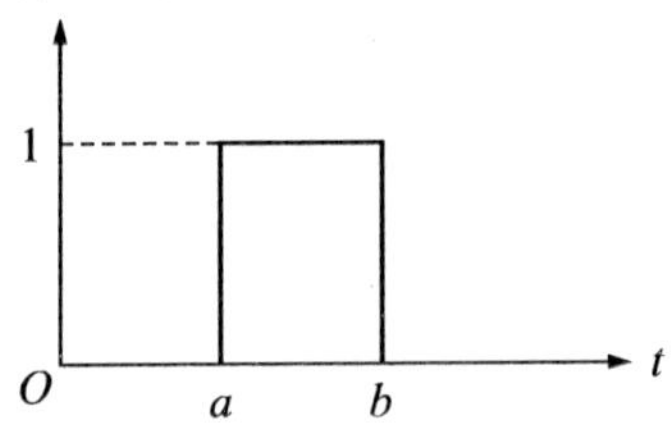

Fig. 2.9. *A gate function*

2.2 DC CIRCUITS

Voltages and currents have transient phenomena insofar as they are dependent upon conditions and change with time in DC circuits. Normally, transient state is for short duration as system is expected to reach steady-state unless it is unstable.

2.2.1 Response of Source-free Circuits

Consider a source free RL circuit of Fig. 2.10.

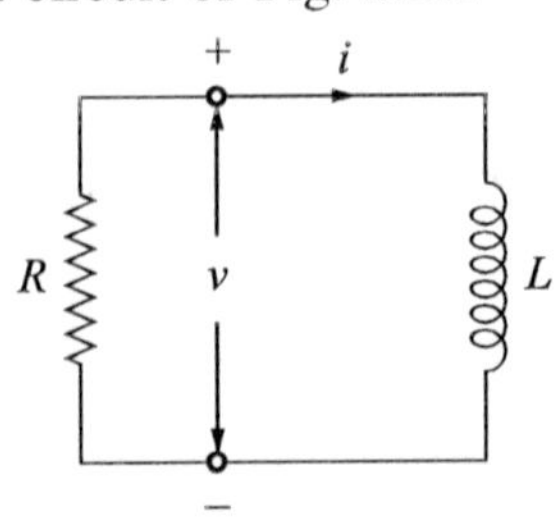

Fig. 2.10. *Source free RL circuit*

KVL gives:

$$L\frac{di}{dt} + Ri = 0$$

or
$$\int_{I_0}^{i} \frac{di}{i} = -\int_{o}^{t} \frac{R}{L}\, dt$$

or
$$i = I_0\, e^{-(R/L)t}$$

Now, consider a source-free LC circuit as shown in Fig. 2.11.

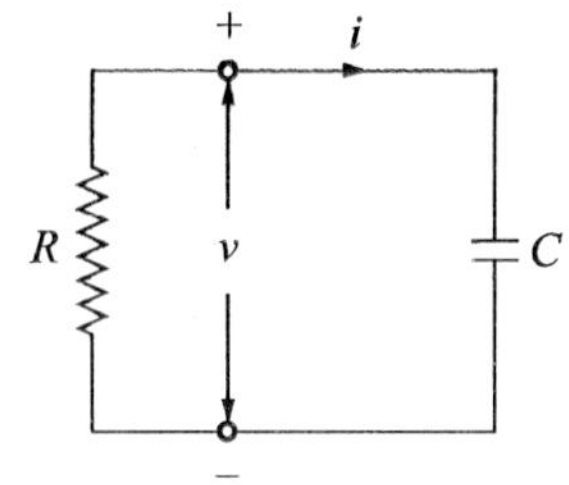

Fig. 2.11. *Source-free RC circuit KVL*

$$C\frac{dv}{dt} + \frac{v}{R} = 0$$

The solution above equation gives :

$$v = V_0\, e^{-t/RC}$$

2.2.2 Complete Response of Circuits

When ac or dc sources are present, the response of a circuit has two components response and steady state response.

2.2.2.1 RL Circuit

Consider a circuit with voltage source as shown in Fig. 2.12.

The voltage V is applied at $t = 0$ by closing the switch S. Consider zero initial conditions *i.e.* $i = 0$ at $t = 0$.

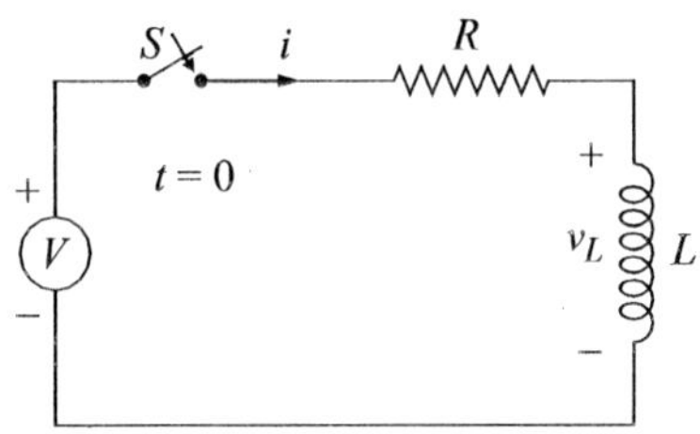

Fig. 2.12. *R_L circuit with voltage source*

KVL gives:

$$L\frac{di}{dt} + Ri = V$$

The solution of the above equation as given by:

$$i = \frac{V}{R}\left(1 - e^{-(R/L)t}\right)$$

or
$$i = \frac{V}{R} - \frac{V}{R}\, e^{-\left(\frac{R}{L}\right)t}$$
$$= i_t + i_s$$

i.e. transient plus steady static components

The response is drawn in Fig. 2.13.

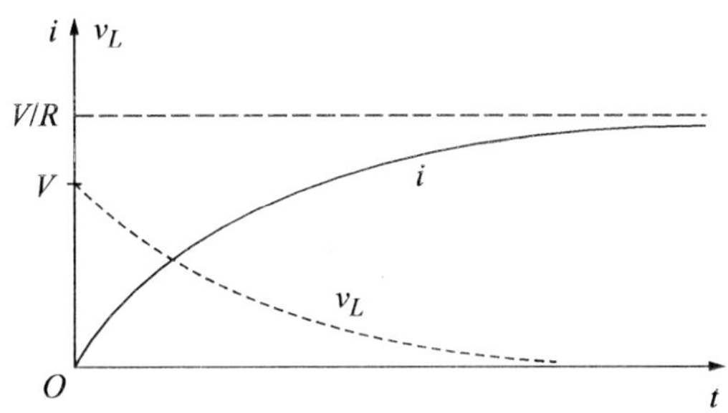

Fig. 2.13. *Response of RL circuit with source voltage*

2.2.2.2 RC Circuit

Now, consider a *RC* circuit with a voltage source as shown in Fig. 2.14.

The switch *S* is closed at $t = 0$ and

$$v = 0 \text{ at } t = 0.$$

KVL gives:

$$Ri + \frac{1}{C}\int_0^t i\,dt = V \text{ for } t > 0$$

We know that:

$$i = \frac{dq}{dt}$$

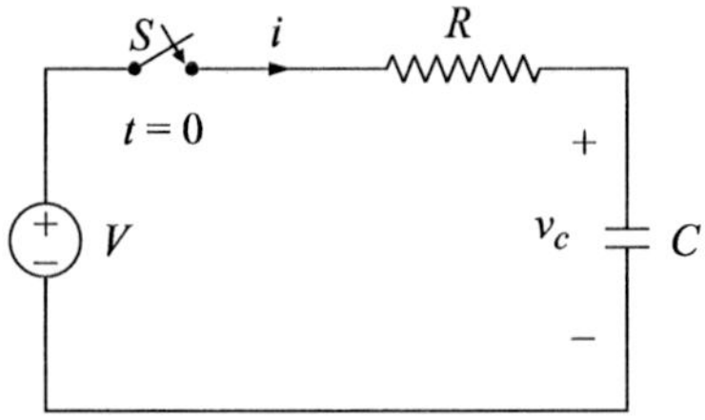

Fig. 2.14. *RC circuit with voltage source*

Hence, the equation can be written as;

$$R\frac{dq}{dt} + \frac{1}{c}q = V$$

The solution of above equation gives:

Now, $$i = \frac{dq}{dt} = \frac{V}{R}e^{-t/RC}$$

and $$v_c = \frac{q}{c} = V(1 - e^{-t/RC})$$

The response is drawn in Fig. 2.15.

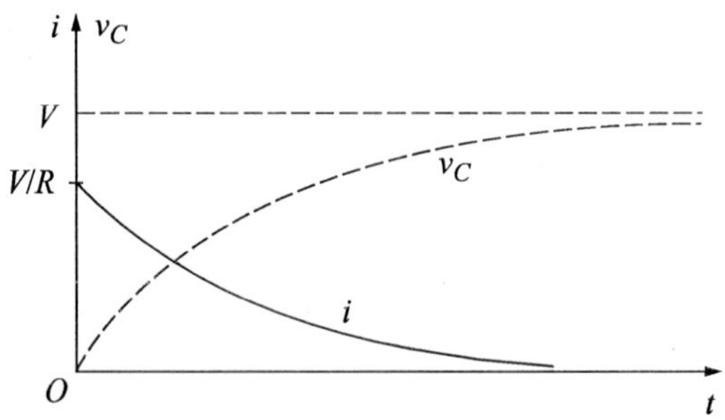

Fig. 2.15. *Response of RC circuit with voltage source*

2.2.2.3 RLC Circuit

Now, consider RLC circuit with a voltage source as shown in Fig. 2.16.

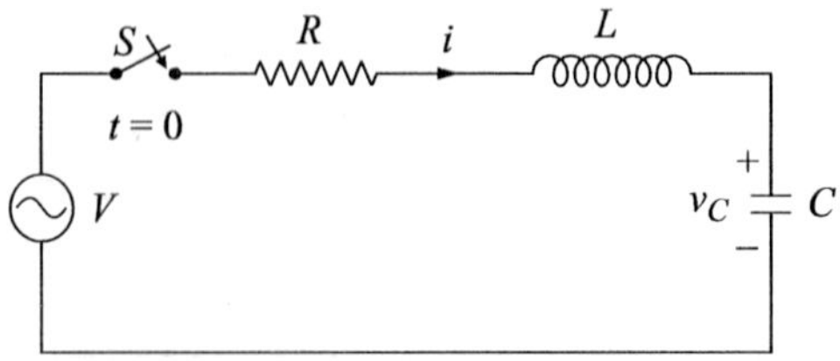

Fig. 2.16. *RLC circuit with voltage source*

By KVL, we get :

$$L\frac{di}{dt} + Ri + \frac{1}{c}\int_0^t idt = V$$

or
$$L\frac{d^2i}{dt^2} + R\frac{di}{dt} + \frac{1}{c}i = 0$$

The solution of above equation gives :

$$i = A_1 e^{s_1 t} + A_2 e^{s_2 t}$$

where
$$s_1 = -\frac{R}{2L} + \sqrt{\left(\frac{R}{2L}\right)^2 - \frac{1}{LC}} = -2 + \beta$$

$$s_2 = -\frac{R}{2L} - \sqrt{\left(\frac{R}{2L}\right)^2 - \frac{1}{LC}} = -2 - \beta$$

The constants A_1 and A_2 can be found out by the following initial conditions:

$$i(0) = 0 \text{ and } L\frac{di}{dt}\bigg|_{t=0} = V$$

Thus,
$$i = A_1\, e^{(-\alpha + \beta)t} + A_2\,{}^{(-\alpha - \beta)t}$$
$$= (A_1 + A_2)\, e^{-\alpha t},\ A_1\, e^{\beta t} + A_2\, e^{-\beta t}$$

Resonant frequency, $\omega_0 = \dfrac{1}{\sqrt{LC}}$

$\therefore$
$$\beta = \sqrt{\alpha^2 - \omega_0^2}$$

The response of the *RLC* circuit is given in Fig. 2.17

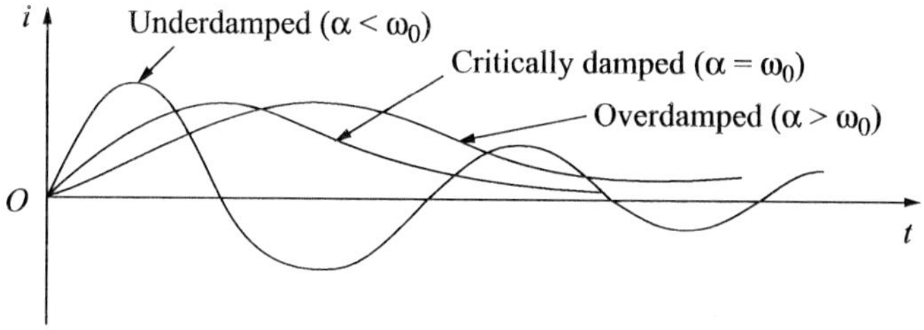

Fig. 2.17. *Response of an RLC circuit with voltage source*

2.3 AC CIRCUITS

Many electrical networks are excited by ac sources. For example, electric power network containing power generating stations, transmission lines, substations, etc. The ac power supplies are basically sinusoidal in nature, therefore, sinusoidal voltages and, currents are considered.

Voltage supply, $v = V_m \sin \omega t$

where V_m = maximum value or amplitude of the sinusoidal wave (*volts*)

ω = angular velocity/frequency (rad/sec).

Also, frequency, $f = \frac{\omega}{2\pi}$ (Hz) or $\omega = 2\pi f$

Time interval, $T = \frac{1}{f} = \frac{2\pi}{\omega}$ sec.

1 Hz = 1 cycle/second.

The sinusoidal voltage supply v is shown graphically in Fig. 2.18.

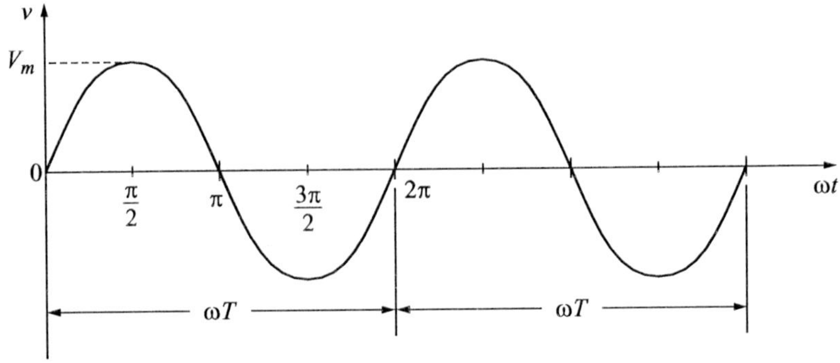

Fig. 2.18. *Sinusoidal voltage*

Similarly, current i will also be sinusoidal and can be expressed as :

$$i = I_m \sin(\omega t + \theta)$$

where I_m = maximum value or amplitude

θ = phase angle between v and i *i.e.* i leads v by θ or v lags i by θ

The relative plot of voltage and current are shown in Fig. 2.19.

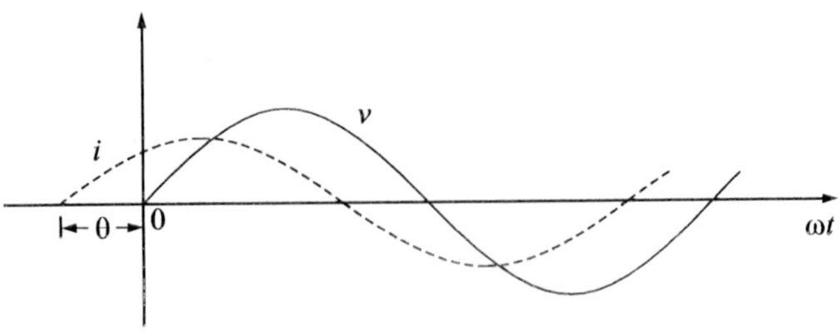

Fig. 2.19. *Sinusoidal current leading voltage v by angle* θ

Euler's formula for sinusoids is:

$$e^{j\omega t} = \cos \omega t + j \sin \omega t$$

i.e., $\cos \omega t = Re(e^{j\omega t}), \ \sin \omega t = I_m(e^{j\omega t})$

However, in practice *Re* and I_m are not written *i.e.*,

$$v = V_m \sin \omega t = V_m\, e^{j\omega t}$$

and $$i = I_m \sin (\omega t + \theta) = I_m\, e^{j(\omega t + \theta)}$$

The above can also be written in the polar, vector or phasor form:

$$\boldsymbol{V} = V_m \underline{|0^\circ}$$

$$\boldsymbol{I} = I_m \underline{|\theta}$$

V and *I* are bold face letters emphasizing that these are complex vectors.

Thus, voltage has amplitude $|V| = V_m$ and angle $\theta = 0$, whereas current has amplitude $|I| = I_m$ and angle θ ant CCW. The phases diagram is given in Fig. 2.20.

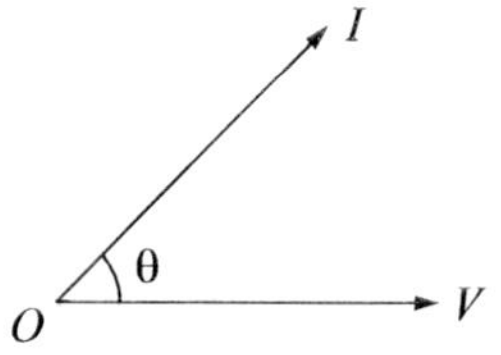

Fig. 2.20. *Voltage and current in phasor form*

A quantity at an instant of time is known as instantaneous value. Average value of a sinusoid is zero and yet any times the average value is that of half cycle in such cases. Root-mean-square (RMS) value of a sinusoid is used quite frequently. It is defined as :

$$V = \sqrt{\frac{1}{T}\int_0^t v^2(t)dt}$$

or $$V = \frac{V_m}{\sqrt{2}}$$

and $$I = \frac{I_m}{\sqrt{2}}$$

2.3.1 Steady-State Responses

Consider instantaneous values and RMS values related with R, L and C elements in an circuit is as follows :

Element circuit		*Instantaneous*	*Steady-state*
Resistor	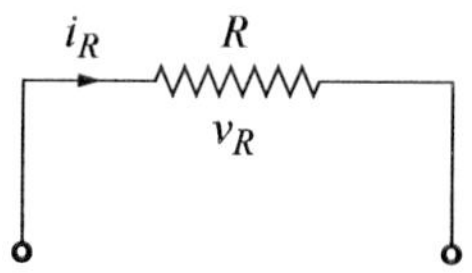	$v_R = Ri_R$	$V_R = RI_R$

Inductor 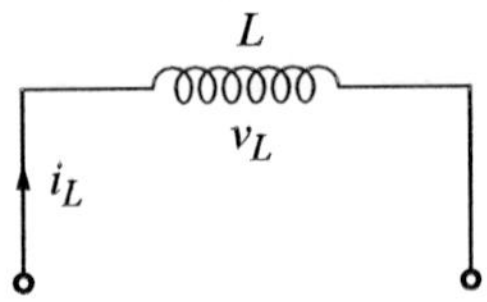

$$v_L = L\frac{d_{iL}}{dt} \qquad V_L = j\omega L\, J_L$$

Capacitor 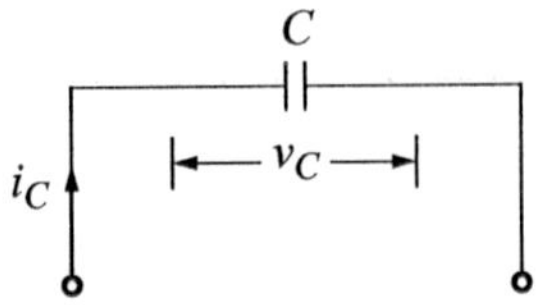

$$v_c = \frac{1}{c}\int i_c dt \qquad V_C = \frac{1}{j\omega c} Ic$$

The phasor plots of R, L, and C circuits is given as follows. (Note that $\omega_L = X_L$ and $\frac{1}{\omega c} = X_C$):

(*a*) Resistor 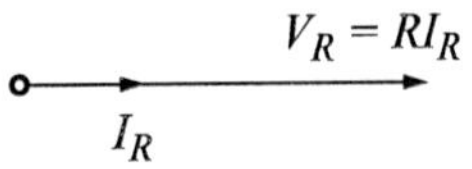

(voltage and current in phase)

(*b*) Inductor 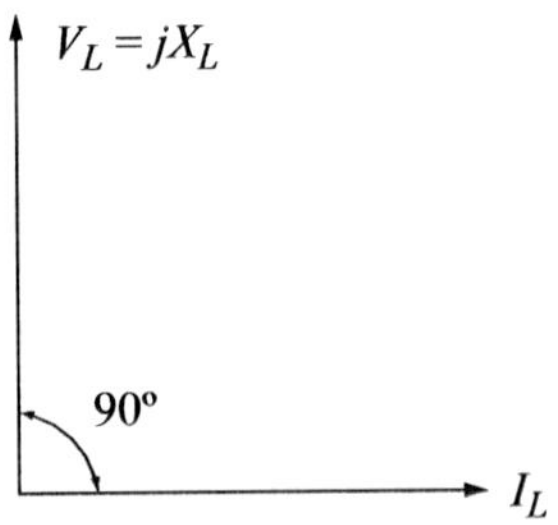

(current lags voltage by 90°)

(*c*) Capacitor 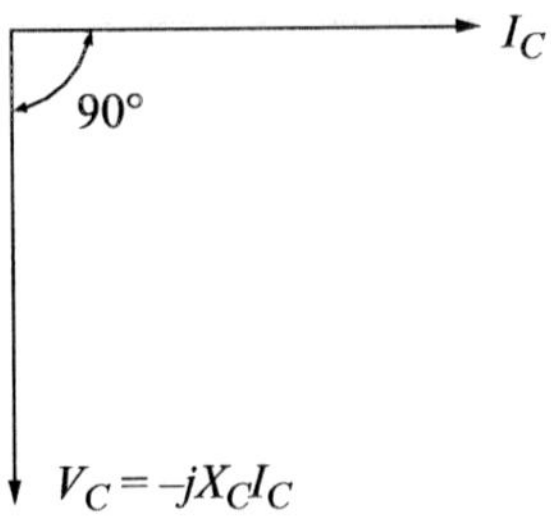

(current leads the voltage by 90°)

2.3.2 Impedance of *R*, *L*, and *C* Series Circuits

Some circuits are considered to demonstrate impedance values as follows:

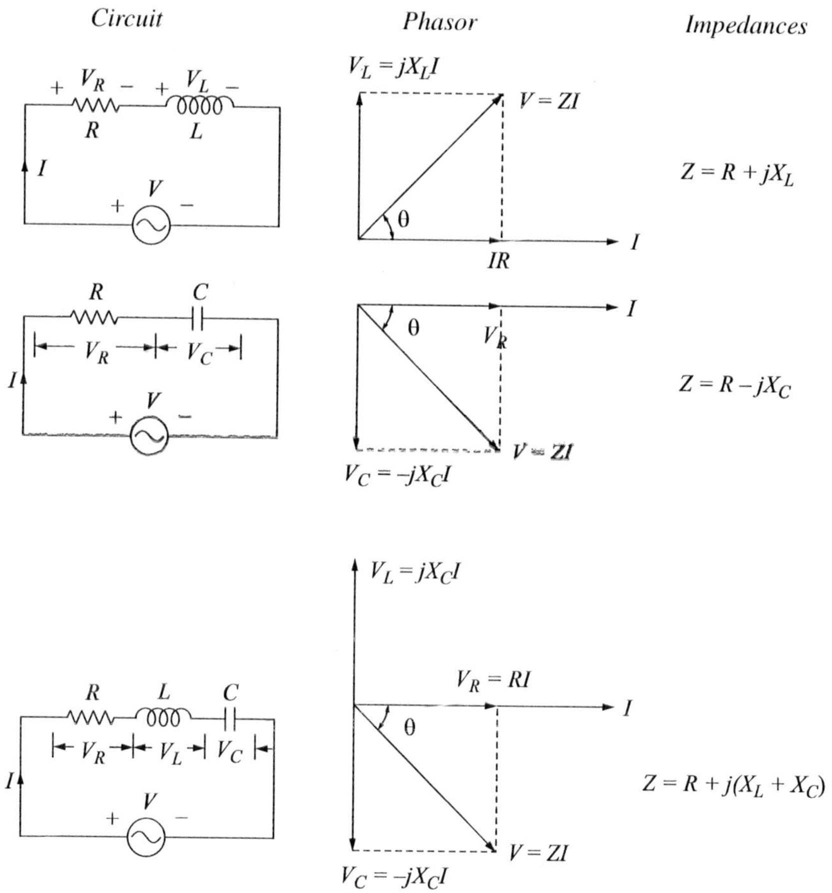

Fig. 2.21. *Impedances of R, L and C series circuits*

2.3.3 Admittance of Parallel *R*, *L* and *C* Circuits

Some general circuits are considered and admittance is as follows :

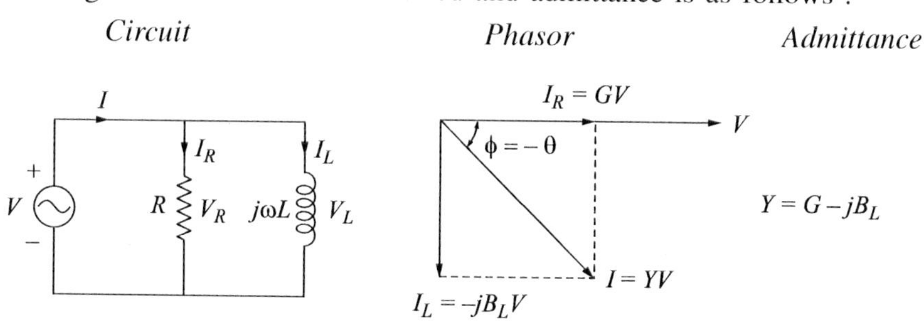

Circuit *Phasor* *Admittance*

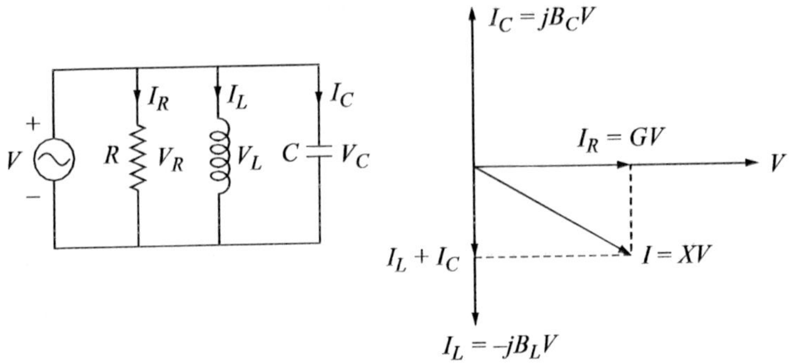

$Y = G + j$

where
$$V_R = V_L = V$$
$$I = I_R + I_L$$
$$G = \frac{1}{R}$$
$$B_L = \frac{1}{X_L}$$
$$IL = -jB_L$$

2.3.4 Power in AC Circuits

Consider that $v = V_m \sin \omega t$ and $i = I_m \sin (\omega t - \theta)$ are the steady-state voltage and current in a circuit.

Instantaneous power, $p = v_i = V_m I_m \sin \omega t \sin (\omega t - \theta)$

$$\text{Average power, } P = \frac{1}{T}\int_0^t pdt = \frac{1}{2}\frac{V_m I_m}{T}\int_0^t \sin \omega t \sin (\omega t - \theta) dt$$

$$= \frac{1}{2} V_m I_m \cos \theta = V_I \cos \theta$$

or
$$P = V_I \cos \theta \text{ watts}$$

where V = rms value of voltage (volts)

I = rms value of current (amperes)

q = Phase angle between voltage and current

T = Time period of the sinusoidal waveforms (seconds)

P is real power and apparent power *i.e.*,

Reactive power, $Q = VI \sin \theta$ voltampere reactive (var)

SUMMARY

1. **Signals for Analysis:** Characteristics of any system can be obtained by observing its response (output) against an excitation (input). These input signals can be in the form of step impulse, ramp, exponential, Sinusoidal, etc.
2. **DC Circuits:** The circuits have transients when dc supply is switch. These are due to the fact that the inductive effect does not allow sudden change in current and the capacitive effect does not allow sudden change in current and the capacitive effect does not allow sudden change of voltage across it. The transients die down with passage of time and steady state condition is achieved.
3. **AC Circuits:** AC supply and distribution is sinusoidal and has 50 Hz frequency. The voltage and current values are RMS values. The voltage and current waves may lead or lag each other depending on the resistive, inductive and capacitive components of the circuit. This lead or large is known as phase angle. The inputs and outputs can be represented as a vector which is known as phasor. AC circuit analysis can be done by using complex algebra or phasor diagrams.
4. **Formulae:** Instantaneous voltage,

$$v(t) = V_m \sin \omega t$$

V_m = maximum instantaneous value of the voltage wave (volts)

ω = angular frequency (rad/sec) = $2\pi f$

f = frequency in Hz (cycles/sec.)

$$T = \frac{1}{f} = \text{time period of the wave.}$$

Inductive reactance, $X_L = \omega L$

Capacitive reactance, $X_C = \dfrac{1}{\omega c}$

Impedance of RLC series network, $Z = R + jX_L - jX_C$

Real power, $P = VI \cos \theta$

Reactive power, $Q = VI \sin \theta$

θ = Phase angle between voltage and current of the element.

EXERCISES

2.1. What do you understand by a Linear Time Invariant (LTI) system? Explain.

2.2. Describe step, impulse and ramp functions.

2.3. Explain exponential, gate function and sinusoidal function.

2.4. What is a transtent in a dc circuit? Explain with an example.

2.5. Describe and plot transients in a series RLC circuit.

2.6. What is phase angle? Explain with an example.

2.7. What are inductive and capacitive reactances?

2.8. What is an impedance? Find out impedance of a series RLC circuit.

❑❑❑

3

Graph Theory

3.1 INTRODUCTION

Graph theory is an analysis technique for a system using matrices representing the system. This technique is very helpful in the analysis of complicated electrical networks. Graph is an interconnection of line segments representing elements of the system.

3.1.1 Network

Any arrangement of the various electrical energy sources along with the different circuit elements is known as an electrical network. Fig. 3.1 shows an electrical network.

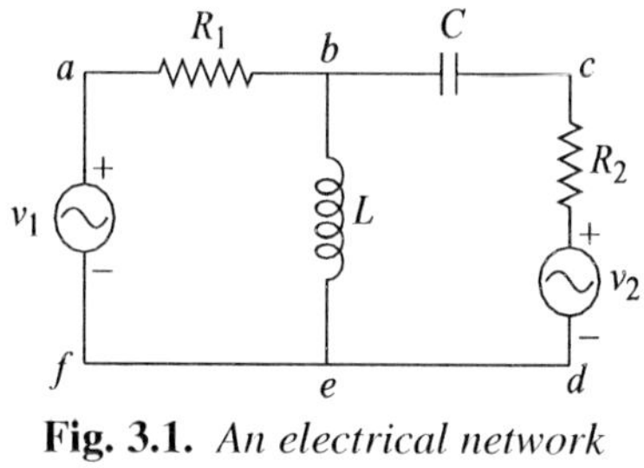

Fig. 3.1. *An electrical network*

3.1.2 Types of Networks

(*i*) **Linear Network:** Behaviour of circuit elements are not dependent on changes in time, voltage, etc.

(*ii*) **Non-linear Network:** Behaviour of circuit elements are dependent on changes in time, voltage, etc.

(*iii*) **Bilateral Network:** Circuit behaviour is same irrespective of the direction of current through various elements.

(*iv*) **Unilateral Network:** Circuit behaviour is dependent on the direction of the current through various elements. Circuits containing diodes, which allow flow of current only in one direction is an unilateral network.

(*v*) **Active Network:** Circuit contains at least one energy source *i.e.*, voltage or current source. Transistors, Op-amps, batteries, signal generators active elements of an active network.

(*vi*) **DC Network:** It consists of dc sources.

(*vii*) **AC Network:** It consists of ac sources.

(*viii*) **Lumped Network:** Circuit elements are physically separable.

(*ix*) **Distributed Network:** Circuit elements like resistance, inductance, etc are not physically separable. Transmission line network is a distributed network.

3.2 GRAPH OF A NETWORK

Given an electrical network, you can easily make its graph. Just replace its each element by a line segment. It is a network that contains an energy source, replace it by its internal impedance.

The internal impedance of a voltage source is zero and to be replaced by short circuit and that of an ideal current source is infinite and, hence to be replaced by an open circuit. A linear graph is defined as a collection of points called nodes, and line segments called branches, the nodes being joined together by the branches. A network is shown in Fig. 3.2 (*a*) and its graph shown in Figs. 3.2 (*b*) and 3.2 (*c*).

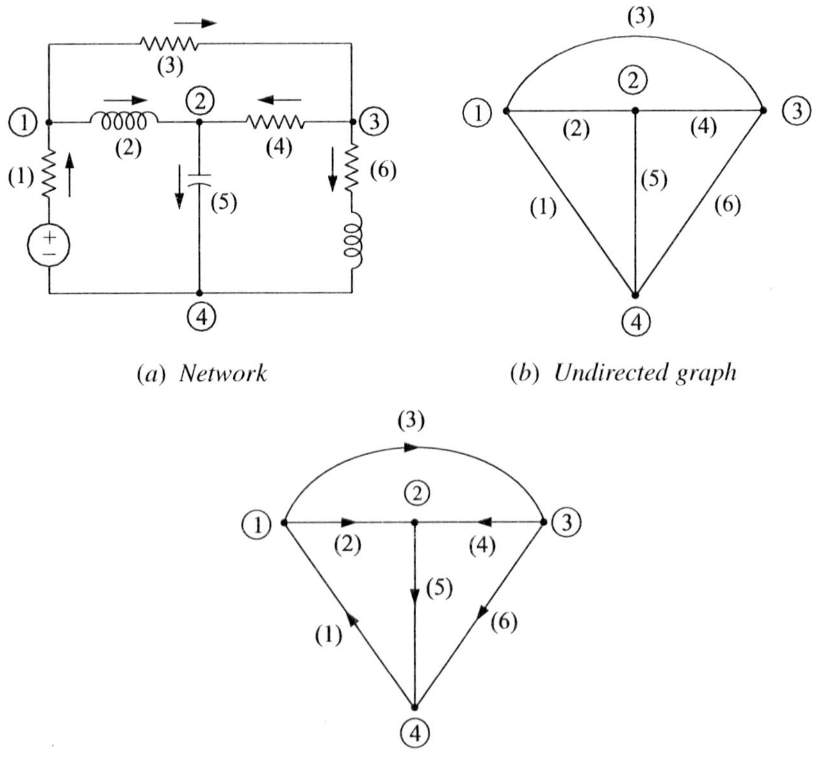

(*a*) *Network* (*b*) *Undirected graph*

(*c*) *Directed graph*

Fig. 3.2. *Network and its graph*

The nodes and branches are numbered. The graph has 6 branches and four nodes. This graph is known as undirected graph Fig. 3.2 (*b*). If each branch carries an arrow to indicate orientation, it is directed or oriented graph Fig. 3.2 (*c*).

Branches whose ends fall on a node are said to be incident at the node. Branches 2, 4, and 5 are incident at node 2.

Rank of graph is given by $n - 1$, where n is the number of nodes of the graph.

3.3 SUBGRAPH

A subgraph of a graph *G*, is a graph whose branches and nodes form a subset of the branch and nodes of a graph *G*. A proper subgraph does not contain all the branches and nodes of the graph *G*. If the subgraph contains all the nodes of the graph *G*, it is called spanning subgraph of *G*. We can get a proper subgraph of a graph *G* by removing branches and nodes from the original graph *G*. For the graph as shown in the Fig. 3.2 (*c*) the spanning subgraphs are as shown in Fig. 3.3 (*a*). The subgraphs of graph *G* are as shown in 3.3 (*b*).

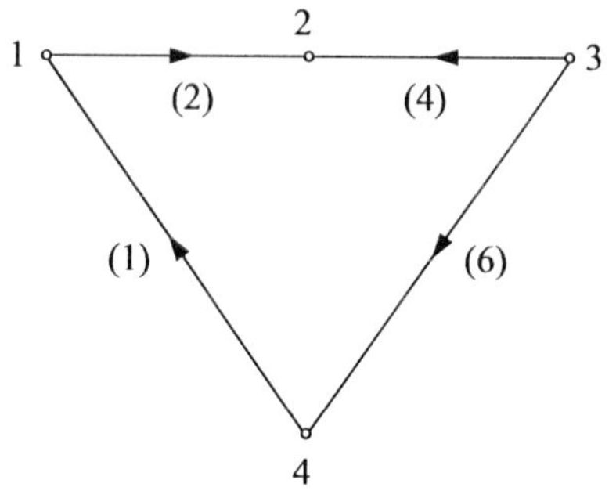

(*a*) *Spanning subgraph*

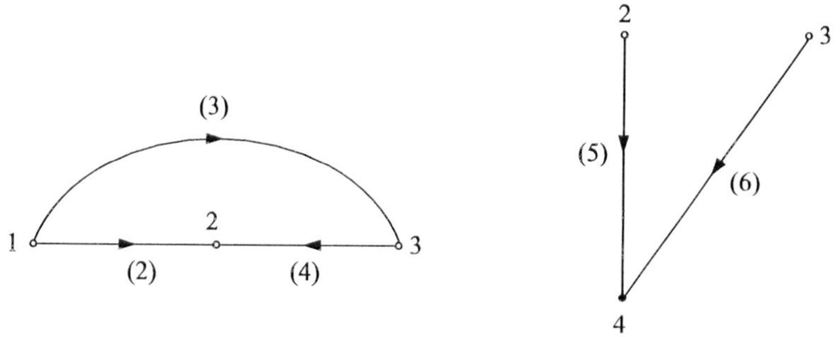

(*b*) *Subgraphs. (do not contain all branches and nodes).*

Fig. 3.3 *Subgraphs of graph G*

3.4 TREE

A tree is connected graph without any circuits (closed loops) of a graph or subgraph.

A tree cannot have a closed path. The trees for the graph G of Fig. 3.2 (c) is shown in Fig. 3.4.

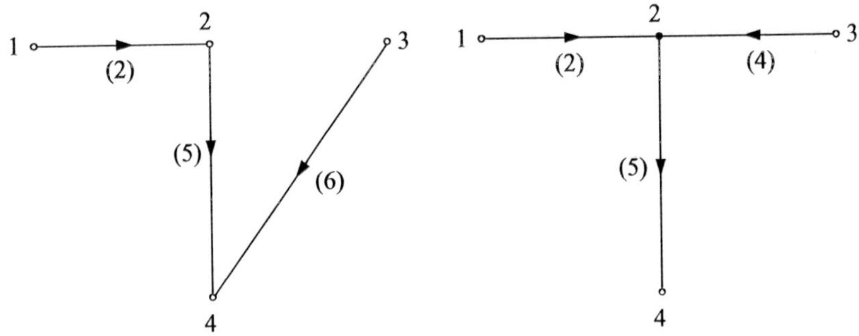

Fig. 3.4. *Trees for a graph G*

3.5 SPANNING TREE OR TREE

If a tree contains all nodes of a given graph G, it is known as spanning tree. It also implies that a tree is normally not expected to have all nodes of a graph G. However, in general concept of graph theory, tree is actually a spanning *tree*.

3.6 CO-TREE, TWIG AND LINKS

The branches of the tree are called as twigs and the branches removed from the graph is forming a tree is called as links (or chords). The set of all links corresponds to a given tree is called as co-tree. For given graph G in Fig. 3.2 (c), the tree, co-tree, twig and links are identified in Fig. 3.5

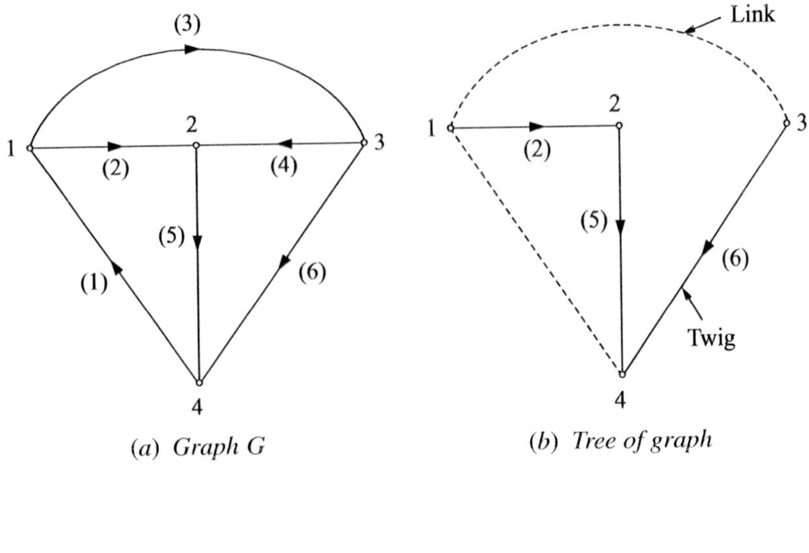

(a) *Graph G* (b) *Tree of graph*

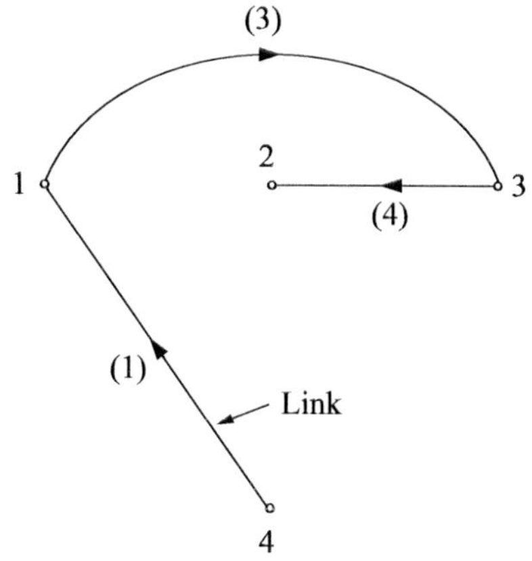

(*c*) *Co-tree*

Fig. 3.5. *Tree of a graph G, co-tree, twigs and links.*

If graph has n nodes, its tree will have $t = n - 1$ branches (or twigs). Its connected graph has b branches and n nodes. The number of links l corresponding to any tree of a graph is given as : $l = b - t = b - (n - 1) = b - n + 1$.

3.7 WALK, EDGE TRAIN, PATH, LOOP

A **walk** is defined as a finite alternating sequence of nodes and branches, beginning and ending with nodes, such that each branch in incident with the nodes preceding and following it. No branch appears (is covered or traversed) more than once in a walk. A node, however, may appear more than once. In Fig. 3.6, for instance, $v_1\ av_2\ bv_3\ cv_3\ dv_4\ ev_2\ fv_5$ is a walk shown with heavy lines. A walk is also referred to as an edge train or a chain. The set of nodes and branches constituting a given walk in graph G is clearly a subgraph of G.

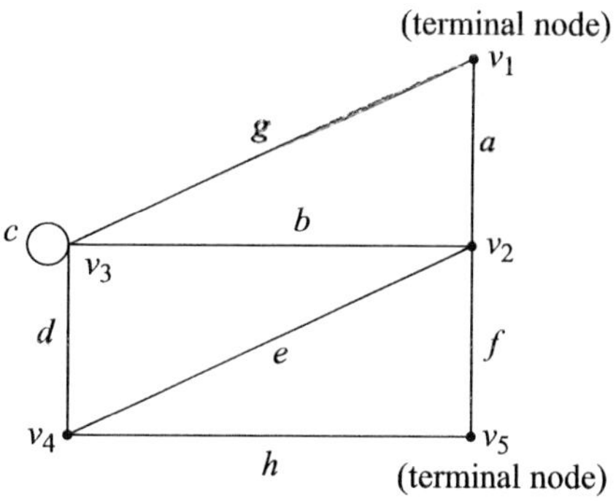

Fig. 3.6. *A walk and a path*

Nodes with which a walk begins and ends are called its terminal nodes. Nodes v_1 and v_2 are the terminal nodes of the walk shown in Fig. 3.6. It is possible for a walk to begin and end at the same node. Such a walk is called a closed walk. A walk that is not closed (*i.e.*, the terminal nodes are distinct) is called an open walk [Fig. 3.6].

An open walk in which no node appears more than once is called a path. In Fig 3.6, $v_1\ a\ v_2\ b\ v_3\ d\ v_4$ is a path. It should be noted that a self loop can be included in a walk but not in a path (Fig. 3.6). A closed walk in which no node (except the initial and the final node) appears more than once is called a circuit or loop.

3.8 PLANAR NETWORK

A network is said to be planar if it can be drawn on a two-dimensional plane, such that no two branches intersect at a point that is not a node.

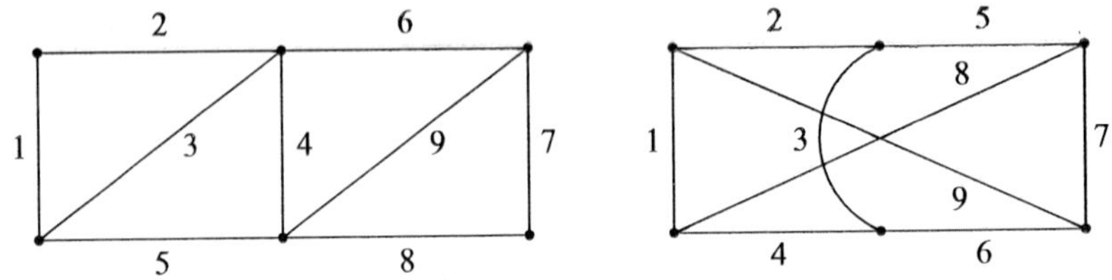

(a) *Planar graph* (b) *Non-planar graph*

Fig. 3.7. *Planar and Non-planar*

Fig. 3.7 (*a*) is a graph of a planar network whereas Fig. 3.7 (*b*) is a graph of a non-planar network, because branch 3 and 8, 9 intersect at non-nodal points. Please note that if branch 3 is withdrawn from the graph, remaining 8 and 9 can be drawn such that they don't intersect *i.e.*, graph becomes planar.

3.9 INCIDENCE MATRIX

Given an oriented graph it is easy to tell as to which nodes and what are their orientations relative to the nodes. This incidence information of the branches of the graph can be given (instead of plotting the graph) in a concise mathematical form by a matrix. For a graph having n nodes and b branches, the complete incidence matrix Aa (subscript a stands for all nodes) is a rectangular matrix of order (n, b). The columns of the matrix correspond to the branches and rows correspond to the nodes. The entry a_{ij} in the ith row and j^{th} column of Aa has the following values (*see* Fig. 3.8):

a_{ij} = i^{th} row and j^{th} column element of incidence matrix.

Incident matrix for a network with n nodes and b branches 10 is a rectangular matrix of order (n, b), rows are nodes and columns are branches.

$a_{ij} = 1$, if branch j is incident at node i and oriented away from it

$a_{ij} = -1$, if branch, is incident at node i and oriented towards it.

$a_{ij} = 0$, if branch j is not incident at node i.

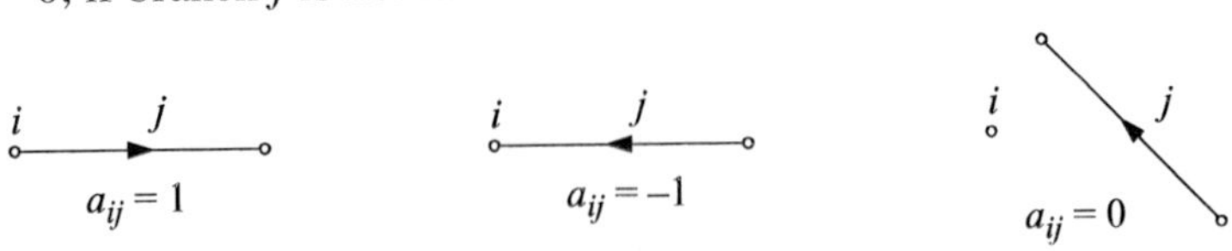

(*a*) (*b*) (*c*)

Fig. 3.8. *Definition of entries of incidence matrix.*

Following the above convention, we can easily write the sample incidence matrix Aa for the graph in Fig. 3.2 (c), as

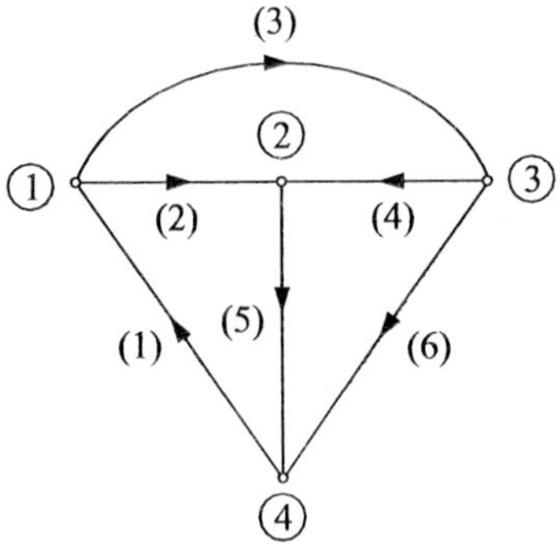

nodes ↓ / branches →	(1)	(2)	(3)	(4)	(5)	(6)
①	–1	1	1	0	0	0
②	0	–1	0	–1	1	0
③	0	0	–1	1	0	1
④	1	0	0	0	–1	–1

A careful look at this matrix indicates that each column has exactly two non-zero elements – a single 1 and a single – 1. This is general property of any linear graph. Each branch is incident at exactly two nodes, and it must be oriented away from one of them and towards the other. If we add all entries of each column of the matrix Aa, the result is row of zeros, indicating that not all the rows are linearly independent. Thus, one of the rows can be eliminated without loosing any information. Any element in the eliminated row can be retrieved, whenever needed, as the negative sum of all the elements in the corresponding column.

3.9.1 Reduced Incidence Matrix

When one of the rows of the matrix Aa is eliminated, the resulting matrix is called a reduced incidence matrix, and is denoted by A. It is of order $(n - 1) \times b$. It is usual practice to eliminate that row from the complete incidence matrix which corresponds to the datum (ground) node of the network. The rank of both the matrices Aa and A in $n - 1$.

It node (4) of graph of Fig. 3.2 (c) is taken as the datum node, the incidence matrix A is obtained from complete incidence matrix Aa, by simply deleting the 4th row. Thus,

nodes ↓ / branches →	(1)	(2)	(3)	(4)	(5)	(6)
(1)	–1	1	1	0	0	0
(2)	0	–1	0	–1	1	0
(3)	0	0	–1	1	0	1

$$A = \begin{bmatrix} -1 & 1 & 1 & 0 & 0 & 0 \\ 0 & -1 & 0 & -1 & 1 & 0 \\ 0 & 0 & -1 & 1 & 0 & 1 \end{bmatrix}$$

Following properties are simple conclusions based on the definition of A :

1. The unit entries (1 or – 1) in a row identify the branches incident at a node. Their number is the degree of the node.
2. Columns of A with unit entries in two identical rows correspond to two branches with same end nodes, and hence they are in parallel.

3. A degree of 1 for a row means that there is only one branch incident at that node. This is possible in a tree.
4. If the degree of a node is two, it shows exactly two branches incident at the node. Hence, these branches are in series.

For incidence matrix of Eq. 3.2 it can be verified that branches (2) and (4) are in series. If the incidence matrix A of a graph is known, then it is possible to find the exact number of trees that can be generated from the graph.

The number of possible trees = det (AA^T) where A^T is the transpose of the matrix A.

3.10 KCL EQUATIONS

The incidence matrix A is useful in writing the KCL (Kirchhoff's Current Law) equations for the graph. Let us consider the following electrical network of Fig. 3.9 (*a*) :

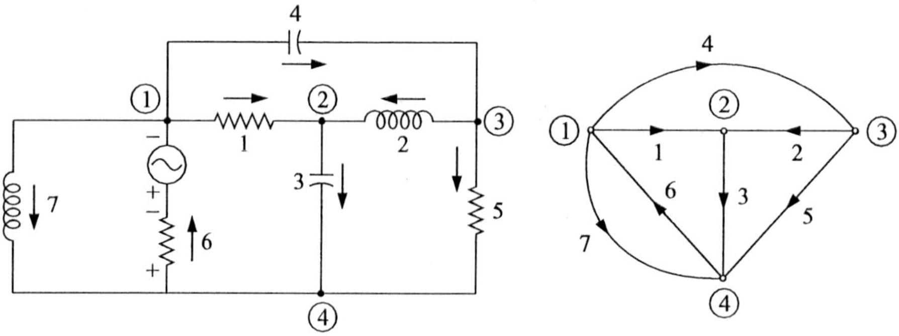

(*a*) (*b*)

Fig. 3.9 (*a*) *An electrical network,* (*b*) *An oriented graph.*

The oriented graph is shown in Fig. 3.9 (*b*). The complete incidence matrix Aa is as follows :

$$Aa = \begin{array}{c} \text{nodes} \\ \downarrow \\ (1) \\ (2) \\ (3) \\ (4) \end{array} \begin{array}{c} \text{branches} \rightarrow \\ \begin{array}{ccccccc} 1 & 2 & 3 & 4 & 5 & 6 & 7 \end{array} \\ \left[\begin{array}{ccccccc} 1 & 0 & 0 & 1 & 0 & -1 & 1 \\ -1 & -1 & 1 & 0 & 0 & 0 & 0 \\ 0 & 1 & 0 & -1 & 1 & 0 & 0 \\ 0 & 0 & -1 & 0 & -1 & 1 & -1 \end{array}\right] \end{array} \quad \text{...(3.1)}$$

The reduced incidence matrix is as follows:

$$A = \begin{array}{c} \text{nodes} \\ \downarrow \\ (1) \\ (2) \\ (3) \end{array} \begin{array}{c} \text{branches} \rightarrow \\ \begin{array}{ccccccc} 1 & 2 & 3 & 4 & 5 & 6 & 7 \end{array} \\ \left[\begin{array}{ccccccc} 1 & 0 & 0 & 1 & 0 & -1 & 1 \\ -1 & -1 & 1 & 0 & 0 & 0 & 0 \\ 0 & 1 & 0 & -1 & 1 & 0 & 0 \end{array}\right] \end{array} \quad \text{....(3.2)}$$

Let $i_1, i_2, i_3, \ldots i_7$ be the branch currents in the seven branches of the graph of Fig. 3.10 (*b*). Let the reference direction of the branch currents correspond to the orientation of the branches. The oriented graph is redrawn with branch currents in Fig. 3.9 (*c*).

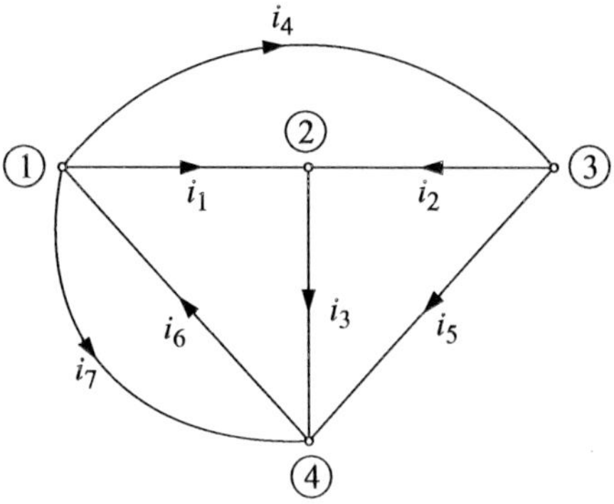

(*c*) *Graph with branch currents*

Fig. 3.9

Applying KCL at nodes (1), (2) and (3), we get

$$i_1 + i_4 - i_6 + i_7 = 0$$
$$-i_1 - i_2 + i_3 = 0$$
$$i_2 - i_4 + i_5 = 0$$

These equations can be written in matrix form as

$$\begin{bmatrix} 1 & 0 & 0 & 1 & 0 & -1 & 1 \\ -1 & -1 & 1 & 0 & 0 & 0 & 0 \\ 0 & 1 & 0 & -1 & 1 & 0 & 0 \end{bmatrix} \begin{bmatrix} i_1 \\ i_2 \\ i_3 \\ i_4 \\ i_5 \\ i_6 \\ i_7 \end{bmatrix} = \begin{bmatrix} 0 \\ 0 \\ 0 \end{bmatrix} \qquad \ldots(3.3)$$

or $$A\,Ib = 0$$

Here I_b represents column matrix or vector of branch currents. A is the incidence matrix as given in Eq. (3.2).

Note that Eq. (3.3) gives a set of $n - 1$ linearly independent equations in branch currents $I_1, I_2, \ldots I_7$. Here, $n = 4$, hence, there are three linearly independent equations.

3.11 FUNDAMENTAL LOOPS OF GRAPH

A tree is obtained from a graph by removing branches form the graph, one by one, till no loops (or closed paths) are left (and, no node is left isolated). It means that it we add a link to the tree a loop will be formed, this loop is called fundamental loop. Consider a connected graph G of Fig. 3.10 (*a*).

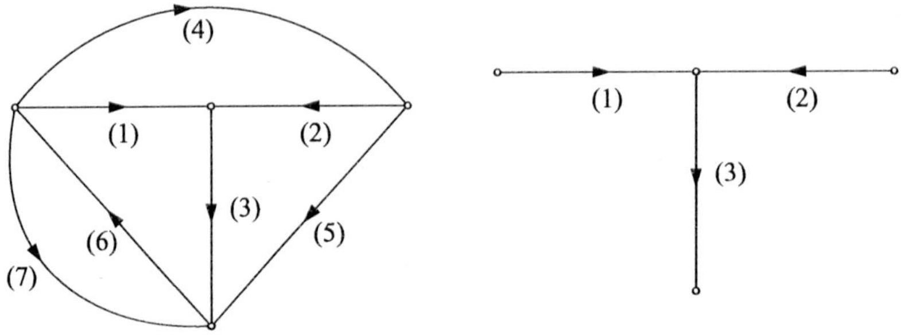

(*a*) *Oriented graph G* (*b*) *Tree T of the graph G*

Fig. 3.10

Let T be a tree of the graph G, as shown in Fig. 3.10 (*b*). The twigs of this tree are branches 1, 2 and 3. The link corresponding to this tree are branches 4, 5, 6 and 7. The fundamental loops are shown in Fig. 3.10 (*c*), (*d*), (*e*) and (*f*).

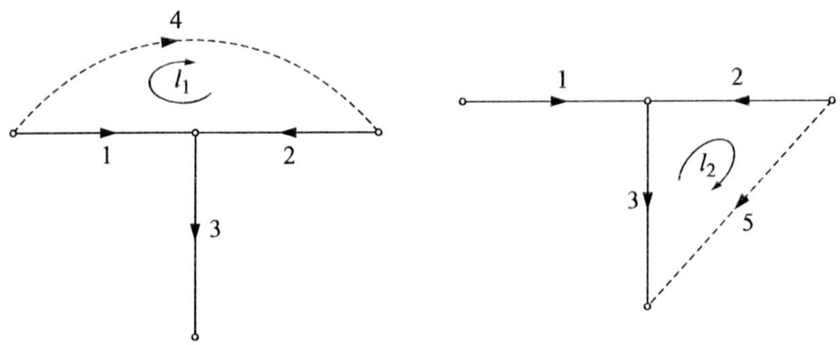

(*c*) (*d*)

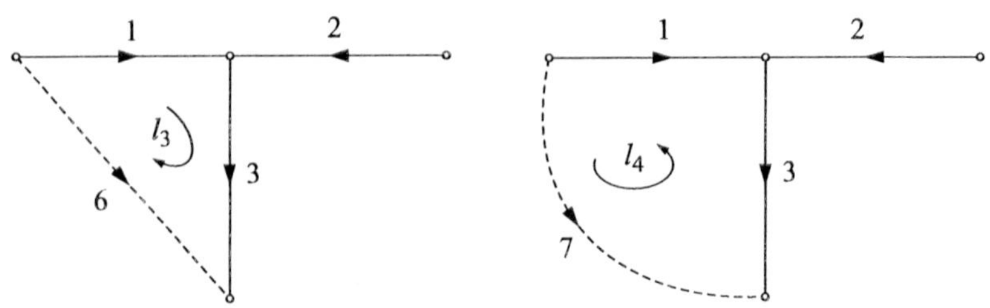

(*e*) (*f*)

Fig. 3.10. *Formation of f-loops*

By convention, f-loop is given the same orientation as its defining link. A fundamental loop is also called a f-circuit or a tie-set.

Corresponding to each link of a tree T, there exists a f-loop, therefore, the number of f-loops is equal to the number of links of the chosen tree, *i.e.*, the number of f-loops $= b - (n - 1) = b - n + 1$.

3.12 KVL EQUATIONS FOR F-LOOPS (TIE-SET)

KVL (Kirchhoff's Voltage Law) can be applied to the fundamental loops to get a set of linearly independent equations. Let us take the example of the graph of Fig. 3.11 (*a*). A tree of the graph G has been shown in Fig. 3.11 (*b*). There are four f-loops l_1, l_2, l_3 and l_4 corresponding to the links 4, 5, 6 and 7 respectively. If v_1, v_2, ..., v_7 are branch voltages, the KVL equations for the four f-loops are:

$$\left.\begin{aligned} l_1 &: -v_1 + v_2 + v_4 = 0 \\ l_2 &: -v_2 - v_3 + v_5 = 0 \\ l_3 &: \ \ v_1 + v_3 + v_6 = 0 \\ l_4 &: -v_1 - v_3 + v_7 = 0 \end{aligned}\right\} \qquad ...(3.4)$$

In writing the above equations, if the branch orientation coincides with that of the f-loop link, the branch voltage is written with a positive sign. If the branch orientation is opposite to that of the f-loops, it is assigned a negative sign. Equation (3.4) can be written in the matrix form as:

$$\begin{bmatrix} -1 & 1 & 0 & 1 & 0 & 0 & 0 \\ 0 & -1 & -1 & 0 & 1 & 0 & 0 \\ 1 & 0 & 1 & 0 & 0 & 1 & 0 \\ -1 & 0 & -1 & 0 & 0 & 0 & 1 \end{bmatrix} \begin{bmatrix} v_1 \\ v_2 \\ v_3 \\ v_4 \\ v_5 \\ v_6 \\ v_7 \end{bmatrix} = \begin{bmatrix} 0 \\ 0 \\ 0 \\ 0 \end{bmatrix}$$

or

$$B_f \, Vb = 0 \qquad ...(3.5)$$

where B_f is an $l \times b$ matrix called fundamental loop matrix or basic loop matrix or tie-set matrix and is defined as follows :

$B_f = [b_{ij}]$ where b_{ij} is the entry of B_f in i^{th} row and j^{th} column. The rows of B_f correspond to the f-loops and the columns to the branches. The entry bij has a value as follows:

$b_{ij} = 1$, if branch j is included in i^{th} f-loop and j^{th} column and their orientations coincide.

$b_{ij} = -1$, if branch j is included in the ith f-loop and their orientations are opposite.

$b_{ij} = 0$, if branch j is not included in th i^{th} f-loop.

V_b represents vector of branch voltage in Eq. (1.121). This equation gives a set of $l = (b - n + 1)$ number of linearly independent equations in the variable v_1, v_2, v_3, v_7. Here, $b = 7, n = 4$, hence, there are 4 linearly independent equations.

3.13 RANK OF FUNDAMENTAL LOOP MATRIX

In order to determine the rank of f-loop matrix let us partition the f-loop matrix of Eq. (3.5) so that the twigs and links are separated. The branches of the graph are numbered such that the twigs of the selected tree are 1, 2, ..., $(n - 1)$, and the links are $n, n + 1, ..., b$ as indicated in Eq. (3.6). The rows of B_f are arranged so that the first row (giving loop 1_1) corresponds to the first link, the second row to the second link and so on.

$$[B_f] = \begin{array}{c} \text{loops} \\ \downarrow \\ l_1 \\ l_2 \\ l_3 \\ l_4 \end{array} \begin{array}{c} \text{twigs} \quad\quad \text{links} \quad\quad \text{branches} \rightarrow \\ \begin{array}{ccc:cccc} 1 & 2 & 3 & 4 & 5 & 6 & 7 \end{array} \\ \begin{bmatrix} -1 & 1 & 0 & 1 & 0 & 0 & 0 \\ 0 & -1 & -1 & 0 & 1 & 0 & 0 \\ 1 & 0 & 1 & 0 & 0 & 1 & 0 \\ -1 & 0 & -1 & 0 & 0 & 0 & 1 \end{bmatrix} \end{array} = [B_t : b_l] \quad \text{...(3.6)}$$

i.e., $$B_f = [B_t : B_l] \quad \text{...(3.7)}$$

where B_t = sub-matrix of B_f corresponding to the twigs of the chosen tree

B_l = Sub-matrix of B_f corresponding to the links of the chosen tree

It is clearly seen that the sub-matrix

$$b_t = \begin{bmatrix} 1 & 0 & 0 & 0 \\ 0 & 1 & 0 & 0 \\ 0 & 0 & 1 & 0 \\ 0 & 0 & 0 & 1 \end{bmatrix} = [U] = \text{Unit matrix}$$

The sub-matrix B_t is a matrix of the order $(b - n + 1) \times (n - 1)$ Thus, the fundamental loop matrix can be written as :

$$B_f = [B_t : B_l]$$

or $$B_f = [B_f : U] \quad \text{...(3.8)}$$

This shows that the rank of B_f is $(b - n + 1)$ since it has non-singular sub matrix U of the order $(b - n + 1) \times (- n + 1)$ consider Eq. (3.5). Let us partition B_f into B_t (corresponding to the twigs) and B_l (corresponding to the links). Also partition the branch voltage vector V_b to conform with the partition of B_f. That is we write :

$$V_b = \begin{bmatrix} V_t \\ V_l \end{bmatrix} \quad \text{...(3.9)}$$

where v_t is set of twig voltages, and v_l the set of link voltages. The KVL Eq. of (3.5) can now be written *as*

$$[B_t\ B_l]\begin{bmatrix}V_t\\V_l\end{bmatrix}=0 \text{ or } [B_t\ U]\begin{bmatrix}V_t\\V_l\end{bmatrix}=0$$

or $$B_t\,v_t+v_l=0$$

which yields $$v_l=-\,b_t\,v_t \qquad ...(3.10)$$

Eq. (3.10) shows that the link voltages can be expresses as linear combination of the twig voltages. In general, the converse is not true, since B^{-1}, may not exist. Hence, we conclude that the link voltage are uniquely determined by the twig voltages. This fact will be used in formation of network equations.

3.14 THE MESH OR LOOP MATRIX

If a graph is planar, there is an alternative way of selecting $(b-n+1)$ independent loops. This is achieved by selecting meshes instead of *f*-loops. By definition, a mesh is a loop in which there are no branches in its interior. Meshes can be defined only for a planar graph. For a non-planar graph, there is no choice but to select a set of *f*-loops.

For a planar graph, there are two choices of getting a set of linearly independent KVL equation. We can either choose a set of *f*-loops or a set of meshes for the graph. For certain planar graphs, it may be possible to select a tree such that the meshes them selves are *f*-loops. But in general, it need not be true. Consider, the planar graph of Fig. 3.11 (*a*). A tree of the graph is shown in Fig. 3.11(b).

(*a*) *Graph* (*b*) *T*

Fig. 3.11. *Meshes in a planar graph*

The Fig. 3.11(*b*) has four meshes, marked as *I*, *II*, *III* and *IV*. It is a general practice to orient all the meshes currents in the direction. Note that the three meshes (*I*, *II* and *III*) are also *f*-loops with respect to the tree *T*. The mesh *IV* differs from *f*-loops 4 of Fig. 3.11.

It may be shown that for a connected planar graph *n* nodes and *b* branches, the number of meshes is $b-n+1$.

For the planar graph of the KVL equations can be written for its meshes n the matrix form as:

$$\begin{array}{c} \text{No. of meshes} \quad \text{braches} \rightarrow \\ \downarrow \begin{array}{ccccccc} 1 & 2 & 3 & 4 & 5 & 6 & 7 \end{array} \\ \begin{array}{c} 1 \\ 2 \\ 3 \\ 4 \end{array} \begin{bmatrix} -1 & 1 & 0 & 1 & 0 & 0 & 0 \\ 0 & -1 & -1 & 0 & 1 & 0 & 0 \\ 1 & 0 & 1 & 0 & 0 & 1 & 0 \\ 0 & 0 & 0 & 0 & 0 & -1 & -1 \end{bmatrix} \end{array} \begin{bmatrix} v_1 \\ v_2 \\ v_3 \\ v_4 \\ v_5 \\ v_6 \\ v_7 \end{bmatrix} = 0 \quad \text{...(3.12)}$$

or $$B_m V_1 = 0 \quad \text{...(3.13)}$$

where B_m is the mesh matrix of the planar graph.

For a connected planar graph G with n nodes and b branches, the mesh matrix can be defined as

$$B_m = [b_{ij}]_{(b-n+1) \times b} \quad \text{...(3.14)}$$

where b_{ij} is given by Eq. (3.14) (where read mesh for loop).

We will make use of mesh matrix while formulating mesh equations for a planar network.

3.15 CUTSET

A connected graph can be broken into two separate parts by removing some branches.

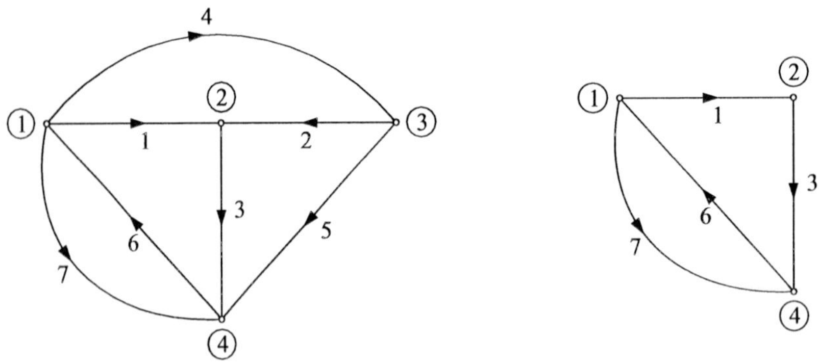

(*a*) *Graph* (*b*) *Parts*

Fig. 3.12. *Graph and parts*

Consider the graph of Fig. 3.12 (*a*). Its branches 4, 2 and 5 are removed we get two distinct parts (Fig. 3.12 (*b*)). One of the parts is just an isolated node (a single node too is the part of the graph). Also note that while removing a branch the end nodes of the branch are not removed. This set of branches form a cutset.

A cutset is a set of minimum number of branches of a connected graph whose removal causes the graph to be cut (separated) into exactly two parts (subgraphs).

A cutset is shown on a graph by a dashed line. The line passes through all the branches defining the cutset. Note that a graph can have more than one cutset. Also, every connected graph, however so, simple, has at least one cutset.

3.16 CUTSET MATRIX

Let us consider the graph of Fig. 3.13 (*a*). It has *f* branches and 4 nodes. The branches have been numbered and their orientations are also marked. As shown in Figs. 3.13 (*b*) – (*h*), the following seven cutsets are possible:

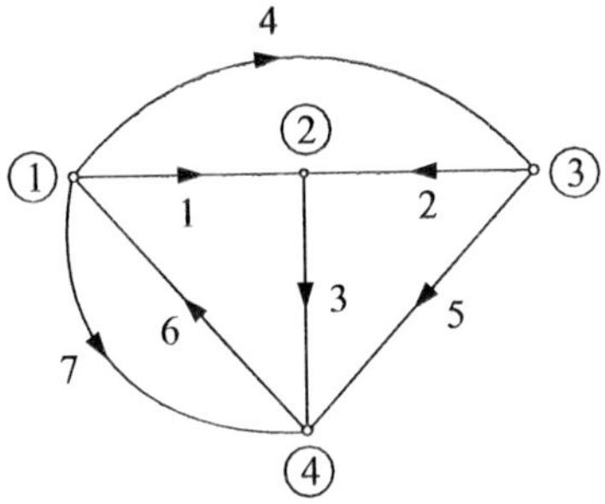

(*a*)

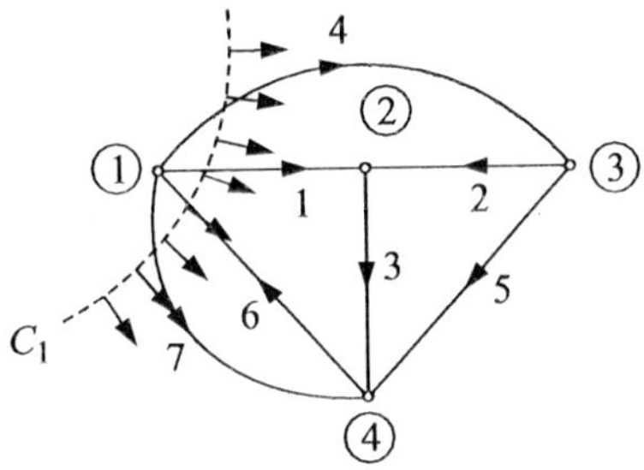

(*b*)

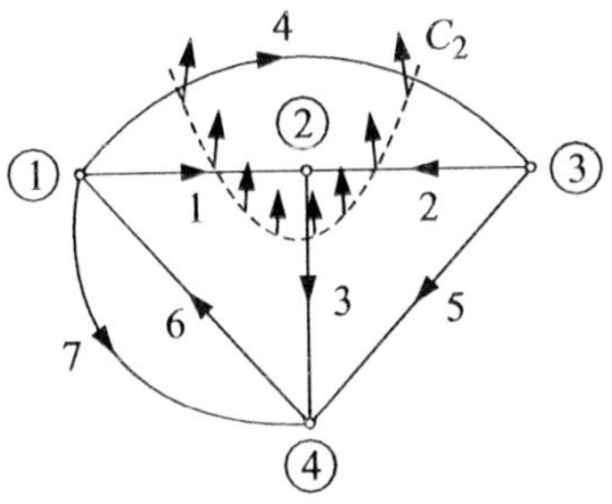

(*c*)

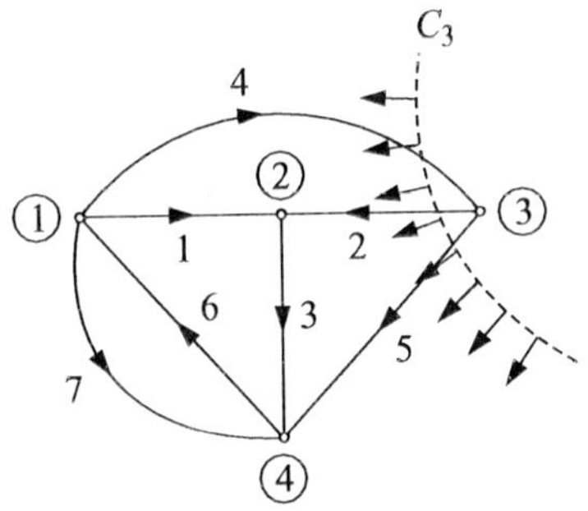

(*d*)

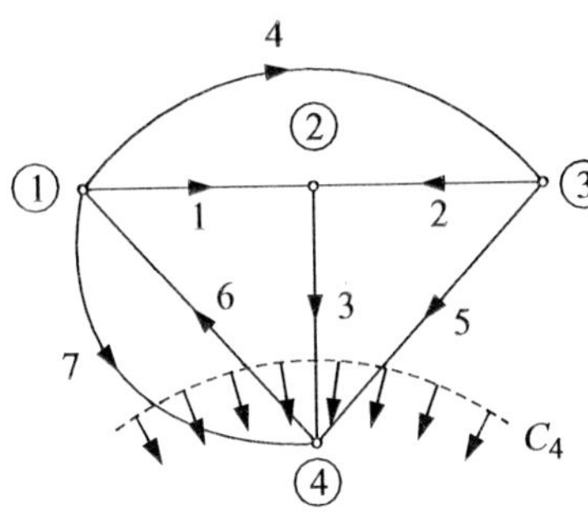

(*e*)

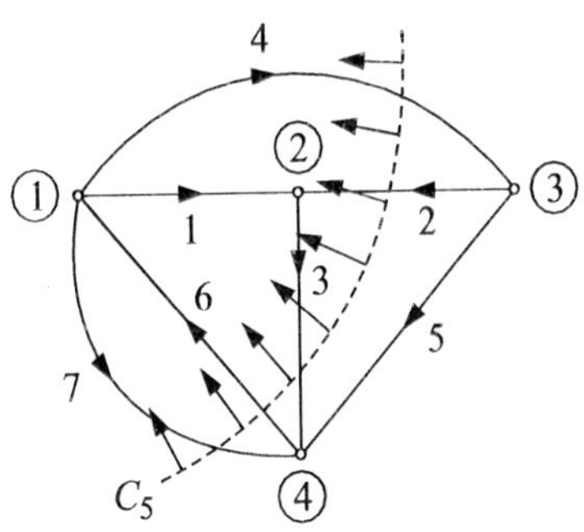

(*f*)

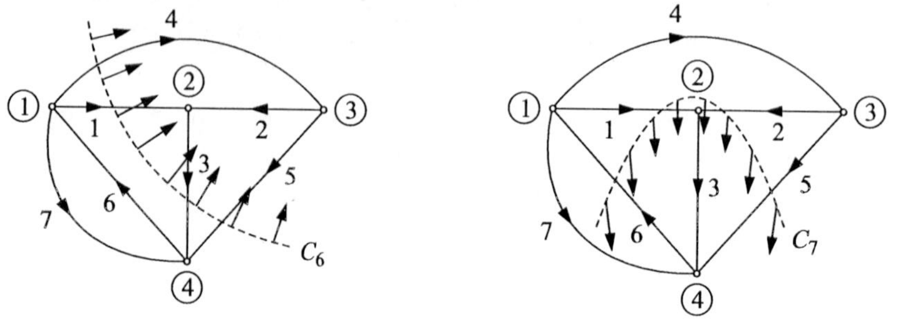

(*g*) (*h*)

Fig. 3.13. *Cutsets*

The orientation of a cutset is arbitrary *i.e.*, from part 1 to *Z* or parts 2 to 4. All cutset matrix *i.e.*, $Q_a = [q_{ij}]$; $q_{ij} = 1$ if branch *j* included in the *i* cutset and branch orientation coincides with that of the cutset.

$$Q_a = \begin{array}{c} \text{cutsets.} \\ \downarrow \\ C_1 \\ C_2 \\ C_3 \\ C_4 \\ C_5 \\ C_6 \\ C_7 \end{array} \begin{array}{c} \qquad\qquad\qquad \text{branches} \rightarrow \\ \begin{array}{ccccccc} 1 & 2 & 3 & 4 & 5 & 6 & 7 \end{array} \\ \begin{bmatrix} 1 & 0 & 0 & 1 & 0 & -1 & 1 \\ 1 & 1 & -1 & 0 & 0 & 0 & 0 \\ 0 & 1 & 0 & -1 & 1 & 0 & 0 \\ 0 & 0 & 1 & 0 & 1 & -1 & 1 \\ 0 & 1 & -1 & -1 & 0 & 1 & -1 \\ 1 & 0 & -1 & 1 & -1 & 0 & 0 \\ 1 & 1 & 0 & 0 & 1 & -1 & 1 \end{bmatrix} \end{array} \qquad \text{...(3.15)}$$

$q_{ij} = -1$ if branch *j* is included in cutset *i* and the branch orientation is opposite to that of the cutset. *qij* = 0 if branch *j* is not present in cutset. For each cutset we can write the KCL eqns.

Cutset

$$C_1 : i_1 + i_4 - i_6 + i_7 = 0$$

$$C_2 : i_1 + i_2 - i_3 = 0$$

$$C_3 : i_2 - i_4 + i_5 = 0$$

$$C_4 : i_3 + i_5 - i_6 + i_7 = 0$$

$$C_5 : \quad i_2 - i_3 - i_4 - i_6 - i_7 = 0$$

$$C_6 : i_1 - i_3 + i_4 - i_5 = 0$$

$$C_7 : i_1 + i_2 + i_5 - i_6 + i_7 = 0$$

These eqns. can be written in matrix form as :

$$\begin{bmatrix} 1 & 0 & 0 & 1 & 0 & -1 & 1 \\ 1 & 1 & -1 & 0 & 0 & 0 & 0 \\ 0 & 1 & 0 & -1 & 1 & 0 & 0 \\ 0 & 0 & 1 & 0 & 1 & -1 & 1 \\ 0 & 1 & -1 & -1 & 0 & 1 & -1 \\ 1 & 0 & -1 & 1 & -1 & 0 & 0 \\ 1 & 1 & 0 & 0 & 1 & -1 & 1 \end{bmatrix} \begin{bmatrix} i_1 \\ i_2 \\ i_3 \\ i_4 \\ i_5 \\ i_6 \\ i_7 \end{bmatrix} = \begin{bmatrix} 0 \\ 0 \\ 0 \\ 0 \\ 0 \\ 0 \\ 0 \end{bmatrix} \qquad ...(3.16)$$

or $$Q\,a\,I_b = 0 \qquad ...(3.17)$$

where I_b is branch current vector

3.17 FUNDAMENTAL CUTSETS (*f*-cutsets)

The basic cutset equations describe the relationship among branch currents, but a closer look at these equations shows that these are not linearly independent equations. Only first three equation of equation 3.16. matrix are linearly independent. The last four equations can be obtained as linear combinations of the first three.

A set of linearly independent equations in branch-current variables can be obtained by using the concept of fundamental cutsets. The *f* cutsets are defined for a given tree of the graph. A fundamental cutset of a graph G with respect to a tree T is a cutset that is formed by one twig and a unique set of links. Thus, corresponding to each twig (of a chosen tree), these will be a fundamental cutset or (*f* cutset). Since there are $n - 1$ twigs in a tree of a connected graph having n nodes, the number of *f* cutsets is also equal to $n - 1$.

As a convention, the orientation of a *f* cutset is chosen to coincide with that of its defining twig.

3.17.1 Fundamental Cutset Matrix

The fundamental cutset matrix Q_f of a graph gives us information about the branches contained in a fundamental cutset. The rows of this matrix correspond to the fundamental cutset and the columns correspond to the branches of the graph. For a graph having n nodes and b branches the fundamental cutset matrix Q_f can be written as

$$Q_f = [q_{ij}]$$

where q_{ij} is the entry in the ith row and j^{th} column as defined in section 3.16.

Let us obtain the fundamental cutset matrix for the network graph of Fig. 3.14 (*a*):

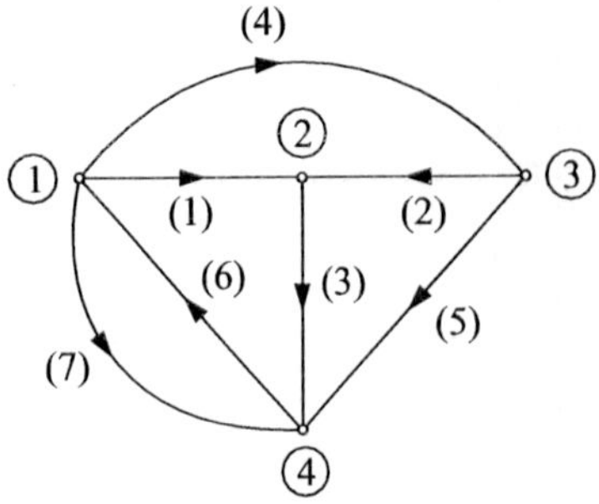

(*a*) *A network graph G*

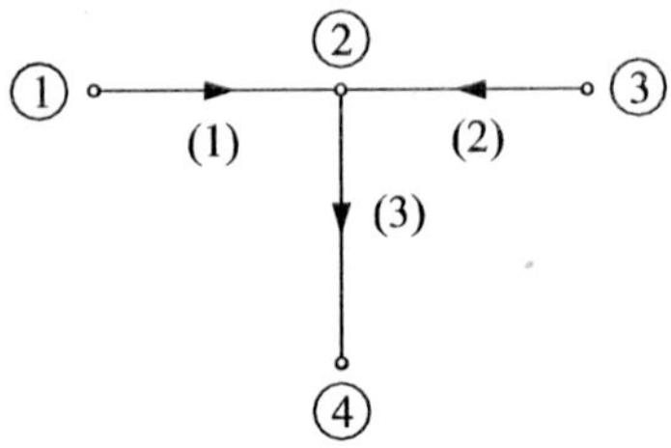

(*b*) *Tree T of the graph G*

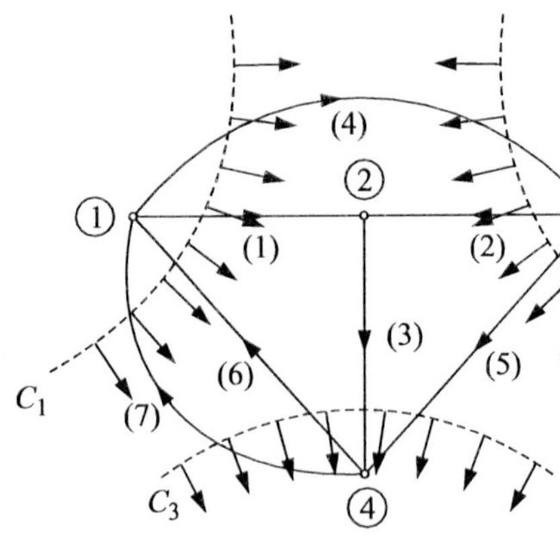

(*c*) *Fundamental cutsets corresponding to a chosen tree T.*

Fig. 3.14. *Fundamental cutsets corresponding to a chosen tree T of the network graph G.*

Let us choose the tree T as given in Fig. 3.14 (*b*). This tree is shown by solid lines and corresponding links by dashed lines in Fig. 3.14 (*c*). The twigs of the chosen tree are {1, 2, 3}. The remaining branches 4, 5, 6 and 7 are the links corresponding to this tree. Let us focus our attention on twig 1. The unique set of links that must be added to this twig to for a cutset C1 is {4, 6, 7}. Thus, corresponding to twig 1, the *f*-cutset C_1 is {1, 4, 6, 7}. Similarly, corresponding to twig 2, we get *f* cutset C_2 {2, 4, 5} and corresponding to twigs we get *f* cut set C_3 { 3, 5, 6, 7}. This, the for cutset matrix is constructed as :

$$
\begin{array}{cc}
 & f \text{ cutsets} \\
 & \downarrow 1 \quad 2 \quad 3 \quad\; 4 \quad 5 \quad 6 \quad 7 \\
 & \quad\;\; \text{twigs} \qquad\quad \text{links} \\
Q_f = \begin{array}{c} C_1 \\ C_2 \\ C_3 \end{array} & \left[\begin{array}{ccc|cccc} 1 & 0 & 0 & 1 & 0 & -1 & 1 \\ 0 & 1 & 0 & -1 & 1 & 0 & 0 \\ 0 & 0 & 1 & 0 & 1 & -1 & 1 \end{array}\right]
\end{array}
$$

3.17.2 Rank of *f* Cutset Matrix

Consider the *f* cutset matrix Q_f obtained as equation (3.17) Partitioning Q_f into submatrices Q_t and Q_l corresponding to twig branches and link branches respectively, we have

$$Q_f = [Q_t : Q_l] \qquad ...(3.19)$$

From Eqn. 3.17 it is seen that sub-matrix Q_t is unit matrix of order $(n - 1)$ and is non-singular. Therefore, the rank of Q_f in $(n - 1)$.

The sub-matrix Q_l is of order $(n - 1) \times (b - n + 1)$. Since the number of twigs for a graph need not be same as no. of links, the sub-matrix Q_l need not be square matrix.

3.17.3 Relation between Branch Voltage and Tree-Twig Voltages

There is another way of getting the *f* cut-set matrix Q_f for a graph. This arises from the fact that it is possible to express branch-voltage in terms of Tree-twig voltages. The tree-branches connect all the nodes in a network graph. It is, therefore, possible to trace a part from one node to any other node by traversing through the tree-branches alone. Hence, potential difference between any two nodes (called the "node pair voltage") can be uniquely determined in terms of the tree-twig voltages. A link-voltage is merely a node-pair voltage. Hence, the link-voltages can be expressed in terms of tree-twig voltages. Let us consider Fig. (3.17 (*c*)). Let branch voltages be denoted by $v_1, v_2, v_3, ..., v_7$. These voltages correspond to the given numbering of the branches. For the chosen tree Fig. 3.17 the branches 1, 2 and 3 are the twigs. Let the twig voltages be denoted by v_{t_1}, v_{t_2} and v_{t_3}. Each branch voltage can be expressed in terms of twig-voltages as follows :

$$v_1 = v_{t_1}$$
$$v_2 = v_{t_2}$$
$$v_3 = v_{t_3}$$
$$v_4 = v_{t_1} - v_{t_2}$$
$$v_5 = v_{t_2} + v_{t_3}$$
$$v_6 = - v_{t_1} - v_{t_3}$$
$$v_7 = v_{t_1} + v_{t_3}$$

The above equations can be written in matrix from as

$$\begin{bmatrix} v_1 \\ v_2 \\ v_3 \\ v_4 \\ v_5 \\ v_6 \\ v_7 \end{bmatrix} = \begin{bmatrix} 1 & 0 & 0 \\ 0 & 1 & 0 \\ 0 & 0 & 1 \\ 1 & -1 & 0 \\ 0 & 1 & 1 \\ -1 & 0 & -1 \\ 1 & 0 & 1 \end{bmatrix} \begin{bmatrix} v_{t_1} \\ v_{t_2} \\ v_{t_3} \end{bmatrix} \qquad \text{...(3.20)}$$

The first matrix on the right hand side of Eq. (3.19) is seen to be the transpose of the f cut-set matrix given in Eqn (3.16) Thus Eqn (3.19) can be written as

$$v_b = Q_f^T \, v_t \qquad \text{...(3.21)}$$

where v_b is the column matrix of branch-voltages and v_t is the column matrix of the twig-voltages corresponding to the chosen tree.

3.18 GENERALISED ELEMENT AND THEIR RELATIONSHIPS

The concept of a generalised element is useful while writing equilibrium equations of a network. The different generalised elements are resistors, inductors, and capacitors. By applying KVL and KCL at different points we can get the mesh equations and nodal equations. But in case of the graph theory the following relationships are used:

Consider the generalised elements as shown in Fig. 3.15 (*a*), (*b*), (*c*) as,

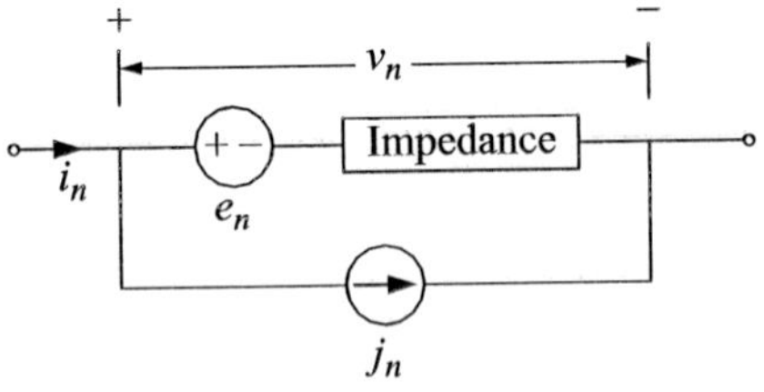

(*a*)

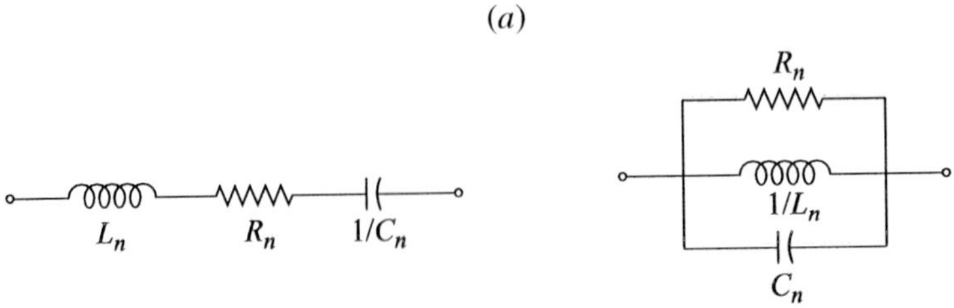

(*b*) *Series* (*c*) *Parallel*

Fig. 3.15. *Generalised elements*

By applying KVL in Fig. 3.18 (*a*) we have for series connection

$$v_n = e_n + R_n (i_n - j_n) + L_n \frac{d}{dt}(i_n - j_n) + \frac{1}{c_n} \int (i_n - j_n)\, dt$$

or

$$v_n = e_n + \left[R_n i_n + L_n \frac{d}{dt} i_n + \frac{1}{c_n} \int i_n \, dt \right]$$

$$- \left[Rn\, j_n + L_n \frac{d}{dt} j_n + \frac{1}{c_n} \int jndt \right]$$

By taking Laplace transform we have

$$v_n(s) = z_n(s)\, I_n(s) + [E_n(s) - Z_n\, J_n(s) - L_n\, iL_n(0) + \frac{1}{s}\; v_{cn}(0)]$$

where $z_n(s) = R_n + sL_n + \dfrac{1}{cn^3}$

This gives the terminal relationship of a genrealised element in Laplace transform. We get such equation for every element of the network. Combining all these elements we can write the overall terminal relationship in matrix form as

$$V_b(s) = z_b(s)\, I_b(s) + E_q(s)$$

where $E_q(s) = E_b(s) - z_b(s)\, J_b(s) - L_b(s)\, L_{Lb}(0) + \dfrac{1}{s}\; V_{cb}(0)$

$$V_b(s) = \text{Branch voltage vector} = \begin{bmatrix} V_1(s) \\ V_2(s) \\ \vdots \\ V_{n(s)} \end{bmatrix}$$

$$I_b(s) = \text{Branch current vector} = \begin{bmatrix} I_1(s) \\ I_2(s) \\ \vdots \\ I_{n(s)} \end{bmatrix}$$

$$E_b(s) = \text{Voltage source vector} = \begin{bmatrix} E_1(s) \\ E_2(s) \\ \vdots \\ E_{n(s)} \end{bmatrix}$$

$$J_b(s) = \text{Current source vector} = \begin{bmatrix} J_1(s) \\ J_2(s) \\ \vdots \\ J_{n(s)} \end{bmatrix}$$

$$i_{Lb}(0) = \text{Initial current vector} = \begin{bmatrix} i_{L_1}(0) \\ i_{L_2}(0) \\ \vdots \\ i_{L_n}(0) \end{bmatrix}$$

$$v_{cb}(0) = \text{Initial voltage vector} = \begin{bmatrix} v_{c_1}(0) \\ v_{c_2}(0) \\ \vdots \\ v_{c_n}(0) \end{bmatrix}$$

where $Z_b\,(s)$ = Branch impedance matrix,

R_b is branch resistance matrix,

L_b is branch incidence matrix and

$\dfrac{1}{cb}$ is the branch capacitance matrix defined by

$$R_b = \text{diagonal } [R_1, R_2, ..., R_n]$$
$$L_b = \text{diagonal } [L_1, L_2, ..., L_n]$$
$$\frac{1}{cb} = \text{diagonal} \left[\frac{1}{c_1}, \frac{1}{c_2}, ..., \frac{1}{c_n}\right]$$

Similarly for parallel connected network, KCL equations give

$$J_b(s) = J_b(s) - y_b(s)\, E_b(s) - C_b\, V_{cb}\,(0) + \frac{1}{3}\, i_{Lb}(0)$$

These can be taought as equivalent current source vector.

3.19 FORMULATION OF NETWORK EQUATIONS

Network equations are set of equations that completely and uniquely determine the state of network at any instant. These equations are written in terms of suitably chosen current variables or voltage variables. For the solution of the equation to be unique, the number of independent variables must be same as the number of independent equations.

3.19.1 Loop Equations and Loop Impedance Matrices

It is convenient to write loop equations for a network containing voltage sources but no current sources. If current sources are present, then these must first be converted into their equivalent voltage sources. The sorting point for writing the loop equations is to write the terminal relationships for the given network in the form of equation.

$$V_b(s) = Z_b(s) \pm (s) + \text{Eq}(s) \qquad ...(3.22)$$

where $V_b(s)$ is the branch voltage vector, $I_b(s)$ is branch current vector $Eq(s)$ is the voltage source vector.

We chose a tree and write KVL equations for the set of *f*-circuits, as

$$B_f V_b(s) = 0$$

Substitution for $V_b(s)$, we have

$$B_f z_b(s) I_b(s) + B_f Eq(s) = 0$$

or $$B_f Z_b(s) I_b(s) = -B_f Eq(s) \qquad ...(3.23)$$

Using link current transformation *i.e.*, $I_b = B_f T_{Il}$, we get

$$B_f Z_b(s) B_f^T I_l(s) = -B_f Eq(s). \qquad ...(3.24)$$

The loop equation then are $Z_l(s) I_l(s) = -B_f Eq(s)$

where $Zl(s) = B_f Z_b(s) B_f^T$ is the loop impedance matrix.

3.19.2 Cutset Equations and Cutset Admittance Matrix

For writing *f*-cutset equations we first ensure that the network contains no voltage sources. If it does, we convert the voltage source into equivalent current source. A tree is then selected so that the twig voltages constitute an independent set of voltages.

The terminal relationship is written in the form

$$I_b(s) = Y_b(s) V_b(s) + J_Q(s) \qquad ...(3.25)$$

where $I(b)(s)$ is the branch current vector, $V_b(s)$ is the branch voltage vector, and $J_Q(s)$ is current source vector.

Corresponding to the chosen tree, *f*-cutset equations are

$$Q_f I_b(s) = 0$$

substituting for $I_b(s)$, we get

$$Q_f [Y_b(s) V_b(s) + J_Q(s)] = 0$$

$$Q_f Y_b(s) V_b(s) = -Q_f V_q(s)$$

Using twig voltage transformation $V_b(s) = Q_f^T V_t(s)$, we get

$$Q_f Y_b(s) Q_f^T v_t(s) = -Q_f Jq(s)$$

or $$[U_{Qe}] \begin{bmatrix} Yt \\ Yl \end{bmatrix} Q_f^T V_t(s) = -Q_f J_q(s)$$

or $$Y_l V_t(s) = -Q_f J_q(s) \qquad ...(3.26)$$

where $$Y_l(s) = Q_f Y_b(s) Q_f^T \qquad ...(3.27)$$

is the cutset admittance matrix. Eqn (3.25) is the required cutset equation for the network.

3.19.3 Mesh Equations and Mesh Impedance Matrix

This method is applicable to planar networks only. To write mesh equations, check that the network does not contain any current source. If it does, first convert it into its equivalent voltage source.

The terminal relationship for the elements of the network are written in the form of

$$V_b(s) = Z_b(s) I_b(s) + E_q(s)$$

The orientation of all mesh currents taken either clockwise or anti clockwise. KVL is applied to each mesh, to give

$$B_m V_b(s) = 0$$

where B_m is mesh matrix, substituting for $V_b(s)$, we have

$$B_m Z_b(s) I_b(s) + B_m Eq(s) = 0$$

or $$B_m Z_b(s), I_b(s) = -B_m Eq(s) \quad ...(3.28)$$

Using the mesh current transformation $I_b(s) = Bm^T I_m(s)$, we get,

$$B_m Z_b(s) Bm^T I_m(s) = -B_m Eq(s)$$

or $$Z_m(s) I_m(s) = -B_m Eq(s) \quad ...(3.29)$$

There are the required mesh equations, here

$$Z_m(s) = B_m Z_b(s) Bm^T$$

is called the mesh impedance matrix.

3.19.4 Node Equations and Node Admittance Matrix

Node equations for a network can be written conveniently, if it contains no voltage source. If the network has a voltage source, the same can be converted into its equivalent current source.

The terminal relationship for the elements of the network is written in the form

$$I_b(s) = Y_b(s) V_b(s) + J_q(s)$$

From KCL, we have

$$A I_b(s) = 0$$

where A is the reduced incidence matrix of the graph, Hence,

$$A Y_b(s) V_b(s) = -A Jq(s)$$

Using the voltage transformation $V_b = A^T V_n(s)$, we get

$$A Y_b(s) A^T V_n(s) = -A Jq(s)$$

or $$Y_n(s) V_n(V) = -A Jq(s) \quad ...(3.30)$$

where $$Y_n(s) = A Y_b(s) A^T \quad ...(3.31)$$

is the node admittance matrix. The eqn. (3.29) is the required node equation. For writing reduced incidence matrix A of the graph, one has to chose a suitable node of the datum node.

3.20 NETWORK ANALYSIS USING GRAPH THEORY IN BRIEF

Graph theory can be used to analyse a network by relating branch current and loop currents or branch voltages and node voltages. Various relationships are given for reference. Normally brackets [ ] are in most of the cases just used to indicate a matrix, however, Capital letter means the related matrix.

3.20.1 Branch Currents and Loop (Link) Currents Relationship

$$I_b = b^T I_l + I_s \quad ...(3.32)$$

where $[I_b]$ = branch current matrix

$[B]$ = fundamental cutset matrix

$[B]^T$ = transpose of $[B]$ matrix

$[I_l]$ = loop current matrix

$[I_s]$ = source current matrix

3.20.2 Branch Voltages and Node Voltages Relationship

$$[V_b] = [Q]^T [V_n] + [V_s] \quad ...(3.33)$$

where $[V_b]$ = branch voltage matrix

$[Q]$ = fundamental cutset matrix

$[Q]$ = transpose of $[Q]$ matrix

$[V_n]$ = node voltage matrix

$[V_s]$ = source voltage matrix.

3.20.3 Loop or Mesh or KVL Equilibrium Equations

$$B Z_b B^T I_l = B (V_S - Z_b I_s) \quad ...(3.34)$$

i.e.,

$$ZI_l = E \quad ...(3.35)$$

where $Z = B Z_b B^T$ = loop impedance matrix of order $(b - n + 1) \times (b - n + 1)$

$E = B (V_s - Z_b I_s)$ = loop voltage coloumn matrix of order $(b - n + 1) \times 1$.

B = f-loop or tie-set matrix of order $(b - n + 1) \times b$.

Z_b = branch impedance matrix of order $b \times b$.

I_l = column matrix of loop currents of order $(b - n + 1) \times 1$.

V_s = column matrix of voltage sources of order $b \times 1$

I_s = column matrix of source currents of order $b \times 1$.

Equation represents a set of $(b - n + 1)$ independent loop equations for a network with n nodes and b branches.

3.20.4 Node or KCL Equilibrium Equations

$$Q Y_b Q^T V_t = Q (I_s - Y_b V_s) \quad ...(3.36)$$

or

$$Y V_t = I \quad ...(3.37)$$

where Y = node admittance matrix of order $(n-1) \times (n-1)$

V_t = column matrix of tree branch voltages of order $(n-1) \times 1$.

I = column matrix of order $(n-1) \times 1$ which represents algebraic sum of current sources.

Q = f-cutset matrix of order $(n-1) \times b$

Y_b = branch admittance matrix of order $b \times b$.

I_s = column matrix of branch current sources of order $b \times 1$.

V_s = column matrix of branch input voltage sources.

3.20.5 General Network Analysis

$$V_b = V_s + Z_b (I_b - I_s) \qquad \text{...(3.38)}$$

$$I_b = I_s + Y_b (V_b - V_s) \qquad \text{...(3.39)}$$

where V_b = branch voltage matrix of order $b \times 1$

I_b = branch current matrix of order $b \times 1$

V_s = voltage source matrix of order $b \times 1$

I_s = current source matrix of order $b \times 1$

Z_b = branch impedance matrix of order $b \times b$

Y_b = branch admittance matrix of order $b \times b$

3.20.5.1 Loop Analysis

Premultiplication of the equation (3.33) by B, we get:

$$B\, V_b = B\, V_s + B\, Z_b (I_b - I_s)$$

where B = reduced incidence matrix of graph of a network

We know that $B\, V_b = 0$ and $I_b = B^T I_l$,

$\therefore \quad B\, V_s + B\, Z_b (I_b - I_s) = 0$

or $$B\, Z_b\, I_s - B\, V_s = B\, Z_b\, I_b$$

or $$B\, Zb\, I_s - B\, V_s = B\, Z_b\, B^T I_l \qquad \text{...(3.40)}$$

If $$V_l = B\, Z_b\, I_s - BV_s \qquad \text{...(3.41)}$$

and $$Z_l = B\, Z_b\, B_T \qquad \text{...(3.42)}$$

Then, the equation (3.39) can be written as:

$$V_l = Z_l\, I_l \qquad \text{...(3.43)}$$

which is known as matrix loop equation, where, Z_1 = loop impedence matrix and I_1 = loop current matrix

3.20.5.2 Nodal Analysis

By premultiplication by A of the equation (3.38) we get:

$$A\, I_b = A\, I_s + A\, Y_b (V_b - V_s)$$

We know that $A\,I_b = 0$ and $V_b = A^T\,V_n$

$$\therefore \quad A\,I_s + A\,Y_b\,V_b = A\,Y_b\,V_s$$

or $$A\,I_s + A\,Y_b\,A^T\,V_n = A\,Y_b\,V_s$$

or $$A\,Y_b\,V_s - A\,I_s = A\,Y_b\,A^T\,V_n \qquad ...(3.44)$$

If $$I_n = A\,Y_b\,V_s - A\,I_s \qquad ...(3.45)$$

and $$Y_n = A\,Y_b\,A^T \qquad ...(3.46)$$

then, the equation (3.43) can be written as:

$$Y_n = Y_n\,V_n \qquad ...(3.47)$$

Which is known as matrix node equation

where, Y_n = node admittance matrix and

V_n = node voltage matrix.

3.20.5.3 Cutset Analysis

Premultiplication of the equation (3.36) by Q, we get :

$$Q\,I_b = Q[I_s + Y_b\,(V_b - V_s)]$$

we know that $$Q\,I_b = 0$$

$$\therefore \quad Q\,[I_s + Y_b\,(V_b - V_s)] = 0$$

or $$Q[I_s + Y_b\,(Q^T\,V_t - V_s)] = 0, \text{ as } V_b = Q^T\,V_t$$

or $$Q\,Y_b\,V_s - Q\,I_s = Q\,Y_b\,Q^T\,V_t \qquad ...(3.48)$$

If $$I = Q\,Y_b\,V_s - Q\,I_s \qquad ...(3.49)$$

and $$Y_t = Q\,Y_b\,Q^T \qquad ...(3.50)$$

then $$I = Y_t\,V_t \qquad ...(3.51)$$

where Y_t = node pair admittance matrix, and

V_t = tree branch voltage matrix.

Example 3.1. Draw the oriented graph, select a tree and draw a tie-set matrix for the circuit given below.

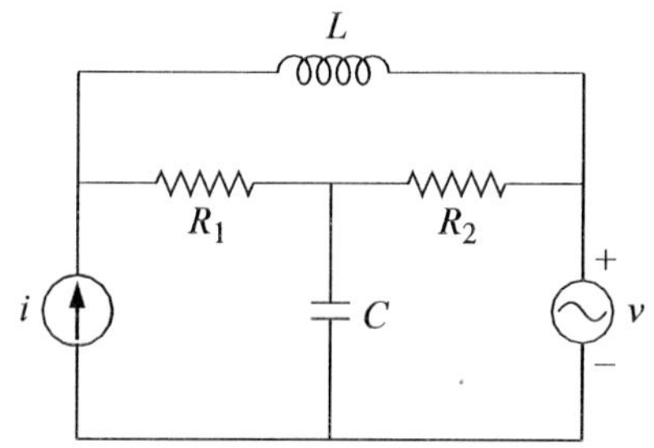

Solution: Consider that direction of currents are as follows:

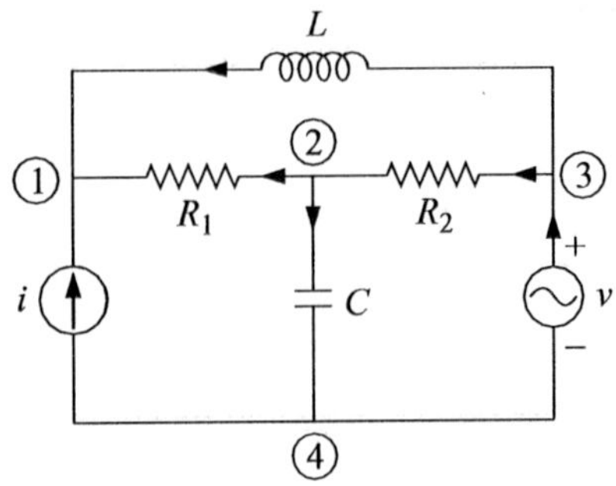

Oriented graph and selected tree are as follows:

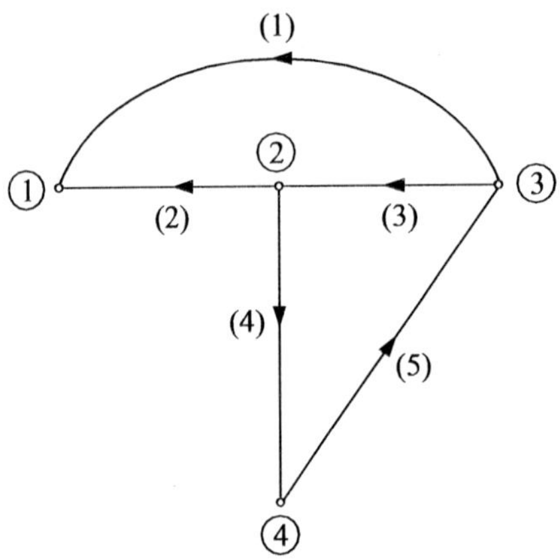

(*a*) *Oriented graph-current source becomes open circuit*

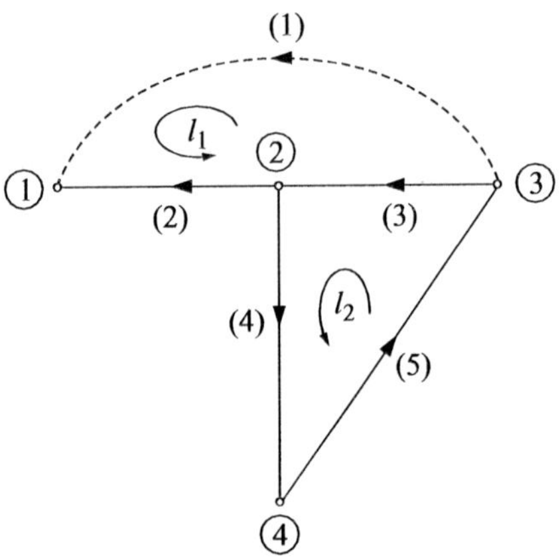

(*b*) *Selected tree*

The number of possible loops in this graph $= b - (n - 1)$

$$\text{Number of branches, } b = 5$$

$$\text{Number of nodes, } n = 4$$

$$\therefore \qquad \text{The number of possible loops} = 5 - (4 - 1) = 2$$

The tie-set matrix is as follows:

Hence, tie-set matrix is given as :

$$\begin{array}{c} \\ \\ B_f = \end{array} \begin{array}{c} f\text{-loops} \\ \\ \begin{array}{c} 1 \\ 2 \end{array} \end{array} \begin{array}{c} \text{twigs} \qquad \text{links} \\ \begin{array}{ccccc} 2 & 3 & 4 & 1 & 5 \end{array} \\ \begin{bmatrix} -1 & -1 & 0 & 1 & 0 \\ 0 & 1 & 1 & 0 & 1 \end{bmatrix} \end{array}$$

Example 3.2. For the network shown in the given figure, draw a directed graph select a suitable tree, write the cut-set and tie-set matrices.

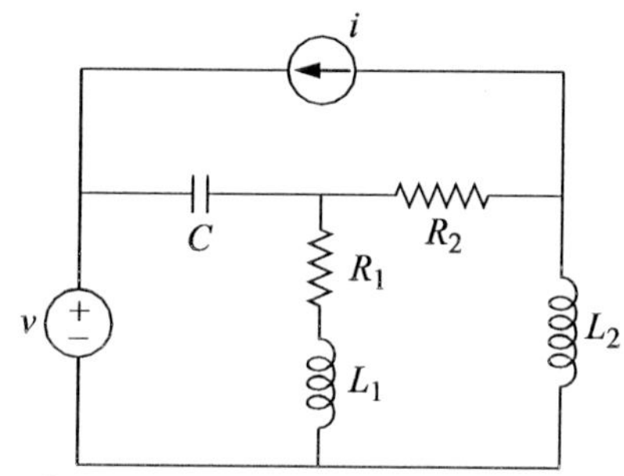

Solution: Consider the circuit with the current directions marked as given below:

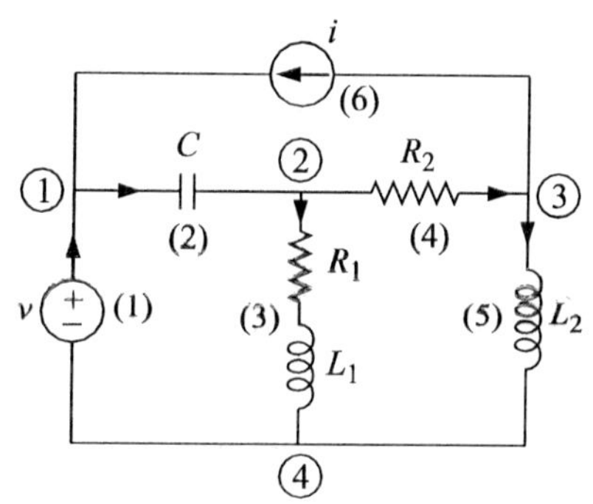

The current source will be taken as open circuit for the graph. Thus, the oriented is as given below:

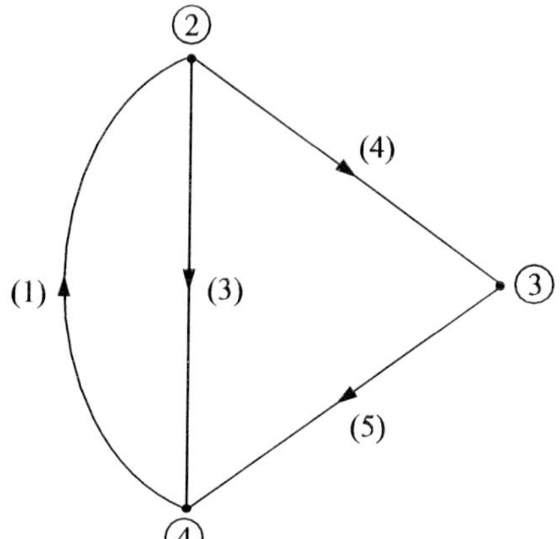

It may be noted that branches (1) and (2) have same direction of current, hence node of can be ignored.

Now, by selecting tree having (4) and (5) branches, we get:

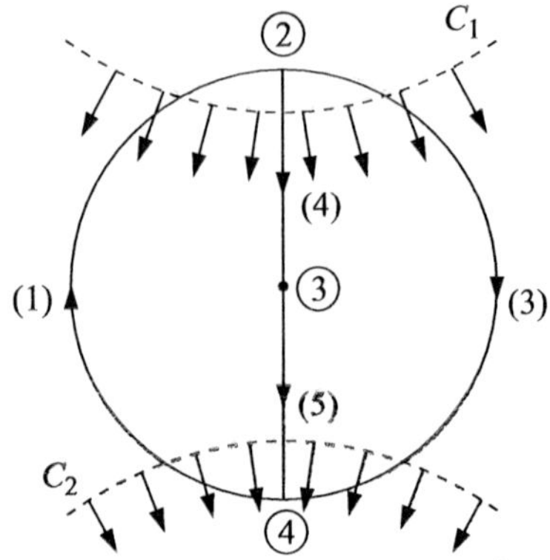

The branches have values 1, if the corresponding branch is in the same direction or else – 1. If the branch is not part of the tree, the value is zero. The arrows of the cuts are taken in direction of the twig. Thus, cutset is given as

$$\begin{array}{c} f\text{-cutssets} \\ \downarrow \text{twigs} \end{array} \quad \begin{array}{c} \text{branches} \rightarrow \\ \text{links} \end{array}$$

$$\begin{array}{c} \\ C_1 \\ C_2 \end{array}\begin{bmatrix} 4 & 5 & 1 & 3 \\ +i & 0 & -1 & +1 \\ 0 & +1 & -1 & +1 \end{bmatrix} \quad i.e., \begin{bmatrix} 1 & 0 & -1 & 1 \\ 0 & 1 & -1 & 1 \end{bmatrix}$$

The direction of loop current is taken in the direction of the link. The tie-set matrix is given for the loops shown in the figure:

$$\begin{array}{l} \text{tie-set} \\ \text{(loop)} \quad \text{branches} \rightarrow \end{array}$$

$$\begin{array}{c} \downarrow \\ i_1 \\ i_2 \end{array}\begin{array}{c} \begin{array}{cccc} 1 & 3 & 4 & 5 \end{array} \\ \begin{bmatrix} 1 & 0 & 1 & 1 \\ 0 & 1 & -1 & -1 \end{bmatrix} \end{array}$$

$$i.e., \quad \begin{bmatrix} 1 & 0 & 1 & 1 \\ 0 & 1 & -1 & -1 \end{bmatrix}$$

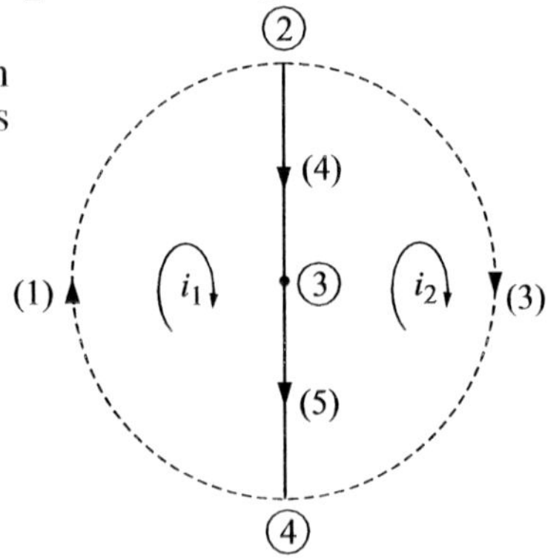

(c) *Tie-sets*

Example 3.3. With reference to the figure, draw the graph and write down the tie-set matrix. Obtain the network equilibrium equations.

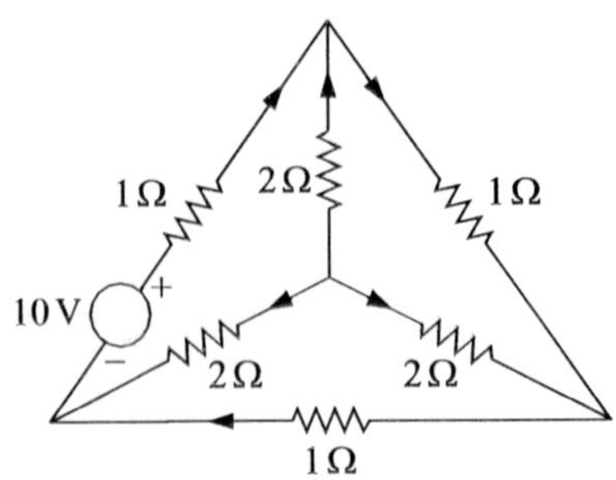

Solution:

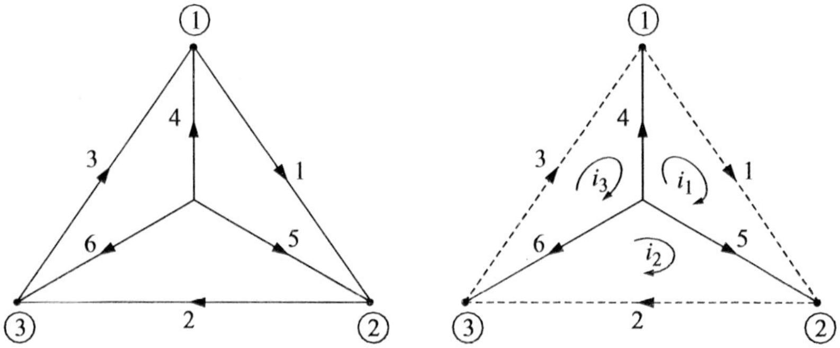

Tie-set matrix is shown below:

$$\begin{array}{c} \text{tie-set(loops)} \quad \text{branches} \rightarrow \\ \begin{array}{c} \downarrow \\ i_1 \\ i_2 \\ i_3 \end{array} \begin{array}{c} \begin{array}{cccccc} 1 & 2 & 3 & 4 & 5 & 6 \end{array} \\ \begin{bmatrix} 1 & 0 & 0 & 1 & -1 & 0 \\ 0 & 1 & 0 & 0 & 1 & -1 \\ 0 & 0 & 1 & -1 & 0 & 1 \end{bmatrix} \end{array} \end{array}$$

Let v_1, v_2, ..., v_6 be branch voltage drops.

Then form the matrix :

$$\left.\begin{aligned} v_1 + v_4 - v_5 &= 0 \\ v_2 + v_5 - v_6 &= 0 \\ v_3 - v_4 + v_6 &= 0 \end{aligned}\right\} \text{ Network equilibrium equation.}$$

Example 3.4. Write incident matrix, cutsets and tie-sets matrices for any selected tree of the circuit given in the figure.

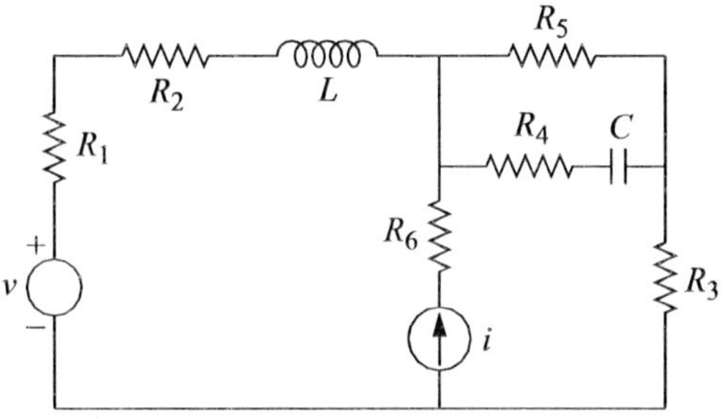

Solution: The directions of currents are chosen based on judgement, therefore, directed graph is taken as per figure (a) and selected tree as per figure (b).

The incidence matrix for the oriented graph of figure (*a*) is as follows:

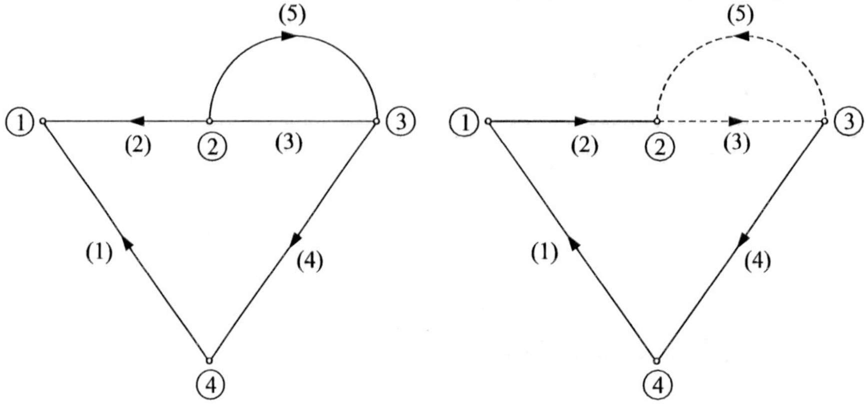

(*a*) *Oriented graph* (*b*) *Selected tree.*

$$\begin{array}{c} \text{Nodes} \\ \downarrow \end{array} \quad \text{Branches} \rightarrow$$

$$A_a = \begin{array}{c} \\ 1 \\ 2 \\ 3 \\ 4 \end{array} \begin{array}{c} \begin{array}{ccccc} 1 & 2 & 3 & 4 & 5 \end{array} \\ \begin{bmatrix} -1 & 1 & 0 & 0 & 0 \\ 0 & -1 & 1 & 0 & -1 \\ 0 & 0 & -1 & 1 & 1 \\ 1 & 0 & 0 & -1 & 0 \end{bmatrix} \end{array}$$

i.e.,

$$A_a = \begin{bmatrix} -1 & 1 & 0 & 0 & 0 \\ 0 & -1 & 1 & 0 & -1 \\ 0 & 0 & -1 & 1 & 1 \\ 1 & 0 & 0 & -1 & 0 \end{bmatrix}$$

If 4th row is eliminated *i.e.* treated as reference node, the reduced incident matrix is given by:

$$A = \begin{bmatrix} -1 & 1 & 0 & 0 & 0 \\ 0 & -1 & 1 & 0 & -1 \\ 0 & 0 & -1 & 1 & 1 \end{bmatrix}$$

The selected tree is shown in figure (*b*). The parameters are :

Number of branches, $b = 5$

Number of nodes, $n = 4$

Hence, number of possible tie-sets $= [b - (n - 1)]$

$$= [5 - (4 - 1)] = 2$$

The possible tie-sets are:

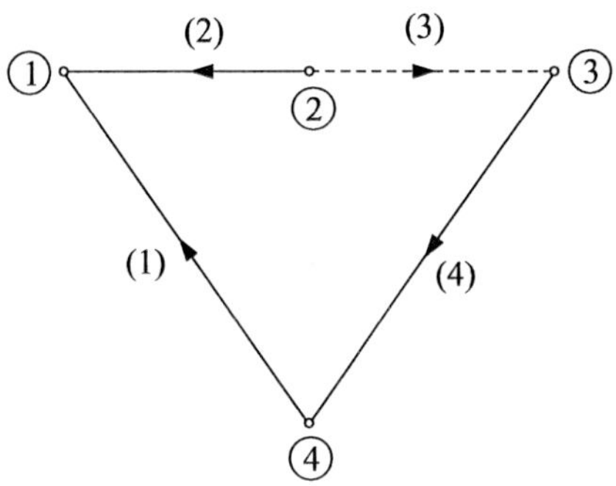

(*c*) *Tie-set 1*

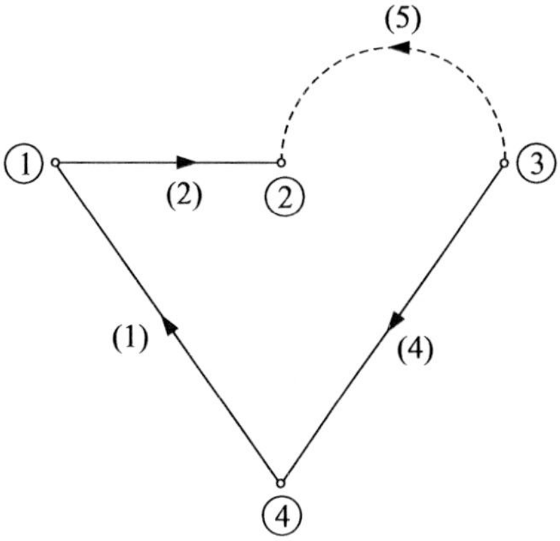

(*d*) *Tie-set 2*

The tie-set matrix is given as:

$$\begin{array}{cc} \textit{f-tie-sets} & \textit{Branches} \rightarrow \\ \downarrow & \begin{array}{ccccc} \textit{links} & & & & \textit{twigs} \\ 3 & 5 & 1 & 2 & 4 \end{array} \\ \begin{array}{c} 1 \\ 2 \end{array} & \begin{bmatrix} 1 & 0 & 1 & 1 & 1 \\ 0 & 1 & -1 & -1 & -1 \end{bmatrix} \end{array}$$

$$\therefore \qquad B = \begin{bmatrix} 1 & 0 & 1 & 1 & 1 \\ 0 & 1 & -1 & -1 & -1 \end{bmatrix}$$

The cut-sets for the chosen tree are :

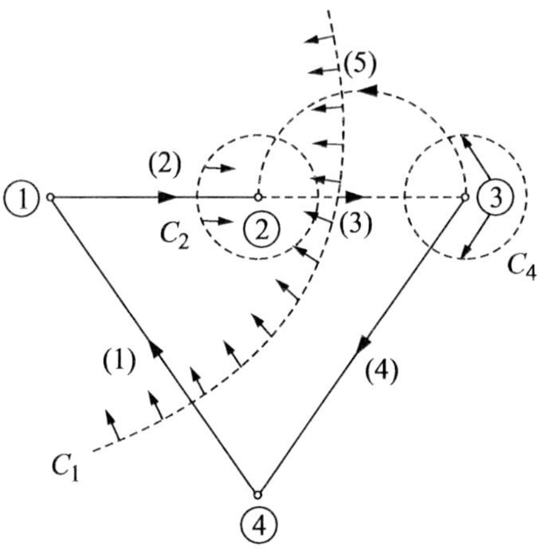

(*e*) *Cut-sets*

The cutset matrix is as follows:

$$\begin{array}{c} f\text{-cut-sets} \\ \downarrow \\ \\ C_1 \\ C_2 \\ C_3 \end{array} \begin{array}{c} \text{Branches} \rightarrow \\ \begin{array}{ccccc} & \text{Twigs} & & \text{Links} & \\ 1 & 2 & 4 & 3 & 5 \end{array} \\ \begin{bmatrix} 1 & 0 & 0 & -1 & 1 \\ 0 & 1 & 0 & -1 & 1 \\ 0 & 0 & 1 & -1 & 1 \end{bmatrix} \end{array}$$

It may be noted that matrix element sign (±) in the case the shown directions toward centre of circle or away from the centre.

$$\therefore \qquad Q = \begin{bmatrix} 1 & 0 & 0 & -1 & 1 \\ 0 & 1 & 0 & -1 & 1 \\ 0 & 0 & 1 & -1 & 1 \end{bmatrix}$$

Example 3.5. An oriented graph of certain circuit is shown in the given figure for the tree [2, 4, 6], find $[Q_f]$ and $[B_f]$ and hence shown that $[Q_f]\,[B_f]^T = 0$

[IPU, 2000]

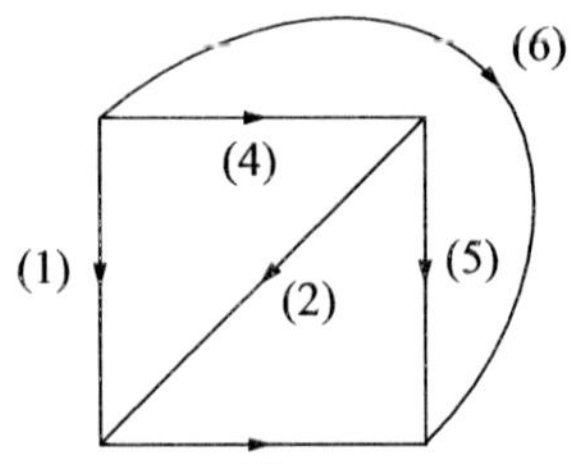

Solution: Tree having (2), (4), (6) as twigs is shown in figure (*a*) and fundamental cutsets are shown in figure (*b*) for the same twigs.

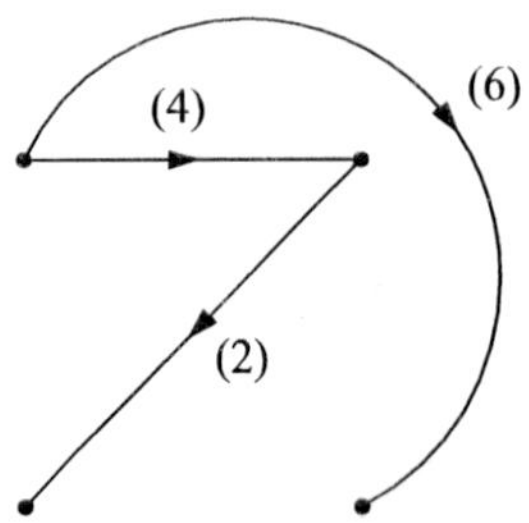

(*a*) *Tree*

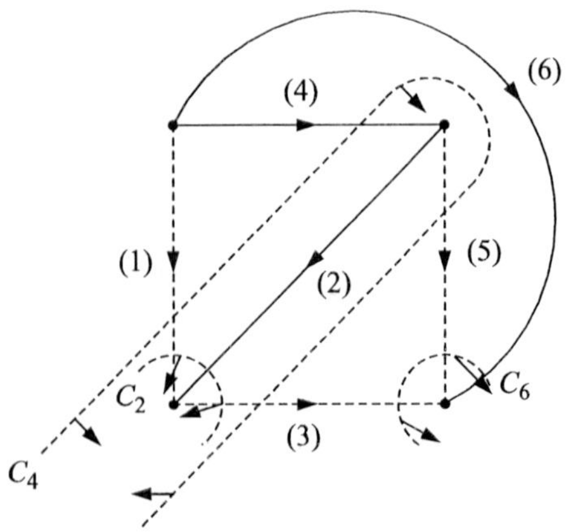

(*b*) *Cutsets*

The f cut-set matrix is given by :

$$\begin{array}{c} f\text{-cut sets} \quad \text{Branches} \rightarrow \\ \downarrow \quad \text{Twigs} \qquad\qquad \text{Links} \end{array}$$

$$\begin{array}{c} \\ C_2 \\ C_4 \\ C_6 \end{array}\begin{array}{c} \begin{array}{cccccc} 2 & 4 & 6 & 1 & 3 & 5 \end{array} \\ \begin{bmatrix} 1 & 0 & 0 & 1 & -1 & 0 \\ 0 & 1 & 0 & 1 & -1 & -1 \\ 0 & 0 & 1 & 0 & 1 & 1 \end{bmatrix} \end{array}$$

i.e.,
$$Q = \begin{bmatrix} 1 & 0 & 0 & 1 & -1 & 0 \\ 0 & 1 & 0 & 1 & -1 & -1 \\ 0 & 0 & 1 & 0 & 1 & 1 \end{bmatrix}$$

Number of f-tie sets = $b - (n - 1) = 5 - (3 - 1) = 3$. The f-tie set graphs are as follows:

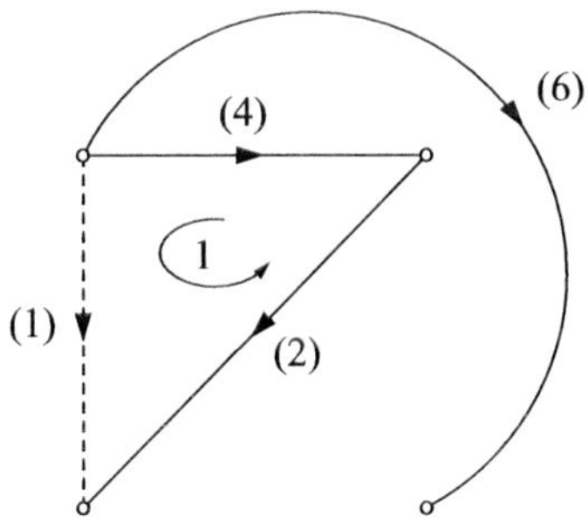

(*c*) *f-tie set 1*

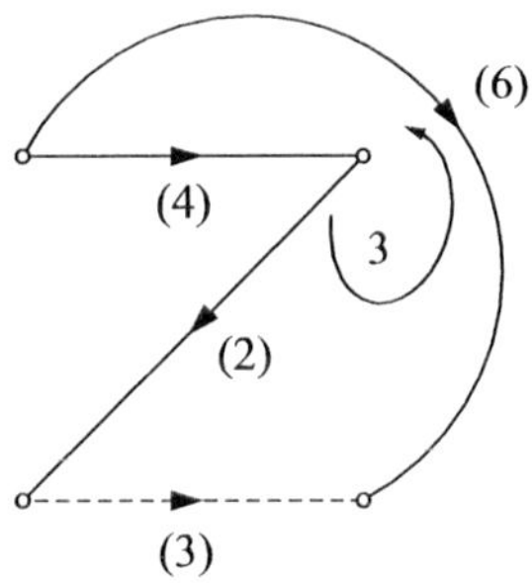

(*d*) *f-tie set 3*

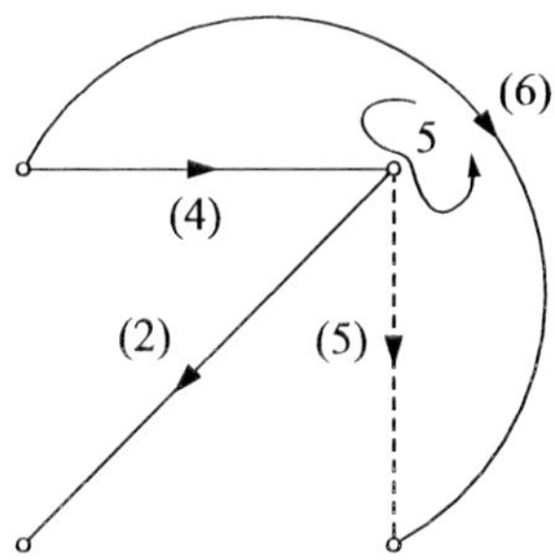

(*e*) *f-tie set 5*

The tie-set matrix is given by :

$$\begin{array}{c} f\text{-tie set} \quad \text{Branches} \\ \downarrow \quad \text{Twigs} \qquad \text{Links} \end{array}$$

$$\begin{array}{c} \\ 2 \\ 3 \\ 5 \end{array}\begin{array}{c} \begin{array}{cccccc} 2 & 4 & 6 & 1 & 3 & 5 \end{array} \\ \begin{bmatrix} -1 & 1 & 0 & 1 & 0 & 0 \\ 1 & 1 & -1 & 0 & 1 & 0 \\ 0 & 1 & -1 & 0 & 0 & 1 \end{bmatrix} \end{array}$$

i.e.,

$$B = \begin{bmatrix} -1 & 1 & 0 & 1 & 0 & 0 \\ 1 & 1 & -1 & 0 & 1 & 0 \\ 0 & 1 & -1 & 0 & 0 & 1 \end{bmatrix}$$

$\therefore$

$$[Q]\,[B]^T = \begin{bmatrix} 1 & 0 & 0 & 1 & -1 & 0 \\ 0 & 1 & 0 & 1 & -1 & -1 \\ 0 & 0 & 1 & 0 & 1 & 1 \end{bmatrix} \begin{bmatrix} -1 & 1 & 0 \\ -1 & 1 & 1 \\ 0 & -1 & -1 \\ 1 & 0 & 0 \\ 0 & 1 & 0 \\ 0 & 0 & 1 \end{bmatrix}$$

$$= \begin{bmatrix} 0 & 0 & 0 \\ 0 & 0 & 0 \\ 0 & 0 & 0 \end{bmatrix}$$

i.e., $[Q]\,[B]^T = 0$

Example 3.6. The reduced incidence matrix of the graph is given by

$$\begin{array}{c} \text{node} \quad \text{branches} \\ \downarrow \end{array}$$

$$\begin{array}{c} \\ a \\ b \\ c \end{array}\begin{array}{c} \begin{array}{cccccc} 1 & 2 & 3 & 4 & 5 & 6 \end{array} \\ \begin{bmatrix} 0 & 1 & 0 & -1 & -1 & 1 \\ 0 & -1 & 1 & 0 & 1 & 0 \\ 1 & 0 & -1 & 0 & 0 & 0 \end{bmatrix} \end{array}$$

Draw the oriented graph. Select a tree and fist f-cutset matrix.

[IPU, 2001]

Solution: Incidence matrix can be written with the principle that algebraic sum of each column should be zero. Hence, incidence matrix is given by:

$$
C \qquad A_a = \begin{array}{c} \text{Nodes,} \\ \downarrow \\ a \\ b \\ c \\ d \end{array}
\begin{array}{c} \\ \begin{array}{cccccc} 1 & 2 & 3 & 4 & 5 & 7 \leftarrow \text{branches} \end{array} \\
\begin{bmatrix} 0 & 1 & 0 & -1 & -1 & 1 \\ 0 & -1 & 1 & 0 & 1 & 0 \\ 1 & 0 & -1 & 0 & 0 & 0 \\ -1 & 0 & 0 & 1 & 0 & -1 \end{bmatrix} \end{array}
$$

Using incidence matrix, the oriented graph is given in figure (*a*)

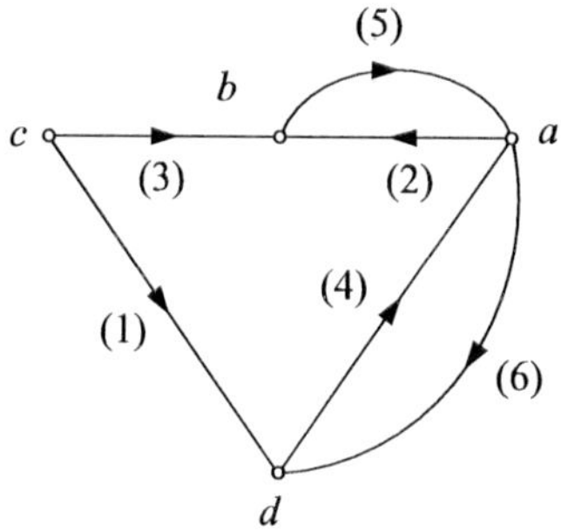

(*a*) *Oriented graph*

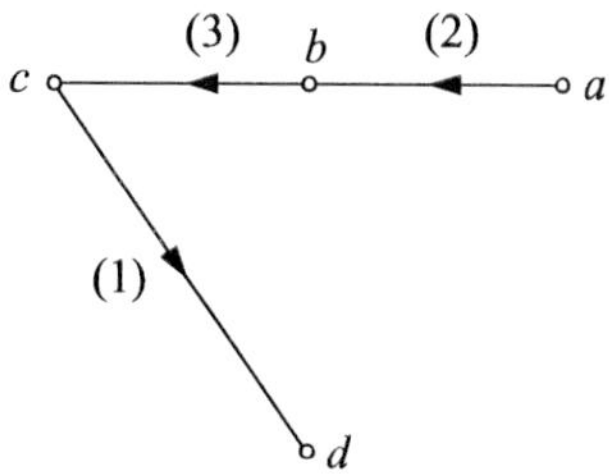

(*b*) *Selected tree*

The selected tree is shown in figure (*b*) and figure (*c*) shown corresponding *f*-cutsets.

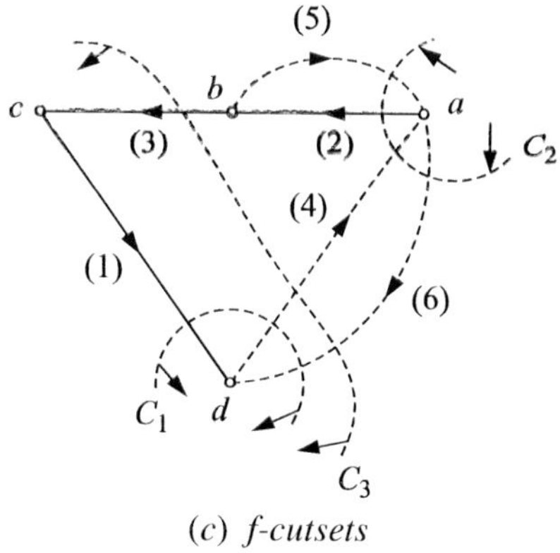

(*c*) *f-cutsets*

Cutset matrix is given by :

$$
\begin{array}{c} f\text{-cutsets} \\ \downarrow \\ \\ C_1 \\ C_2 \\ C_3 \end{array}
\begin{array}{c} \text{branches} \rightarrow \\ \text{Twigs} \qquad\qquad \text{Links} \\ \begin{array}{cccccc} 1 & 2 & 3 & 4 & 5 & 6 \end{array} \\
\begin{bmatrix} 1 & 0 & 0 & -1 & 0 & 1 \\ 0 & 1 & 0 & -1 & -1 & 1 \\ 0 & 0 & 1 & -1 & 0 & 1 \end{bmatrix} \end{array}
$$

$$\text{i.e.,} \qquad Q = \begin{bmatrix} 1 & 0 & 0 & -1 & 0 & 1 \\ 0 & 1 & 0 & -1 & -1 & 1 \\ 0 & 0 & 1 & -1 & 0 & 1 \end{bmatrix}$$

Example 3.7. For the network shown in the figure given, drawn network graph selecting elements 1, 2, 3 as the tree elements. Obtain basic cutsets and write basic cutset matrix. **[UPTU, 2002]**

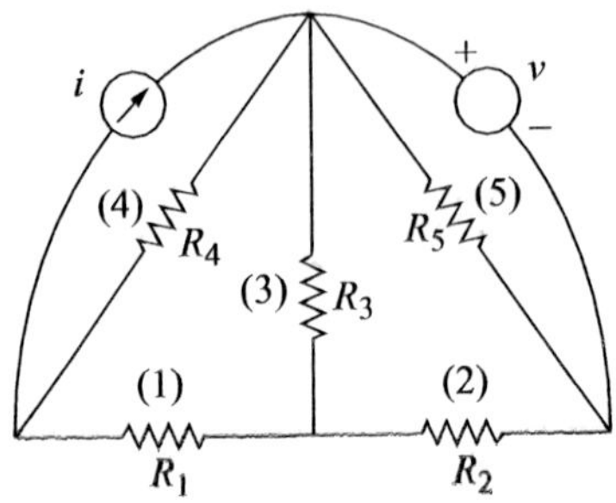

Solution: The voltage source is short circuited and current source is open circuited. A oriented graph of the circuit is drawn in figure (*a*).

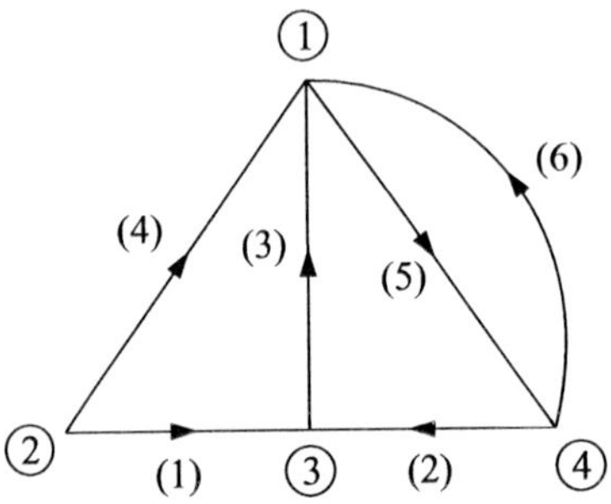

(*a*) *Oriented graph*

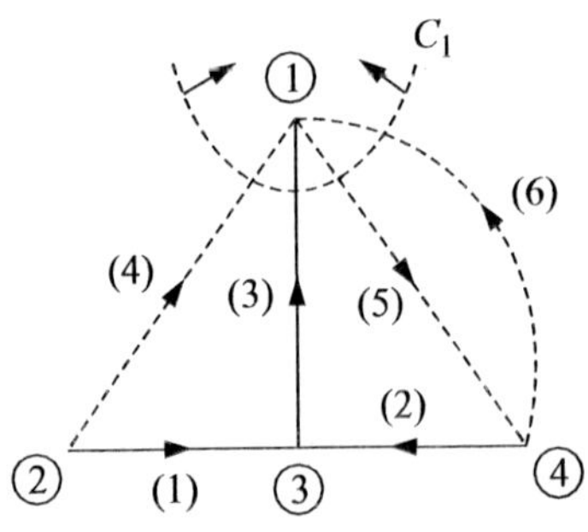

(*b*) *f-cutsets*

The *f*-cutset matrix is given by :

f-cutsets branches →

↓ twig links

$$\begin{matrix} & 1 & 2 & 3 & 4 & 5 & 6 \\ C_1 & 1 & 0 & 0 & 1 & 0 & 0 \\ C_2 & 0 & 1 & 0 & 0 & -1 & 1 \\ C_3 & 0 & 0 & 1 & 1 & -1 & 1 \end{matrix}$$

$$\text{i.e.,} \qquad Q \begin{bmatrix} 1 & 0 & 0 & 1 & 0 & 0 \\ 0 & 1 & 0 & 0 & -1 & 1 \\ 0 & 0 & 1 & 1 & -1 & 1 \end{bmatrix}$$

Example 3.8. A network graph has three basic cutsels and six basic loops. Draw :

(*i*) The oriented network graph having all the nodes in one line.

(*ii*) All the six basic loops. **[UPTU, 2002]**

Solution: (*i*) We know that the number of twigs equals number basic cutsets. Further, the number of nodes equal number of twigs plus one. Hence, the graph is given by:

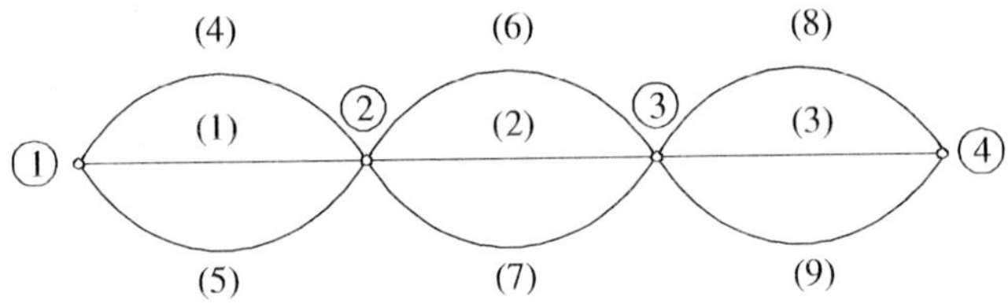

(*ii*) All the six basic loops

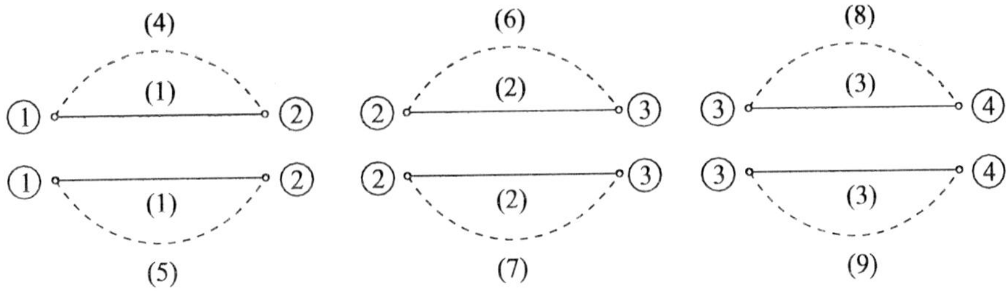

Example 3.9. For the graph shown in the figure,

(*i*) Write down augmented incidence matrix.

(*ii*) Indicate the number of fundamental loops. **[UPTU, 2004]**

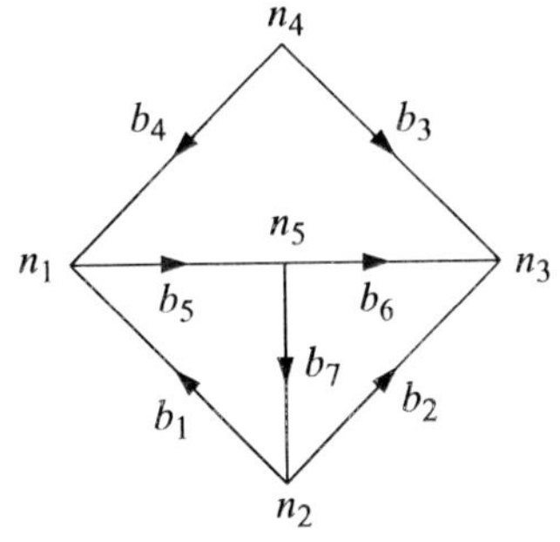

Solution: (*i*) Complete or augmented incidence matrix for the graph is as under:

$$A_a = \begin{array}{c} \\ n_1 \\ n_2 \\ n_3 \\ n_4 \\ n_5 \end{array} \begin{array}{c} \text{nodes} \downarrow \quad \text{branches} \rightarrow \\ \begin{bmatrix} b_1 & b_2 & b_3 & b_4 & b_5 & b_6 & b_7 \\ -1 & 0 & 0 & 1 & 1 & 0 & 0 \\ 1 & 1 & 0 & 0 & 0 & 0 & -1 \\ 0 & -1 & -1 & 0 & 0 & -1 & 0 \\ 0 & 0 & 1 & -1 & 0 & 0 & 0 \\ 0 & 0 & 0 & 0 & -1 & 1 & 1 \end{bmatrix} \end{array}$$

(*ii*) Number of nodes $n = 5$, number of branch, $b = 7$

$\therefore$ Number fundamental loops $=$ Number of links

$$= b - (n - 1) = 7 - (5 - 1)$$

$$= 7 - 4 = 3$$

Example 3.10. Draw the network graph of a network shown the figure given and reduce in the simplest form eliminating temporary nodes. Divide the reduced network graph into two connected sub-graphs eliminating minimum number of elements. **[UPTU, 2003]**

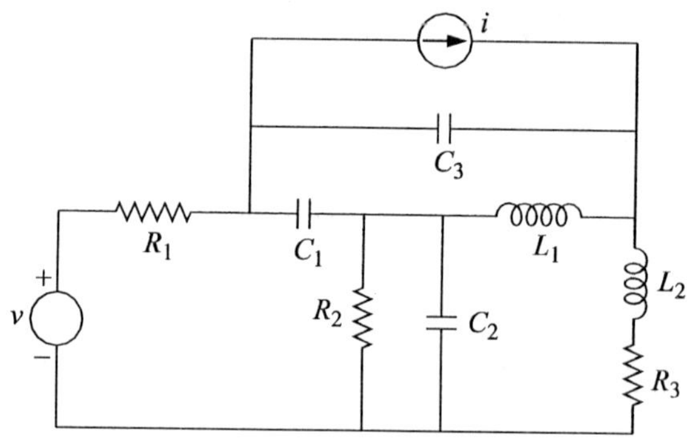

Solution: The voltage source is shorted and current source is open circuited and then graph is as shown figure (*a*).

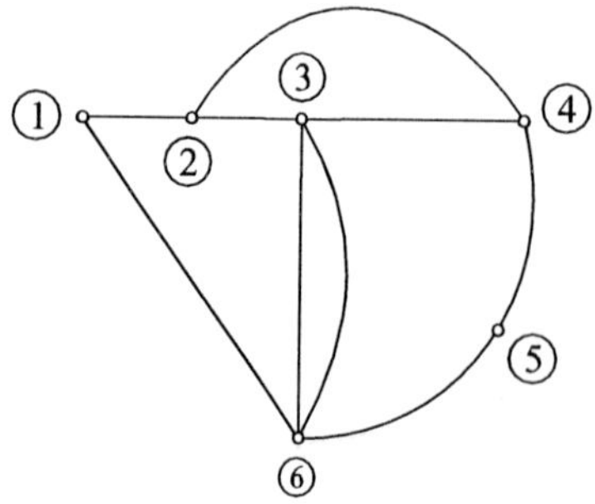

(*a*) *Graph of the network*

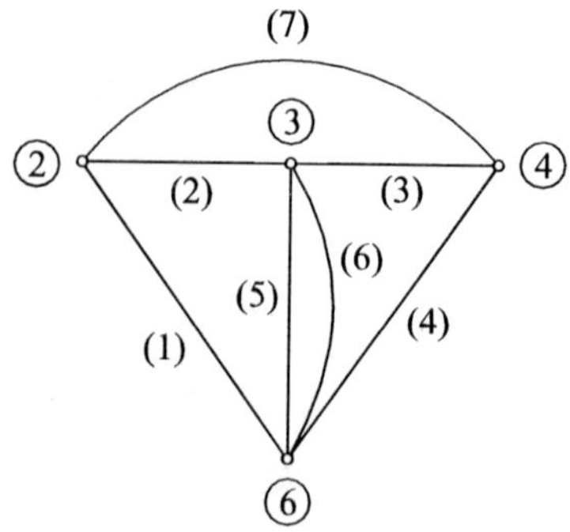

(*b*)

R_1, and v can be treated as single branch, L_2 and R_3 can be treated in single branch, therefore, simplest form of graph is shown figure (*b*). The two connected subgraphs eliminating minimum number of elements are shown in figure (*c*).

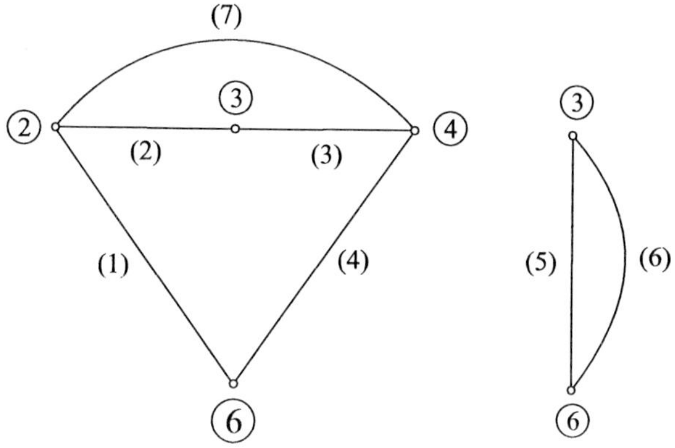

(*c*) *Two subgraphs*

Example 3.11. Draw the graph of the network shown in the figure given, Find the number of possible trees and drain them. **[UPTU, 2005]**

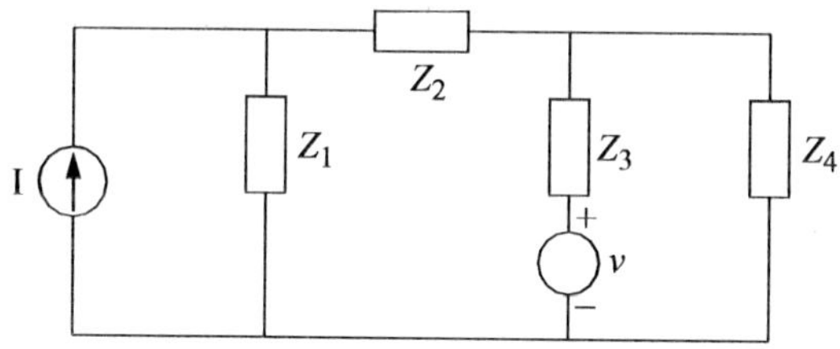

Solution: The current sour is taken as open circuit and voltage source as short circuit. Then the graph of the network is given in figure (*a*).

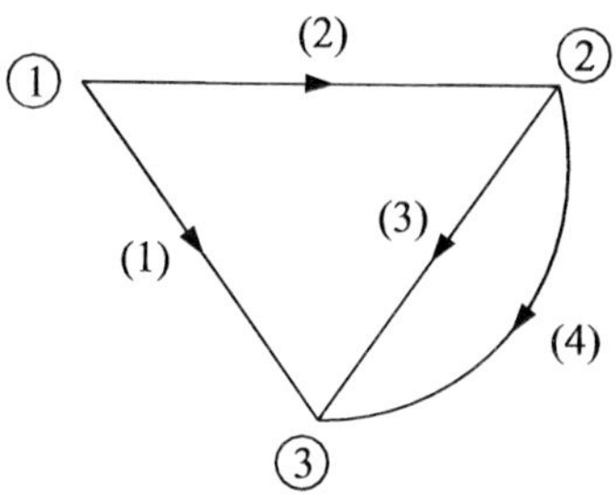

(*a*) *oriented, graph of the network*

Node us can be taken as reference, therefore, reduced incident matrix is given by,

$$\begin{array}{c} \text{nodes} \\ \downarrow \\ (1) \\ (2) \end{array} \begin{array}{c} \text{branches} \rightarrow \\ \begin{array}{cccc} 1 & 2 & 3 & 4 \end{array} \\ \begin{bmatrix} 1 & 1 & 0 & 0 \\ 0 & -1 & 1 & 1 \end{bmatrix} \end{array}$$

i.e.
$$A = \begin{bmatrix} 1 & 1 & 0 & 0 \\ 0 & -1 & 1 & 1 \end{bmatrix}$$

Number of trees can be determined by value of det $[A\,A^T]$

$$AA^T = \begin{bmatrix} 1 & 1 & 0 & 0 \\ 0 & -1 & 1 & 1 \end{bmatrix} \begin{bmatrix} 1 & 0 \\ 1 & -1 \\ 0 & 1 \\ 0 & 1 \end{bmatrix} = \begin{bmatrix} 2 & -1 \\ -1 & 3 \end{bmatrix}$$

$$\therefore \text{Number of trees} = |A\,A^T| = \begin{vmatrix} 2 & -1 \\ -1 & 3 \end{vmatrix} = 6 - 1 = 5$$

The corresponding trees are as follows:

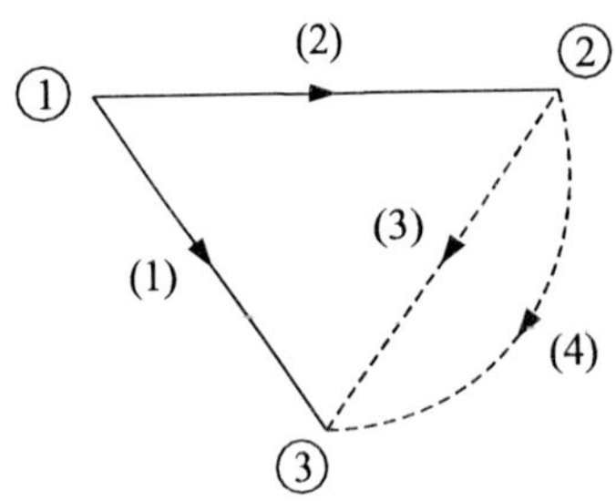

(*i*)

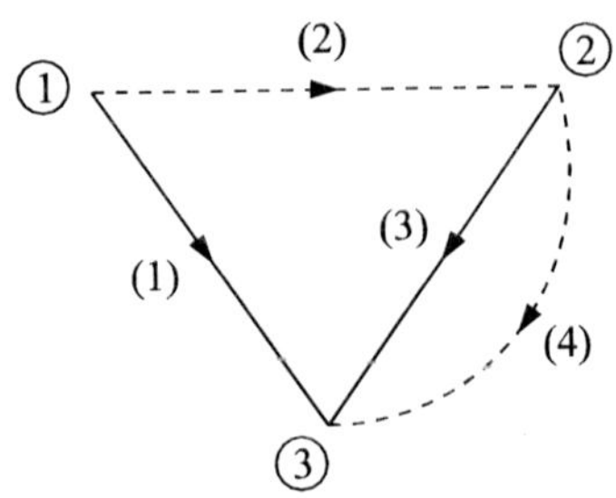

(*ii*)

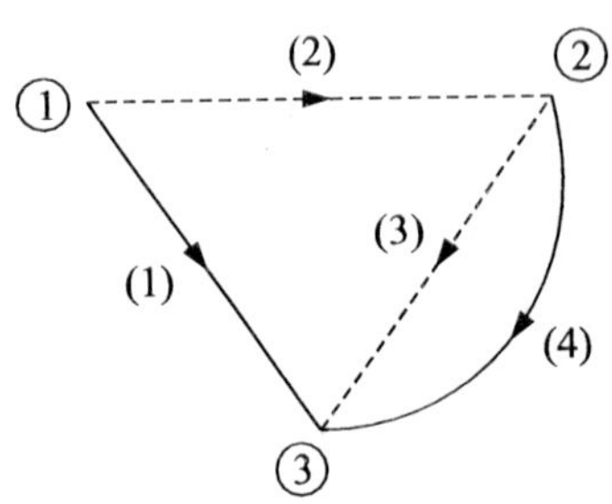

(*iii*)

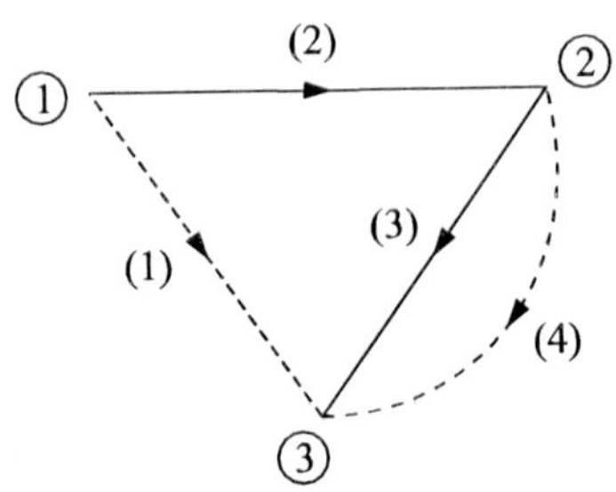

(*iv*)

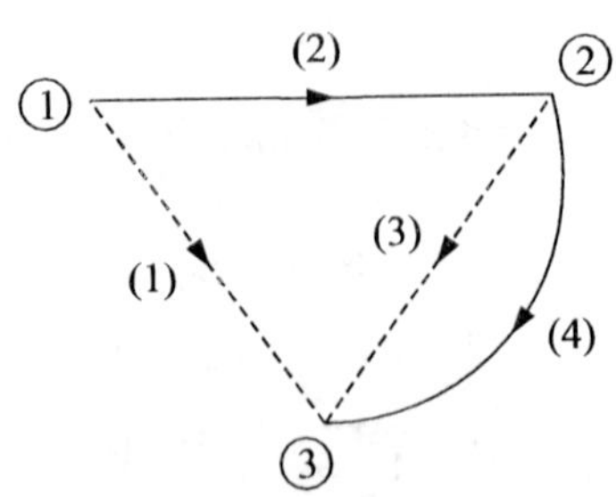

(*v*)

Example 3.12. A reduced incidence matrix of a graph is given by :

$$[A] = \begin{bmatrix} 1 & 1 & 0 & 0 & 0 & 1 \\ 0 & -1 & 1 & -1 & 0 & 0 \\ -1 & 0 & -1 & 0 & -1 & 0 \end{bmatrix}$$

Obtain the number of possible trees. **[UPTU, 2006]**

Solution: Number of possible trees = det $[A\, A^T]$

From given incidence matrix $[A]$, we get

$$A\,A^T = \begin{bmatrix} 1 & 1 & 0 & 0 & 0 & 1 \\ 0 & -1 & 1 & -1 & 0 & 0 \\ -1 & 0 & -1 & 0 & -1 & 0 \end{bmatrix} \begin{bmatrix} 1 & 0 & -1 \\ 1 & -1 & 0 \\ 0 & 1 & -1 \\ 0 & -1 & 0 \\ 0 & 0 & -1 \\ 1 & 0 & 0 \end{bmatrix}$$

$$= \begin{bmatrix} 3 & -1 & -1 \\ -1 & 3 & -1 \\ -1 & -1 & 3 \end{bmatrix}$$

$$\therefore \qquad \det[A\,A^T] = \begin{vmatrix} 3 & -1 & -1 \\ -1 & 3 & -1 \\ -1 & -1 & 3 \end{vmatrix}$$

$$= 3(9-1) - (-1)(-3-1) - 1\,(1+3) = 16$$

Hence, number of possible trees are 16.

Example 3.13. Develop the tie-set matrix of the circuit given in the figure.

[UPTU, 2006]

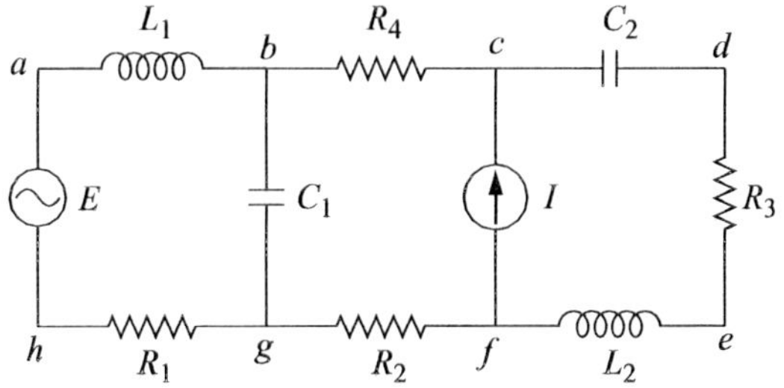

Solution: An oriented graph of the circuit is given in figure (a).

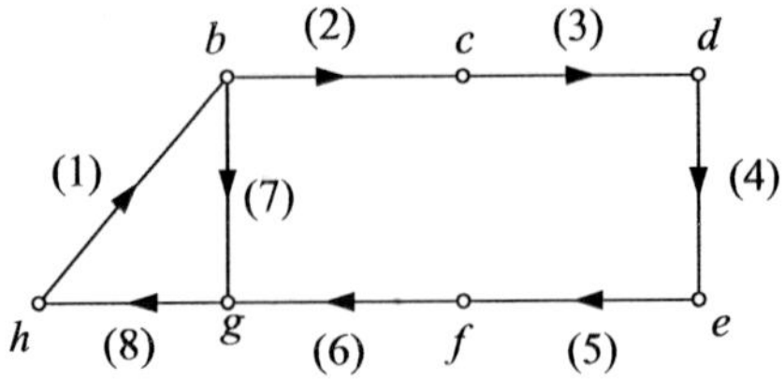

(*a*) *Oriented graph of the circuit*

A tree of the graph and *f*-loops are shown in figure (*b*).

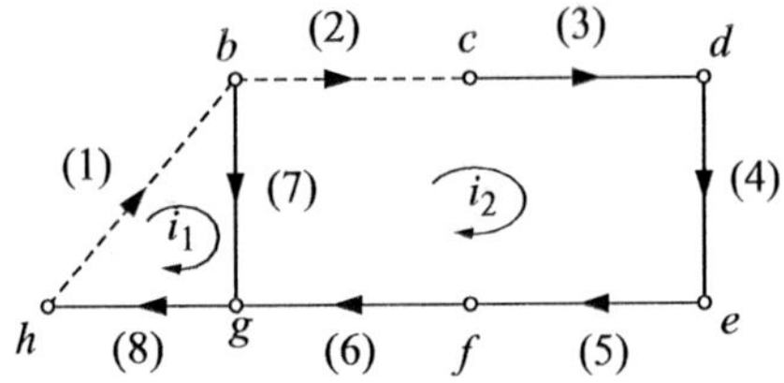

(*b*) *Tree with f-loops*

The tie-set matrix is given by :

$$
\begin{array}{c} f\text{-loops} \\ \downarrow \\ i_1 \\ i_2 \end{array}
\quad
\begin{array}{c} \text{branches} \rightarrow \\ \begin{array}{cccccccc} 1 & 2 & 3 & 4 & 5 & 6 & 7 & 8 \end{array} \\ \begin{bmatrix} 1 & 0 & 0 & 0 & 0 & 0 & 1 & 1 \\ 0 & 1 & 1 & 1 & 1 & 1 & -1 & 0 \end{bmatrix} \end{array}
$$

i.e.
$$B = \begin{bmatrix} 1 & 0 & 0 & 0 & 0 & 0 & 1 & 1 \\ 0 & 1 & 1 & 1 & 1 & 1 & -1 & 0 \end{bmatrix}$$

Example 3.14. Draw a graph of the network shown in the figure. Select a suitable tree and obtain tie-set matrix. **[UPTU, 2006]**

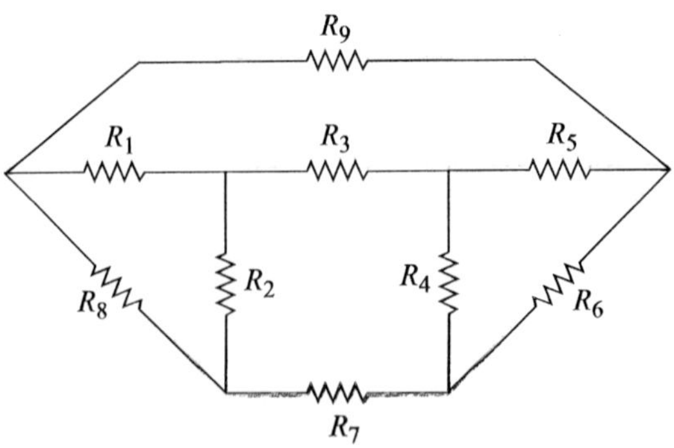

Solution: An oriented graph of the network is shown in figure (*a*).

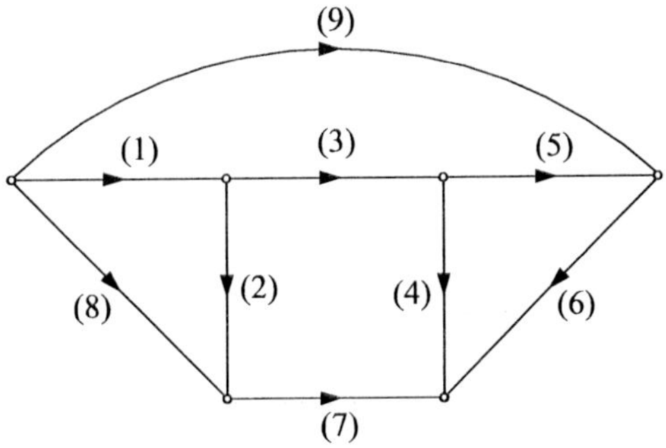

(*a*) *Oriented of the given network*

The chosen tree is shown by dark lines and corresponding loops are shown in figure (*b*).

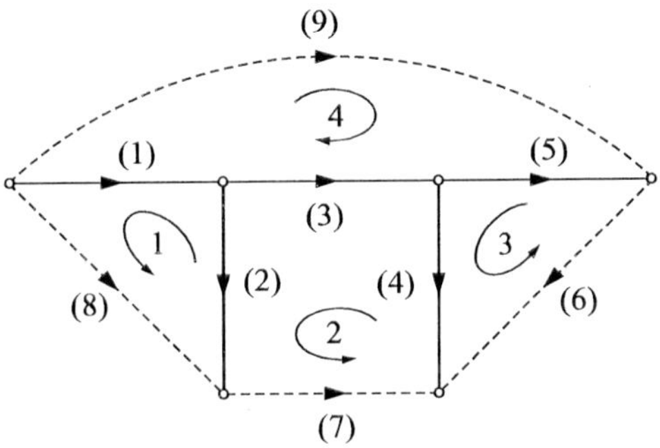

The tie-set matrix is given by :

$$\begin{array}{c} f\text{-loops} \\ \downarrow \\ 1 \\ 2 \\ 3 \\ 4 \end{array} \begin{array}{c} \text{branches} \rightarrow \\ \begin{array}{ccccccccc} 1 & 2 & 3 & 4 & 5 & 6 & 7 & 8 & 9 \end{array} \\ \begin{bmatrix} 0 & 0 & 0 & -1 & 1 & 1 & 0 & 0 & 0 \\ 0 & 1 & -1 & -1 & 0 & 0 & 1 & 0 & 0 \\ -1 & -1 & 0 & 0 & 0 & 0 & 0 & 1 & 0 \\ -1 & 0 & -1 & 0 & -1 & 0 & 0 & 0 & 1 \end{bmatrix} \end{array}$$

$$\therefore \qquad B = \begin{bmatrix} 0 & 0 & 0 & -1 & 1 & 1 & 0 & 0 & 0 \\ 0 & 1 & -1 & -1 & 0 & 0 & 1 & 0 & 0 \\ -1 & -1 & 0 & 0 & 0 & 0 & 0 & 1 & 0 \\ -1 & 0 & -1 & 0 & -1 & 0 & 0 & 0 & 1 \end{bmatrix}$$

Example 3.15 Find Norton's current I_N for the network shown in the given figure using graph theory. **[UPTU, 2006]**

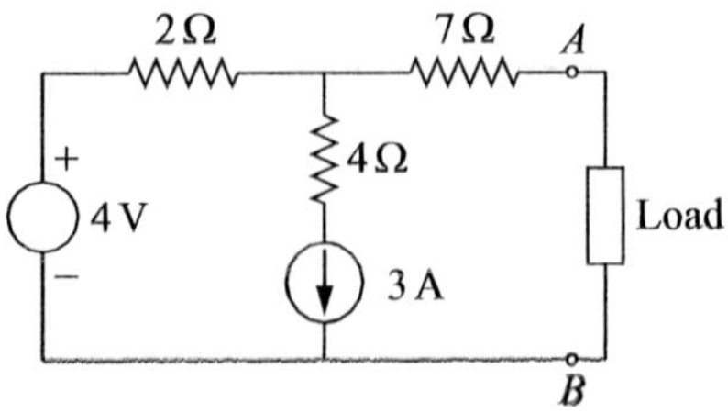

Solution: Norton current I_N is obtained by shorting the load, hence, network is show in figure (*a*).

Consider that voltage source has zero resistance and voltage source has zero admittance. An oriented graph of the network is shown in figure (*a*).

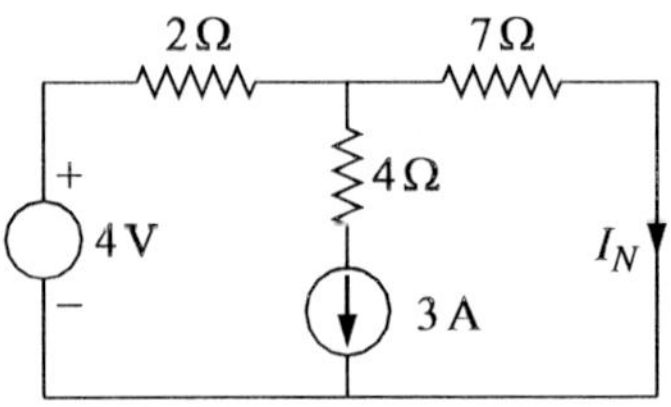

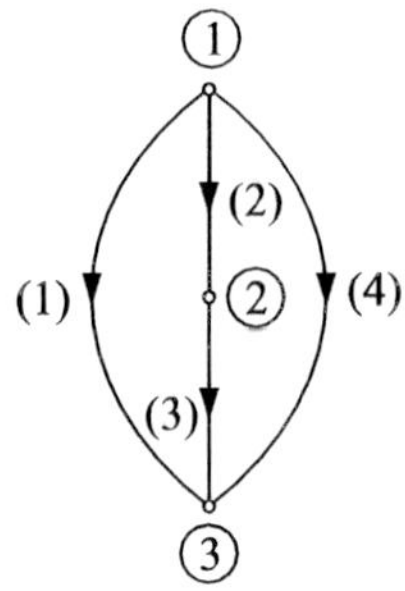

(*a*) (*b*) *Obtained graph of the given network*

Reduced incidence matrix is given by :

nodes branches

$$\begin{array}{c} \downarrow \\ 1 \\ 2 \end{array} \begin{array}{c} \begin{array}{cccc} 1 & 2 & 3 & 4 \end{array} \\ \begin{bmatrix} 1 & 1 & 0 & 1 \\ 0 & -1 & 1 & 0 \end{bmatrix} \end{array}$$

i.e.,

$$A = \begin{bmatrix} 1 & 1 & 0 & 1 \\ 0 & -1 & 1 & 0 \end{bmatrix}$$

The branch admittance matrix is given by :

$$Y_b = \begin{bmatrix} 1/2 & 0 & 0 & 0 \\ 0 & 1/4 & 0 & 0 \\ 0 & 0 & 0 & 0 \\ 0 & 0 & 0 & 1/7 \end{bmatrix}$$

Node admittance matrix is given by:

$$Y_n = A\, Y_b\, A^T$$

$$= \begin{bmatrix} 1 & 1 & 0 & 1 \\ 0 & -1 & 1 & 0 \end{bmatrix} \begin{bmatrix} 1/2 & 0 & 0 & 0 \\ 0 & 1/4 & 0 & 0 \\ 0 & 0 & 0 & 0 \\ 0 & 0 & 0 & 1/7 \end{bmatrix} \begin{bmatrix} 1 & 0 \\ 1 & -1 \\ 0 & 1 \\ 1 & 0 \end{bmatrix}$$

$$= \begin{bmatrix} \frac{1}{2} & \frac{1}{4} & 0 & \frac{1}{7} \\ 0 & -\frac{1}{4} & 0 & 0 \end{bmatrix} \begin{bmatrix} 1 & 0 \\ 1 & -1 \\ 0 & 1 \\ 1 & 0 \end{bmatrix}$$

$$= \begin{bmatrix} \left(\frac{1}{2}+\frac{1}{4}+\frac{1}{7}\right) & -\frac{1}{4} \\ -\frac{1}{4} & \frac{1}{4} \end{bmatrix} = \begin{bmatrix} \frac{25}{28} & -\frac{1}{4} \\ -\frac{1}{4} & \frac{1}{4} \end{bmatrix}$$

or
$$Y_n = \begin{bmatrix} \frac{25}{28} & -\frac{1}{4} \\ -\frac{1}{4} & \frac{1}{4} \end{bmatrix}$$

$$Y_n\, V_n = \mathrm{A}\,[I_s + Y_b\, V_s]$$

where
$$I_s = \begin{bmatrix} 0 \\ 0 \\ 3 \\ 0 \end{bmatrix}$$

and
$$V_s = \begin{bmatrix} 4 \\ 0 \\ 0 \\ 0 \end{bmatrix}$$

$$Y_n\, V_n = -AI_s + A\, Y_b\, V_s$$

$$= \begin{bmatrix} 1 & 1 & 0 & 1 \\ 0 & -1 & 1 & 0 \end{bmatrix} \begin{bmatrix} 0 \\ 0 \\ 3 \\ 0 \end{bmatrix} + \begin{bmatrix} 1 & 1 & 0 & 1 \\ 0 & -1 & 1 & 0 \end{bmatrix} \begin{bmatrix} 1/2 & 0 & 0 & 0 \\ 0 & 1/4 & 0 & 0 \\ 0 & 0 & 0 & 0 \\ 0 & 0 & 0 & 1/2 \end{bmatrix} \begin{bmatrix} 4 \\ 0 \\ 0 \\ 0 \end{bmatrix}$$

$$= -\begin{bmatrix} 0 \\ 3 \end{bmatrix} + \begin{bmatrix} \frac{1}{2} & \frac{1}{4} & 0 & \frac{1}{7} \\ 0 & -\frac{1}{4} & 0 & 0 \end{bmatrix} \begin{bmatrix} 4 \\ 0 \\ 0 \\ 0 \end{bmatrix}$$

$$= -\begin{bmatrix} 0 \\ 3 \end{bmatrix} + \begin{bmatrix} 2 \\ 0 \end{bmatrix} = \begin{bmatrix} 2 \\ -3 \end{bmatrix}$$

$\therefore$ $$Y_n V_n = \begin{bmatrix} 2 \\ -3 \end{bmatrix}$$

i.e, $$\begin{bmatrix} \frac{25}{28} & -\frac{1}{4} \\ -\frac{1}{4} & \frac{1}{4} \end{bmatrix} \begin{bmatrix} V_1 \\ V_2 \end{bmatrix} = \begin{bmatrix} 2 \\ -3 \end{bmatrix}$$

or $$\frac{25}{28} V_1 - \frac{1}{4} V_2 = 2 \quad ...(i)$$

and $$-\frac{1}{4} V_1 + \frac{1}{4} V_2 = -3 \quad ...(ii)$$

By simplifying Eqn (*ii*), we get:

$$V_2 = V_1 - 12 \quad ...(iii)$$

By substituting V_2 from Eqn. (*iii*) into Eqn. (*i*), we get :

or $$\frac{25}{28} V_1 - \frac{1}{4}(V_1 - 12) = 2 \quad ...(i)$$

and $$-\frac{1}{4} V_1 + \frac{1}{4} V_2 = -3 \quad ...(ii)$$

By simplifying Eqn (*ii*), we get :

$$V_2 = V_1 - 12 \quad ...(iii)$$

By substituting V_2 from Eqn. (*iii*) into Eqn. (*i*), we get :

$$\frac{25}{28} V_1 - \frac{1}{4}(V_1 - 12) = 2$$

or $$\frac{25}{28} V_1 - \frac{1}{4} V_1 + 3 = 2$$

or $$\frac{9}{14} V_1 = -1$$

or $$V_1 = -\frac{14}{9} V$$

$$I_N = \frac{V_1}{7} = -\frac{14}{9} \times \frac{1}{7}$$

or $$I_N = -\frac{2}{9} A$$

Example 3.16. Draw the oriented graph of a network with the fundamental cut-set matrix as shown below :

Twigs				Links		
1	2	3	4	5	6	7
1	0	0	0	–1	0	0
0	1	0	0	1	0	1
0	0	1	0	0	1	1
0	0	0	1	0	1	0

Also find numbers of fundamental cutsets and draw them. **[UPTU, 2006]**

Solution: The fundamental cutset corresponds to a twig, therefore, fundamental cutsets can be first drawn.

(*i*) Twig (1) Links (5) : C

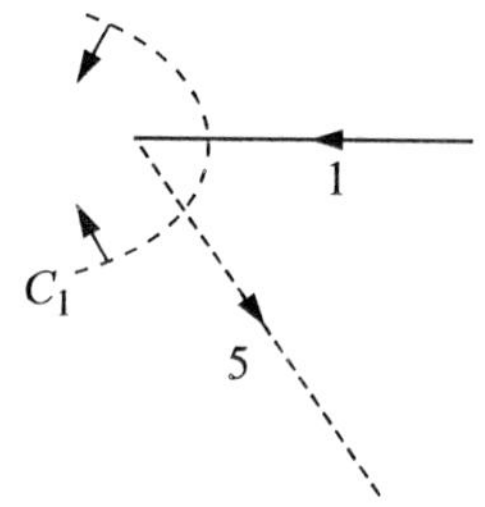

(*ii*) Twig (2) and Links (5) and (4): C_2

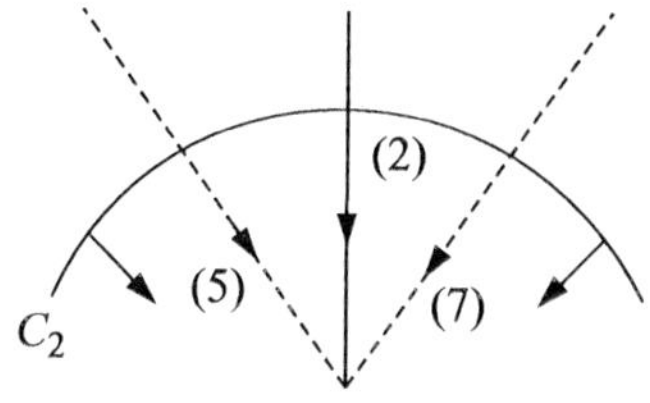

(*iii*) Twig (3) and Links (6), (7) : C_3

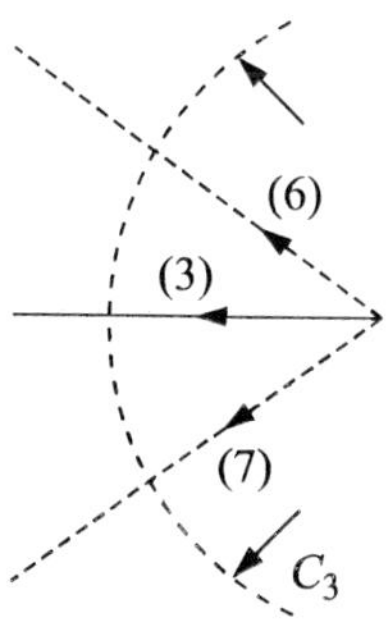

(*iv*) Twig (4) and Link (6): C_4

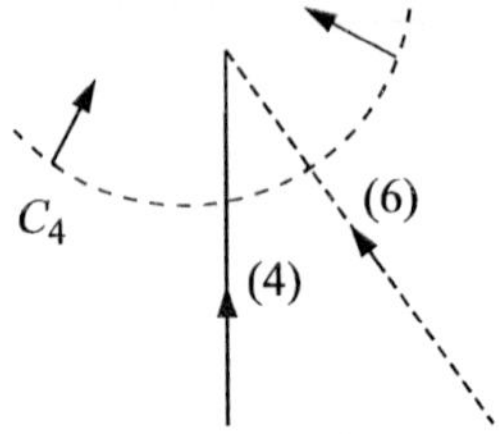

The oriented graph is combination of all four *f*-cutsets hence, the graph is given by:

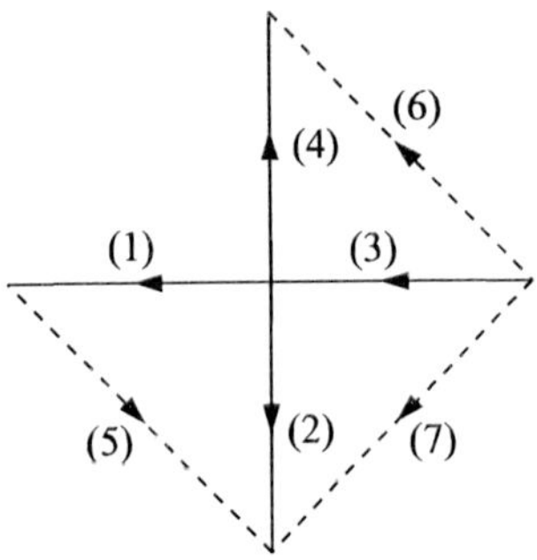

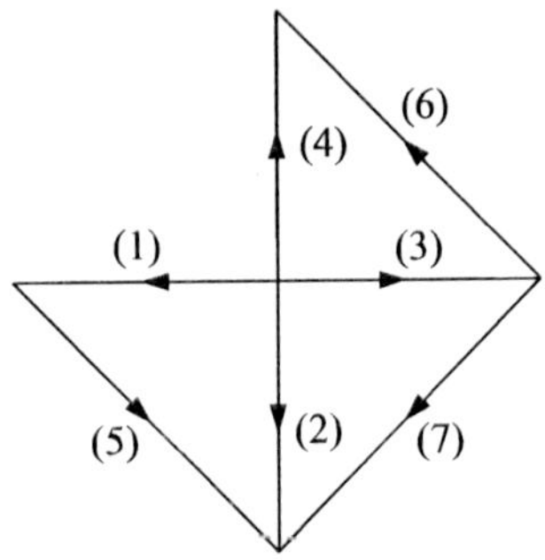

It is important to note that node corresponding to (1), (2), (3) and (4) is not essential as per given not *f*-cutset. Hence, more oriented graphs are given by:

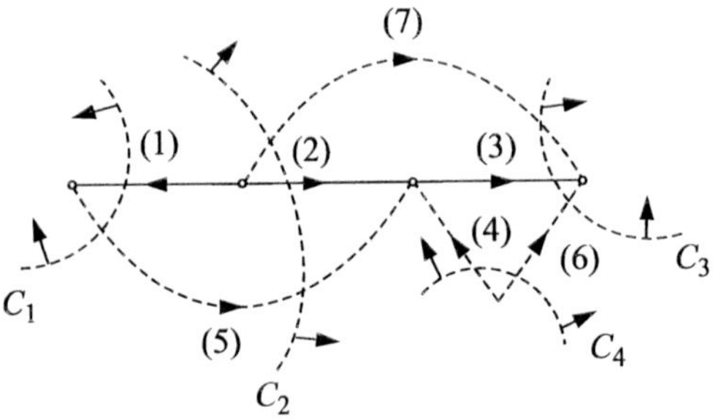

i.e.

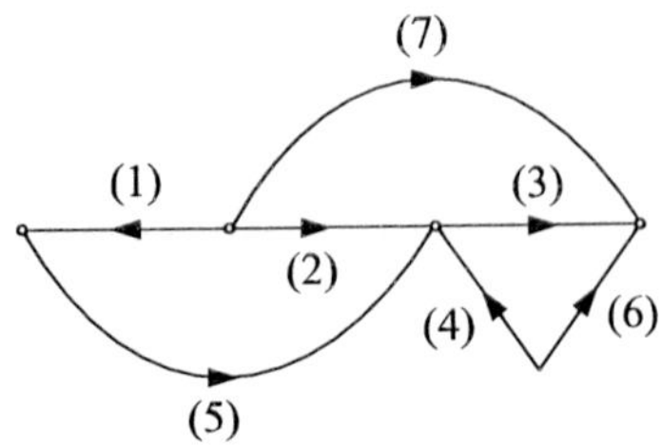

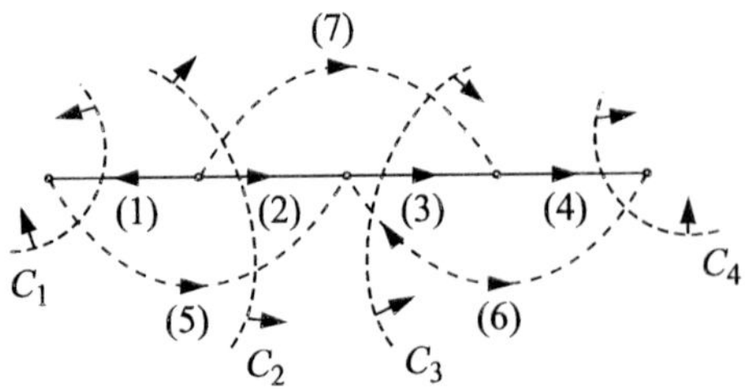

i.e.

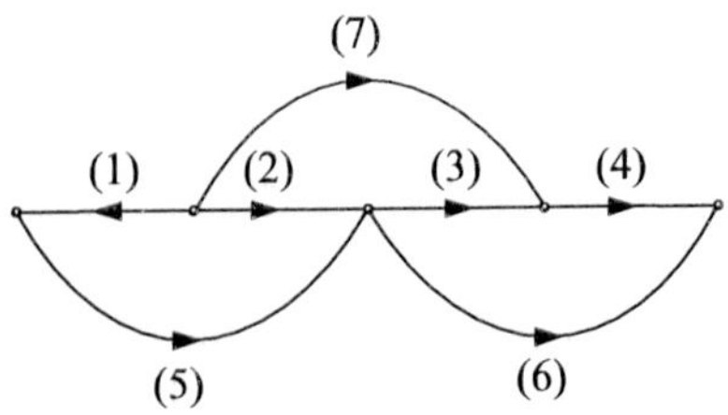

Example 3.17. Draw the graph and find a tree for the network shown in the figure. Considering '0' as datum node and assuming elements BD and BC as links, determine the tie-set schedule, branch impedance matrix and source voltage matrix. Obtain the loop equations using the above said matrices.

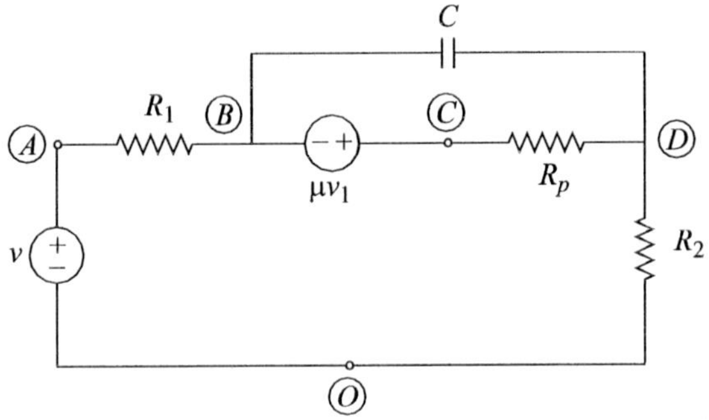

[UPTU, 2001]

Solution: The voltage sources are taken as short circuited. The voltage source V_1 and R_1 can be treated as single branch and UV_1 and R_p can be treated also as a single branch. Then oriented graph is as given in figure (a). '0' is considered as datum BD and BC are taken as links *i.e.*, branches (2) and (3) act as links. In such case, the tree is as shown in figure (b) and loops are shown in figures (c) and (d).

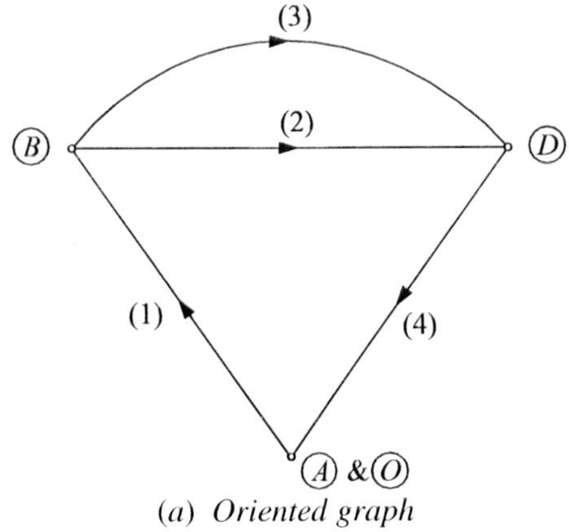

(a) *Oriented graph*

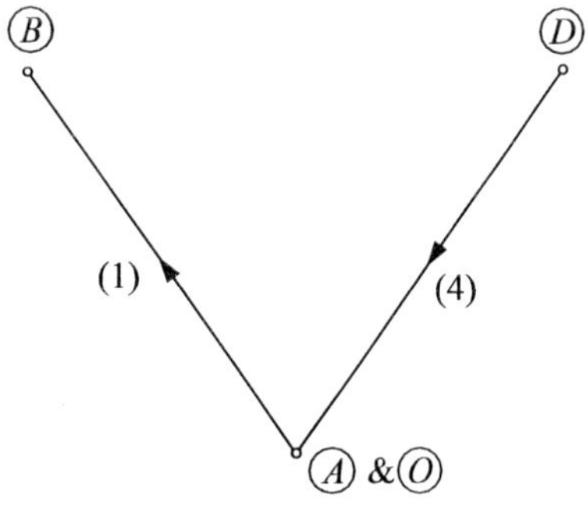

(b) *Tree*

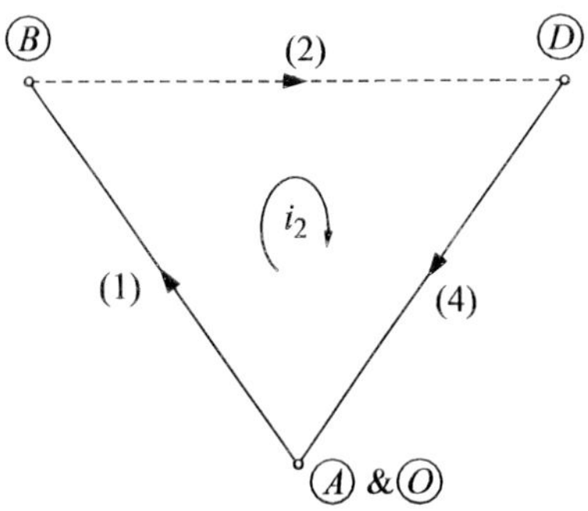

(c) *Loop 1*

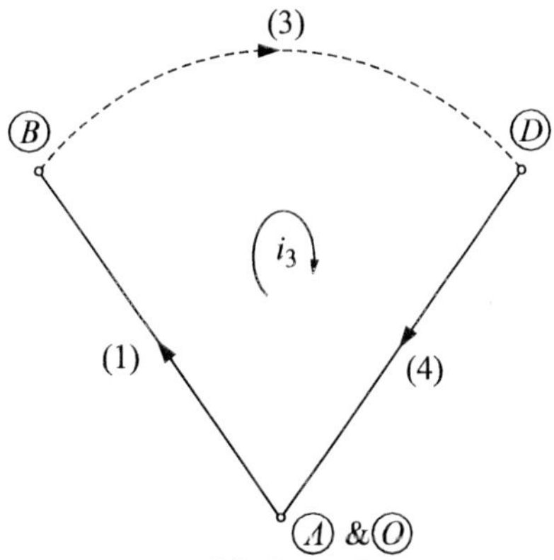

(d) *Loop 2*

The tie-set matrix is given by:

$$\begin{array}{c} \text{loops} \quad \text{branches} \rightarrow \\ \downarrow \quad 1 \;\; 2 \;\; 3 \;\; 4 \end{array}$$

$$\begin{array}{c} i_2 \\ i_3 \end{array} \begin{bmatrix} 1 & 1 & 0 & 1 \\ 1 & 0 & 1 & 1 \end{bmatrix}$$

i.e., $$B = \begin{bmatrix} 1 & 1 & 0 & 1 \\ 1 & 0 & 1 & 1 \end{bmatrix} \qquad \text{...}(i)$$

The branch impedance matrix is given by:

$$Z_b = \begin{bmatrix} R_1 & 0 & 0 & 0 \\ 0 & R_p & 0 & 0 \\ 0 & 0 & \dfrac{1}{cs} & 0 \\ 0 & 0 & 0 & R_2 \end{bmatrix} \qquad \text{...}(ii)$$

The voltage source matrix, $V_s = \begin{bmatrix} V_1 \\ -\mu V_1 \\ 0 \\ 0 \end{bmatrix}$ ...(*iii*)

The loop current matrix, $I_l = \begin{bmatrix} i_2 \\ i_3 \end{bmatrix}$ ...(*iv*)

The equilibrium equation is given by :

$$BZ_bB^TI_l = (BV_s - Z_bI_s) \qquad \text{...}(v)$$

RHS of Eqn (*iv*) is given by:

$$BZ_bB^TI_l = \begin{bmatrix} 1 & 1 & 0 & 1 \\ 1 & 0 & 1 & 1 \end{bmatrix} \begin{bmatrix} R_1 & 0 & 0 & 0 \\ 0 & Rp & 0 & 0 \\ 0 & 0 & \dfrac{1}{cs} & 0 \\ 0 & 0 & 0 & R_2 \end{bmatrix} \begin{bmatrix} 1 & 1 \\ 1 & 0 \\ 0 & 1 \\ 1 & 1 \end{bmatrix} \begin{bmatrix} i_2 \\ i_3 \end{bmatrix}$$

$$BZ_bB^TI_l = \begin{bmatrix} 1 & 1 & 0 & 1 \\ 1 & 0 & 1 & 1 \end{bmatrix} \begin{bmatrix} R_1 & R_1 \\ R_b & 0 \\ 0 & \dfrac{1}{cs} \\ R_2 & R_2 \end{bmatrix} \begin{bmatrix} i_2 \\ i_3 \end{bmatrix}$$

or

$$BZ_b\,B^T I_l = \begin{bmatrix} R_1 + R_b + R_2 & R_1 + R_2 \\ R_1 + R_2 & R_1 + \dfrac{1}{cs} + R_2 \end{bmatrix}$$

RHS of Eqn (v) is given by :

$$BV_s - Z_b\,I_s = B\,V_s \text{ as } I_s = 0$$

$$= \begin{bmatrix} 1 & 1 & 0 & 1 \\ 1 & 0 & 1 & 1 \end{bmatrix} \begin{bmatrix} V_1 \\ \mu V_1 \\ 0 \\ 0 \end{bmatrix}$$

or

$$BV_s - Z_b\,I_s = \begin{bmatrix} V_1 - \mu V_1 \\ V_1 \end{bmatrix}$$

$$\therefore \qquad \begin{bmatrix} V_1 - \mu V_1 \\ V_1 \end{bmatrix} = \begin{bmatrix} R_1 + R_p + R_z & R_1 + R_2 \\ R_1 + R_2 & R_1 + \dfrac{1}{cs} + R_2 \end{bmatrix} \begin{bmatrix} i_2 \\ i_3 \end{bmatrix}$$

Example 3.18. Determine currents in all branches of the network shown in the given figure using node analysis method.

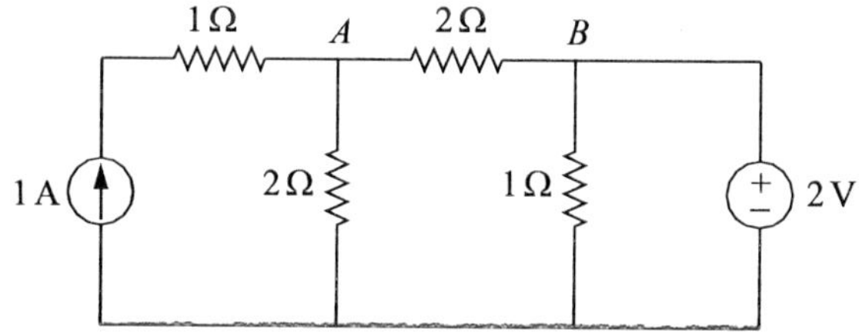

[UPTU, 2002]

Solution: The $2V$ source fixes the voltage at node B at 2 volts, hence, 1 Ω branch across $2V$ source need not be included in the graph. Further, current source is open circuited and voltage source is short circuited, thus an oriented graph is shown in figure (a). A tree and f-loop is shown in figure (b).

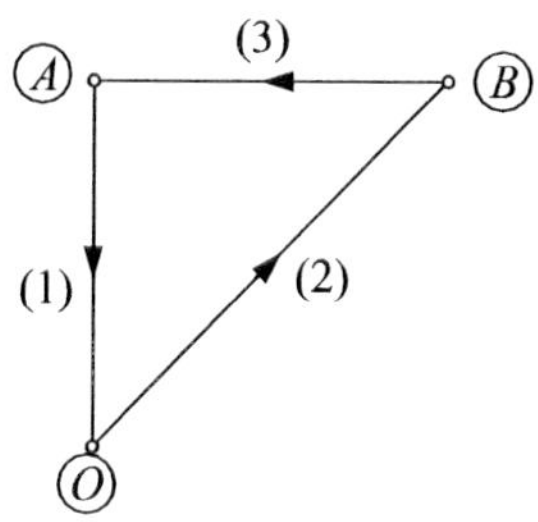

(a) *Oriented graph*

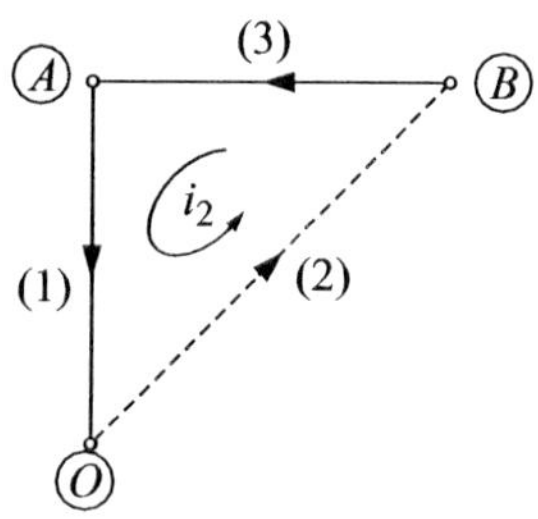

(b) *Tree and f-loop*

The incidence matrix is given by :

$$B = [1\ 1\ 1]$$

From $BV_b = 0$, we get:

$$[\,1\ 1\ 1]\begin{bmatrix} v_1 \\ v_2 \\ v_3 \end{bmatrix} = 0$$

or $$v_1 + v_2 + v_3 = 0 \qquad ...(i)$$

The branch currents are same *i.e.*,

$$i_1 = i_2 = i_3 = i$$

$$R_1 = 2\Omega,\ R_2 = 1\Omega,\ R_3 = 2\Omega$$

∴ $$v_1 = (i_1 + 1)\,R_1 = 2\,(i + 1) \qquad ...(ii)$$

$$v_2 = -\,2 \qquad ...(iii)$$

and $$v_3 = i_3\,R_3 = 2i \qquad ...(iv)$$

By substituting values of v_1, v_2 and v_3 from Eqns. (*ii*), (*iii*) and (*iv*) respectively into Eqn (*i*); we get :

$$2(i + 1) - 2 + 2i = 0$$

or $$i = 0$$

Thus, currents in all the branches shall be:

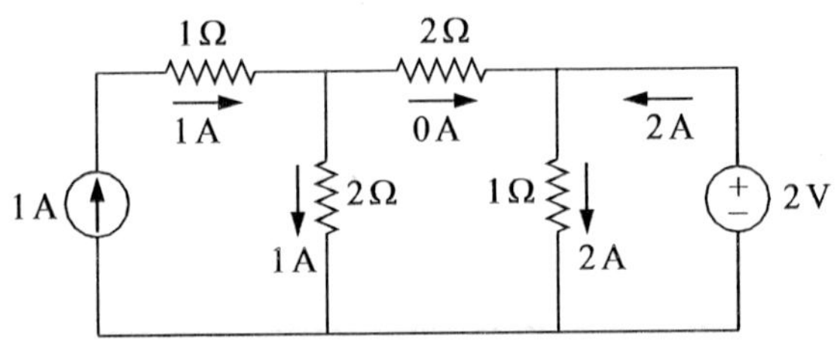

Example 3.19. Consider a network shown in the figure. Using loop method of analysis, determine currents in all the branches indicating their directions.

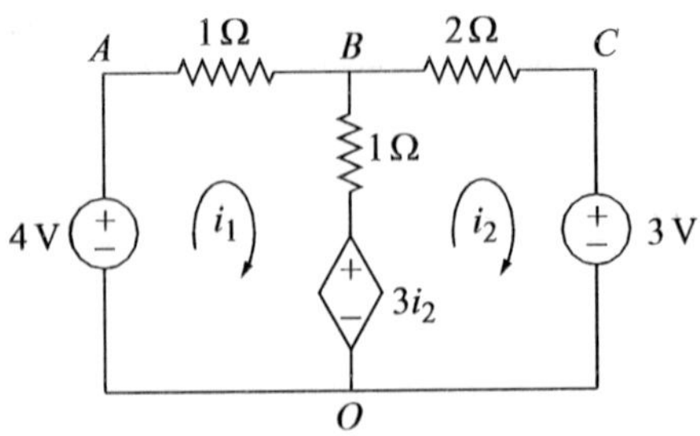

Solution: An oriented graph of the network is shown in the figure (*a*) and the chosen tree with *f*-loops is shown in figure (*b*). The voltage sources are taken as short circuit.

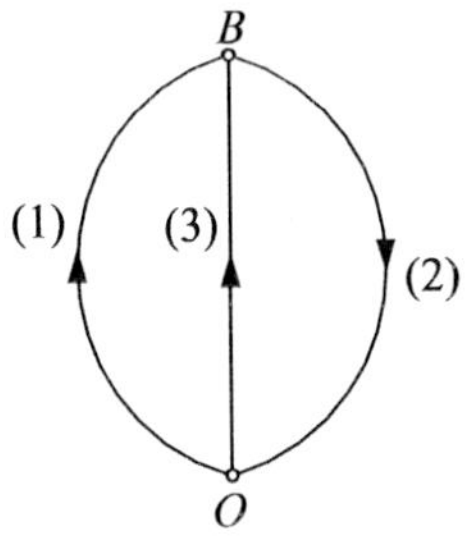

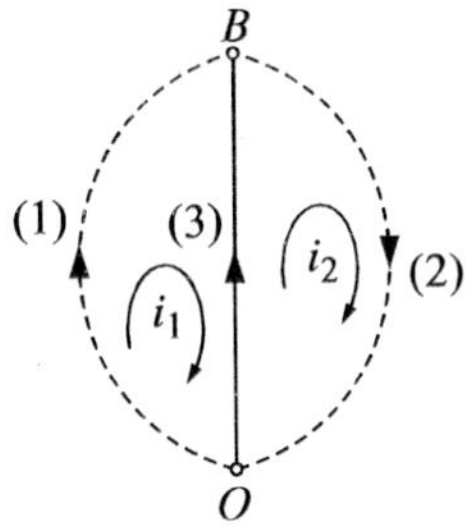

The f-loop/tie set matrix is given by :

$$\begin{array}{c} f\text{-loop} \quad \text{branches} \rightarrow \\ \downarrow \quad 1 \quad 2 \quad 3 \end{array}$$

$$\begin{array}{c} i_1 \\ i_2 \end{array} \begin{bmatrix} 1 & 0 & 1 \\ 0 & 1 & -1 \end{bmatrix}$$

i.e., $$B = \begin{bmatrix} 1 & 0 & 1 \\ 0 & 1 & -1 \end{bmatrix} \qquad ...(i)$$

The branch impedance matrix is given by :

$$Z_b = \begin{bmatrix} 1 & 0 & 0 \\ 0 & 2 & 0 \\ 0 & 0 & 1 \end{bmatrix} \quad (\text{columns } 1 \; 2 \; 3) \qquad ...(ii)$$

The voltage source matrix, $V_s = \begin{bmatrix} 4 \\ -3 \\ -3\,i_2 \end{bmatrix}$ ...(iii)

The branch current source, $I_s = 0$

The loop current matrix, $I_l = \begin{bmatrix} i_1 \\ i_2 \end{bmatrix}$ ...(iv)

The loop equilibrium equation in matrix form is as follows :

$$B\,Z_b\,B^T\,I_l = B\,(V_s - Z_b\,I_s)$$

i.e., $$B\,Z_b\,B^T\,I_l = BV_s \text{ since } I_s = 0 \qquad ...(v)$$

Thus, $$B\,Z_b\,B^T\,I_l = \begin{bmatrix} 1 & 0 & 1 \\ 0 & 1 & -1 \end{bmatrix} \begin{bmatrix} 1 & 0 & 0 \\ 0 & 2 & 0 \\ 0 & 0 & 1 \end{bmatrix} \begin{bmatrix} 1 & 0 \\ 0 & 1 \\ 1 & -1 \end{bmatrix} \begin{bmatrix} i_1 \\ i_2 \end{bmatrix}$$

or $$BZ_b\,B^T\,I_l = \begin{bmatrix} 1 & 0 & 1 \\ 0 & 2 & -1 \end{bmatrix} \begin{bmatrix} 1 & 0 \\ 0 & 1 \\ 1 & -1 \end{bmatrix} \begin{bmatrix} i_1 \\ i_2 \end{bmatrix} \begin{bmatrix} 2 & -1 \\ -1 & 3 \end{bmatrix} \begin{bmatrix} i_1 \\ i_2 \end{bmatrix} \qquad ...(vi)$$

and $$BV_s = \begin{bmatrix} 1 & 0 & 1 \\ 0 & 1 & -1 \end{bmatrix} \begin{bmatrix} 4 \\ -3 \\ -3i_2 \end{bmatrix} = \begin{bmatrix} 4 - 3i_2 \\ -3 + 3i_2 \end{bmatrix} \quad ...(vii)$$

From Eqns. (*vi*) and (*vii*), we get :

$$\begin{bmatrix} 2 & -1 \\ -1 & 3 \end{bmatrix} \begin{bmatrix} i_1 \\ i_2 \end{bmatrix} = \begin{bmatrix} 4 - 3i_2 \\ -3 + 3i_2 \end{bmatrix}$$

The above give the following equations:

$$2i_1 - i_2 = 4 - 3i_2$$
$$-i_1 + 3i_2 = -3 + 3i_2$$

Thus, above equations give :

$$i_1 - i_2 = 2$$
$$i_1 = 3$$

$\therefore$ $$i_1 = 3A$$

and $$i_2 = -1\ A$$

Thus, branch currents are given as :

Branch (1) current, $i_1 = 3A$ *i.e.*, $3A$ from A to B

Branch (2) current, $i_2 = -1\ A$ *i.e.*, A from C to B

and branch (3) current $= i_2 - i_1 = -1 - 3 = -4\ A$

i.e., $4A$ from B to 0.

Example 3.20. For the network shown in the figure given, draw the network graph. Select 2, 4, 5 as the branches. Obtain loop incidence matrix and loop equations.

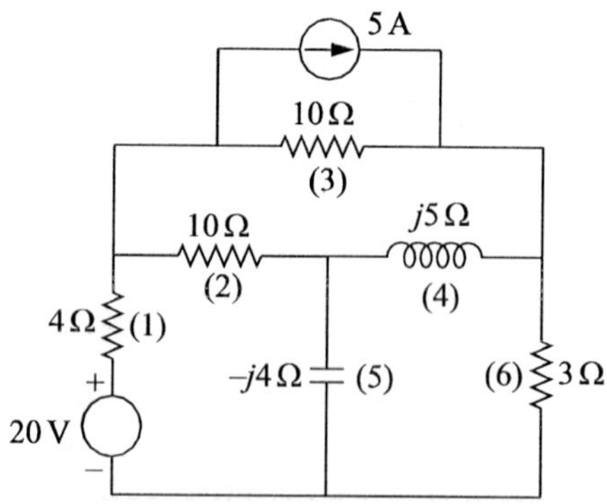

[UPTU, 2003]

Solution: The graph of the network is shown in figure (*a*) where directions are also marked for analysis purposes. Note that voltage source is short circuited and current source open circuited.

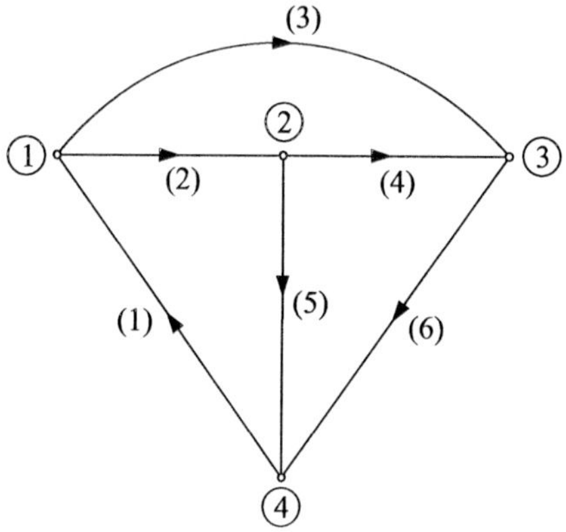

(a) *Directed graph*

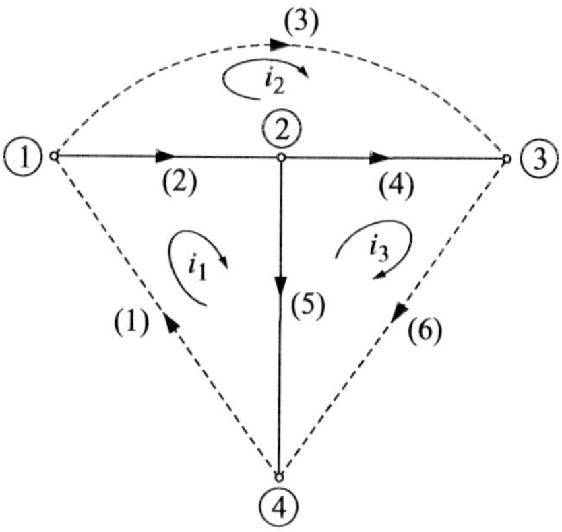

(b) *Tree and loops*

The loop incidence matrix is as follows:

$$\begin{array}{c} f\text{-loops} \\ \downarrow \\ i_1 \\ i_2 \\ i_3 \end{array} \begin{array}{c} \text{branches} \rightarrow \\ \begin{array}{cccccc} 1 & 2 & 3 & 4 & 5 & 6 \end{array} \\ \begin{bmatrix} 1 & 1 & 0 & 0 & 1 & 0 \\ 0 & -1 & 1 & -1 & 0 & 0 \\ 0 & 0 & 0 & 1 & -1 & 1 \end{bmatrix} \end{array}$$

i.e.,

$$B = \begin{bmatrix} 1 & 1 & 0 & 0 & 1 & 0 \\ 0 & -1 & 1 & -1 & 0 & 0 \\ 0 & 0 & 0 & 1 & -1 & 1 \end{bmatrix} \quad ...(i)$$

The branch impedance matrix Z_b is given as :

$$\begin{array}{c} \text{branches} \rightarrow \\ \begin{array}{cccccc} 1 & 2 & 3 & 4 & 5 & 6 \end{array} \end{array}$$

$$Z_b = \begin{bmatrix} 4 & 0 & 0 & 0 & 0 & 0 \\ 0 & 10 & 0 & 0 & 0 & 0 \\ 0 & 0 & 10 & 0 & 0 & 0 \\ 0 & 0 & 0 & j5 & 0 & 0 \\ 0 & 0 & 0 & 0 & -j4 & 0 \\ 0 & 0 & 0 & 0 & 0 & 3 \end{bmatrix} \quad ...(ii)$$

The column matrix of voltage and current source are given as :

$$V_s = \begin{bmatrix} 20 \\ 0 \\ 0 \\ 0 \\ 0 \\ 0 \end{bmatrix} \quad ...(iii)$$

$$I_s = \begin{bmatrix} 0 \\ 0 \\ -5 \\ 0 \\ 0 \\ 0 \end{bmatrix} \quad ...(iv)$$

The loop equilibrium equations in matrix form are given as :

$$B\,Z_b\,B^T\,I_l = B\,(V_s - Z_b\,I_s) \quad ...(v)$$

for

$$I_e = \begin{bmatrix} i_1 \\ i_2 \\ i_3 \end{bmatrix} \quad ...(vi)$$

LHS of Eqn. (v) can be written after substitution as :

$$B\,Z_b\,I_l = \begin{bmatrix} 1 & 1 & 0 & 0 & 1 & 0 \\ 0 & -1 & 1 & -1 & 0 & 0 \\ 0 & 0 & 0 & 1 & -1 & 1 \end{bmatrix}$$

$$\begin{bmatrix} 4 & 0 & 0 & 0 & 0 & 0 \\ 0 & 10 & 0 & 0 & 0 & 0 \\ 0 & 0 & 10 & 0 & 0 & 0 \\ 0 & 0 & 0 & j5 & 0 & 0 \\ 0 & 0 & 0 & 0 & -j4 & 0 \\ 0 & 0 & 0 & 0 & 0 & 3 \end{bmatrix} \begin{bmatrix} 1 & 0 & 0 \\ 1 & -1 & 0 \\ 0 & 1 & 0 \\ 0 & -1 & 1 \\ 1 & 0 & -1 \\ 0 & 0 & 1 \end{bmatrix} \begin{bmatrix} i_1 \\ i_2 \\ i_3 \end{bmatrix}$$

or

$$B\,Z_b\,I_l =$$

$$\begin{bmatrix} 4 & 10 & 0 & 0 & -j4 & 0 \\ 0 & -10 & 10 & -j5 & 0 & 0 \\ 0 & 0 & 0 & j4 & j4 & 3 \end{bmatrix} \begin{bmatrix} 1 & 0 & 0 \\ 0 & 1 & 0 \\ 0 & -1 & 1 \\ 1 & 0 & -1 \\ 0 & 0 & 1 \end{bmatrix} \begin{bmatrix} i_1 \\ i_2 \\ i_3 \end{bmatrix}$$

or

$$B\,Z_b\,I_l = \begin{bmatrix} (14 - j4) & -10 & j4 \\ -10 & (20 + j5) & -j5 \\ j4 & -j5 & (3 + j) \end{bmatrix} \begin{bmatrix} i_1 \\ i_2 \\ i_3 \end{bmatrix} \quad ...(vii)$$

RHS of Eqn. (*v*) can be written after substitution as :

$B(V_s - Z_b I_s)$

$$= \begin{bmatrix} 1 & 1 & 0 & 0 & 1 & 0 \\ 0 & -1 & 1 & -1 & 0 & 0 \\ 0 & 0 & 0 & 1 & -1 & 1 \end{bmatrix} \left\{ \begin{bmatrix} 20 \\ 0 \\ 0 \\ 0 \\ 0 \\ 0 \end{bmatrix} - \begin{bmatrix} 4 & 0 & 0 & 0 & 0 & 0 \\ 0 & 10 & 0 & 0 & 0 & 0 \\ 0 & 0 & 10 & 0 & 0 & 0 \\ 0 & 0 & 0 & j5 & 0 & 0 \\ 0 & 0 & 0 & 0 & -j4 & 0 \\ 0 & 0 & 0 & 0 & 0 & 3 \end{bmatrix} \begin{bmatrix} 0 \\ 0 \\ -5 \\ 0 \\ 0 \\ 0 \end{bmatrix} \right\}$$

or $$B(V_s - Z_b I_s) = \begin{bmatrix} 1 & 1 & 0 & 0 & 1 & 0 \\ 0 & -1 & 1 & -1 & 0 & 0 \\ 0 & 0 & 0 & 1 & -1 & 1 \end{bmatrix} \left\{ \begin{bmatrix} 20 \\ 0 \\ 0 \\ 0 \\ 0 \\ 0 \end{bmatrix} - \begin{bmatrix} 0 \\ 0 \\ -50 \\ 0 \\ 0 \\ 0 \end{bmatrix} \right\}$$

or $$B(V_s - Z_b I_s) = \begin{bmatrix} 1 & 1 & 0 & 0 & 1 & 0 \\ 0 & -1 & 1 & -1 & 0 & 0 \\ 0 & 0 & 0 & 1 & -1 & 1 \end{bmatrix} \begin{bmatrix} 20 \\ 0 \\ 50 \\ 0 \\ 0 \\ 0 \end{bmatrix} = \begin{bmatrix} 20 \\ 50 \\ 0 \end{bmatrix} \quad \text{...(viii)}$$

Thus, Eqn. (*v*) becomes :

$$\begin{bmatrix} (14 + j4) & -10 & j4 \\ -10 & (20 + j5) & -j5 \\ j4 & -j5 & (3 + i) \end{bmatrix} \begin{bmatrix} i_1 \\ i_2 \\ i_3 \end{bmatrix} = \begin{bmatrix} 20 \\ 50 \\ 0 \end{bmatrix}$$

Example 3.21. For the network shown in the figure, drawn the graph, write the tie-set schedule and hence obtain the equilibrium equation on loop basis. Calculate the values of branch currents and branch voltages.

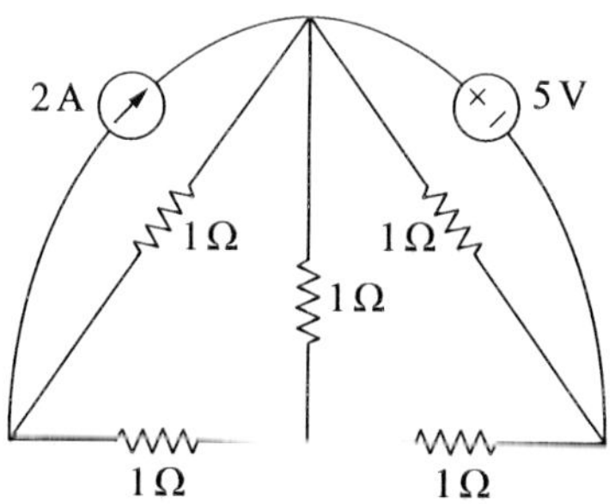

[UPTU, 2004]

Solution: The current source is open circuited and voltage source is short circuited. An oriented graph of the network is shown in figure (*a*) and a free is shown in figure (*b*).

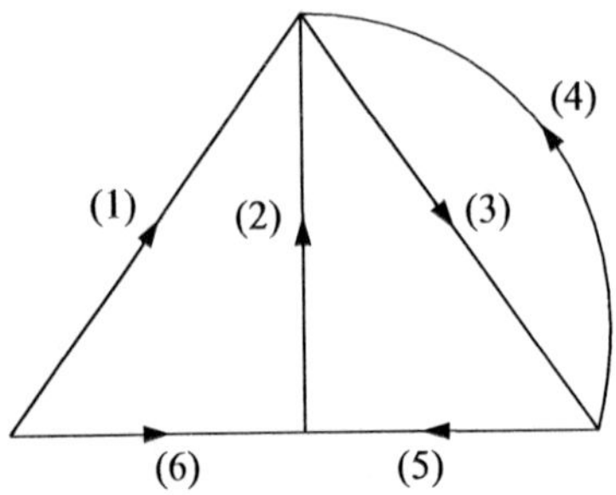

(*a*) *An oriented graph*

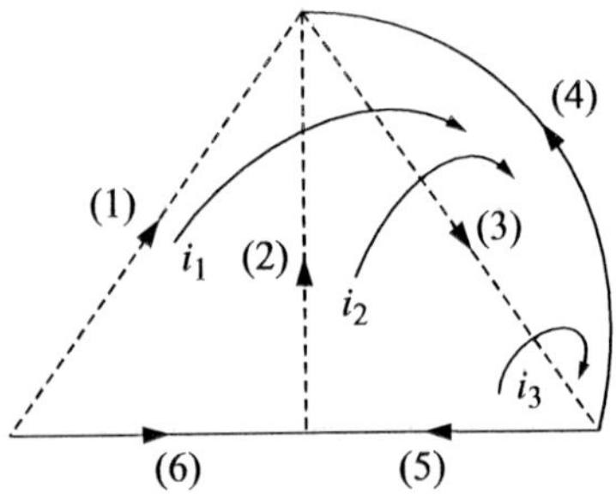

(*b*) *A tree and f-loops*

Tie-set matrix is given by:

$$\begin{array}{c} f\text{-loops} \quad \text{branches} \rightarrow \\ \begin{array}{c|cccccc} \downarrow & 1 & 2 & 3 & 4 & 5 & 5 \\ 1 & 1 & 0 & 0 & -1 & 1 & -1 \\ 2 & 0 & 1 & 0 & -1 & 1 & 0 \\ 3 & 0 & 0 & 1 & 1 & 0 & 0 \end{array} \end{array}$$

i.e.,
$$B = \begin{bmatrix} 1 & 0 & 0 & -1 & 1 & -1 \\ 0 & 1 & 0 & -1 & 1 & 0 \\ 0 & 0 & 1 & 1 & 0 & 0 \end{bmatrix} \qquad ...(i)$$

The branch impedance matrix is given by:

$$Z_b = \begin{array}{c} \begin{array}{cccccc} 1 & 2 & 3 & 4 & 5 & 6 \end{array} \\ \begin{bmatrix} 1 & 0 & 0 & 0 & 0 & 0 \\ 0 & 1 & 0 & 0 & 0 & 0 \\ 0 & 0 & 1 & 0 & 0 & 0 \\ 0 & 0 & 0 & 0 & 0 & 0 \\ 0 & 0 & 0 & 0 & 1 & 0 \\ 0 & 0 & 0 & 0 & 0 & 1 \end{bmatrix} \end{array} \qquad ...(ii)$$

$\therefore$
$$B\,Z_b\,B^T = \begin{bmatrix} 1 & 0 & 0 & -1 & 1 & -1 \\ 0 & 1 & 0 & -1 & 1 & 0 \\ 0 & 0 & 1 & 1 & 0 & 0 \end{bmatrix}$$

$$\begin{bmatrix} 1 & 0 & 0 & 0 & 0 & 0 \\ 0 & 1 & 0 & 0 & 0 & 0 \\ 0 & 0 & 1 & 0 & 0 & 0 \\ 0 & 0 & 0 & 0 & 0 & 0 \\ 0 & 0 & 0 & 0 & 1 & 0 \\ 0 & 0 & 0 & 0 & 0 & 1 \end{bmatrix} \begin{bmatrix} 1 & 0 & 0 \\ 0 & 1 & 0 \\ 0 & 0 & 1 \\ -1 & -1 & 1 \\ 1 & 1 & 0 \\ -1 & 0 & 0 \end{bmatrix}$$

or
$$B\,Z_b\,B^T = \begin{bmatrix} 3 & 1 & 0 \\ 1 & 2 & 0 \\ 0 & 0 & 1 \end{bmatrix} = Z_l \qquad ...(iii)$$

$$V_s = \begin{bmatrix} 0 \\ 0 \\ 0 \\ -5 \\ 0 \\ 0 \end{bmatrix}; \; I_s = \begin{bmatrix} 2 \\ 0 \\ 0 \\ 0 \\ 0 \\ 0 \end{bmatrix}$$

$$B\,Z_b\,I_s - B\,V_s = (B\,Z_b)\,I_s - B\,V_s$$

$$= \begin{bmatrix} 1 & 0 & 0 & 0 & 1 & -1 \\ 0 & 1 & 0 & 0 & 1 & 0 \\ 0 & 0 & 1 & 0 & 0 & 0 \end{bmatrix} \begin{bmatrix} 2 \\ 0 \\ 0 \\ 0 \\ 0 \\ 0 \end{bmatrix} - \begin{bmatrix} 1 & 0 & 0 & -1 & 1 & -1 \\ 0 & 1 & 0 & -1 & 1 & 0 \\ 0 & 0 & 1 & 1 & 0 & 0 \end{bmatrix} \begin{bmatrix} 0 \\ 0 \\ 0 \\ -5 \\ 0 \\ 0 \end{bmatrix}$$

$$= \begin{bmatrix} 2 \\ 0 \\ 0 \end{bmatrix} - \begin{bmatrix} 5 \\ 5 \\ -5 \end{bmatrix} = \begin{bmatrix} -3 \\ -5 \\ 5 \end{bmatrix} = V_l \qquad ...(iv)$$

The equilibrium equation for a loop is given by :

$$Z_l\,I_l = V_l$$

$$\therefore \qquad \begin{bmatrix} 3 & 1 & 0 \\ 1 & 2 & 0 \\ 0 & 0 & 1 \end{bmatrix} \begin{bmatrix} i_1 \\ i_2 \\ i_3 \end{bmatrix} = \begin{bmatrix} -3 \\ -5 \\ 5 \end{bmatrix}$$

Hence, the loop currents are given by:

$$i_1 = \frac{\begin{vmatrix} -3 & 1 & 0 \\ -5 & 2 & 0 \\ 5 & 0 & 1 \end{vmatrix}}{\begin{vmatrix} 3 & 1 & 0 \\ 1 & 2 & 0 \\ 0 & 0 & 1 \end{vmatrix}} = \frac{\Delta_1}{\Delta}$$

$$\Delta_1 = -\,3(2-0) - 1\,(-\,5-0) + 0 = -1$$

and $$\Delta = 3\,(2-0) - (1-0) + 0 = 5$$

$\therefore$ $$i_1 = \frac{\Delta_1}{\Delta} = \frac{-1}{5} = -\frac{1}{5}A$$

$$i_2 = \frac{\begin{vmatrix} 3 & -3 & 0 \\ 1 & -5 & 0 \\ 0 & 5 & 1 \end{vmatrix}}{\Lambda} = \frac{\Delta_2}{\Delta}$$

and $$\Delta_2 = 3\,(-\,5-0) - (-\,3)\,(1-0) + 0 = -\,12$$

$\therefore$ $$i_2 = \frac{\Delta_2}{\Delta} = \frac{-12}{5} = -\frac{12}{5}A$$

and $$i_3 = \frac{\begin{vmatrix} 3 & 1 & -3 \\ 1 & 2 & -5 \\ 0 & 0 & 5 \end{vmatrix}}{\Delta} = \frac{\Delta_3}{\Delta}$$

and $$\Delta_3 = 3\,(10-0) - (5-0) + (-\,3)\,(0)$$
$$= 30 - 5 = 25$$

$\therefore$ $$i_3 = \frac{\Delta_3}{\Delta} = \frac{25}{5} = 5A$$

The equation for branch currents is :

$$I_b = B^T I_l - I_s$$

$\therefore$ $$\begin{bmatrix} i_1 \\ i_2 \\ i_3 \\ i_4 \\ i_5 \\ i_6 \end{bmatrix} = \begin{bmatrix} 1 & 0 & 0 \\ 0 & 1 & 0 \\ 0 & 0 & 1 \\ -1 & -1 & 1 \\ 1 & 1 & 0 \\ -1 & 0 & 0 \end{bmatrix} \begin{bmatrix} -\frac{1}{5} \\ -\frac{12}{15} \\ 5 \end{bmatrix} \begin{bmatrix} 2 \\ 0 \\ 0 \\ 0 \\ 0 \\ 0 \end{bmatrix}$$

or

$$\begin{bmatrix} i_1 \\ i_2 \\ i_3 \\ i_4 \\ i_5 \\ i_6 \end{bmatrix} = \begin{bmatrix} -\frac{1}{5} \\ -\frac{12}{5} \\ 5 \\ \frac{38}{5} \\ -\frac{13}{5} \\ \frac{1}{5} \end{bmatrix} - \begin{bmatrix} 2 \\ 0 \\ 0 \\ 0 \\ 0 \\ 0 \end{bmatrix} = \begin{bmatrix} -\frac{11}{5} \\ -\frac{12}{5} \\ 5 \\ \frac{38}{5} \\ -\frac{13}{5} \\ \frac{1}{5} \end{bmatrix}$$

The branch voltages are given by:

$$Z_b I_b + V_s = V_b$$

$$Z_b I_b + \mathrm{V_s} = \begin{bmatrix} 1 & 0 & 0 & 0 & 0 & 0 \\ 0 & 1 & 0 & 0 & 0 & 0 \\ 0 & 0 & 1 & 0 & 0 & 0 \\ 0 & 0 & 0 & 0 & 0 & 0 \\ 0 & 0 & 0 & 0 & 1 & 0 \\ 0 & 0 & 0 & 0 & 0 & 1 \end{bmatrix} \begin{bmatrix} -\frac{11}{5} \\ -\frac{12}{5} \\ 5 \\ \frac{3}{5} \\ -\frac{13}{5} \\ \frac{1}{5} \end{bmatrix} + \begin{bmatrix} 0 \\ 0 \\ 0 \\ -5 \\ 0 \\ 0 \end{bmatrix}$$

$$= \begin{bmatrix} -\frac{11}{5} \\ -\frac{12}{5} \\ 5 \\ 0 \\ -\frac{13}{5} \\ \frac{1}{5} \end{bmatrix} + \begin{bmatrix} 0 \\ 0 \\ 0 \\ -5 \\ 0 \\ 0 \end{bmatrix} = \begin{bmatrix} -\frac{11}{5} \\ -\frac{12}{5} \\ 5 \\ -5 \\ -\frac{13}{5} \\ \frac{1}{5} \end{bmatrix}$$

$$\therefore \qquad V_b = \begin{bmatrix} -\frac{11}{5} \\ -\frac{12}{5} \\ 5 \\ -5 \\ -\frac{13}{5} \\ \frac{1}{5} \end{bmatrix}$$

Example 3.22. Derive relation between branch voltage matrix, Twig voltage matrix and Node voltage matrix of a network. **[UPTU, 2006]**

Solution: A network graph with identified tree twigs and links in shown in figure (*a*).

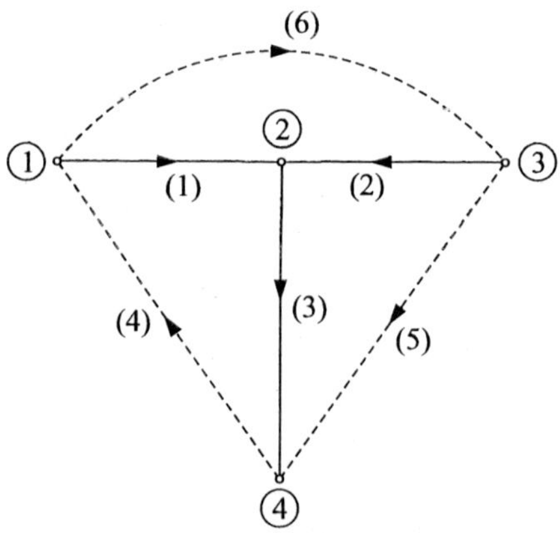

(*i*) Relation between branch voltages and twig voltages:

Consider that twig voltages are v_{t_1}, v_{t_2} and v_{t_3}. The branch voltages in terms of twig voltages are :

$$v_1 = v_{t_1}$$
$$v_2 = v_{t_2}$$
$$v_3 = v_{t_3}$$
$$v_4 = -v_{t_1} - v_{t_3}$$
$$v_5 = v_{t_2} + v_{t_3}$$
$$v_6 = v_{t_1} - v_{t_2}$$

i.e.,
$$\begin{bmatrix} v_1 \\ v_2 \\ v_3 \\ v_4 \\ v_5 \\ v_6 \end{bmatrix} = \begin{bmatrix} 1 & 0 & 0 \\ 0 & 1 & 0 \\ 0 & 0 & 1 \\ -1 & 0 & -1 \\ 0 & 1 & 0 \\ 1 & -1 & 0 \end{bmatrix} \begin{bmatrix} v_{t_1} \\ v_{t_2} \\ v_{t_3} \end{bmatrix}$$

i.e., $V_b = Q^T V_t$ where Q is *f*-cutset matrix.

(*ii*) Relation between branch voltages and node voltages:

Consider the node voltage as v_{n_1}, v_{n_2}, v_{n_3} and v_{n_4} for the nodes ①, ②, ③, and ④ respectively. The following relationships exist between branch voltages and node voltages for the given graph:

$$v_1 = v_{n_1} - v_{n_2}$$
$$v_2 = -v_{n_2} + v_{n_3}$$
$$v_3 = v_{n_2} - v_{n_4} = v_{n_2} \qquad \text{by taking } v_{n_4} \text{ as reference.}$$
$$v_4 = v_{n_4} - v_{n_1} = -v_{n_1}$$
$$v_5 = v_{n_3} - v_{n_4} = v_{n_3}$$
$$v_6 = v_{n_1} - v_{n_3}$$

Hence, the relationship becomes :

$$\begin{bmatrix} v_1 \\ v_2 \\ v_3 \\ v_4 \\ v_5 \\ v_6 \end{bmatrix} = \begin{bmatrix} 1 & -1 & 0 \\ 0 & -1 & 1 \\ 0 & 1 & 0 \\ -1 & 0 & 0 \\ 0 & 0 & 1 \\ 1 & 0 & -1 \end{bmatrix} \begin{bmatrix} v_{n_1} \\ v_{n_2} \\ v_{n_3} \end{bmatrix}$$

i.e., $V_b = A^T V_n$ where A is reduced incidence matrix.

3.21 DUALITY

Circuits where the cut-set matrix of one corresponds to tie-set matrix of the other are the duals of each other. Consider circuit of figures 3.16(*a*) and 3.16(*b*).

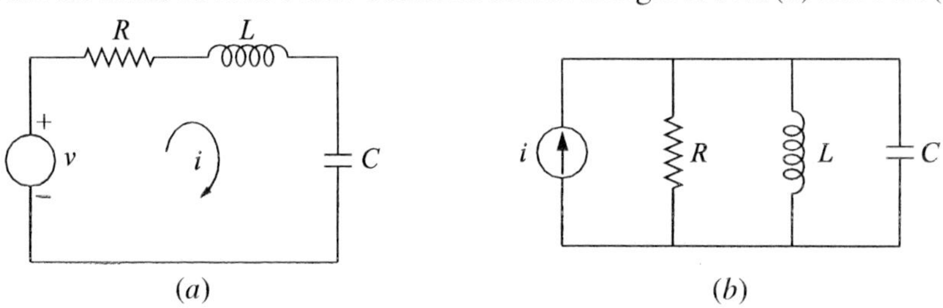

Fig. 3.16 (*a*) *Series RLC circuit,* (*b*) *Parallel RLC circuit.*

KVL gives : (Fig. 3.16(*a*)) $v = iR + L\frac{di}{dt} + \frac{1}{c}\int_a^t i dt$ ...(3.51)

KCL gives (Fig. 3.16(*b*)): $i = vG + c\frac{dv}{dt} + \frac{1}{L}\int_0^t v\, dt$ ...(3.52)

Note that KVL and KCL of either figures are identical, therefore, the voltage source *v* of the RLC circuit corresponds to current source *I* in the parallel circuit. In addition, the current in first circuit corresponds to voltage in second circuit. However, the dual circuits are not equivalent of each other.

3.21.1 The Dual Relationships are as Follows

S. No.	*Element*	↔	*Dual element*
1.	Resistance	↔	Conductance
2.	Capacitance	↔	Inductance
3.	Voltage	↔	Current
4.	Impedance	↔	Admittance
5.	Voltage source	↔	Current source
6.	KVL	↔	KCL
7.	Mesh	↔	Node
8.	Mesh current	↔	Node voltage
9.	Mesh equations	↔	Nodal equations
10.	Thevenin's equivalent	↔	Norton's equivalent
11.	Voltage division	↔	Current division
12.	Tree	↔	Co-tree
13.	Twig	↔	Link
14.	Cut-set	↔	Loop
15.	Closed switch	↔	Open switch
16.	Series branch	↔	Parallel branch

3.21.2 Graphical Method of Constructing the Dual Network

(*i*) A node is placed in the centre of each mesh and labelled.

(*ii*) An extra node is placed out side the network to serve as reference node.

(*iii*) Dashed lines are drawn between the nodes in such a way that each line crosses only one element and looped to reference node.

(*iv*) Dashed lines from centre node to remaining elements leading to another node is drawn.

(*v*) A convention positive direction for mesh current is clockwise and correspondingly the node voltage is positive with respect to the reference node or *vice-versa*.

(*vi*) The dashed lines along with the dual elements constitute the dual network.

Example 3.23. Draw a dual network of the network given in the figure.

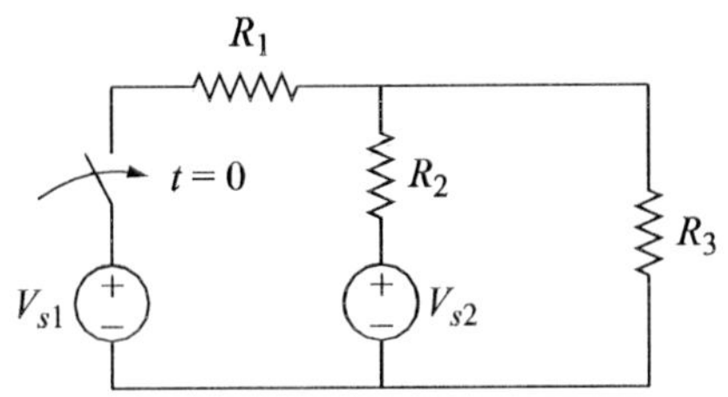

Solution:

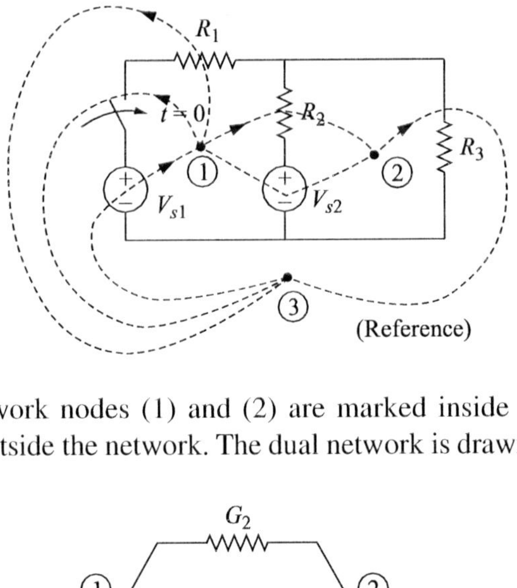

In the above network nodes (1) and (2) are marked inside the meshes. Node (3) is marked outside the network. The dual network is drawn as follows:

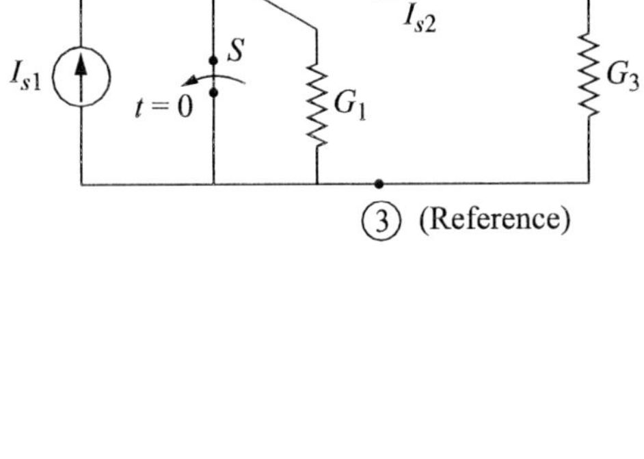

Example 3.24. Construct dual network of the circuit shown in the given figure.

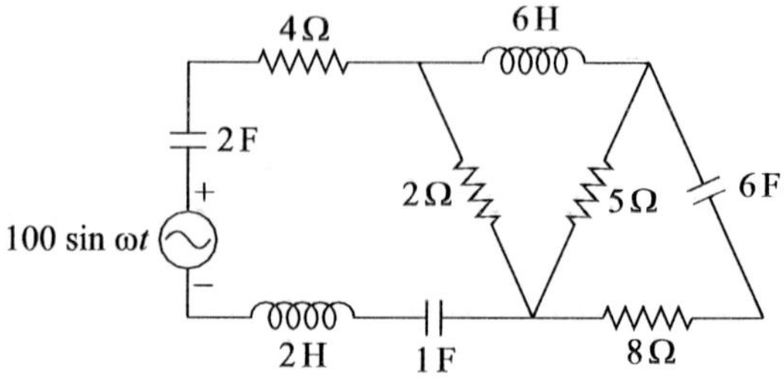

Solution: A node is placed in each mesh and a reference node is placed outside; and the given network is marked as follows:

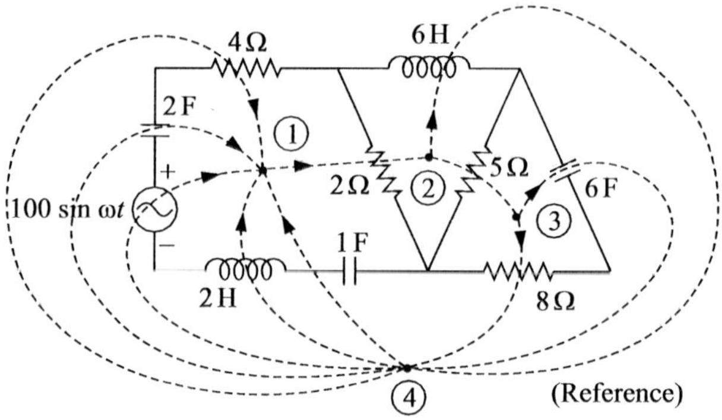

The dual network is as follows:

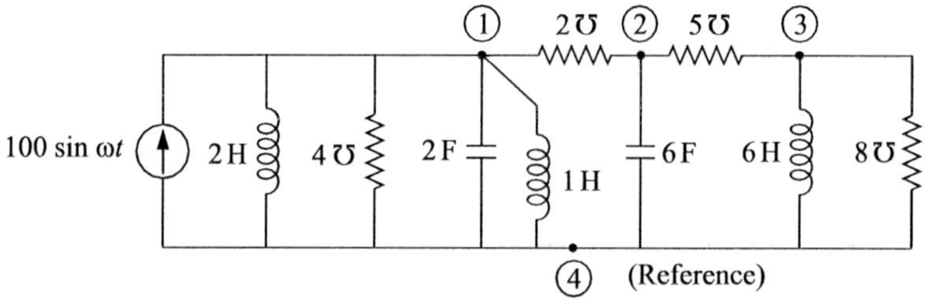

Example 3.25. Draw the dual of the network shown in the given figure.

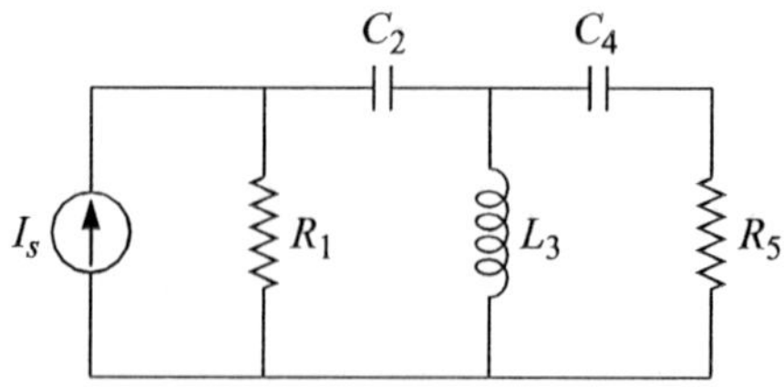

[UPTU, 2005]

Solution:

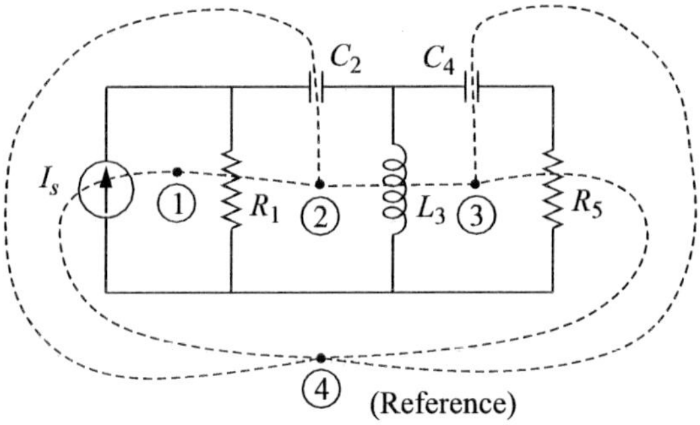

(*a*) *Technique for dual network*

A node is fixed in each loop and one reference node is taken outside the circuit. Join all the tree nodes through elements and Join nodes (1), (2), (3) and (4) (reference) through each element as shown in figure (*a*).

The dual network is drawn with dual elements w.r.t the circuit given. Thus, the dual network for the given network is given by:

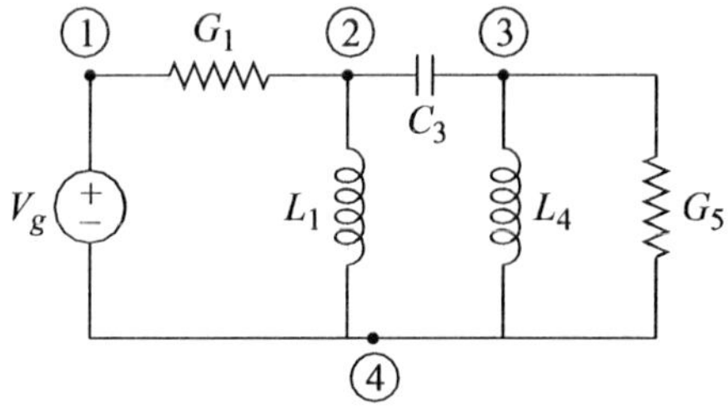

3.22 NETWORK WITH MUTUAL INDUCTANCE

Let M be the mutual inductance between the coupled coils as shown by dotted terminals in the Fig. 3.20.

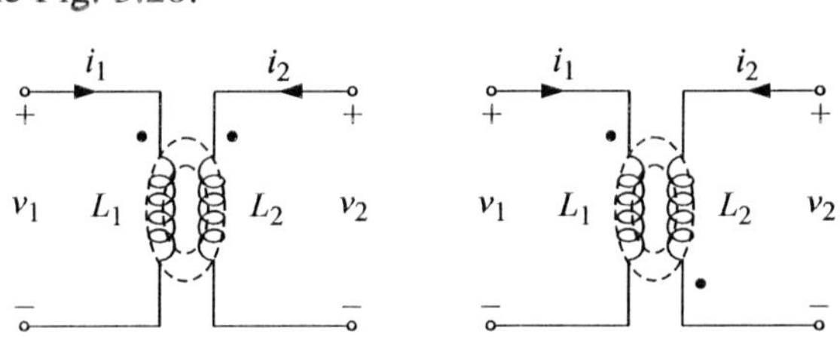

Fig. 3.20. *A pair of coupled coils.*

The branch inductance gets modified by :

$$L_b = \begin{bmatrix} L_1 & \pm M \\ \pm M & L_2 \end{bmatrix}$$

Example 3.26. Write loop equations for the network shown in the figure.

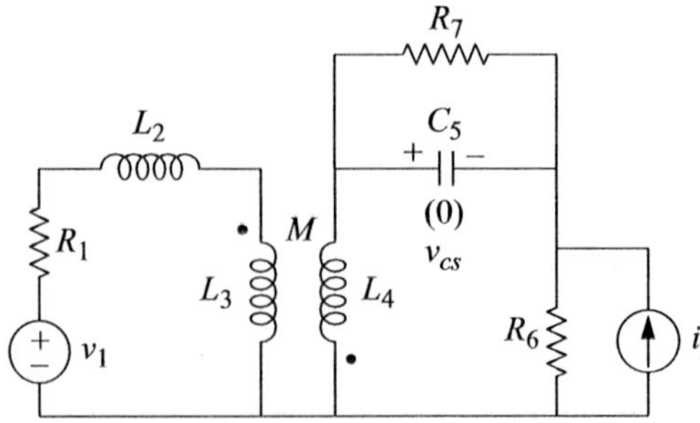

Solution: Current source is to be converted into voltage for writing loop equations.

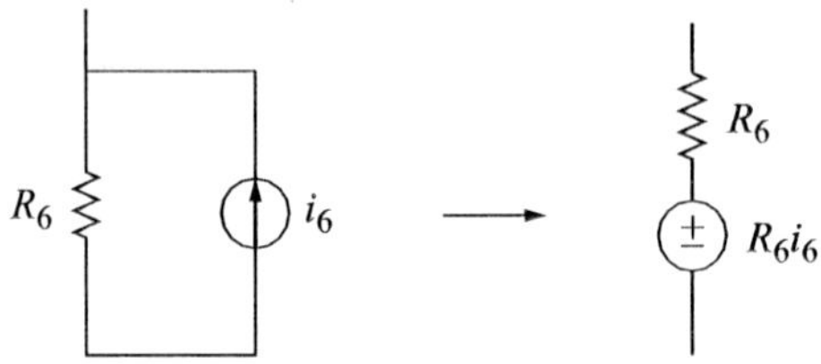

An oriented graph of the network is shown in figure (a) and corresponding trees with loop currents are shown in figure (b).

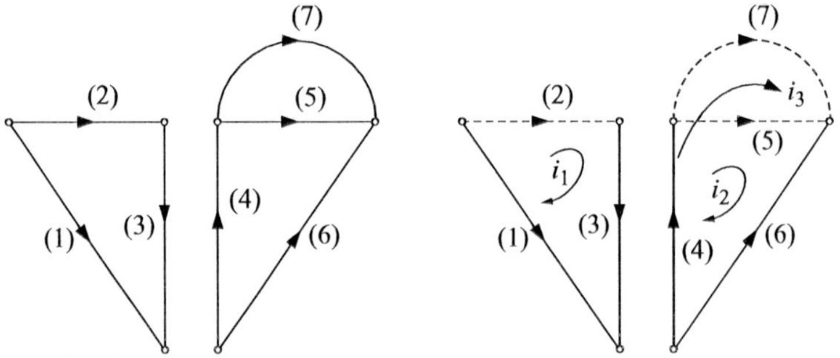

The loop currents are shown corresponding to the links namely (2), (5) and (7). Currents i_2 and i_5 are entering their respective dots in the two coils. Hence, the sign of the term corresponding to the mutual inductance would be positive. The branches (3) and (4) are coupled by M. Therefore, the branch impedance matrix $Z_b\,(s)$ and loop (tie-set) matrix B are:

$$Z_b\,(s) = \begin{bmatrix} R_1 & 0 & 0 & 0 & 0 & 0 & 0 \\ 0 & sL_2 & 0 & 0 & 0 & 0 & 0 \\ 0 & 0 & sL_3 & sM & 0 & 0 & 0 \\ 0 & 0 & sM & sL_4 & 0 & 0 & 0 \\ 0 & 0 & 0 & 0 & \dfrac{1}{sC_5} & 0 & 0 \\ 0 & 0 & 0 & 0 & 0 & R_6 & 0 \\ 0 & 0 & 0 & 0 & 0 & 0 & R_7 \end{bmatrix}$$

$$B = \begin{array}{c} \text{loops} \\ \downarrow \\ 1 \\ 2 \\ 3 \end{array} \begin{array}{c} \text{branches} \rightarrow \\ \begin{array}{ccccccc} 1 & 2 & 3 & 4 & 5 & 6 & 7 \end{array} \\ \begin{bmatrix} -1 & 1 & 1 & 0 & 0 & 0 & 0 \\ 0 & 0 & 0 & 1 & 1 & -1 & 0 \\ 0 & 0 & 0 & 0 & -1 & 0 & 1 \end{bmatrix} \end{array}$$

The loop impedance matrix is given by :

$$Z_l = B\,Z_b\,B^T = \begin{bmatrix} R_1 + sL_2 + sL_3 & sM & 0 \\ sM & sL_4 + \dfrac{1}{sc_5} + R_6 & -\dfrac{1}{sc_5} \\ 0 & -\dfrac{1}{sc_5} & \dfrac{1}{sc_5} + R_7 \end{bmatrix}$$

The equivalent voltage source matrix is given by :

$$V_q(s) = \begin{bmatrix} v_1 \\ 0 \\ 0 \\ 0 \\ \dfrac{v_{c_5}(c)}{s} \\ -R_6\, i_6 \\ 0 \end{bmatrix}$$

The loop equations are :

$$\begin{bmatrix} R_1 + sL_2 + sL_3 & sM & 0 \\ sM & R_6 + sL_4 + \dfrac{1}{sc_5} & -\dfrac{1}{sc_5} \\ 0 & -\dfrac{1}{sc_5} & \dfrac{1}{sc_5} + R_7 \end{bmatrix} \begin{bmatrix} i_1 \\ i_2 \\ i_3 \end{bmatrix} = \begin{bmatrix} v_1 \\ -\left[\dfrac{v_{cs}(0)}{s} + R_6 i_6\right] \\ \dfrac{v_{c_5}(0)}{s} \end{bmatrix}$$

SUMMARY

1. **Graph Theory:** Graph theory is an analysis technique for a system using matrices representing the system.
2. **Types of Networks:** (*i*) Linear Network; (*ii*) Non-linear Network; (*iii*) Bilateral Network; (*iv*) Unilateral Network; (*v*) Active Network; (*vi*) DC Network; (*vii*) AC Network; (*viii*) Lumped Network; (*ix*) Distributed Network.
3. **Subgraph:** A subgraph of a graph *G*, is a graph whose branches and nodes form a subset of the branch and nodes of a graph *G*.
4. **Tree:** A tree is connected graph without any circuits (closed loops) of a graph or subgraph.
5. **Spanning Tree or Tree:** If a tree contains all nodes of a given graph *G*, it is known as spanning tree.
6. **Co-tree, Twig and Links:** The branches of the tree are called as twigs and the branches removed from the graph is forming a tree is called as links (or chords). The set of all links corresponds to a given tree is called as co-tree.
7. **Planar Network:** A network is said to be planar if it can be drawn on a two-dimensional plane, such that no two branches intersect at a point that is not a node.

EXERCISES

3.1. Draw the nodes and branches elements shown in the figures (*a*) to (*f*).

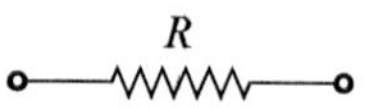

(*a*)

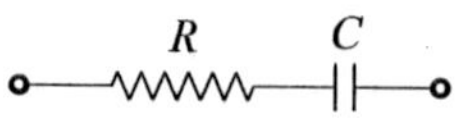

(*b*)

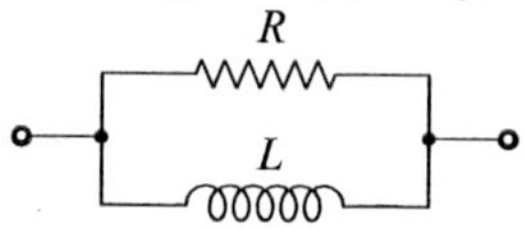

(*c*)

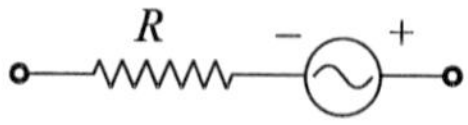

(*d*)

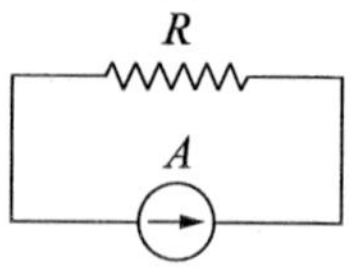

(*e*)

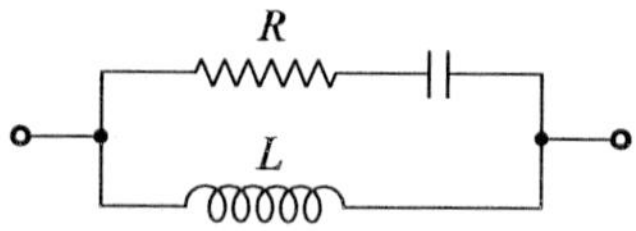

(*f*)

3.2. Draw the graphs of the Network shown in given figures (*a*) to (*d*).

(*a*)

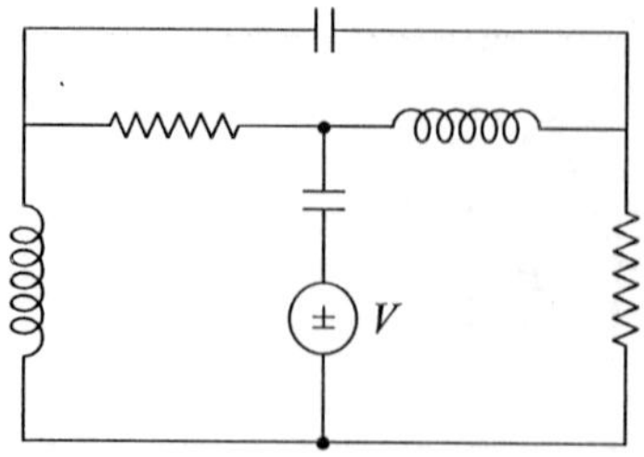

(*b*)

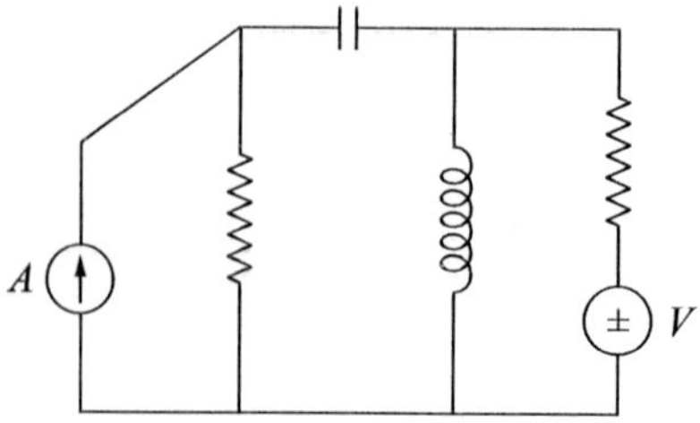

(c)

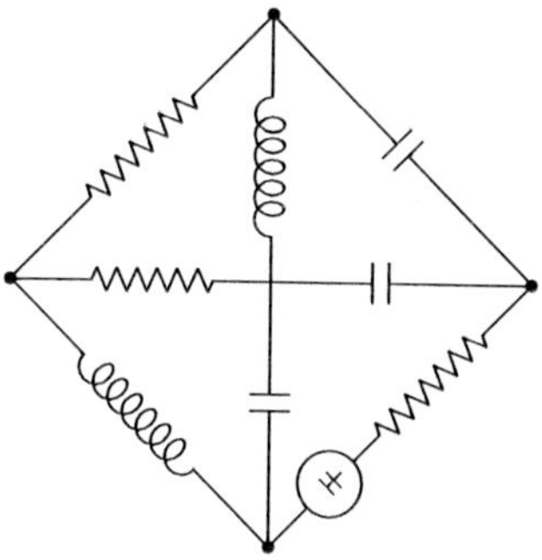

(d)

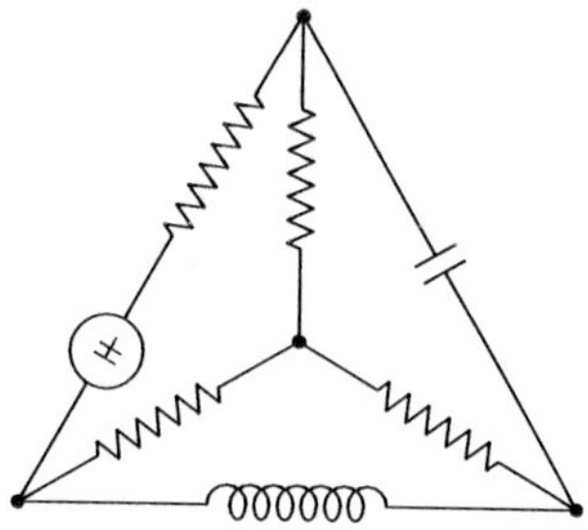

3.3. For the Network shown in given figure. Determine all the tree and corresponding co-trees for the graph of the Network. Then, consider the tree formed by branches (1, 2, 5), using tree write A, B_f and Q_f.

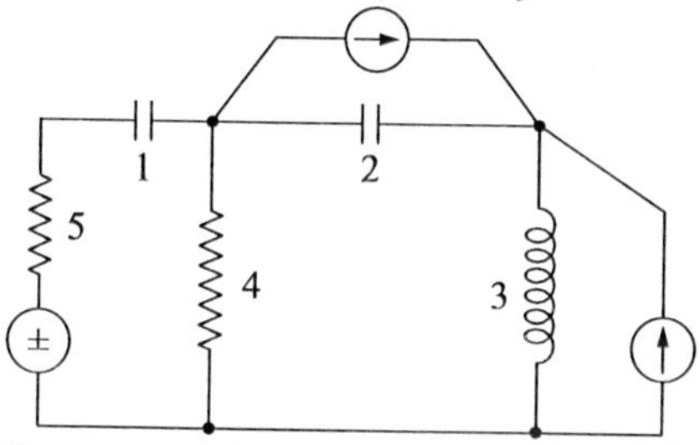

3.4. For the shown in figure, consider the tree formed by branches (2, 3, 4), using this tree write A, B, and Q_f.

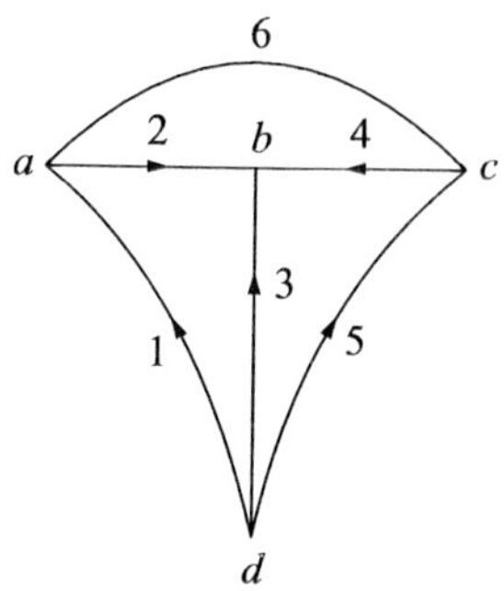

3.5. Consider the graph shown in figure. Select a tree with twigs (4, 5, 6) for this tree write down the tie set matrix.

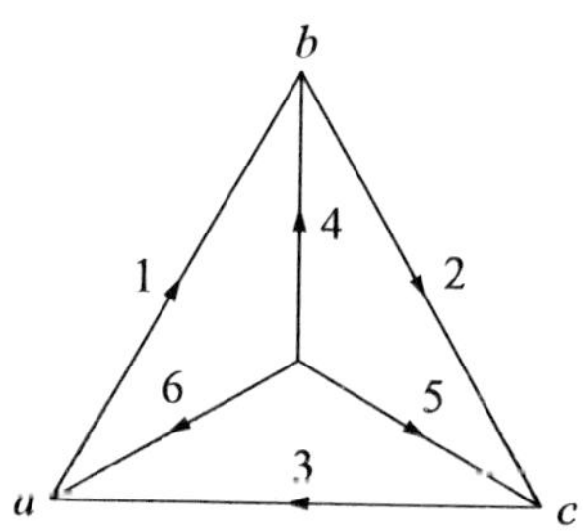

3.6. The reduced incidence matrix of a graph is given by Nodes/Branches →

$$\begin{array}{c} \downarrow \\ a \\ b \\ c \end{array} \begin{array}{c} \begin{matrix} 1 & 2 & 3 & 4 & 5 & 6 \end{matrix} \\ \begin{bmatrix} 0 & 1 & 0 & -1 & -1 & 1 \\ 0 & -1 & 1 & 0 & 1 & 0 \\ 1 & 0 & -1 & 0 & 0 & 0 \end{bmatrix} \end{array}$$

Draw the oriented graph. Select a tree and find *f* cutset matrix.

3.7. Write the matrix loop equation for the Network shown in given figure, using the loop analysis.

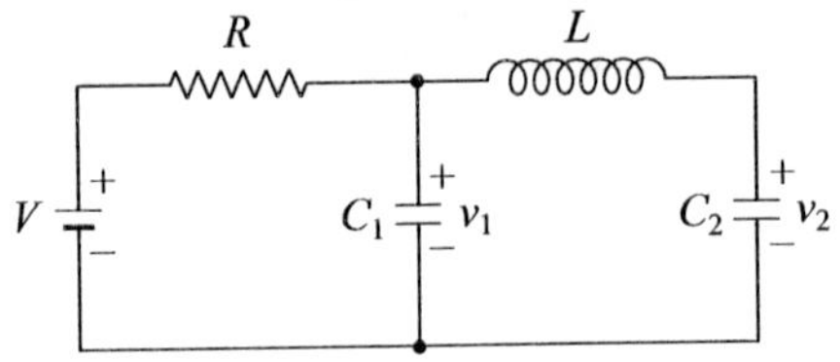

3.8. For the network shown in figure obtain the incidence matrix, the node admittance matrix and the matrix node equation.

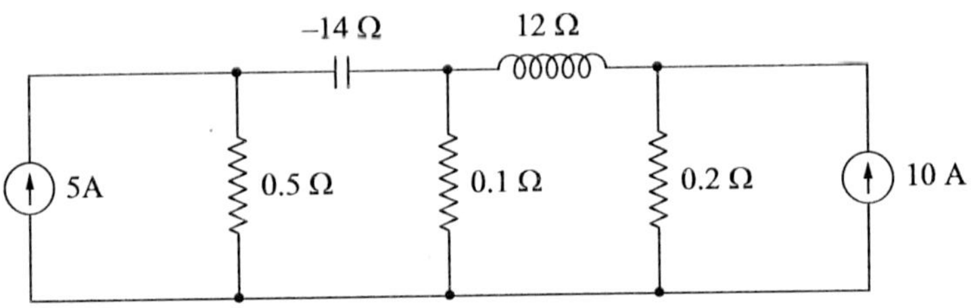

3.9. The reduce incidence matrix of a graph is given. Draw the graph and obtain the *f*-loop and *f*-cutset matrices. Also verify the results by obtaining mathematically.

$$A = \begin{array}{c} \\ a \\ b \\ c \end{array} \begin{array}{c} \begin{matrix} 1 & 2 & 3 & 4 & 5 & 6 & 7 \end{matrix} \\ \begin{bmatrix} 0 & 0 & 1 & 1 & 1 & 0 & -1 \\ 0 & 1 & 0 & 0 & -1 & 1 & 1 \\ -1 & 0 & -1 & 0 & 0 & -1 & 0 \end{bmatrix} \end{array}$$

3.10. For the Network shown in figure write down the *f*-cutset matrix for the tree having 3H inductor and 2F capacity as twigs and write the matrix cutset equations.

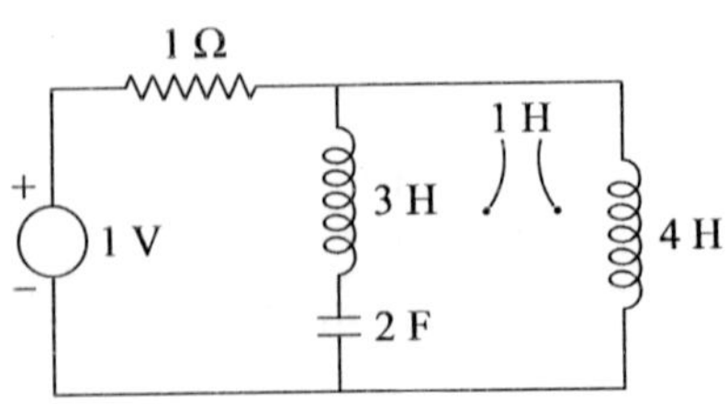

3.11. For the network shown in figure. Determine the f-loop matrix and f-loop equations.

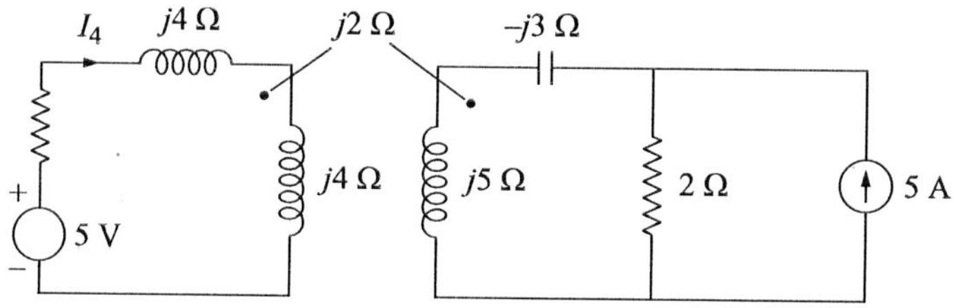

3.12. The fundamental at set matrix Q_f of a circuit is given, obtain the link voltages in terms of the tree branch voltages. The tree branches and C, V_S and R_1.

$$Q_f = \begin{array}{c} \begin{array}{ccccccc} C & V_s & R_1 & R_2 & R_3 & X_3 & L \end{array} \\ \begin{bmatrix} 0 & 0 & 1 & 1 & 1 & 0 & -1 \\ 0 & 1 & 0 & 0 & -1 & 1 & 1 \\ -1 & 0 & -1 & 0 & 0 & -1 & 0 \end{bmatrix} \end{array}$$

3.13. An oriented graph of a certain circuit is shown in figure for the tree [2, 4, 6], find $[Q_f]$ and $[B_f]$ and hence shown let $[Q_f]\,[B_f]^T = 0$.

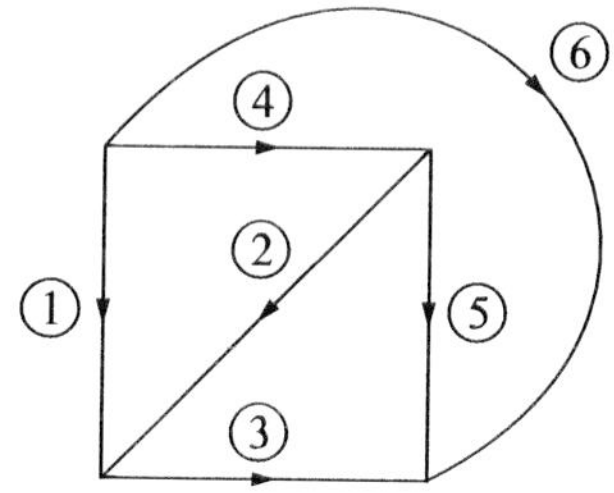

3.14. Draw the graph and find a tree for the network shown in figure. Considering 'O' as a datum mode and assuming elements BD and BC as links determine the tie-set schedule. Branch impedance matrix and source voltage matrix. Obtain the loop equations using the above said matrices.

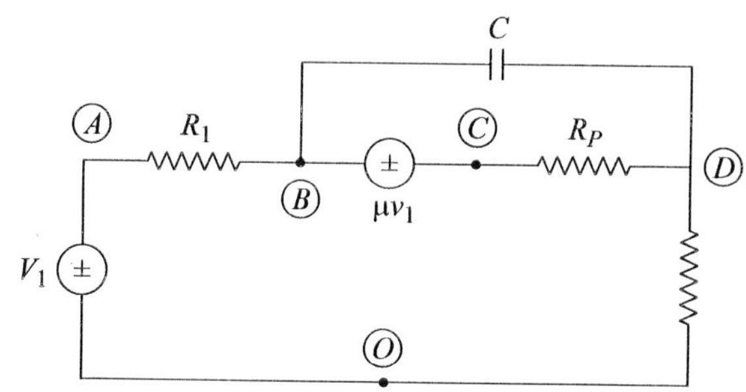

3.15. Draw the directed graph of the network of figure, choose a tree, write the cutset matrix, branch admittance and node-pair equations.

❑❑❑

4

Network Theorems

4.1 INTRODUCTION

Ohm's law, Kirchhoff's Current Law (KCL) and Kirchhoff's Voltage Law (KVL) can solve any network problem. These methods are laborious and time-consuming, therefore, Network Theorems are very handy to solve very difficult problems.

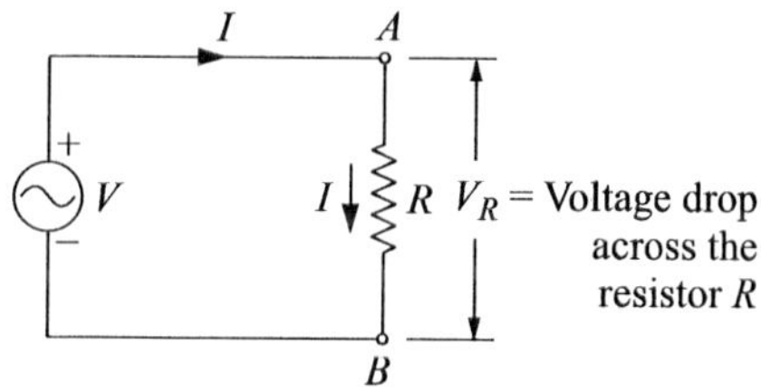

Fig. 4.1. *Conventions of voltage and current directions*

We have already covered Ohm's law in the basic concepts, therefore, KCL and KVL need to be discussed before taking up the network theorems. The current directions and polarity of voltages are important in solving any network. The +*ve* terminal of a voltage source is at higher potential with respect to – *ve* terminal of the voltage source. The + *ve* current flow direction is from + *ve* terminal to – *ve* terminal through the load. Consider the circuit of Fig. 4.1.

The current flows from higher potential to lower potential. In the figure, the current from voltage source + *ve* terminal through load resistor *R* to – *ve* terminal is + *ve* direction current flow. The point *A* is at higher potential with respect to point *B* across the load resistor *R*. There is a voltage drop from point *A* to point *B*.

The higher potential at point *A* is known as + *ve* with respect to potential at point *B i.e.*, – *ve* potential. *IR* voltage drop is equal and opposite of voltage source.

4.1.1 First Law or Point Law or Kirchoff's Current Law (KCL)

KCL is also known as Kirchoff's first law or point law. KCL states that algebraic sum of currents at a point or node in a network is zero. In Fig. 4.2, I_1 to I_5 are shown at point or node with their respective directions. As per KCL at a node:

$\Sigma I = 0$ *i.e.*, $I_1 + I_2 - I_3 - I_4 + I_5 = 0$

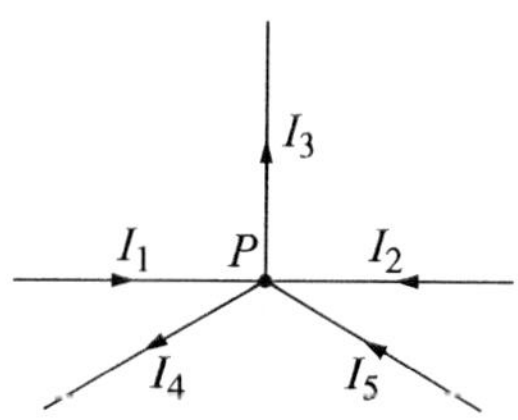

Fig. 4.2. *KCL illustration*

or $$(I_1 + I_2 + I_5) = (I_3 + I_4)$$

Please note that current flow toward point P is + *ve* and away from point P is – *ve* as per normal convention.

Example 4.1. A node of a network is shown in the figure with currents flowing in the direction shown. Determine the magnitude and direction of I.

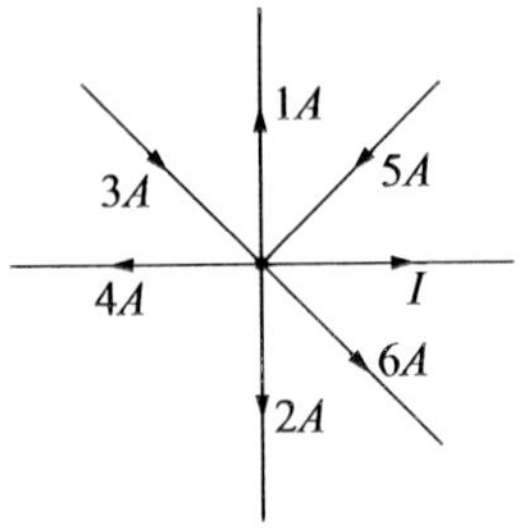

Solution: $-4 + 3 - 1 + 5 - I - 6 - 2 = 0$

or $I = -5$ A *i.e.*, the current is towards the node.

4.1.2 Second Law or Mesh Law or Kirchoff's Voltage Law (KVL)

In any closed network or mesh, the algebraic sum of products of current and resistances in each conductor plus the algebraic sum of emf in that path is zero.

i.e., $$\Sigma IR + \Sigma \text{emf} = 0$$

As per convention a voltage rise is given a + *ve* sign and a voltage fall is given – *ve* sign. The current flows from higher potential to lower potential.

Example 4.2. Determine the value of current in each branch for the network shown in the figure given below using Kirchhoff's Laws.

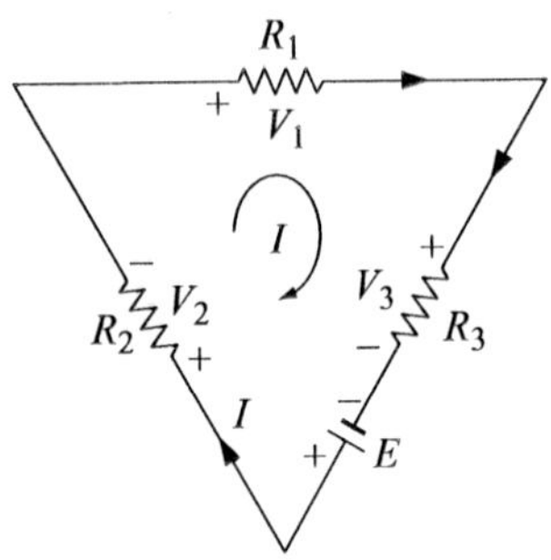

Solution: As per KVL for triangular circuit we get,

$$V_2 + V_1 + V_3 - E = 0$$

or $$E = IR_2 + IR_1 + IR_3$$

or $$I = \frac{E}{R_1 + R_2 + R_3}$$

4.2 SIMULTANEOUS EQUATIONS SOLUTION USING MATRIX

We already know that the solution of following equation using matrix is to be found out

$$a_1 x_1 + b_1 x_2 = p \quad \text{... (i)}$$

$$a_2 x_1 + b_2 x_2 = q \quad \text{... (ii)}$$

The above can be written in matrix form as :

$$\begin{bmatrix} a_1 & b_1 \\ a_2 & b_2 \end{bmatrix} \begin{bmatrix} x_1 \\ x_2 \end{bmatrix} = \begin{bmatrix} p \\ q \end{bmatrix}$$

The values of x_1 and x_2 as per Crammer's rule are as follows:

$$x_1 = \frac{\Delta_1}{\Delta}$$

where, $$\Delta_1 = \begin{bmatrix} p & b_1 \\ q & b_2 \end{bmatrix} = (p\,b_2 - q)$$

and $$\Delta = \begin{bmatrix} a_1 & b_1 \\ a_2 & b_2 \end{bmatrix} = (a_1\,b_2 - b_1\,a_2)$$

$\therefore$ $$x_1 = \frac{pb_2 - b_1 q}{a_1 b_2 - b_1 a_2}$$

Similarly, $$x_2 = \frac{\Delta^2}{\Delta}$$

where $$\Delta_2 = \begin{bmatrix} a_1 & p \\ a_2 & q \end{bmatrix} = a_1 q - p a_2$$

$\therefore$ $$x_2 = \frac{a_1 q - p a_2}{a_1 b_2 - b_1 a_2}$$

Now, if the simultaneous equations are three as follows:

$$a_1\,x_1 + b_1\,x_2 + c_1\,x_3 = p$$
$$a_2\,x_1 + b_2\,x_2 + c_2\,x_3 = q$$
$$a_3\,x_1 + b_3\,x_2 + c_3\,x_3 = r$$

The above equation can be written in matrix form as follows:

$$\begin{bmatrix} a_1 & b_1 & c_1 \\ a_2 & b_2 & c_2 \\ a_3 & b_3 & c_3 \end{bmatrix} \begin{bmatrix} x_1 \\ x_2 \\ x_3 \end{bmatrix} = \begin{bmatrix} p \\ q \\ r \end{bmatrix}$$

The values of x_1, x_2 and x_3 are given by:

$$x_1 = \frac{\Delta_1}{\Delta},\ x_2 = \frac{\Delta_2}{\Delta} \text{ and } x_3 = \frac{\Delta_3}{\Delta}$$

where the determinant values are:

$$\Delta = \begin{vmatrix} a_1 & b_1 & c_1 \\ a_2 & b_2 & c_2 \\ a_3 & b_3 & c_3 \end{vmatrix}$$

$$= a_1(b_2\,c_3 - c_2\,b_3) - b_1\,(a_2\,c_3 - c_2\,a_3) + c_1\,(a_2\,b_3 - b_2\,a_3)$$

$$\Delta_1 = \begin{vmatrix} p & b_1 & c_1 \\ q & b_2 & c_2 \\ r & b_3 & c_3 \end{vmatrix}$$

$$= p(b_2\,c_3 - c_2\,b_3) - b_1\,(q\,c_3 - c_2\,r) + c_1\,(q\,b_3 - b_2\,r)$$

$$\Delta_2 = \begin{vmatrix} a_1 & p & c_1 \\ a_2 & q & c_2 \\ a_3 & r & c_3 \end{vmatrix}$$

$$= a_1(q\,c_3 - c_2\,r) - p\,(a_2\,c_3 - c_2\,a_3) + c_1\,(a_2 r - q\,a_3)$$

Example 4.3. The figure below shown two batteries connected in parallel each represented by an emf along with its internal resistance. A load resistance of 6 Ω is connected across the ends of the batteries. Calculate the current through each battery and the load.

[UPTU, 2002]

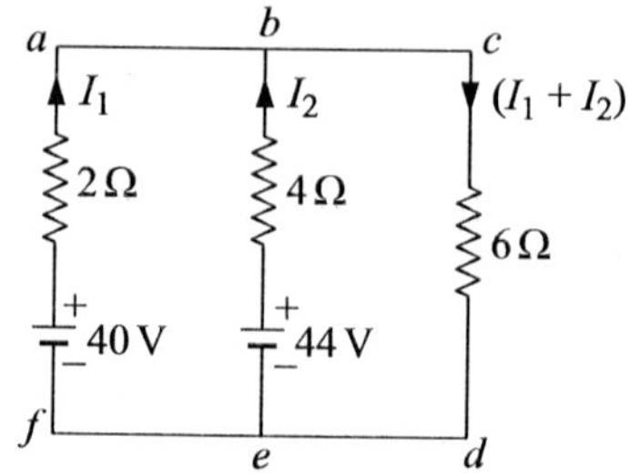

Solution: KVL in mesh or loops *a b e f a* and *a b c d e f a*. The currents towards a node + *ve* and away is negative and clockwise voltage drops and sources are positive directions.

$$\therefore \quad 2I_1 - 4I_2 + 44 - 40 = 0$$

or $$2I_1 - 4I_2 = 40 - 44$$

or $$I_1 - 2I_2 = 2 \qquad \text{...(i)}$$

The other mesh *a c d e f a*, we can write:

$$2I_1 + 6(I_1 + I_2) - 40 = 0$$

or $$I_1 + 3(I_1 + I_2) = 20 \qquad \text{...(ii)}$$

or $$4I_1 + 3I_2 = 20 \qquad \text{...(iii)}$$

From equations (*i*) and (*ii*), we get

$$I_1 = \frac{\begin{vmatrix} -2 & -2 \\ 20 & 3 \end{vmatrix}}{\begin{vmatrix} 1 & -2 \\ 4 & 3 \end{vmatrix}} = \frac{-2\times 3-(-2)\times 20}{1\times 3-(-2\times 4)} = \frac{-6+40}{3+8} = \frac{34}{11}$$

or $$I_1 = \frac{34}{11} \text{ Amp.}$$

and $$I_2 = \frac{\begin{vmatrix} 1 & -2 \\ 4 & 20 \end{vmatrix}}{\begin{vmatrix} 1 & -2 \\ 4 & 3 \end{vmatrix}} = \frac{1\times 20-(-2)\times 4}{1\times 3-(-2)\times 4} = \frac{20+8}{3+8} = \frac{28}{11} \text{ Amp.}$$

Current through load is $I_1 + I_2$

$$= \frac{34}{11} + \frac{28}{11} = \frac{62}{11} \text{ Amps.}$$

Example 4.4. Two batteries A and B are connected in parallel and a load of 10 Ω is connected across their terminals. A has an emf of 12V and an internal resistance of 2Ω and B has an emf of 8V and an internal resistance of 1W. Use Kirchhoff's laws to determine the magnitude of currents and also the direction in each of the batteries. Also determine *pd* across external resistance.

Solution: By Applying KVL in Mesh $a\,b\,e\,f\,a$, we get

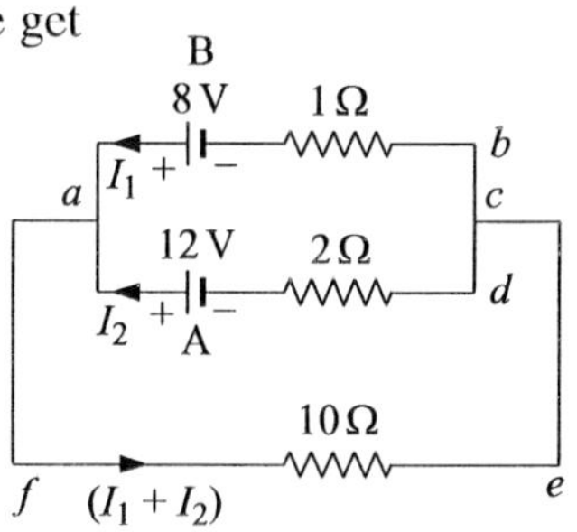

$$I_1 + 10\,(I_1 + I_2) - 8 = 0$$

or $$11I_1 + 10I_2 = 8 \qquad ...(i)$$

From mesh $a\,b\,c\,d\,a$ using KVL, we get

$$I_1 - 2I_2 - 8 + 12 = 0$$

or $$I_1 - 2I_2 = -4 \qquad ...(ii)$$

Using matrix techniques, we get:

$$I_1 = \frac{\begin{vmatrix} 8 & 10 \\ -4 & -2 \end{vmatrix}}{\begin{vmatrix} 11 & 10 \\ 1 & -2 \end{vmatrix}} = \frac{8(-2) - 10\times(-4)}{11(-2) - 10\times 1} = \frac{-16+40}{-22-10} = \frac{24}{-32} = -\frac{3}{4}$$

or $$I_1 = -0.75 \text{ A}$$

and $$I_2 = \frac{\begin{vmatrix} 11 & 8 \\ 1 & -4 \end{vmatrix}}{\begin{vmatrix} 11 & 10 \\ 1 & -2 \end{vmatrix}} = \frac{-44-8}{(11\times -2) - 10\times 1} = \frac{-52}{-22-10} = \frac{52}{32} = \frac{13}{8}$$

or $$I_2 = 1.625 \text{ A}$$

pd across external resistance

$$= (I_1 + I_2)\,10 = (-0.75 + 1.625)10$$
$$= 0.875 \times 10 = 8.75 \text{ Volts.}$$

Example 4.5. In figure given here the potential of point A = –30V. Using Kirchhoff's laws find (a) value of V and (b) power dissipated by 5Ω resistance.

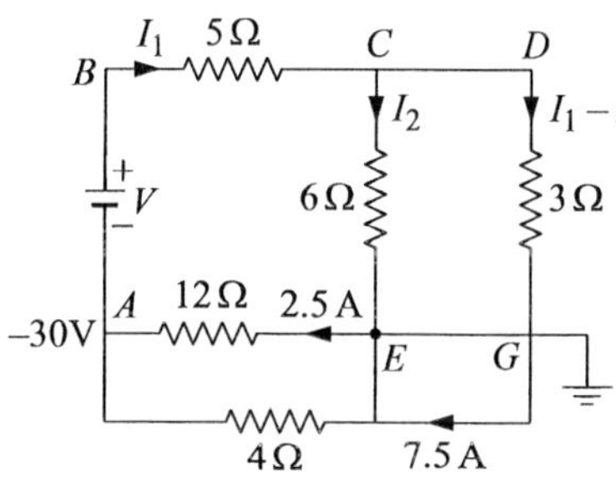

Solution: The currents I_1, I_2 with directions have been marked to solve the problem. The currents 2.5 A and 7.5 A are calculated on the basis the point A is at – 30 V and G at zero potential.

Using KVL in the loop $A\ B\ C\ E\ A$, we get

$$5I_1 + 6I_2 + 30 - (V + 30) = 0$$

or $$5I_1 + 6I_2 = V \qquad ...(i)$$

From loop $ABCDG\ A$ using KVL, we get

$$5I_1 + 3(I_1 - I_2) + 30 - (V + 30) = 0$$

or $$8I_1 - 3I_2 = V \qquad ...(ii)$$

Substract equation (*i*) from (*ii*), we get

$$3I_1 - 9I_2 = 0$$

or $$I_1 - 3I_2 = 0 \qquad ...(iii)$$

Using KVL at EG point, we get

$$-2.5 - 7.5 + I_2 + (I_1 - I_2) = 0$$

or $$-10 + I_1 = 0$$

or $$I_1 = 10\text{A} \qquad ...(iv)$$

Substituting value of I_1 in equation (*iii*), we get

$$3I_2 = 10$$

or $$I_2 = \frac{10}{3}A$$

From equation. (*ii*) after substituting values of I_1 and I_2, we get

$$V = 8 \times 10 - 3 \times \frac{10}{3} = 80 - 10$$

(*a*) or $V = 70$ Volts.

(*b*) Power dissipated by 5 W resistance

$$= I_1^2 R = (10)^2 \times 5 = 500 \text{ W}$$

Example 4.6. Applying Kirchhoff's current law, determine current I_S in the electric circuit of figure given below. Take $V_0 = 16$V. **[UPTU, 2001]**

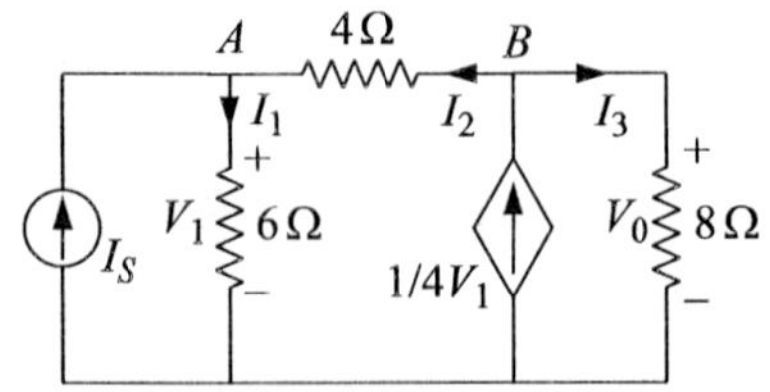

Solution: Apply KCL at node, A, we get

$$I_s - I_1 + I_2 = 0$$

$$I_s = I_1 - I_2 \qquad ...(i)$$

Apply KCL at *B*, we get

$$-I_2 - I_3 + \frac{1}{4}V_1 = 0$$

or $$V_1 = 4\,(I_2 + I_3) \qquad \text{...}(ii)$$

As $$V_0 = 16\text{V}$$

$\therefore$ $$I_3 = \frac{16}{8} = 2\text{ A}$$

From equation (*ii*), we get

$$V_1 = 4(I_2 + 2) \qquad \text{...}(iii)$$

From 6 Ω branch, we get

$$V_1 = 6I_1 \qquad \text{...}(iv)$$

Also *B* to *A*, we get

$$I_2 = \frac{16 - V_1}{4} \text{ as } V_B = V_0 = 16\text{ V}$$

or $$4I_2 = 16 - V_1$$

By substituting value of V_1 from equation (*iii*), we get

$$4I_2 = 16 - 4(I_2 + 2)$$

or $$4I_2 = 16 - 4I_2 - 8$$

or $$8I_2 = 8$$

or $$I_2 = 1\text{ A}$$

$\therefore$ $$V_1 = 6I_1$$

Now, from equation (*ii*), we get

$$6I_1 = 4(I_2 + I_3)$$

or $$6I_1 = 4(1 + 2)$$

or $$I_1 = \frac{12}{6} = 2\text{ A}.$$

From equation (*i*) by substituting values, we get

$$I_s = 2 - 1 = 1\text{ A}.$$

Example 4.7. Find the value of resistance R and current I_R in figure given below when the branch AD carries no current.

Solution: Branch AD carries no current, hence branch CD and DB have same current I_2. Also branch CA and AB carry same current I_1.

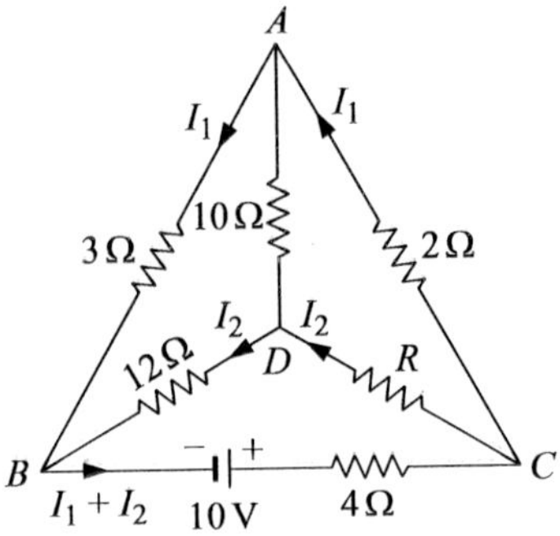

Applying KVL in loop $CADC$, we get

$$2I_1 + 10 \times 0 - I_2 \times R = 0$$

or $$2I_1 = I_2 R$$

or $$R = \frac{2I_1}{I_2} \qquad ...(i)$$

Applying KVL in loop $ABDA$, we get

$$3I_1 - 12I_2 + 10 \times 0 = 0$$

$$3I_1 = 12I_2$$

or $$\frac{I_1}{I_2} = 4 \qquad ...(ii)$$

Substituting $\frac{I_1}{I_2}$ in equation (i), we get

$$R = 2 \times 4 = 8\ \Omega.$$

Applying KVL in loop $BCDB$, we get

$$4(I_1 + I_2) + I_2 R + 12I_2 - 10 = 0$$

or $$4I_1 + 4I_2 + 8I_2 + 12\,I_2 = 10$$

or $$4I_1 + 24I_2 = 10$$

or $$4\frac{I_1}{I_2} + 24 = \frac{10}{I_2}$$

or $$4 \times 4 + 24 = \frac{10}{I_2}$$

or $$40 = \frac{10}{I_2}$$

or $$I_2 = 0.25\text{ A.}$$

We know that $$I_R = I_2 = 0.25\text{ A.}$$

Example 4.8. In the circuit given below, find the value of I_s for $I = 0$.

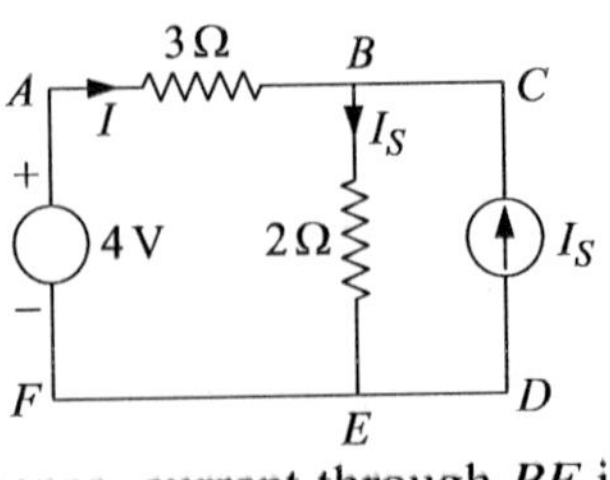

Solution: Current $I = 0$, hence, current through BE is I_s.

$\therefore$ $$4\text{V} = 2I_s\text{ V}$$

or $$I_s = 2A.$$

Example 4.9. Find the total power delivered to the circuit by two sources in the figure given below:

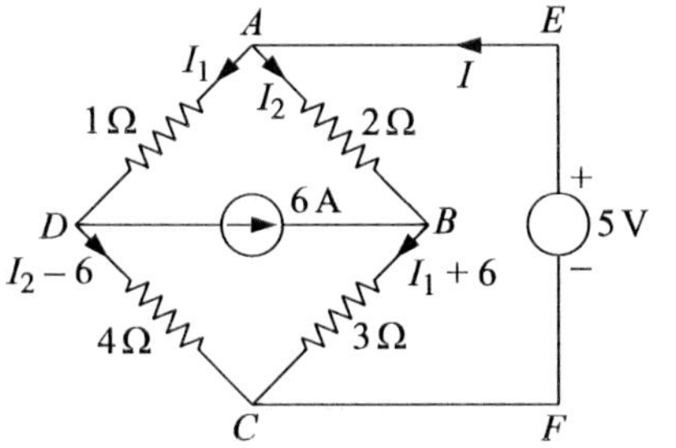

Solution: The currents I_1 and I_2 are being marked as shown and according currents in other branches are deduced.

Applying KVL in mesh *AEFCDA*, we get:

$$I_2 + 4(I_2 - 6) - 5 = 0$$

or $$5I_2 - 24 - 5 = 0$$

or $$5I_2 = 29$$

or $$I_2 = \frac{29}{5} = 5.8 \text{ A.}$$

From mesh *AEFCBA* by applying KVL, we get:

$$2I_1 + 3(I_1 + 6) - 5 = 0$$

or $$5I_1 + 18 - 5 = 0$$

or $$5I_1 = -13$$

or $$I_1 = -\frac{13}{5} = -2.6 \text{ A.}$$

Total power delivered to the circuit by two sources

$= (I_1)^2 \times 2 + (I_2)^2 \times 1 + (I_2 - 6)^2 \times 4 + (I_1 + 6)^2 \times 3$

$= (2.6)^2 \times 2 + (5.8)^2 \times 1 + (5.8 - 6)^2 \times 4 + (-2.6 + 6)^2 \times 3$

$\simeq 82$ W.

Example 4.10. Calculate the current through the galvanometer in the following bridge of the figure: **[UPTU, 2003]**

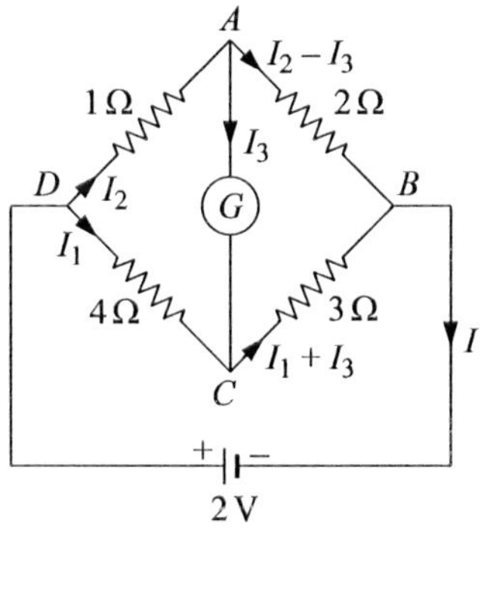

Solution: The currents and directions have been marked as shown, and nodes have also been marked.

KVL for *ADCA*:

$$I_2 \times 1 + I_3 \times 4 - I_1 \times 2 = 0$$

or $$2I_1 + I_2 - 4I_3 = 0 \quad ...(i)$$

KVL for *ABCA*:

$$2(I_2 - I_3) - 3(I_1 + I_3) - 4I_3 = 0$$

or $$2I_2 - 2I_3 - 3I_1 - 3I_3 - 4I_3 = 0$$

or $$-3I_1 + 2I_2 - 9I_3 = 0$$

or $$3I_1 - 2I_2 + 9I_3 = 0 \quad ...(ii)$$

KVL for *DCBD*:

$$2I_1 + 3(I_1 + I_3) - 2 = 0$$

or $$2I_1 + 3I_1 + 3I_3 = 2$$

or $$5I_1 + 3I_3 = 2$$

or $$5I_1 = 2 - 3I_3 \text{ or } I_1 = \frac{2 - 3I_3}{5} \quad ...(iii)$$

By substituting value of I_1 in equation (*i*), we get

$$2\left(\frac{2 - 3I_3}{5}\right) - I_2 - 4I_3 = 0$$

$$4 - 6I_3 \; 5I_2 - 20I_3 = 0$$

$$4 - 5I_2 - 26I_3 = 0 \quad ...(iv)$$

By substituting value of I_1 in equation (*i*), we get

$$3\left(\frac{2 - 3I_3}{5}\right) - 2I_2 + 9I_3 = 0$$

$$6 - 9I_3 - 10I_2 + 45I_3 = 0$$

or $$6 - 10I_2 + 36I_3 = 0$$

or $$10I_2 = 6 + 36I_3 \quad ...(v)$$

or $$I_2 = \left(\frac{6 + 36I_3}{10}\right) = \left(\frac{3 + 18I_3}{5}\right)$$

By substituting I_2 in equation (*iv*), we get

$$4 - 5\left(\frac{3 + 18I_3}{5}\right) - 26\,I_3 = 0$$

or $$4 - 3 - 18I_3 - 26I_3 = 0$$

or $$1 - 44I_3 = 0$$

or $$I_3 = \frac{1}{44} A$$

Current flowing through galvanometer

$$= I_3 = \frac{1}{44} \text{ A.}$$

Example 4.11. Determine the current in 4 Ω resistance of the circuit shown.

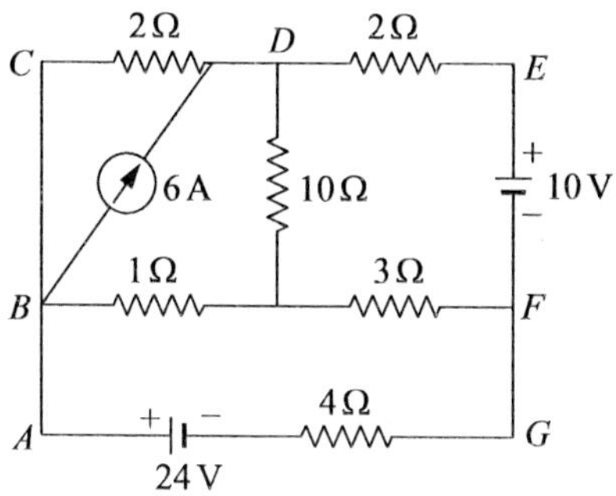

Solution:

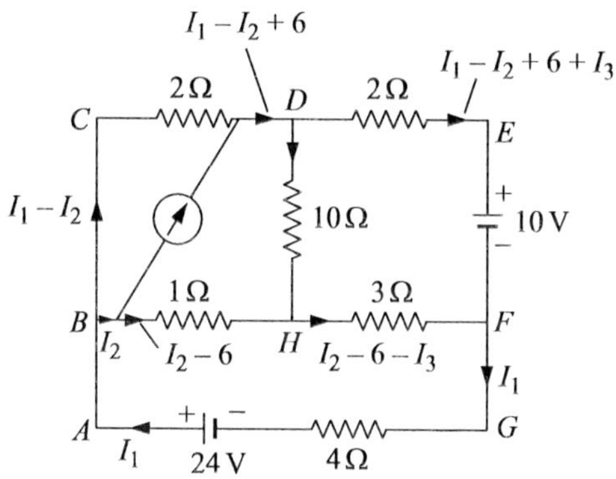

KVL in the loop *BHFGAB*:

$$(I_2 - 6) + 3\,(I_2 - 6 - I_3) + 4I_1 + 24 = 0$$

or $$I_2 - 6 + 3I_2 - 18 - 3I_3 + 4I_1 = 24$$

or $$4I_1 + 4I_2 - 3I_3 = 48 \qquad ...(i)$$

KVL in the loop *DEFHD*:

$$2(I_1 - I_2 + 6 + I_3) + 10 - 3(I_2 - 6 - I_3) + 10I_3 = 0$$

or $$2I_1 - 5I_2 + 15I_3 = 40 \qquad ...(ii)$$

KVL in the loop *BCDHB*:

$$2(I_1 - I_2) - 10I_3 - (I_2 - 6) = 0$$

or $$2I_1 - 3I_2 - 10I_3 = -6 \qquad ...(iii)$$

The three simultaneous equations are:

$$4I_1 + 4I_2 - 3I_3 = 48$$
$$2I_1 - 5I_2 + 15I_3 = 40$$
$$2I_1 - 3I_2 - 10I_3 = -6$$

In matrix form:

$$\begin{bmatrix} 4 & 4 & -3 \\ 2 & -5 & 15 \\ 2 & -3 & -10 \end{bmatrix} \begin{bmatrix} I_1 \\ I_2 \\ I_3 \end{bmatrix} = \begin{bmatrix} 48 \\ 40 \\ -6 \end{bmatrix}$$

$$\text{Common determinant, } \Delta = \begin{vmatrix} 4 & 4 & -3 \\ 2 & -5 & 15 \\ 2 & -3 & -10 \end{vmatrix} = -568$$

$$\Delta_1 = \begin{vmatrix} 48 & 4 & -3 \\ 40 & -5 & 15 \\ -6 & -3 & -10 \end{vmatrix} = -2330$$

$$\therefore \quad I_1 = \frac{\Delta_1}{\Delta} = \frac{-2330}{-568} = 4.1\text{A}.$$

4.3 MAXWELL'S MESH CURRENT METHOD

The concept of mesh current reduces the number of equations required to be solved. Consider the Fig. 4.3. A mesh is a loop which does not contain any other loop.

The procedure for writing the equations as per mesh current method is as under :

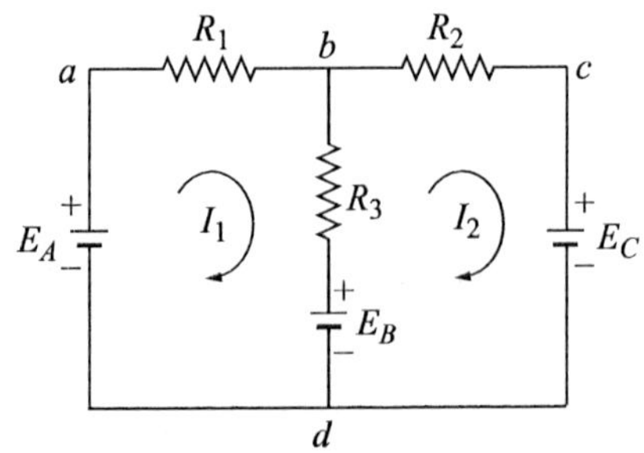

Fig. 4.9. *Network*

(*i*) The mesh currents are usually presumed in clockwise direction.

(*ii*) The number of mesh currents should be as lower as possible.

(*iii*) The KVL equations are written for each mesh.

KVL for mesh *abda:*

$$I_1 R_1 + (I_1 - I_2) R_3 = E_A - E_B \quad \text{...}(i)$$

KVL for *bcdb:*

$$(I_2 - I_1) R_3 + I_2 R_2 = E_B - E_C \quad \text{...}(ii)$$

The equations (*i*) and (*ii*) can be rewritten as :

$$I_1 (R_1 + R_3) - I_2 (R_3) = E_A - E_B \quad \text{...}(iii)$$

$$-I_1 (R_3) + I_2 (R_2 + R_3) = E_B - E_C \quad \text{...}(iv)$$

(*iv*) The above three two equations if solved give values of I_1 and I_2.

The equations (*iii*) and (*iv*) can be written on the following principle:

Mesh current × Sum of resistances in the mesh + Algebraic sum of multiplication of common other mesh currents and respective resistances = Sum of voltage rise in the mesh.

Example 4.12. Determine the loop currents network shown in the following figure:

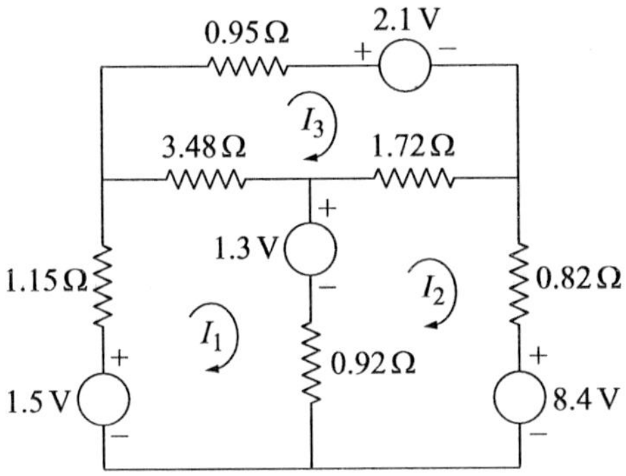

Solution: Using mesh current principle for mesh 1:

$I_1\,(1.5 + 3.48 + 0.92) - I_2(0.92) - I_3(3.48) = 1.5 - 1.3$

or $\quad 5.55I_1 - 0.92I_2 - 3.48I_3 = 0.2 \qquad$ ...(*i*)

Equation for mesh 2:

$I_2(1.72 + 0.82 + 0.92) - I_1(0.92) - I_3(1.72) = 1.3 - 8.4$

or $\quad 3.46I_2 - 0.92I_1 - 1.72I_3 = -7.1$

or $\quad 0.92I_1 - 3.46I_2 + 1.72I_3 = 7.1 \qquad$ ...(*ii*)

Equation for mesh 3:

$I_3(0.95 + 1.72 + 3.48) - I_1(3.48) - I_2(1.72) = -2.1$

or $\quad 3.48\,I_1 + 1.7\,2I_2 - 6.15\,I_3 = 2.1 \qquad$ ...(*iii*)

The simultaneous equations are:

$$5.55I_1 - 0.92I_2 - 3.48I_3 = 0.2$$
$$0.92I_1 - 3.46I_2 + 1.72I_3 = 7.1$$
$$3.48I_1 + 1.72I_2 - 6.15I_3 = 2.1$$

The matrix form is:

$$\begin{bmatrix} 5.55 & -0.92 & -3.48 \\ 0.92 & -3.46 & 1.72 \\ 3.48 & 1.72 & -6.15 \end{bmatrix} \begin{bmatrix} I_1 \\ I_2 \\ I_3 \end{bmatrix} = \begin{bmatrix} 0.2 \\ 1.7 \\ 2.1 \end{bmatrix}$$

Common determinant,

$$\Delta = \begin{vmatrix} 5.55 & -0.92 & -3.48 \\ 0.92 & -3.46 & 1.72 \\ 3.48 & 1.72 & -6.15 \end{vmatrix}$$

or $\quad \Delta = 43.56$

For current I_1, determinant $\Delta_1 = \begin{bmatrix} 0.2 & -0.92 & -3.48 \\ 1.7 & -3.46 & 1.72 \\ 2.1 & 1.72 & -6.15 \end{bmatrix}$

or $\Delta_1 = -107.61$

For current I_2, determinant $\Delta_2 = \begin{bmatrix} 5.55 & 0.2 & -3.48 \\ 0.92 & 1.7 & 1.72 \\ 3.48 & 2.1 & -6.15 \end{bmatrix}$

or $\Delta_2 = -180.8$

For current I_3, determinant $\Delta_3 = \begin{bmatrix} 5.55 & -0.92 & 0.2 \\ 0.92 & -3.46 & 1.7 \\ 3.48 & 1.72 & 2.1 \end{bmatrix}$

or $\Delta_3 = -126.33$

∴ Current, $I_1 = \dfrac{-107.61}{43.56} = -2.47$ A

Current, $I_2 = \dfrac{-180.8}{43.56} = -4.15$ A

Current, $I_3 = \dfrac{-126.33}{43.56} = -2.9$ A

The presumption was CW for each current,

Now since all are having *–ve* sign, hence, all currents I_1,I_2 and I_3 will be in CCW direction.

$$\therefore \quad \left.\begin{aligned} I_1 &= 2.47 \text{ A} \\ I_2 &= 4.15 \text{ A} \\ I_3 &= 2.9 \text{ A} \end{aligned}\right\}$$

Example 4.13. Find mesh currents i_1 and i_2 in the electric circuit.

[UPTU, 2001]

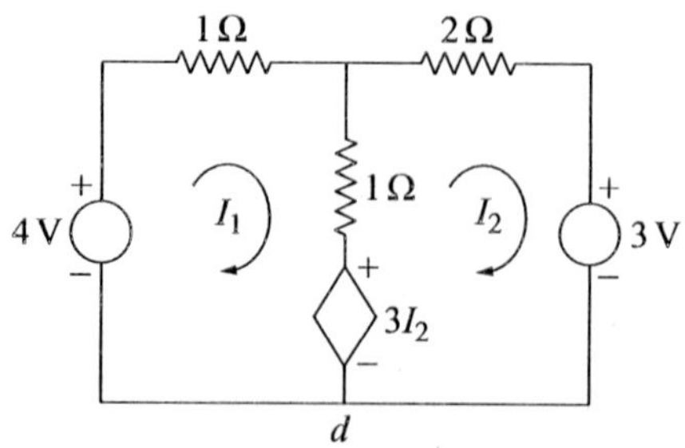

Solution: Equation for mesh 1:

$I_1(1) + (I_1 - I_2)(1) + 3I_2 - 4 = 0$

or $$I_1 + I_2 = 2 \qquad ...(i)$$

Equation for mesh 2:

$I_2(2) + 3 - 3I_2 + (I_2 - I_1)(1) = 0$

or $$I_1 = 3 \text{ A in CW direction}$$

By substituting I_1 in equation (i), we get

$$I_2 = -1 \text{ A} = 1 \text{ A in CCW direction.}$$

Example 4.14. Using loop current method, find the currents I_1 and I_2 in the figure in the right. **[UPTU, 2003]**

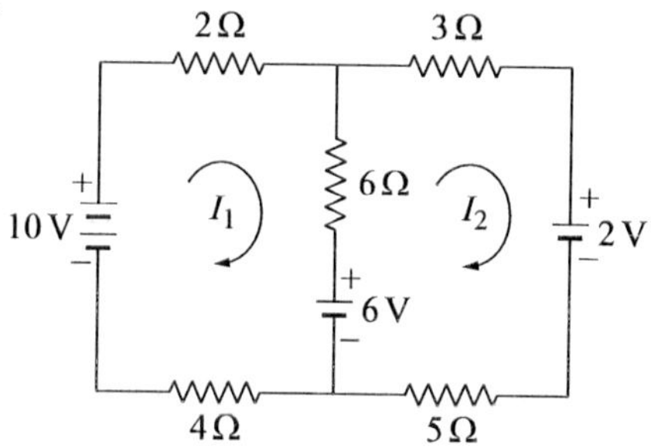

Solution: Mesh current equation for loop 1:

$$I_1(2 + 6 + 4) - I_2(6) = 10 - 6$$

or $$12I_1 - 6I_2 = 4$$

or $$6I_1 - 3I_2 = 2 \qquad ...(i)$$

Mesh current equation for loop 2:

$$I_2(3 + 6 + 5) - I_1(6) = 6 - 2$$

or $$14I_2 - 6I_1 = 4$$

or $$-6I_1 + 14I_2 = 4 \qquad ...(ii)$$

$$I_1 = \frac{\begin{vmatrix} 2 & -3 \\ 4 & 14 \end{vmatrix}}{\begin{vmatrix} 6 & -3 \\ -6 & 14 \end{vmatrix}} = \frac{28 + 12}{84 - 18} = \frac{40}{66}$$

or $$I_1 = \frac{20}{33} \text{ A.}$$

$$I_2 = \frac{\begin{vmatrix} 6 & 2 \\ -6 & 4 \end{vmatrix}}{\begin{vmatrix} 6 & -3 \\ -6 & 14 \end{vmatrix}} = \frac{24 + 12}{66} = \frac{36}{66} = \frac{6}{11} \text{ A.}$$

4.4 NODAL VOLTAGE METHOD

A node is a junction of two or more circuit elements. One of the nodes acts as reference and voltages of remaining independent nodes are presumed. Equations are written as per KCL. These equations are solved for its node voltages. Once, node voltages are known, branch current can be calculated.

Consider the circuit of Fig. 4.3. It has two independent nodes b and d. Node d is taken as reference node.

The voltage at node $b = V_b$

By KCL: $I_{ab} + I_{cb} = I_{bd}$

Using Ohm's law:

$$I_{ab} = \frac{E_A - N_b}{R_1},\; I_{cb} = \frac{E_C - V_b}{R_2}$$

and

$$I_{bd} = \frac{V_b - E_b}{R_3}$$

Hence, $\dfrac{E_A - V_b}{R_1} + \dfrac{E_C - V_b}{R_2} = \dfrac{V_b - E_b}{R_3}$

or

$$V_b\left[\frac{1}{R_1} + \frac{1}{R_2} + \frac{1}{R_3}\right] - \frac{E_A}{R_1} - \frac{E_B}{R_3} - \frac{E_C}{R_2} = 0$$

∴ Node voltage × sum of conductances connected to node – node voltages to other end × conductance of the branch = 0.

Example 4.15. Find the currents I_1 and I_2 in the passive elements of the network given in the figure.

Solution. Using nodal method:

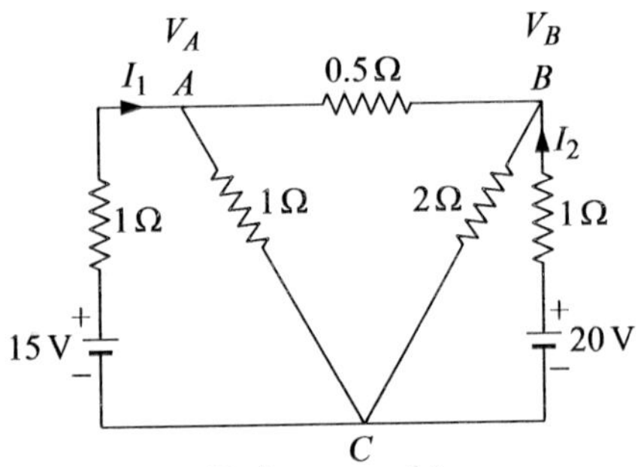

$$\frac{V_A - V_B}{0.5} + \frac{V_A}{1} = \frac{15 - V_A}{1}$$

or $V_A\left[\dfrac{1}{1} + \dfrac{1}{1} + \dfrac{1}{0.5}\right] - \dfrac{15}{1} - \dfrac{V_B}{0.5} = 0$

or $V_A\,[1 + 1 + 2] - 15 - 2V_B = 0$

or $4V_A - 2V_B = 15$

Similarly, $V_B\left[\dfrac{1}{0.5} + \dfrac{1}{2} + \dfrac{1}{1}\right] - \dfrac{20}{1} - \dfrac{V_B}{0.5} = 0$

or $V_B\,[2 + 0.5 + 1] - 20 - 2V_A = 0$

or $3.5\,V_B - 2V_A = 20$

or $7V_B - 4V_A = 40$ ... (*ii*)

By addition of equations (*i*) and (*ii*)

we get, $5\,V_B = 55$

or $V_B = 11$ V

By substituting value of V_B in equation (*i*), we get

$$4V_A - 22 = 15$$

or

$$V_A = \frac{37}{4} = 9.25 \text{ V}$$

Hence, $$I_1 = \frac{15 - V_A}{1} = \frac{15 - 9.25}{1} = 5.75$$

and $$I_2 = \frac{20 - V_B}{1} = \frac{20 - 11}{1} = 9\text{ A}$$

Example 4.16. Find the voltage across 12 Ω resistor in the following figure using nodal method.

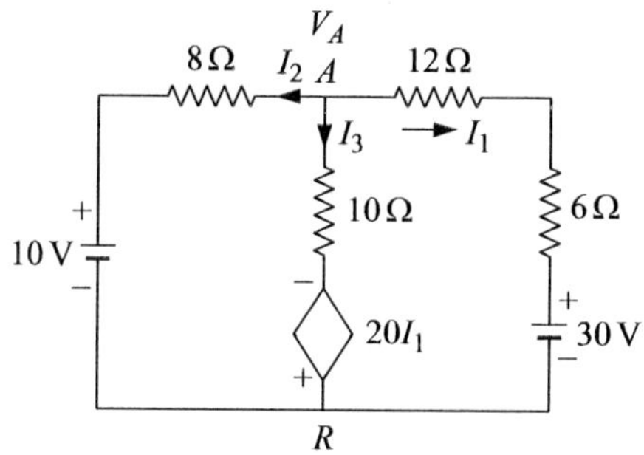

Solution: Using nodal method at node A,

$$I_1 + I_2 + I_3 = 0$$

$$\frac{V_A - 10}{8} + \frac{V_A - 30}{18} + \frac{V_A - 20I_1}{10} = 0$$

or $$V_A\left[\frac{1}{8} + \frac{1}{10} + \frac{1}{18}\right] - \frac{10}{8} - \frac{(-20I_1)}{10} - \frac{30}{18} = 0$$

or $$V_A\left[\frac{9\times 5 + 36 + 20}{72\times 5}\right] - \frac{450 - 720I_1 + 600}{72\times 5} = 0$$

or $$V_A[45 + 36 + 20] = 1050 - 720\, I_1$$

or $$101V_A = 1050 - 720I_1$$

From figure $$I_1 = \frac{V_A - 30}{18}$$

∴ $$I_1 = \frac{V_A - 30}{18} = \frac{\frac{750}{47} - 30}{18} = \frac{750}{47\times 18} - \frac{30}{18}$$

$$= \frac{750 - 47\times 30}{47\times 18} = -\frac{110}{141}$$

∴ Current through 12Ω resistor is $\frac{110}{141}$ A towards node A.

Example 4.17. By applying Nodal method of network analysis, find the current in 15Ω resistor the network shown in figure given below:

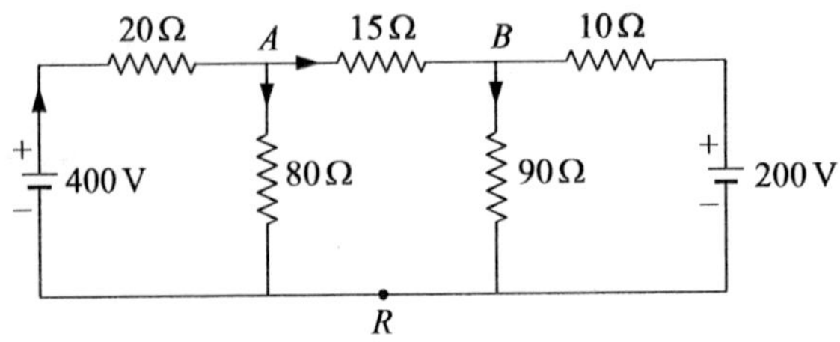

Solution: A and B nodes have voltages V_A and V_B respectively and node R acts as reference.

Using nodal method at node A:

$$\frac{400 - V_A}{20} = \frac{V_A}{80} + \frac{V_A - V_B}{15}$$

or $$V_A\left[\frac{1}{20} + \frac{1}{80} + \frac{1}{15}\right] - \frac{400}{20} - \frac{V_B}{15} = 0$$

or $$V_A\left[\frac{12 + 3 + 16}{240}\right] = 20 + \frac{V_B}{15}$$

or $$\frac{31}{240}V_A - \frac{V_B}{15} = 20$$

or $$31V_A - 16V_B = 4800 \quad ...(i)$$

Using nodal method at node B:

$$\frac{V_A - V_B}{15} + \frac{200 - V_B}{10} = \frac{V_B}{90}$$

or $$V_B\left[\frac{1}{15} + \frac{1}{90} + \frac{1}{10}\right] - \frac{200}{10} - \frac{V_A}{15} = 0$$

or $$V_B\left[\frac{6 + 1 + 9}{90}\right] - 20 - \frac{V_A}{15} = 0$$

or $$\frac{8}{45}V_B - \frac{V_A}{15} = 0$$

or $$3V_A - 8V_B = -900 \quad ...(ii)$$

Equations (i) and (ii) in matrix form can be written as:

$$\begin{bmatrix} 31 & -16 \\ 3 & -8 \end{bmatrix}\begin{bmatrix} V_A \\ V_B \end{bmatrix} = \begin{bmatrix} 4800 \\ -900 \end{bmatrix}$$

$$\therefore \qquad V_A = \frac{\begin{vmatrix} 4800 & -16 \\ -900 & -8 \end{vmatrix}}{\begin{vmatrix} 31 & -16 \\ 3 & -8 \end{vmatrix}} = 264 \text{ V}$$

and

$$V_B = \frac{\begin{vmatrix} 31 & 4800 \\ 3 & -900 \end{vmatrix}}{\begin{vmatrix} 31 & -16 \\ 3 & -8 \end{vmatrix}} = 211.5 \text{ V}$$

$$\text{Current in } 15\ \Omega \text{ resistor} = \frac{V_A - V_B}{15} = \frac{264 - 211.5}{15} = 3.5 \text{ A.}$$

Example 4.18. Use the node voltage method to solve the mesh currents in the network shown in the figure given below:

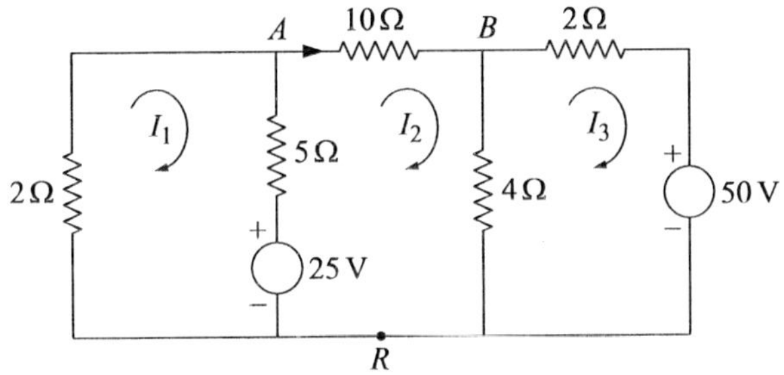

Solution: A, B and R nodes are meshed to facilitate nodal method, R is reference node.

$$\frac{V_A}{2} + \frac{V_A - V_B}{10} = \frac{25 - V_A}{5}$$

$$\text{or} \quad V_A\left[\frac{1}{2} + \frac{1}{5} + \frac{1}{10}\right] - \frac{25}{5} - \frac{V_B}{10} = 0$$

$$\text{or} \qquad V_A\left[\frac{5+2+1}{10}\right] - 5 - \frac{V_B}{10} = 0$$

$$\text{or} \qquad 8V_A - V_B = 50 \qquad \ldots(i)$$

$$\text{and} \qquad \frac{V_A - V_B}{10} = \frac{V_B}{4} + \frac{V_B - 50}{2}$$

$$\text{or} \quad V_B\left[\frac{1}{10} + \frac{1}{4} + \frac{1}{2}\right] - \frac{50}{2} - \frac{V_A}{10} = 0$$

$$\text{or} \qquad V_B\left[\frac{2+5+10}{20}\right] - 25 - \frac{V_A}{10} = 0$$

or $$17V_B - 2V_A = 500$$

or $$2V_A - 17V_B = 500 \qquad ...(ii)$$

$$\therefore \qquad V_A = \frac{\begin{vmatrix} 50 & -1 \\ -500 & -17 \end{vmatrix}}{\begin{vmatrix} 8 & -1 \\ 2 & -17 \end{vmatrix}} = 10.07\text{V}$$

and $$V_B = \frac{\begin{vmatrix} 8 & 50 \\ 2 & -500 \end{vmatrix}}{\begin{vmatrix} 8 & -1 \\ 2 & -17 \end{vmatrix}} = 30.60 \text{ V}$$

Hence, $$I_1 = \frac{-V_A}{2} = \frac{-10.07}{2} = -5.035 \text{ A.}$$

$$I_2 = \frac{V_A - V_B}{10} = \frac{10.075 - 30.60}{10} = -2.05 \text{ A.}$$

$$I_3 = \frac{V_B - 50}{10} = \frac{30.60 - 50}{10} = -9.7 \text{ A.}$$

Example 4.19. Using Nodal analysis, determine current in each branch of the network shown below. Also find total power loss in the network.

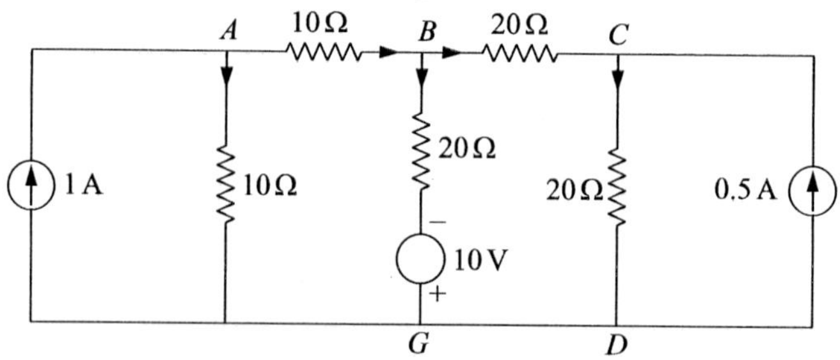

[UPTU,2002]

Solution: Using nodal method at node A:

$$1 = \frac{V_A}{10} + \frac{V_A - V_B}{10}$$

or $$2V_A - V_B = 10 \qquad ...(i)$$

At node B: $$\frac{V_A - V_B}{10} = \frac{V_B + 10}{20} + \frac{V_B - V_C}{20} = 0$$

or $$V_B\left[\frac{1}{10} + \frac{1}{20} + \frac{1}{20}\right] + \left(\frac{10}{20}\right) - \frac{V_A}{10} - \frac{V_C}{20} = 0$$

or $$V_B[2 + 1 + 1] + 10 - 2V_A - V_C = 0$$

or $$4V_B + 10 - 2V_A - V_C = 0$$

or $$2V_A - 4V_B + V_C = 10 \qquad ...(ii)$$

At node C: $$\frac{V_B - V_C}{20} + 0.5 = \frac{V_C}{20}$$

or $$10 = 2V_C - V_B$$

or $$-V_B + 2V_C = 10 \qquad ...(iii)$$

Hence, $$V_A = \frac{\begin{vmatrix} 10 & -1 & 0 \\ 10 & -4 & 1 \\ 10 & -1 & 2 \end{vmatrix}}{\begin{vmatrix} 2 & -1 & 0 \\ 2 & -4 & 1 \\ 0 & -1 & 2 \end{vmatrix}} = 6 \text{ V}$$

$$V_B = \frac{\begin{vmatrix} 2 & 10 & 0 \\ 2 & 10 & 1 \\ 0 & 10 & 2 \end{vmatrix}}{\begin{vmatrix} 2 & -1 & 0 \\ 2 & -4 & 1 \\ 0 & -1 & 2 \end{vmatrix}} = 2 \text{ V}$$

and $$V_C = \frac{\begin{vmatrix} 2 & -1 & 10 \\ 2 & -4 & 10 \\ 0 & -1 & 10 \end{vmatrix}}{\begin{vmatrix} 2 & -1 & 0 \\ 2 & -4 & 1 \\ 0 & -1 & 2 \end{vmatrix}} = 6 \text{ V.}$$

Hence, currents are:

$$I_{AG} = \frac{V_A}{10} = \frac{6}{10} = 0.6 \text{ A}$$

$$I_{AB} = \frac{V_A - V_B}{10} = \frac{6-2}{10} = 0.4 \text{ A}$$

$$I_{BG} = \frac{2-(-10)}{20} = \frac{12}{20} = 0.6 \text{ A}$$

$$I_{CB} = \frac{V_C - V_B}{20} = \frac{6-2}{20} = 0.2 \text{ A}$$

$$I_{CD} = \frac{V_C}{20} = \frac{6}{20} = 0.3 \text{ A}$$

Total power loss in the network

$= (I_{AG})^2 \times 10 + (I_{AB})^2 \times 10 + (I_{BG})^2 \times 20 + (I_{CB})^2 \times 20 + (I_{CD})^2 \times 20$

$= (0.6)^2 \times 10 + (0.4)^2 \times 10 + (0.6)^2 \times 20 + (0.2)^2 \times 20 + (0.3)^2 \times 20$

$= 3.6 + 1.6 + 7.2 + 0.8 + 1.8$

$= 15$ W

Example 4.20. For the circuit shown in figure given under, find (*a*) voltage *v* (*b*) current through the 2*A* – resistor using nodal method.

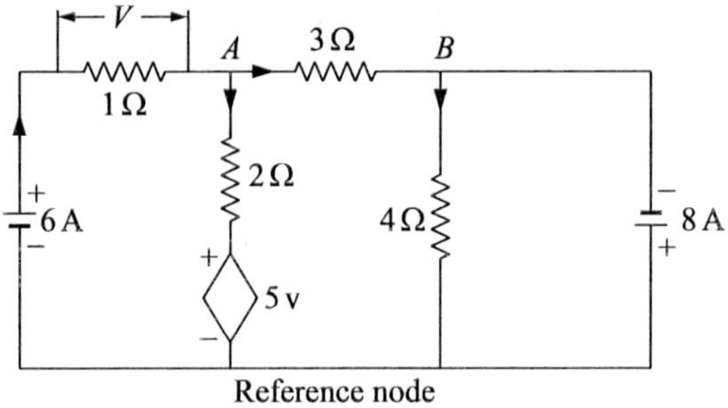

Solution: At node *A*:

$$\frac{V_A - 6}{1} + \frac{V_A - 5v}{2} + \frac{V_A - V_B}{3} = 0$$

or $$V_A\left[\frac{1}{1} + \frac{1}{2} + \frac{1}{3}\right] - \frac{6}{1} - \frac{5v}{2} - \frac{V_B}{3} = 0$$

At node *B*: $$V_B = -8 \text{ V}$$

$$\therefore\ V_A\left[\frac{6+3+2}{6}\right] - 6 - \frac{5v}{2} + \frac{8}{3} = 0$$

$$v = 6 - V_A$$

$$\therefore\ 11\,V_A - 36 - 15\,(6 - V_A) + 16 = 0$$

or $$11V_A - 36 - 90 + 15V_A + 16 = 0$$

or $$26V_A - 110 = 0$$

or $$V_A = \frac{55}{13} \text{V}$$

$\therefore$ $$v = 6 - V_A = 6 - \frac{55}{13} = \frac{78 - 55}{13}$$

or $$v = \frac{23}{13} \text{V}$$

Current through 2 Ω resistor

$$= \frac{5v - V_A}{2} = \frac{5\times\frac{23}{13} - \frac{55}{13}}{2}$$

$$= \frac{115-55}{13\times 2} = \frac{60}{13\times 2}$$

$$= \frac{30}{13}\text{A}$$

Example 4.21. Use nodal analysis method to find currents in the various resistors of the circuit shown in figure given under.

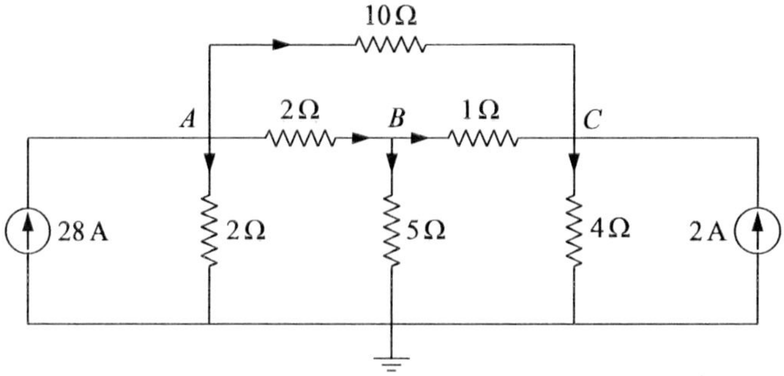

Solution: At node *A*:

$$28 = \frac{V_A}{2} + \frac{V_A - V_B}{2} + \frac{V_A - V_C}{10}$$

or $$280 = 5V_A + 5V_A - 5V_B + V_A - V_C$$

or $$11V_A - 5V_A - V_C = 280 \qquad ...(i)$$

or $$11V_A - 5V_B - V_C = 280$$

At node *B*:

$$\frac{V_A - V_B}{2} = \frac{V_B}{5} + \frac{V_B - V_C}{1}$$

or $$V_B\left[\frac{1}{2}+\frac{1}{5}+\frac{1}{1}\right] - \frac{V_A}{2} - \frac{V_C}{1} = 0$$

The $$V_B\left[\frac{5+2+10}{10}\right] - \frac{V_A}{2} - V_C = 0$$

or $$17V_B - 5V_A - 10V_C = 0$$

or $$5V_A - 17V_B + 10V_C = 0 \qquad ...(ii)$$

At node C:

$$\frac{V_B - V_C}{1} + \frac{V_A - V_C}{10} = \frac{V_C}{4} + 2$$

or $$20V_B - 20V_C - 2V_A - 2V_C = 5V_C + 40$$

or $$2V_A + 20V_B - 27V_C = 40 \quad \text{...(iii)}$$

Hence, $$V_A = \frac{\begin{vmatrix} 280 & -5 & -1 \\ 0 & -17 & 10 \\ 40 & 20 & -27 \end{vmatrix}}{\begin{vmatrix} 11 & -5 & -1 \\ 5 & -17 & 10 \\ 2 & 20 & -27 \end{vmatrix}} = 36 \text{ V}$$

$$V_B = \frac{\begin{vmatrix} 11 & 280 & -1 \\ +5 & 0 & 10 \\ 20 & 40 & -27 \end{vmatrix}}{\begin{vmatrix} 11 & -5 & -1 \\ 5 & -17 & 10 \\ 2 & 20 & -27 \end{vmatrix}} = 20 \text{ V}$$

$$V_C = \frac{\begin{vmatrix} 11 & -5 & 280 \\ 5 & -17 & 0 \\ 2 & 20 & 40 \end{vmatrix}}{\begin{vmatrix} 11 & -5 & -1 \\ 5 & -17 & 10 \\ 2 & 20 & -27 \end{vmatrix}} = 16 \text{ V}$$

Hence, currents are:

$$I_{AG} = \frac{V_A}{2} = \frac{36}{2} = 18 \text{ A}$$

$$I_{AB} = \frac{V_A - V_B}{2} = \frac{36 - 20}{2} = \frac{16}{2} = 8 \text{ A}$$

$$I_{AC} = \frac{V_A - V_C}{10} = \frac{36 - 10}{10} = \frac{20}{10} = 2 \text{ A}$$

$$I_{BC} = \frac{V_B - V_C}{1} = \frac{20 - 16}{1} = 4\text{A}$$

$$I_{CG} = \frac{V_C}{4} = \frac{16}{4} = 4 \text{ A}$$

4.5 SUPERPOSITION THEOREM

In a linear resistive circuit which contains voltage and/or current sources, the current through or the voltage across any element can be found out by the algebraic addition of the individual currents or voltages due to each independent voltage or current sources acting alone. The independent sources which are non-acting are replaced by their internal resistance or short circuits for ideal sources, and independent non-acting current sources are replaced by their internal conductances or open circuits for ideal current sources.

Example 4.22. Using "superposition theorem" or "Norton's theorem", find the current I in the figure given under. **[UPTU, 2003]**

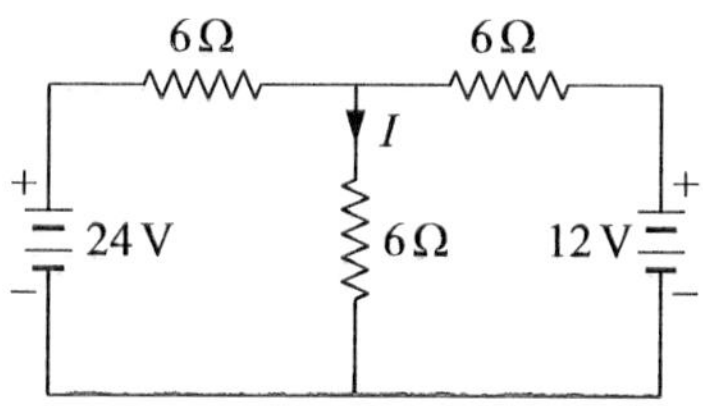

Solution: Using superposition theorem solution is being found out. Consider 24 V source is removed:

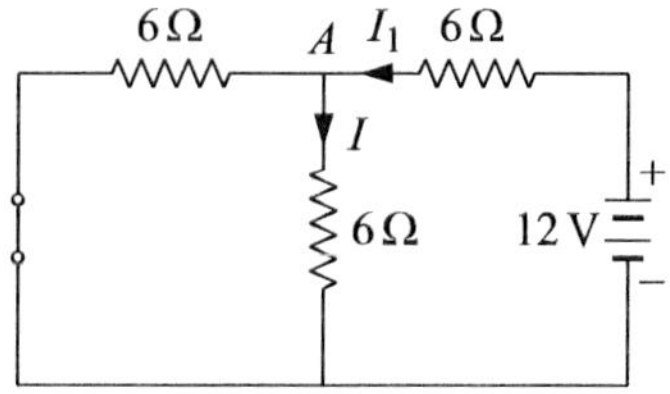

$$I_1 = \left[12 \div 6 + \frac{1}{\left(\frac{1}{6}+\frac{1}{6}\right)}\right]$$

$$= \left[12 \div 6 + \frac{6}{2}\right] = \frac{12}{9} = \frac{4}{3}\text{A.}$$

$$I' = \frac{I_1}{2}$$

$$\therefore \quad I' = \left(\frac{4}{3}\right) \times \frac{1}{2} = \frac{2}{3}\text{A}$$

Consider 12 V is eliminated:

$$I'' = \left[24 \div 6 + \frac{1}{\left(\frac{1}{6}+\frac{1}{6}\right)}\right]$$

$$= \left[24 \div 6 + \frac{6}{2}\right] = \frac{24}{9} = \frac{8}{3}\text{A}$$

$$\therefore \quad I'' = \left(\frac{8}{3}\right) \times \frac{1}{2} = \frac{4}{3}\text{A.}$$

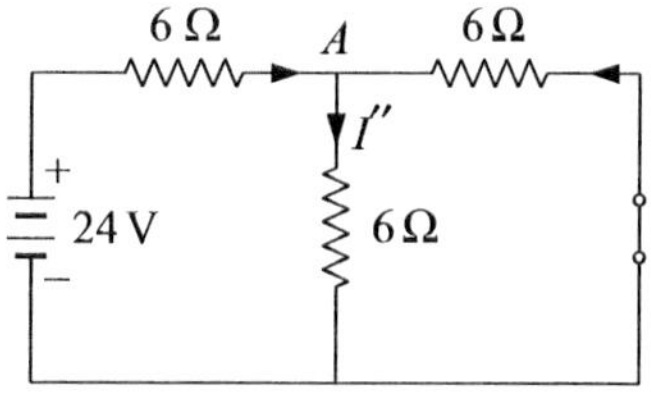

$$\therefore \quad I = I' + I'' = \frac{2}{3} + \frac{4}{3} = \frac{6}{3} = 2\text{ A.}$$

Example 4.23. For the circuit of figure given under, find I using superposition theorem. **[UPTU, 2001]**

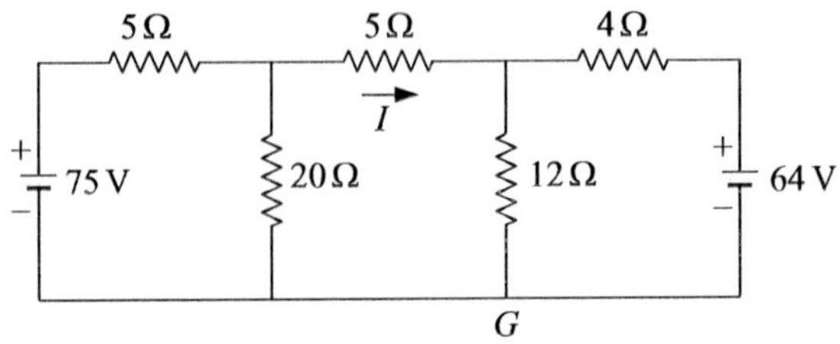

Solution: Remove 75 V from circuit:

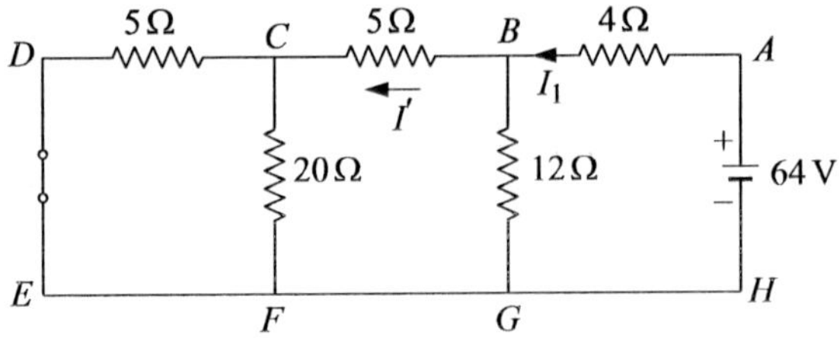

$$R_{BG} = 4 + 12 \,||\left[5 + \frac{1}{\frac{1}{5}+\frac{1}{20}}\right]$$

$$= 4 + 12 \,||\left[5 + \frac{20}{5}\right] = 4 + 12 \,||\, 9$$

$$= 4 + \frac{12 \times 9}{12+9} = 4 + \frac{12 \times 9}{21} = 4 + \frac{36}{7} = \frac{64}{7}\,\Omega.$$

$$I_1 = \frac{64}{\left(\frac{64}{7}\right)} = 7\,A \quad R_{BCDEF} = 5 + \frac{5 \times 20}{5+20}$$

$$= 5 + \frac{5 \times 20}{25} = 9\,\Omega.$$

$$I' = I_1 \times \frac{12 \times 9}{12+9} \times \frac{1}{9} = 7 \times \frac{12}{21} = 4\text{A}.$$

Remove 64 Battery:

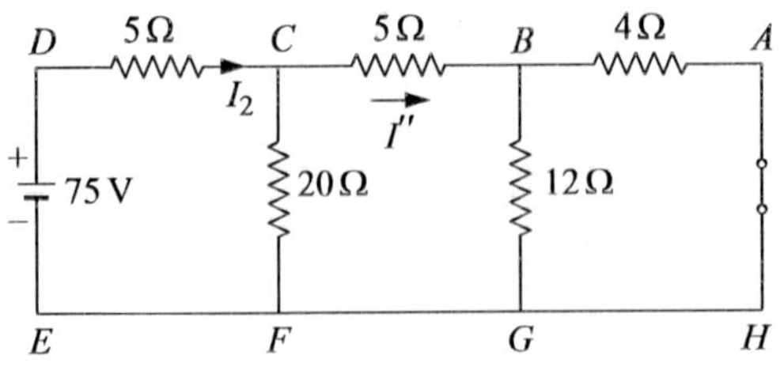

Equivalent resistive load across 75 V.

$$= 5 + [20 \parallel \{5 + (12 \parallel 4)\}]$$

$$= 5 + \left[20 \parallel \left\{5 + \frac{12 \times 4}{12 + 4}\right\}\right]$$

$$= 5 + [20 \parallel \{5 + 3\}]$$

$$= 5 + \left[\frac{20 \times 8}{20 + 8}\right] = 5 + \frac{5 \times 8}{7}$$

$$= \frac{35 + 40}{7} = \frac{75}{7}\,\Omega.$$

$$\therefore \qquad I_2 = \frac{75}{\left(\frac{75}{7}\right)} = 7\text{ A}$$

$$R_{CBAHG} = 5 + (12 \parallel 4) = 5 + \frac{12 \times 4}{10}$$

$$= 5 + \frac{12 \times 4}{16} = 8\Omega$$

$$\therefore \qquad I'' = I_2 \times \frac{20 \times 8}{20 + 8} \times \frac{1}{8} = 7 \times \frac{20}{28} = 5\text{ A.}$$

$$I = I'' - I' = 5 - 4 = 1\text{ A.}$$

4.6 THEVENIN'S THEOREM

At a pair of terminals of any linear network can be replaced by an equivalent circuit composed of a voltage source V_{TH} in series with a resistance Z_{TH}. The voltage V_{TH} is the voltage across the open circuited terminals considered and Z_{TH} is the equivalent resistance of the network as seen from these terminals with all independent sources removed. Thevenin's equivalent is described in Fig. 4.4 for the given networks.

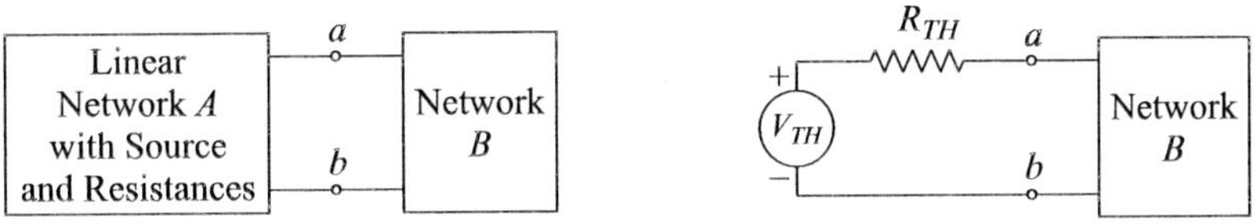

(i) Networks *(ii) Thevenin's equivalent*

Fig. 4.4 *Networks and Thevenin's equivalent.*

Consider circuit given in Fig. 4.5. Thevenin's equivalent is to be found across the load.

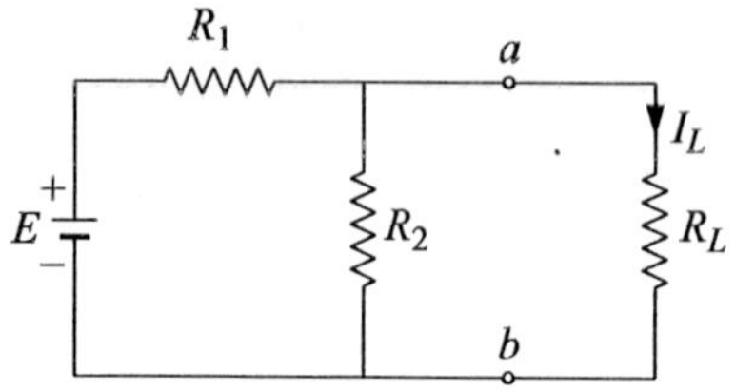

Fig. 4.5. *Circuit with load RL*

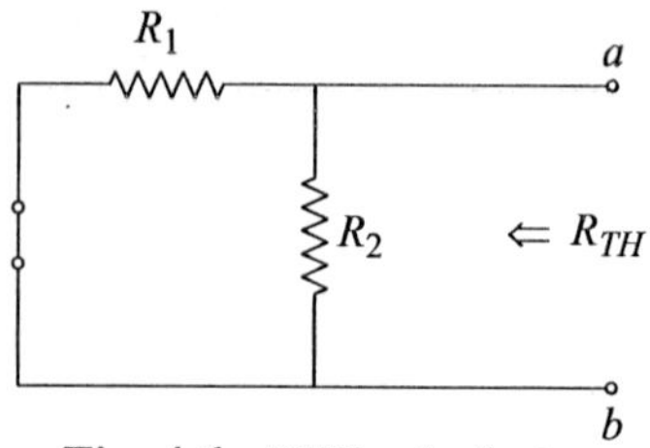

Fig. 4.6. *RTH calculation*

To find R_{TH}, R_L is to be removed *i.e.*, *ab* made open, also *E* is to be removed, *i.e.*, a short circuit is to be shown in its place. The circuit becomes as shown in Fig. 4.6.

$$\therefore \quad R_{TH} = R_1 \parallel R_2 = \frac{R_1 R_2}{R_1 + R_2}$$

To find V_{TH}, again R_L is to be removed *i.e.*, *ab* made open, also *E* remain in the circuit. The circuit becomes as shown in Fig. 4.7.

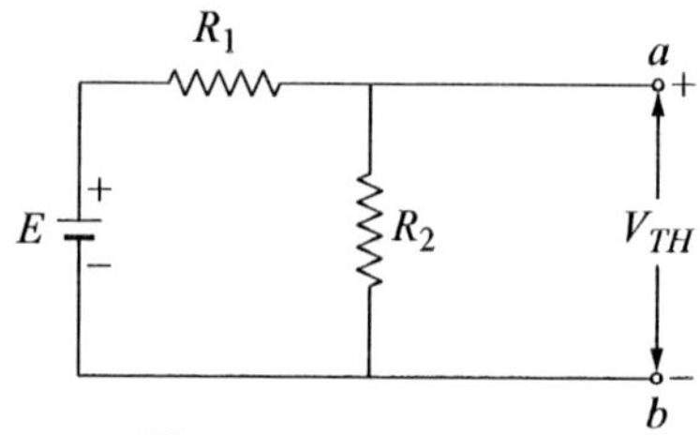

Fig. 4.7. *V_{TH} calculation*

$$\therefore \quad V_{TH} = \frac{E R_2}{R_1 + R_2}$$

The Thevenin's equivalent is shown in Fig. 2.14.

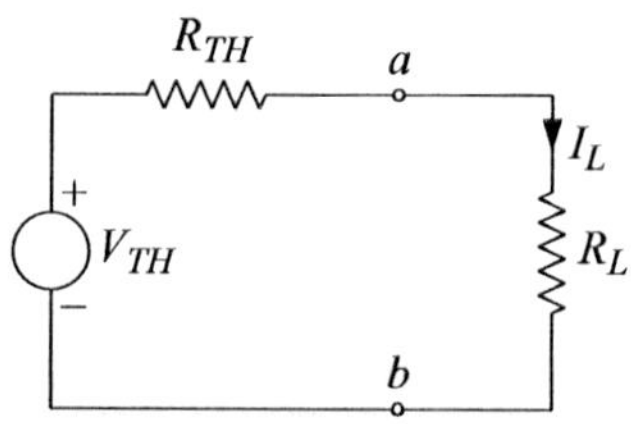

Fig. 4.8. *Thevenin's equivalent with load R_L*

$$\text{The current, } I_L = \frac{V_{TH}}{R_{TH} + R_L} = \frac{E\left(\dfrac{R_2}{R_1 + R_2}\right)}{\dfrac{R_1 R_2}{R_1 + R_2} + R_L}$$

or
$$I_1 = \frac{E R_2}{R_1 R_2 + R_L(R_1 + R_2)}$$

The value of IL can be found directly from the circuit as under:

$$I_L = \frac{E}{R_1 + (R_2 \parallel R_L)} \times (R_2 \parallel R_L) \times \frac{1}{R_L}$$

$$= \frac{E}{R_1 + \dfrac{R_2 R_L}{R_2 + R_L}} \times \frac{R_2 R_L}{R_2 + R_L} \times \frac{1}{R_L}$$

$$= \frac{E}{R_1(R_2 + R_L) + R_2 R_L} \times R_2$$

$$= \frac{E R_2}{R_1 R_2 + R_L (R_1 + R_2)}$$

The value of current is same as calculated by using Thevenin's method.

Example 4.24. Find the Thevenin's equivalent circuit at terminals *BC* of the figure given under. Hence, determine the current through resistor $R = 1\ \Omega$.

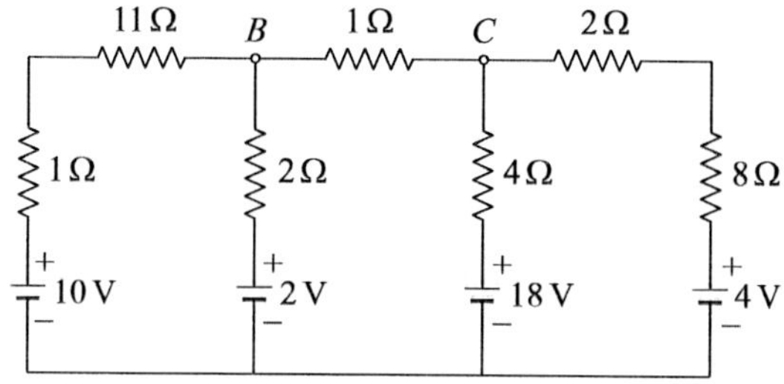

Solution: Thevenin's equivalent is to be found across *BC* and $R = 1\ \Omega$ series as load.

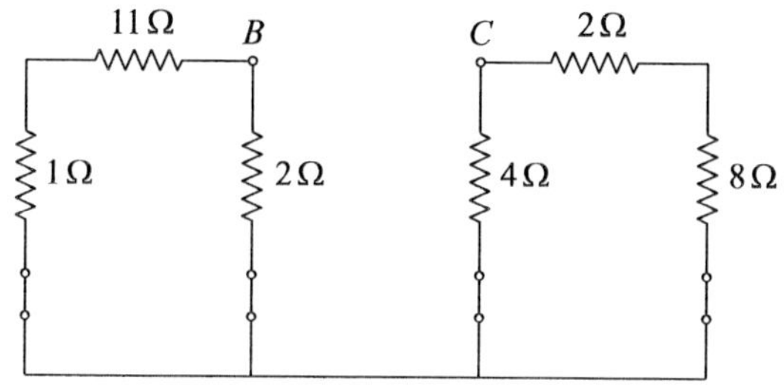

For Thevenin's resistance, voltages are eliminated:

$$R_{TH} = R_{BC} = [(1 + 11) \parallel 2] + [(2 + 8) \parallel 4]$$

$$= \frac{12+2}{12+2} + \frac{10+4}{10+4} = \frac{24}{14} + \frac{40}{14} = \frac{64}{14} = \frac{32}{7}\ \Omega.$$

For Thevenin's voltage, circuit is as follows:

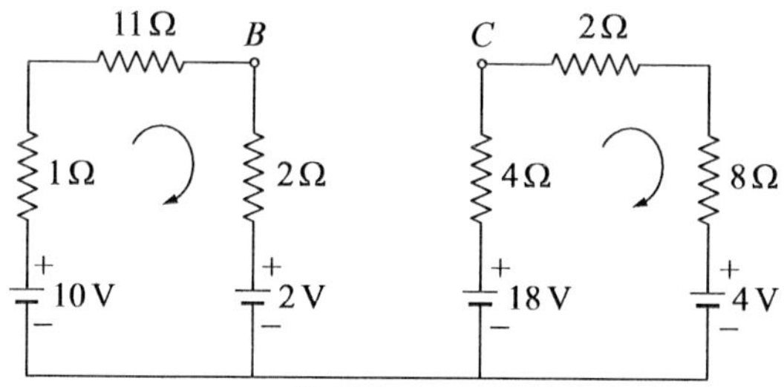

$$V_{TH} = V_B - V_C$$

$$V_B = \left[\left(\frac{10-2}{1+11+2} \times 2\right) + 2\right]$$

$$= \frac{8+14}{7} = \frac{22}{7}$$

$$\therefore \quad V_C = \left[-\frac{18\times4}{4+2+8}\times4+18\right]$$

$$= \left[18-\frac{14\times4}{14}\right] = 14 \text{ V}$$

$$\therefore \quad V_{TH} = V_B - V_C = \frac{22}{7} - 14 = \frac{22-98}{7} = -\frac{76}{7} \text{ V.}$$

The equivalent circuit with load is

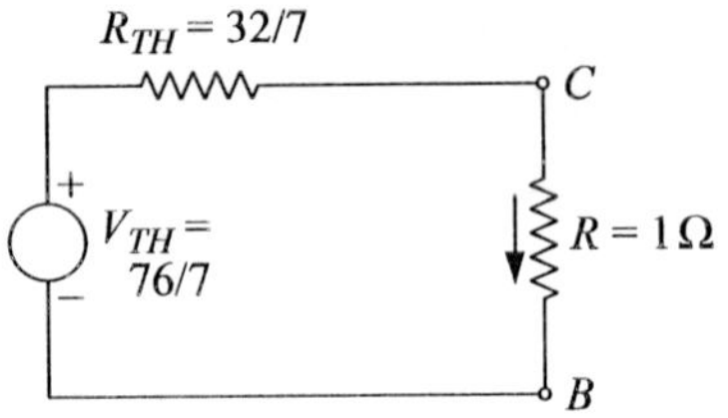

$$\therefore \quad I_R = \frac{V_{TH}}{R_{TH}+R_L} = \frac{\left(\frac{76}{7}\right)}{\left(\frac{32}{7}\right)+1}$$

$$= \left(\frac{76}{7}\right)\times\frac{7}{(32+7)}$$

or current through 1 Ω-resistor $= I_R = \frac{76}{39}$ A.

The current flows form C to B.

Example 4.25. Determine current through 6 Ω resistance connected across A B terminals in the electric circuit of figure given under using Thevenin's theorem. **[UPTU, 2001]**

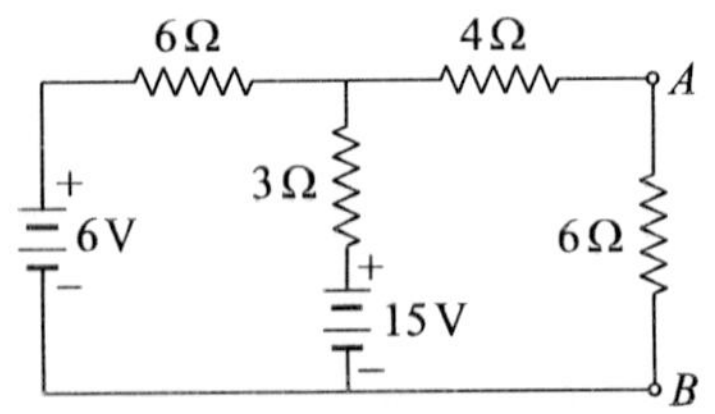

Solution: R_{TH} can be found from the following circuit:

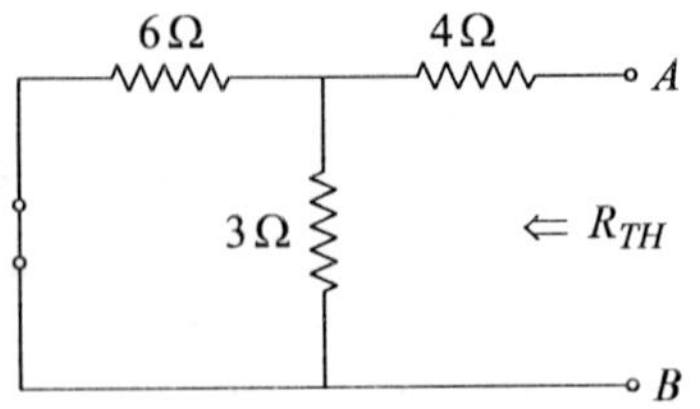

$$\therefore \qquad R_{TH} = 4 + \frac{6 \times 3}{6+3} = 4 + \frac{18}{9} = 6\ \Omega$$

V_{TH} can be found from the following circuit:

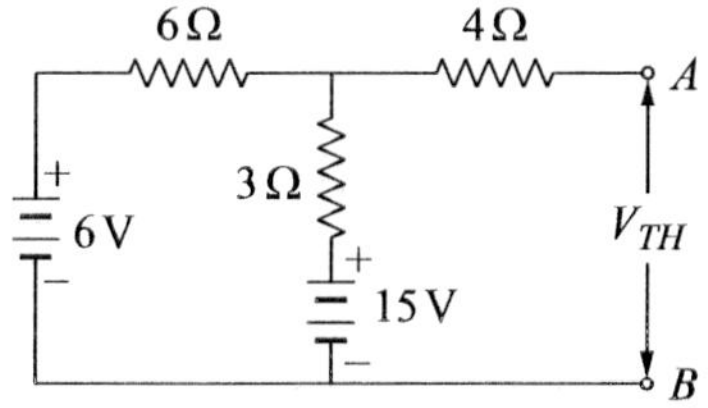

$$V_{TH} = \left[-\left(\frac{15-6}{6+3} \right) \times 3 + 15 \right] \left[15 - \frac{9}{9} \times 3 \right]$$

$$= 15 - 3 = 12\ \text{V}$$

The Thevenin's equivalent circuit is as under:

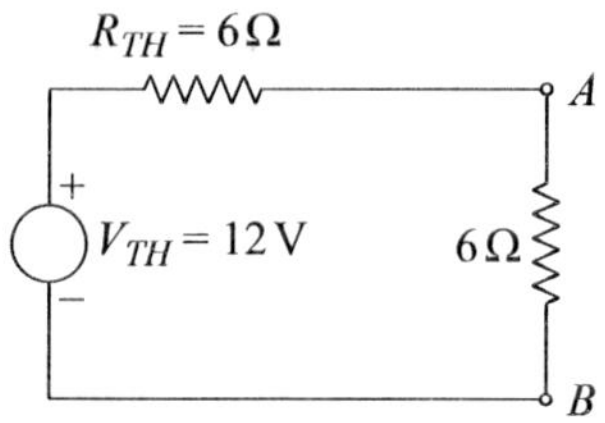

$$\text{Current through } 6\ \Omega = I_{AB} = \frac{12}{6+6} = 1\ \text{A}.$$

Example 4.26. Find Thevenin's equivalent circuit of the network in the following figure at terminals *A*, *B*. Determine the current through the load resistance of 4 Ω connected across terminals *AB*.

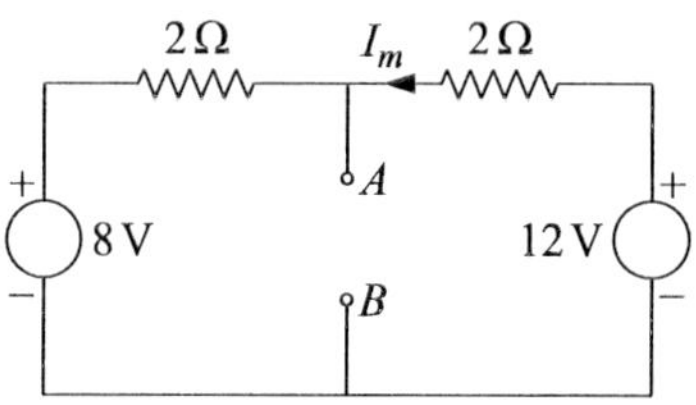

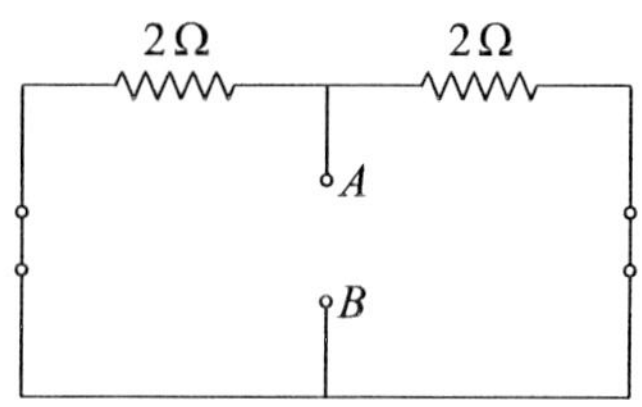

Solution: The current through 2 Ω resistors in CCW direction

$$I_m = (12 - 8) \div (2 + 2) = \frac{4}{4} = 1\ \text{A}$$

$$\therefore \qquad V_{TH} = (12 - 2 \times 1) = 10\ \text{V}$$

or $\quad (8 + 2 \times 1) = 10\ \text{V}$

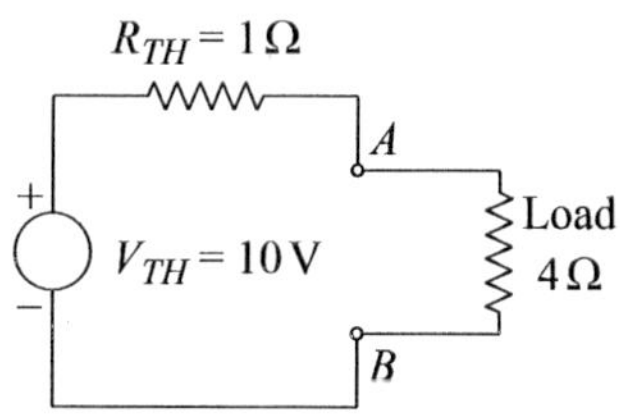

For R_{TH}, the circuit shall be as follows:

$$R_{TH} = R_{AB} = (2 \parallel 2) = \frac{2 \times 2}{2+2} = 1\ \Omega.$$

The equivalent circuit shall be:

Current through 6 Ω resistor

$$= \frac{10}{1+4} = 2\ \text{A}.$$

Example 4.27. Draw the Thevenin's equivalent of the circuit across AB shown in figure.

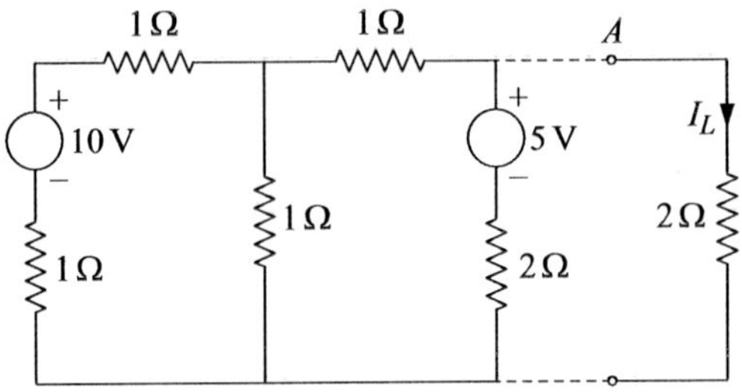

Solution: For R_{TH}.

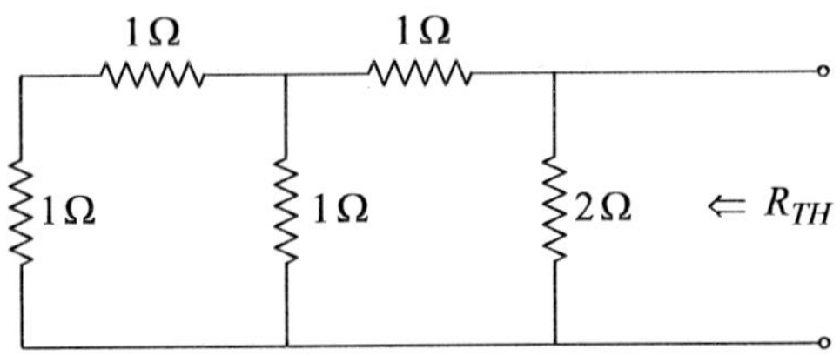

$$\therefore \quad R_{TH} = [2 \parallel \{1 + 1 \parallel (1+1)\}] = \left[1 \parallel \left\{1 + \frac{1 \times 2}{1+2}\right\}\right]$$

$$= \left[1 \parallel \frac{5}{3}\right] = \frac{2 \times \frac{5}{3}}{2 + \frac{5}{3}} = \frac{5 \times 2}{6+5} = \frac{10}{11}\ \Omega$$

For V_{TH};s

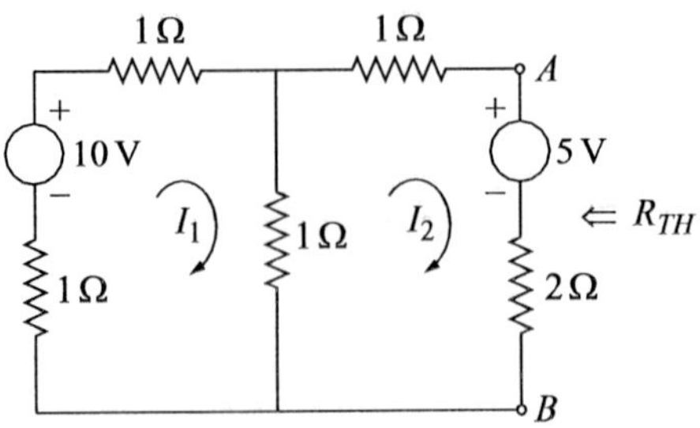

For mesh 1:

$$I_1 + (I_1 - I_2) + I_1 = 10$$

or $$3I_1 - I_2 = 10 \quad ...(i)$$

For mesh 2:

$$I_2 + 2I_2 + (I_2 - I_1) = -5$$

$$4I_2 - I_1 = -5$$

or $$I_1 - 4I_2 = 5 \quad ...(ii)$$

$\therefore$ $$I_1 = (5 + 4I_2)$$

By substituting above in equation (*i*), we get

$$3(5 + 4I_2) - I_2 = 10$$

$$15 + 12I_2 - I_2 = 10$$

$$11I_2 = -5$$

or $$I_2 = -\frac{5}{11}\text{ A}$$

From equation (*i*), we get

$$3I_1 - \left(-\frac{5}{11}\right) = 10$$

or $$3I_1 + \frac{5}{11} = 10$$

or $$33I_1 + 5 = 110$$

or $$33I_1 = 105 \text{ or } I_1 = \frac{105}{33} = \frac{35}{11}\text{ A.}$$

$$V_{TH} = V_{AB} = 5 + 2 \times -\frac{5}{11} = \frac{55 - 10}{11}$$

or $$V_{TH} = \frac{45}{11}\text{ V.}$$

Thevenin's equivalent across *AB* is as under:

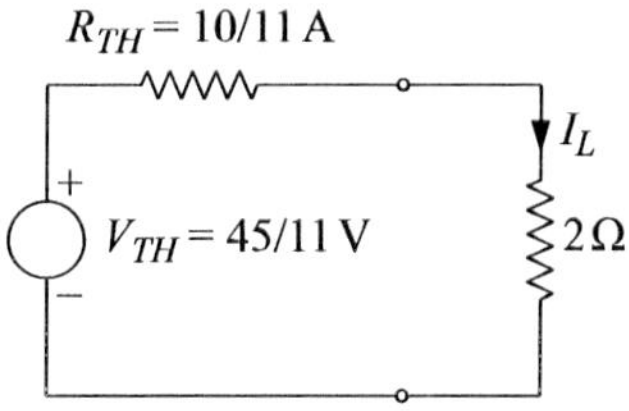

$\therefore$ $$I_L = \left(\frac{45}{11}\right) \div \left(\frac{10}{11} + 2\right)$$

$$= \left(\frac{45}{11}\right) \div \left(\frac{10 + 22}{11}\right)$$

$$= \frac{45}{11} \div \frac{32}{11} = \frac{45}{11} \times \frac{11}{32}$$

$$= \frac{45}{32}\text{ A}$$

Example. 4.28. Obtain Thevenin's equivalent circuit at AB, as shown in figure given under

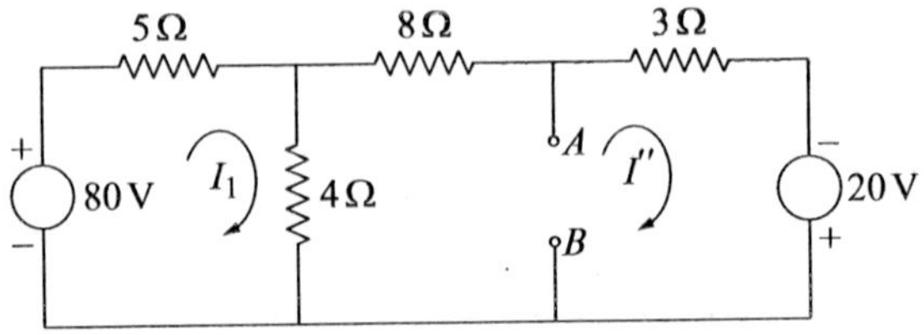

Solution: For R_{TH};

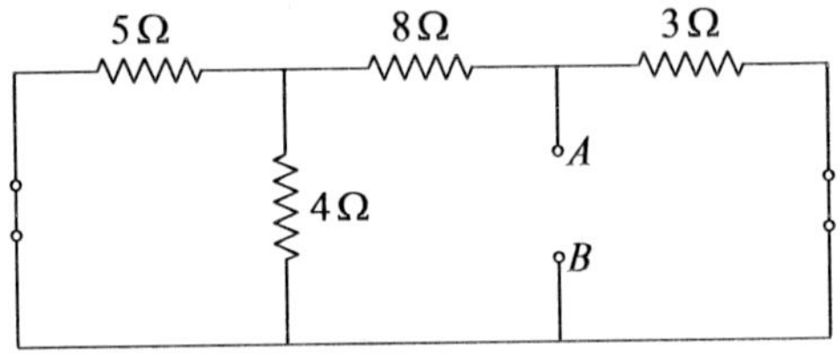

$$R_{TH} = R_{AB} = [\{3 \parallel (8 + 5 \parallel 4)\}] = 3 \parallel \left(8 + \frac{5 \times 4}{5+4}\right)$$

$$= 3 \parallel \left(8 + \frac{20}{9}\right) = 3 \parallel \left(\frac{72+20}{9}\right) = 3 \parallel \frac{92}{9}$$

or
$$R_{TH} = \frac{3 \times \frac{92}{9}}{3 + \frac{92}{9}} = \frac{3 \times 92}{27 + 92} = \frac{276}{119}\,\Omega$$

KVL for mesh 1:

$$5I'' + 4(I' - I'') = 80$$

or
$$9I' - 4I'' = 80$$

KVL for mesh 2:

$$8I'' + 3I'' + 4(I'' - I') = 20$$

or
$$15I'' - 4I' = 20$$

or
$$15I'' = (20 + 4I')$$

Substituting I'' in equation (i), we get

$$9I' - 4\left(\frac{20 + 4I'}{15}\right) = 80$$

$$135I' - 80 - 16I' = 1200$$

$$119I' = 1280$$

$$I'' = \frac{1280}{119}\text{ A}$$

By substituting I' in equation (i), we get

$$9I' - 4\left(\frac{20+4I'}{15}\right) = 80$$

$$135I' - 80 - 16I' = 1200$$

$$119I' = 1280$$

or $$I'' = \frac{1280}{119}\text{ A}$$

By substituting I' in equation (i), we get

$$9\times\frac{1280}{119} - 4I'' = 80$$

or $$I'' = \frac{500}{119}\text{ A}$$

$\therefore$ $$V_{TH} = V_{AB} = 3I'' - 20 = 3\times\frac{500}{119} - 20$$

$$= \frac{1500-2380}{119} = -\frac{880}{119}\text{ V}$$

The Thevenin's equivalent circuit is given as under:

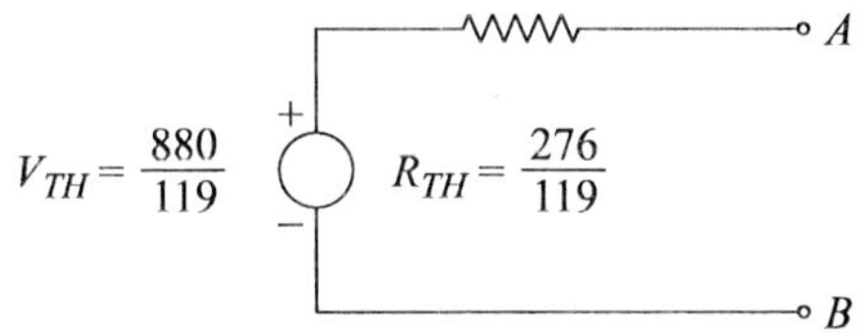

Example 4.29. The resistance of various arms of a Wheatstone bridge are shown in the figure given below. The battery has an emf 2 V and negligible internal resistance. Using Thevenin's theorem, determine the value and direction of the current in the galvanometer circuit BD. **[UPTU, 2002]**

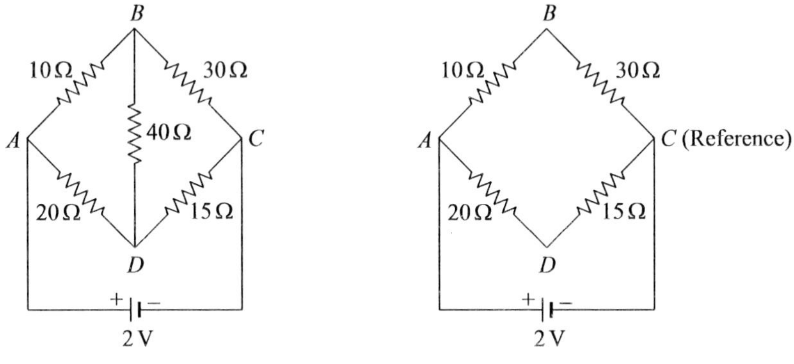

Solution: 40 Ω resistance is to be a opened for Thevenin's equivalent:

$$V_{TH} = V_{BC} - V_{DC}$$

$$V_{BC} = \frac{30}{30+110} \times 2 = \frac{3}{2}\text{ V}$$

$$V_{DC} = \frac{15}{20+15} \times 2 = \frac{6}{7}\text{ V}$$

$$V_{TH} = \frac{6}{7} = \frac{21-12}{14} = \frac{19}{4}$$

or $$V_{TH} = \frac{9}{14}\text{ V}.$$

For R_{TH};

$$R_{TH} = R_{BD} = \left[\frac{10\times 30}{10+30} + \frac{20\times 15}{20+15}\right]$$

$$= \frac{300}{40} + \frac{300}{35} = 7.5 + \frac{60}{7}$$

$$= \frac{5.25+60}{7} = \frac{112.5}{7}$$

or $$R_{TH} = \frac{225}{14}\ \Omega$$

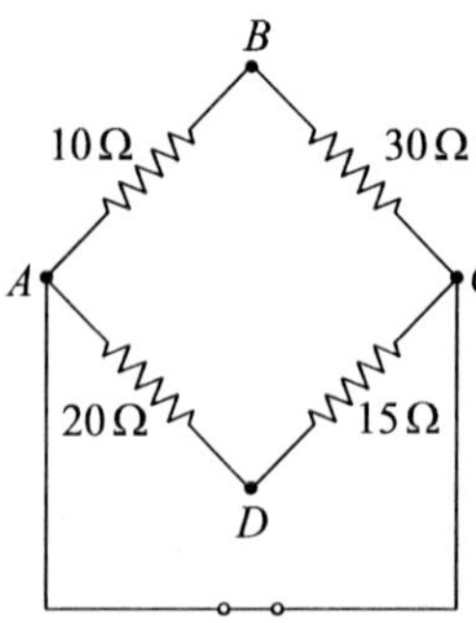

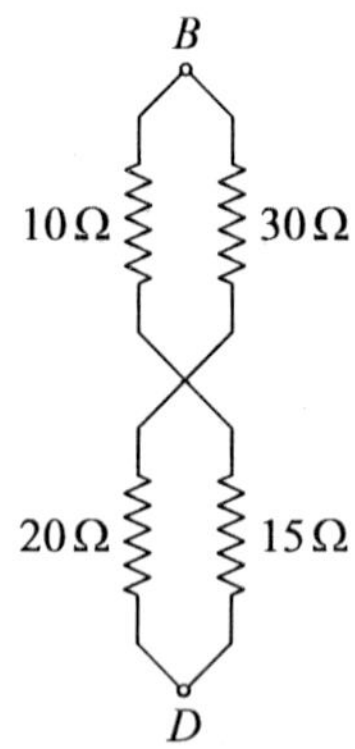

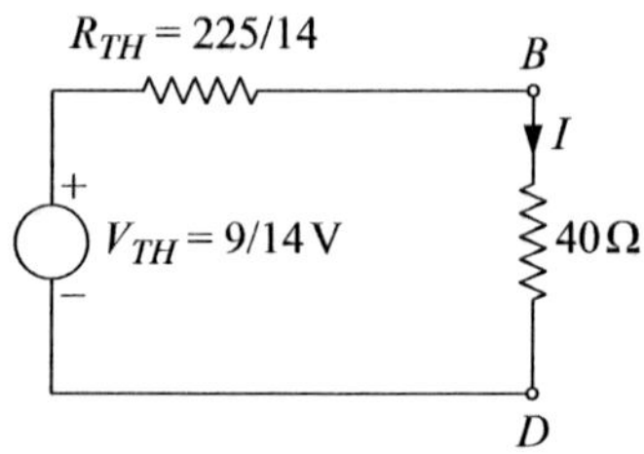

Thevenin's equivalent is as follows: Current through 10 W from B to D

$$= \frac{9}{14} \times \frac{1}{\left(\frac{225}{14} + 40\right)}$$

$$= \frac{1}{14} \times \frac{14}{225+560} = \frac{9}{785}\text{ A}$$

Example 4.30. In the circuit of figure find the branch current I_2 which flows through R_2, when R_2 has the following values 5 Ω, 15 Ω and 50 Ω.

[UPTQ, 2001]

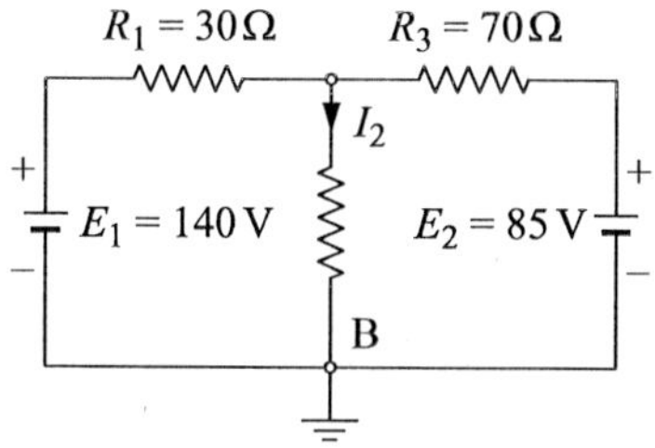

Solution: For R_{TH}:

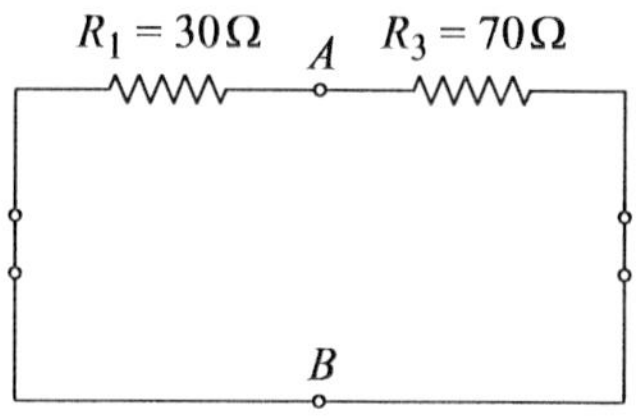

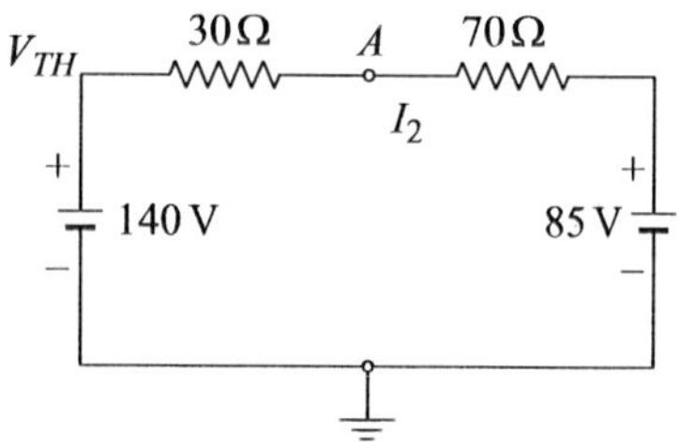

$$R_{TH} = \frac{30 \times 70}{30 + 70} = 21\ \Omega$$

for V_{TH}

$$V_{TH} = V_{AB} = \left[-\left(\frac{140 - 85}{30 + 75} \right) \times 30 + 140 \right]$$

$$= \left[140 - 30 \times \frac{55}{100} \right] = 123.5\ \text{V}$$

The Thevenin's equivalent circuit is as follows:

The currents through R_2 is as under:

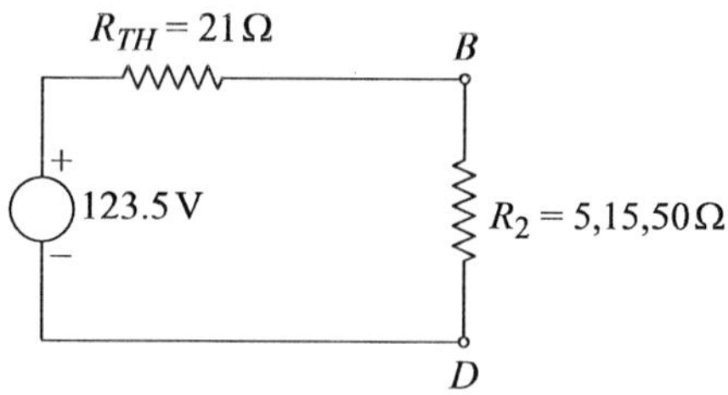

$$I_{R_2 = 5} = \frac{123.5}{21 + 5} = 4.75\ \text{A}$$

$$I_{R_2 = 15} = \frac{123.5}{21 + 15} = 3.43\ \text{A}$$

$$I_{R_2 = 50} = 1.74\ \text{A}$$

Example 4.31. Using Thevenin's theorem, find the current I in the following figure: **[UPTU, 2003]**

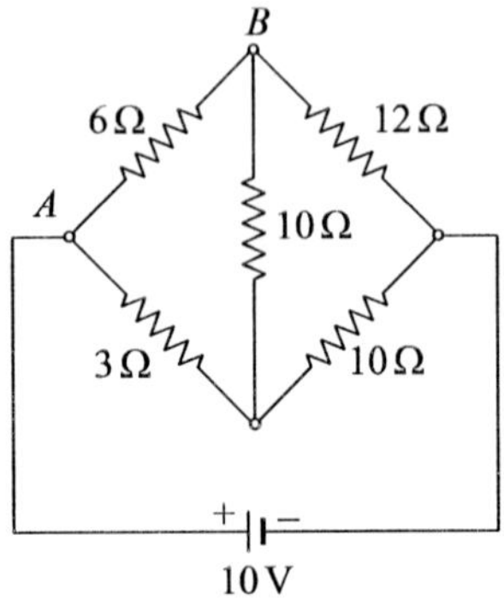

Solution: For R_{TH}:

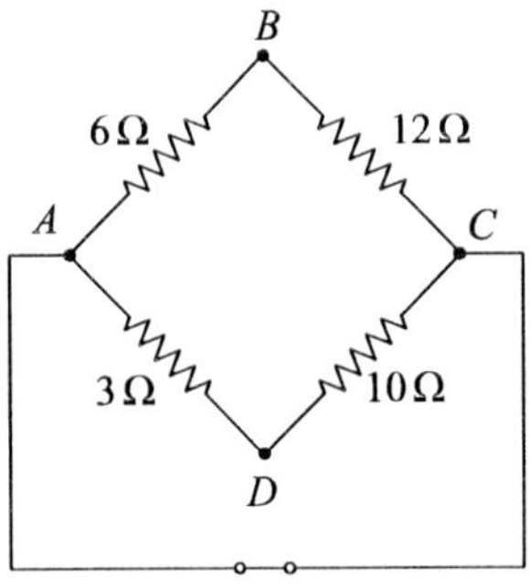

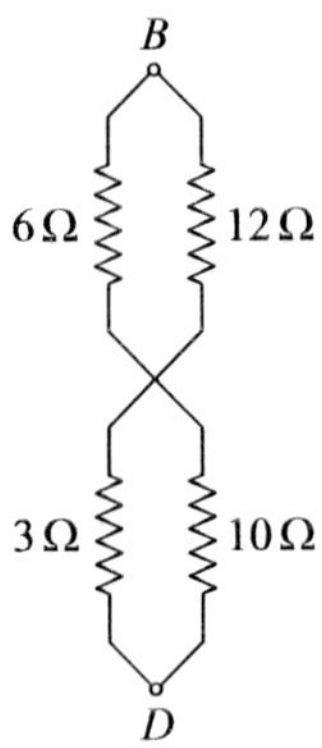

$$R_{TH} = R_{BD} = \left[\frac{6\times 12}{6+12} + \frac{3\times 10}{3+10}\right] = \frac{72}{18} + \frac{30}{13}$$

$$= 4 + \frac{30}{13} = \frac{82}{13}\ \Omega$$

For V_{TH}:

$$V_{TH} = V_{BC} - V_{DC}$$

$$V_{BC} = \frac{10}{6+12} \times 12 = \frac{120}{18} = \frac{20}{3}\ \text{V}$$

$$V_{DC} = \frac{10}{3+10} \times 10 = \frac{100}{13}\ \text{V}$$

$$V_{TH} = V_{BC} - V_{DC} = \frac{20}{3} - \frac{100}{13}$$

$$= \frac{260 - 300}{39} = -\frac{40}{39}\ \text{V}$$

∴ Thevenin's equivalent is:

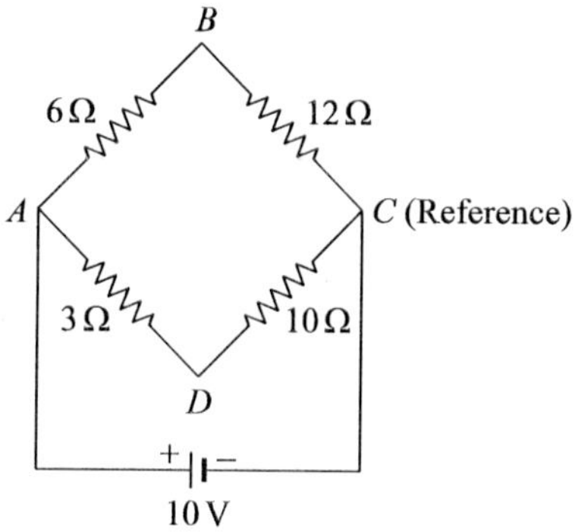

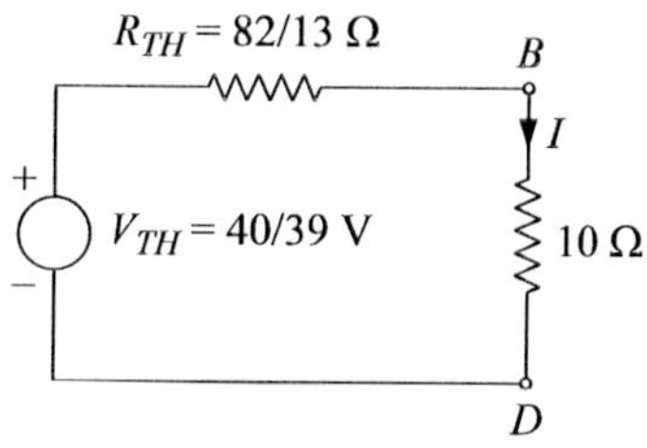

$$\therefore \qquad I = \frac{\left(\frac{40}{39}\right)}{\left(\frac{82}{13}+10\right)} = \frac{40}{39} \times \frac{13}{82+130}$$

$$= \frac{40}{3} \times \frac{1}{212} = \frac{10}{159}\ \text{A}$$

The current direction is opposite of the direction method in the given figure *i.e.*, from *D* to *B*.

Example 4.32. Calculate current in a 1,000 Ω resistor connected between terminals *A* and *B* as shown in the following figure using Thevenin's theorem.

[UPTU, 2001]

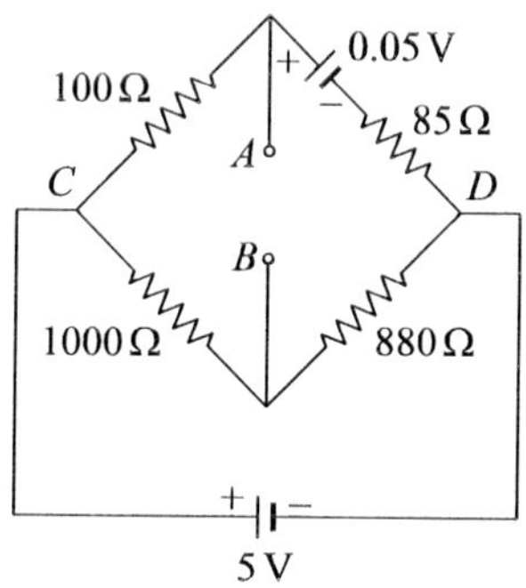

Solution: For R_{TH}:

$$R_{TH} = R_{AB} = \left[\frac{100 \times 85}{100+85} + \frac{1000 \times 880}{1000+880}\right] = 514\ \Omega$$

For V_{TH}, the given figure is used

$$V_{TH} = V_{AD} - V_{BD}$$

$$V_{AC} = \left(\frac{5-0.5}{100+85}\right) \times 895 + 0.5 = 2.32\ \text{V}$$

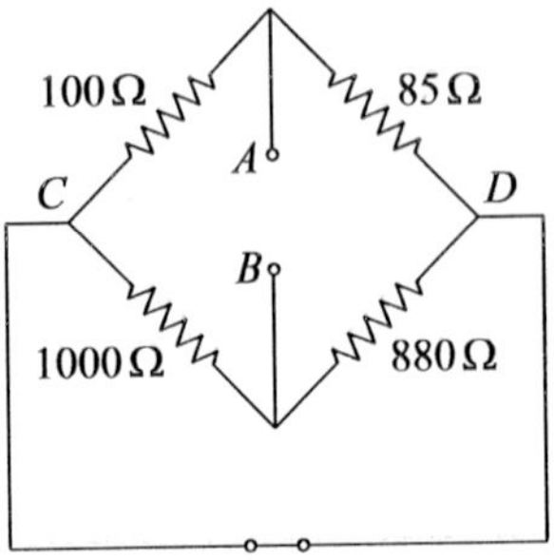

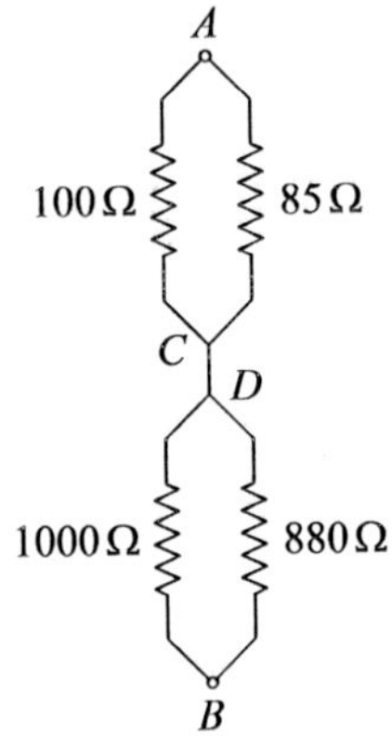

$$V_{BC} = \frac{5}{1000+880} \times 880 = 2.34 \text{ V}$$

$$\therefore \quad V_{TH} = V_{AD} - V_{BD} = 2.32 - 2.34$$

or $\quad V_{TH} = -0.02$ V

The Thevenin's equivalent is as follows:

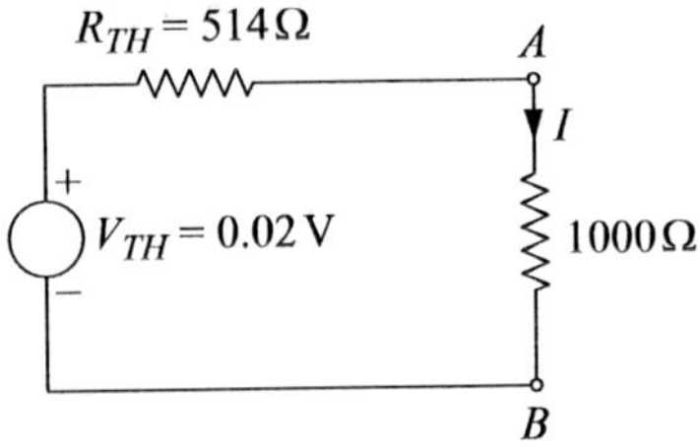

$$\therefore \quad I = \frac{0.02}{514+1000} = -10.6 \ \mu\text{A}.$$

or $\quad I = 10.6$ mA from B to A.

Example 4.33. Find the Thevenin's equivalent circuit for terminal pair AB of the network shown in figure given below:

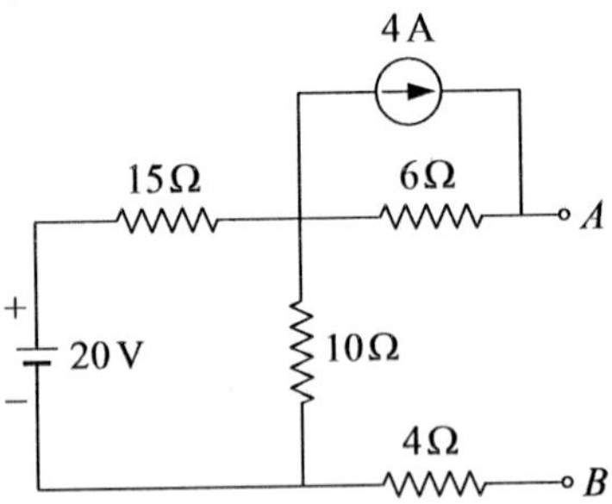

Solution: For R_{TH}:

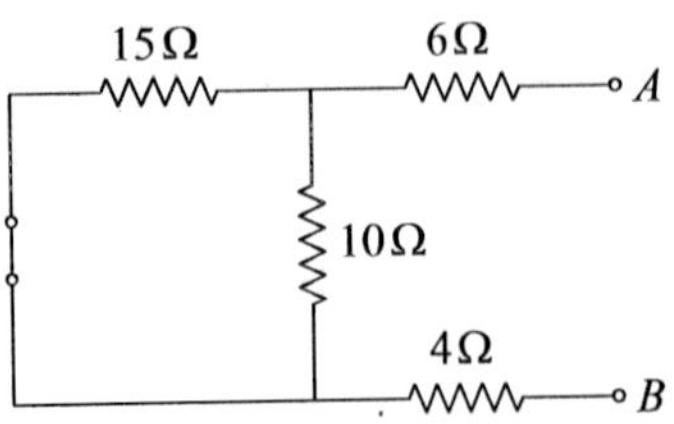

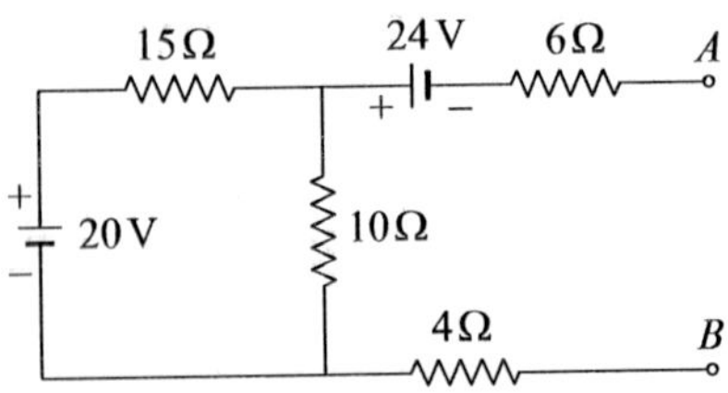

$$R_{TH} = R_{AB} = 6 + \frac{15 \times 10}{15+10} + 4 = 16 \ \Omega$$

For V_{TH}: The current source is converted in voltage source; hence, circuit becomes:

$$V_{TH} = V_{AB} = \frac{20}{(15+10)} \times 10 - 24$$

or $$V_{TH} = -16\text{V}.$$

The Thevenin's equivalent is as follows:

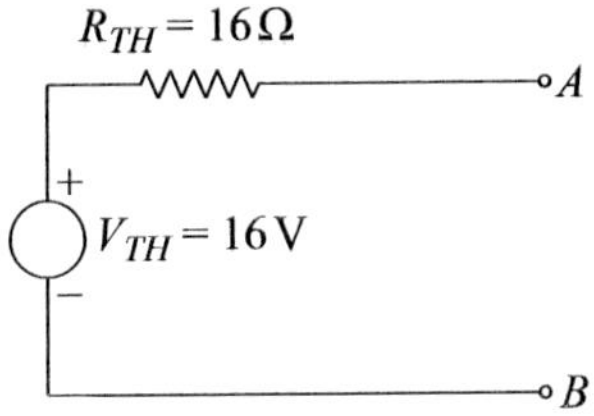

Example 4.34. Find the magnitude and direction of the current in the 2Ω resistor by using Thevenin's theorem for the circuit shown in figure as under:

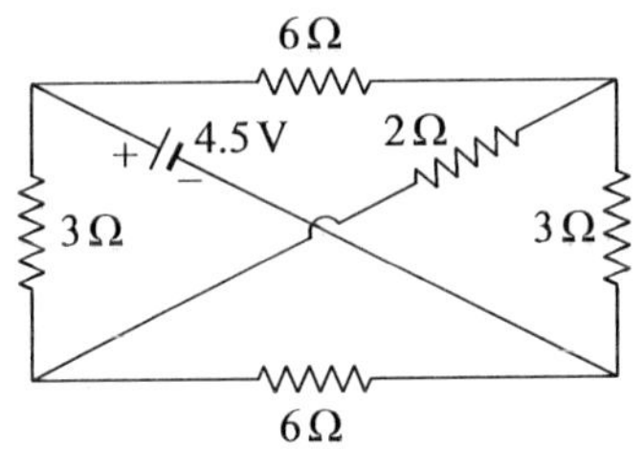

Solution: For R_{TH}:

$$R_{TH} = R_{AB} = \left[\frac{6\times 3}{6+3} + \frac{3\times 6}{3+6}\right] = 4\ \Omega$$

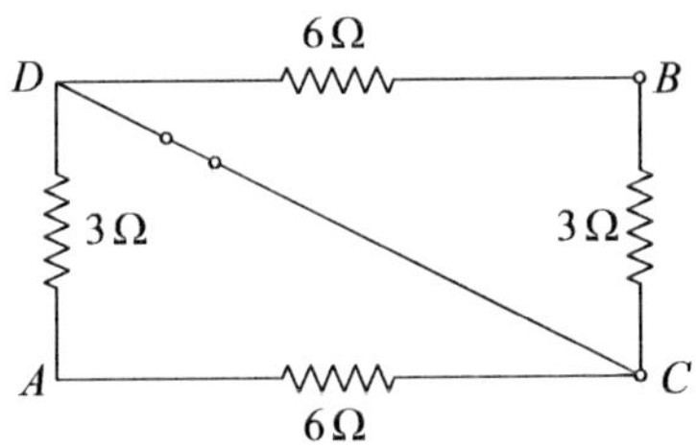

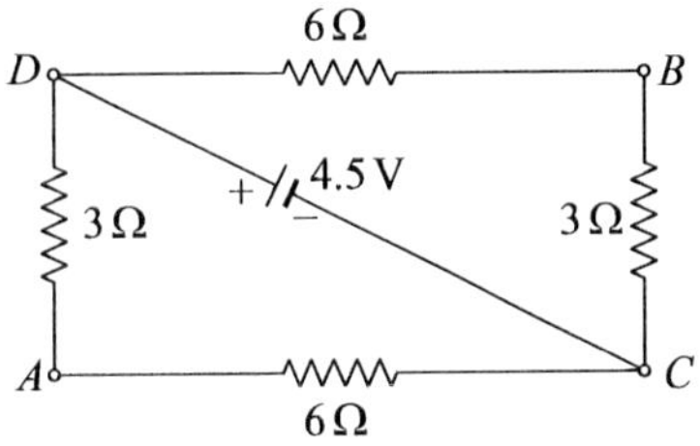

For V_{TH}:

$$V_{TH} = V_{AB} = V_{AC} - V_{BC}$$

$$V_{AC} = \frac{4.5}{3+6} \times 6 = 3\ \text{V}$$

$$V_{BC} = \frac{4.5}{6+3} \times 3 - 1.5\ \text{V}$$

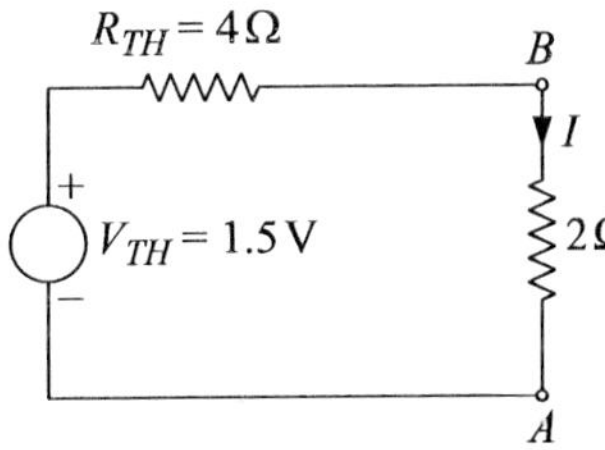

$$V_{TH} = V_{AC} - V_{BC} = 3 - 1.5 = 1.5 \text{ V}$$

The Thevenin's equivalent is:

$$\therefore \qquad I = \frac{1.5}{4+2}$$

$$= 0.25 \text{ A from } A \text{ to } B.$$

Example 4.35. Determine the value of R in the figure shown below so that a power loss of 400 W takes place in it, using Thevenin's theorem.

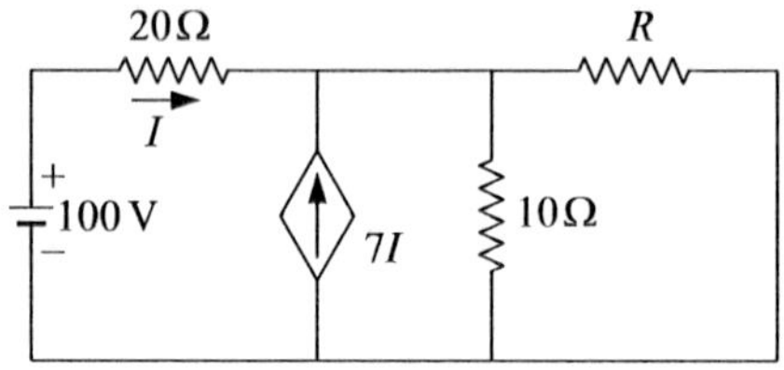

Solution: For V_{TH}:

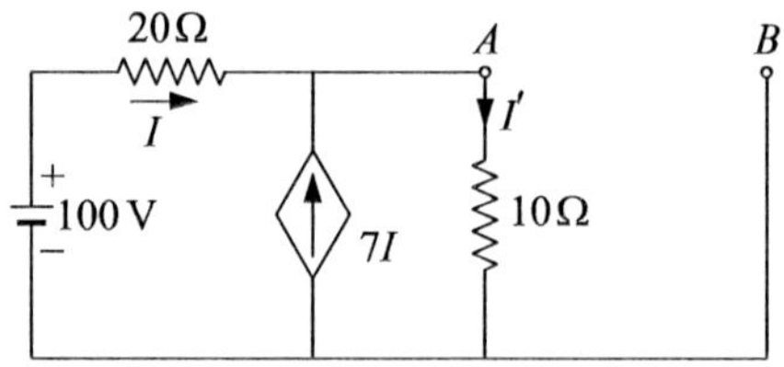

$V_{TH} = V_{AB}$ *i.e.*, voltage across 10 Ω resistor.

At node A:

$$I = \frac{10 - V_{AB}}{20}$$

$$\frac{100 - V_{AB}}{20} + 7\left(\frac{100 - V_{AB}}{20}\right) = \frac{V_{AB}}{10}$$

or $$8(100 - V_{AB}) = 2V_{AB}$$

or $$800 - 8V_{AB} = 2V_{AB}$$

or $$10\ V_{AB} = 800$$

or $$V_{AB} = 80 \text{ V}$$

$$V_{TH} = V_{AB} = 80 \text{ V}$$

For value of R_{TH};

I_{sc} in the current flowing through terminals A and B when they are short circuited as shown in the figure

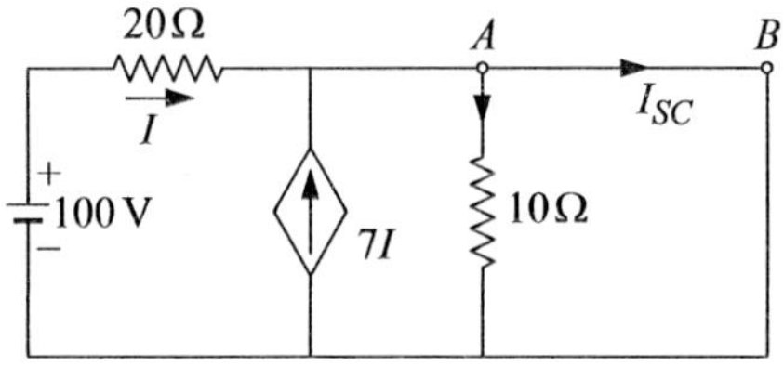

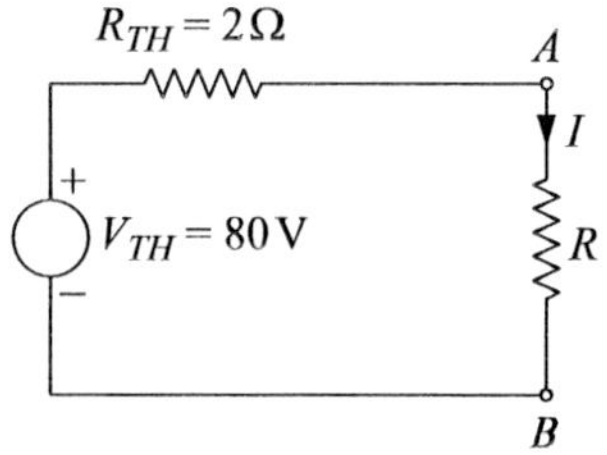

$$I_{sc} = I + 7I = 8I,$$

but $$I = \frac{100}{20} = 5 \text{ A}$$

∴ $$I_{sc} = 8 \times 5 = 40 \text{ A}$$

Hence, $$R_{TH} = \frac{V_{TH}}{I_{sc}} = \frac{80}{40} = 2\Omega$$

Current flowing through R:

$$I = \frac{80}{2+R}$$

$$\therefore \text{ Power loss in } R = I_2\,R = \left(\frac{80}{2+R}\right)^2 R = 400 \text{ W}$$

$$R = 11.66\ \Omega.$$

Example 4.36. Determine the current in 1Ω resistor connected across AB of the network of the figure as under using Thevenin's theorem.

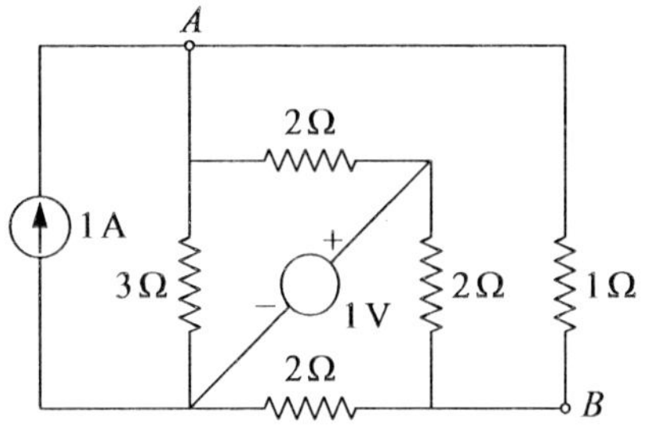

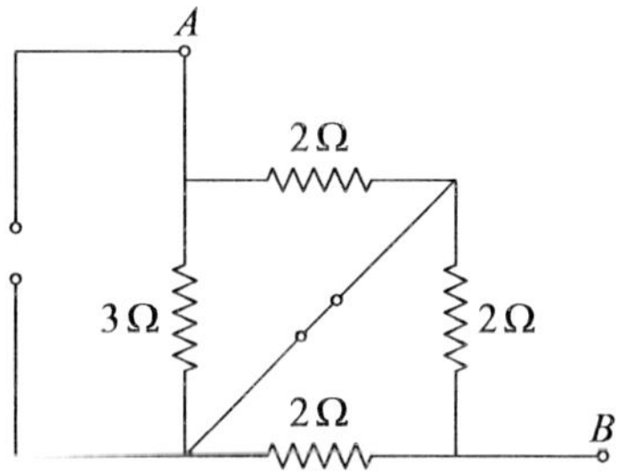

Solution: For R_{TH}:

The 1 V dc supply is across 3 Ω and 2 Ω resistor on both sides, hence, current through resistors 3 W and 2 W is equal to

$$\frac{1V}{5\Omega} = \frac{1}{5}\text{A}$$

$$\therefore \quad V_A = \frac{1}{5} \times 3 = \frac{1}{5} \times 3 = \frac{3}{5}\text{ V}$$

Similarly, $V_B = \frac{1}{2}$ V.

Now, by superposition of both values *i.e.*, objective sums *i.e.*,

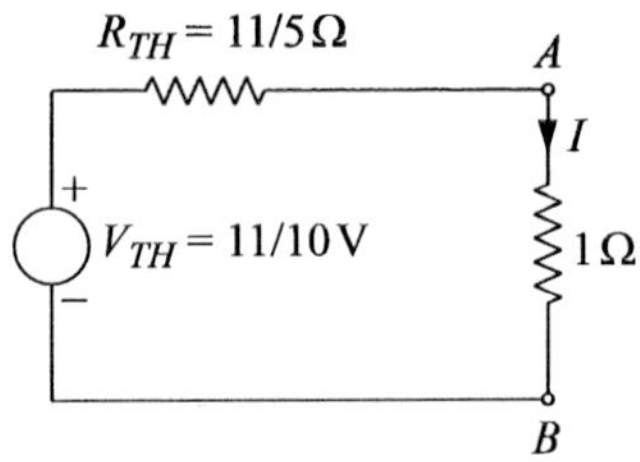

$$V_A = 1 + \frac{3}{5} = \frac{8}{5}$$

$$V_B = 0 + \frac{1}{2} = \frac{1}{2}\text{ V}$$

$$\therefore \quad V_{AB} = \frac{8}{5} - \frac{1}{2} = \frac{16-5}{5 \times 2} = \frac{11}{10}\text{ V}$$

$$\therefore \quad V_{TH} = V_{AB} = \frac{11}{10}\text{ V}$$

The Thevenin's equivalent is as follows:

Current through 1 Ω resistor between A and B

$$I = \frac{\left(\frac{11}{10}\right)\text{V}}{\left(\frac{11}{5} + 1\right)\Omega} = \frac{\left(\frac{11}{10}\right)}{\frac{11+5}{5}}$$

or $$I = \frac{11}{10} \times \frac{8}{16} = \frac{11}{32}\text{ A}$$

4.7 NORTON'S THEOREM

Norton's equivalent can be easily obtained from the Thevenin's equivalent circuit. Norton's Theorem states that any linear network containing resistors and energy source, can be replaced by an equivalent circuit consisting of a current source in parallel with a conductance. See Fig. 4.9.

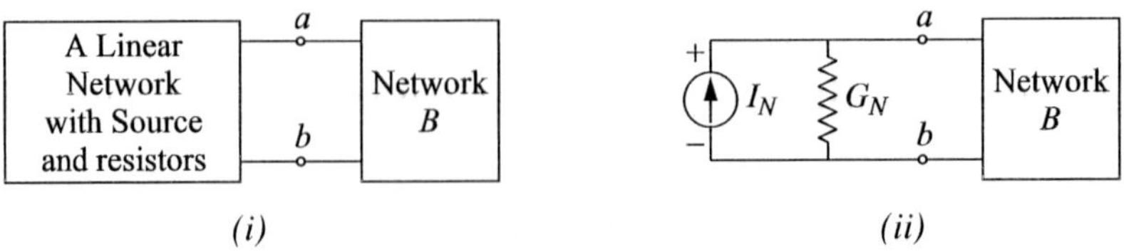

Fig. 4.9. *Norton's equivalent is shown in (ii) for a circuit given in (i).*

The current can be found out by shorting a and b, whereas G_N is the ratio of the short circuit current to the open circuit voltage across a and b. One can notice that

$$V_{TH} = I_N \times \frac{1}{G_N} \text{ and } R_{TH} = \frac{1}{G_N}$$

Consider circuit given in Fig. 4.10 for detailed discussion of Norton's theorem.

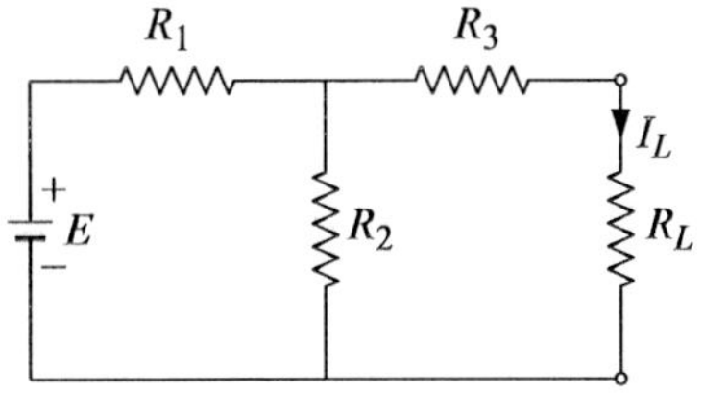

Fig. 4.10. *Norton's equivalent and its proof.*

The Norton's equivalent current I_N can be calculated as per Fig. 4.11.

$$I_1 = E \div \left[R_1 + \frac{R_2 \times R_3}{R_2 + R_3} \right]$$

$$= E \div \left[\frac{R_1(R_2 + R_3) + R_2 R_3}{(R_2 + R_3)} \right]$$

$$= \left[\frac{E(R_2 + R_3)}{(R_1 + R_2 + R_2 R_3 + R_1 R_3)} \right]$$

$$I_N = I_{SC}$$

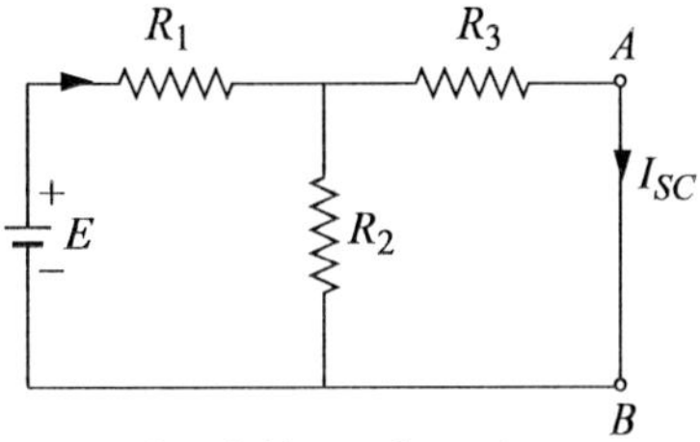

Fig. 4.11. *Value of I_N*

$$\therefore \quad I_N = I_1 \times (R_2 \parallel R_3) \times \frac{1}{R_3}$$

$$= I_1 \times \frac{R_2 \, R_3}{R_2 + R_3} \times \frac{1}{R_3}$$

$$= I_1 \times \frac{R_2}{R_2 + R_3}$$

or $$I_N = \left[\frac{E\,(R_2 + R_3)\, R_2}{(R_2 + R_3)\,(R_1 R_2 + R_2 \, R_3 + R_1 R_3)} \right]$$

or $$I_N = \frac{E \, R_2}{(R_1 R_2 + R_2 R_3 + R_1 R_3)} \qquad \text{... } (i)$$

The Norton's equivalent conductance can be calculated as per Fig. 4.12.

$$R_{TH} = R_3 + \frac{R_1 R_2}{R_1 + R_2}$$

$$= \frac{R_3(R_1 + R_2) + R_1 R_2}{R_1 + R_3}$$

$$\therefore \quad G_N = \frac{R_1 + R_2}{R_1 R_2 + R_2 R_3 + R_1 R_3}$$

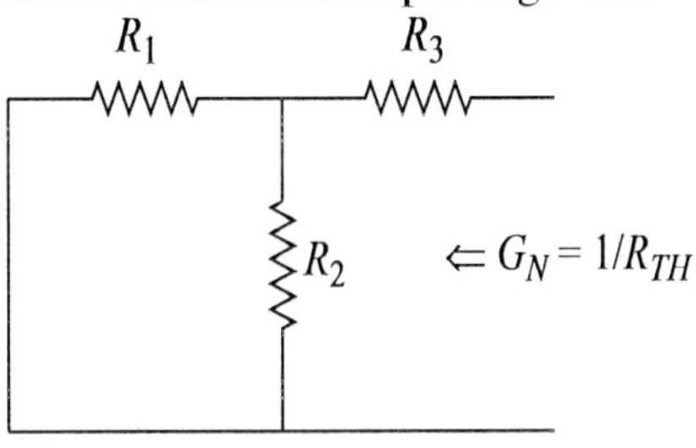

Fig. 4.12. *G_N of a Norton's equivalent*

or $$R_{TH} = \frac{R_1R_2 + R_2R_3 + R_1R_3}{R_1 + R_2} \qquad \text{... (ii)}$$

The Norton's equivalent is as follows in Fig 4.13.

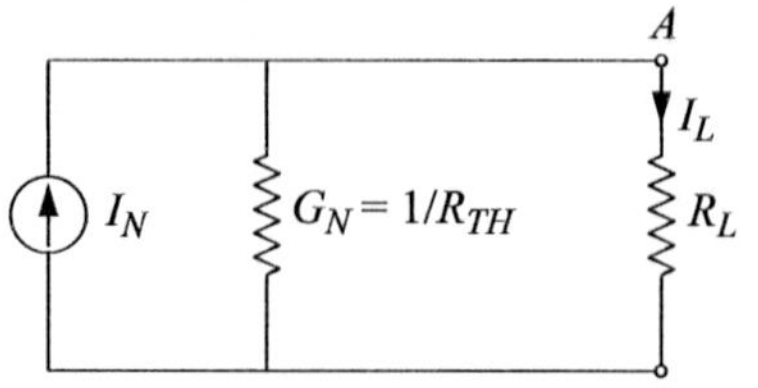

Fig. 4.13. *Norton's equivalent circuit.*

$$I_L = \frac{I_N\ R_{TH}}{R_{TH} + R_L} = \frac{E\ R_2 \times (R_1R_2 + R_2R_3 + R_1R_3)}{(R_1R_2 + R_2R_3 + R_1R_3)\,(R_1 + R_2)} \times \frac{1}{R_{TH} + R_L}$$

$$= \frac{E\ R_2}{R_1 + R_2} \times \frac{1}{\left(\frac{R_1R_2 + R_2R_3 + R_1R_3}{R_1 + R_2}\right) + R_L}$$

or $$I_L = \frac{R\ R_2}{(R_1R_2 + R_2R_3 + R_1R_3) + R_L(R_1 + R_2)}$$

Example 4.37. Solve for the current in the 20 Ω resistor of the network of figure given under by applying Norton's Theorem.

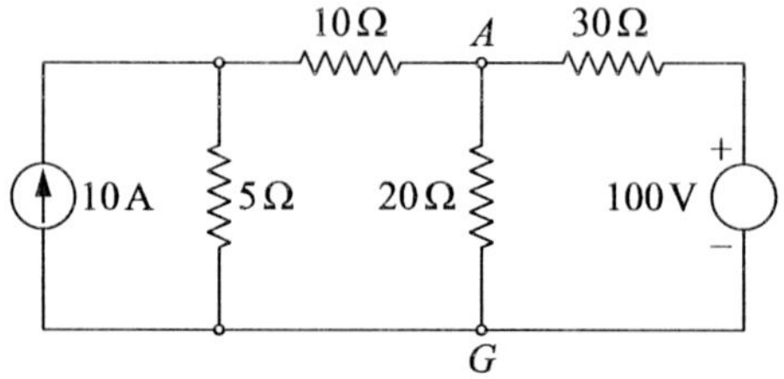

Solution: For R_N:

$$R_N = R_{AG} = (5 + 10) \,||\, 30 = \frac{15 \times 30}{15 + 30}$$

or $$R_N = \frac{15 \times 30}{45} = 10\ \Omega.$$

For I_N:

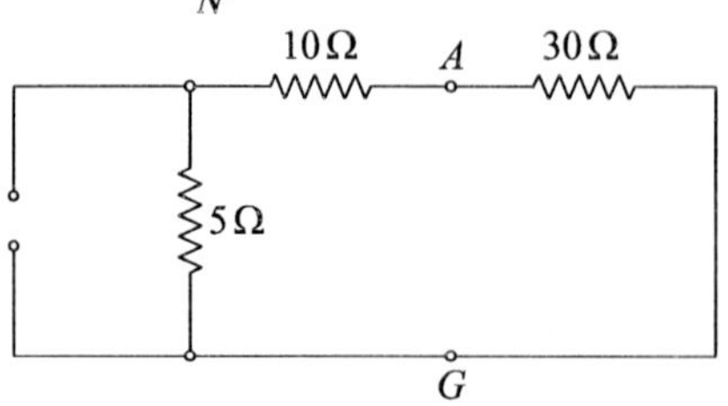

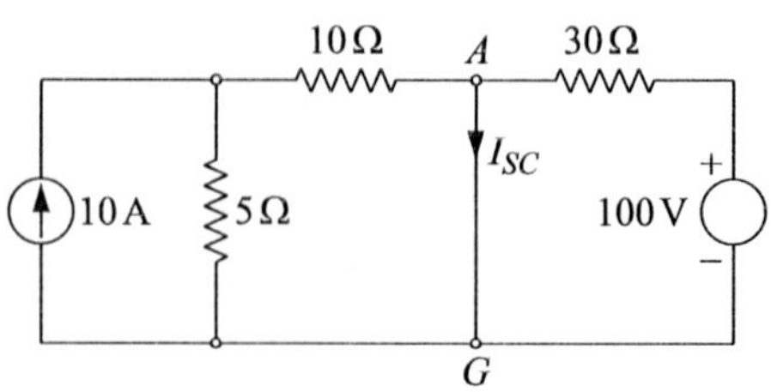

Using superposition theorem:

Use only current source:

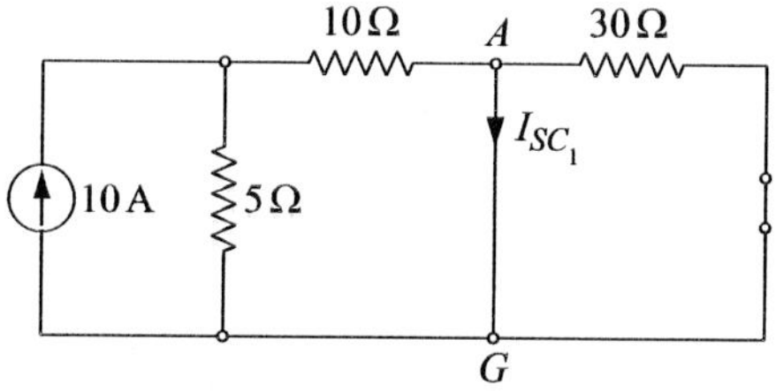

No current flows through 30 Ω resistor as AG is short circuited.

$$\therefore \quad I_{SC_1} = \frac{5\times 10}{5+10}\times 10\times\frac{1}{10}$$

$$= \frac{50}{10} = \frac{10}{3}\text{ A.}$$

Use only voltage source:

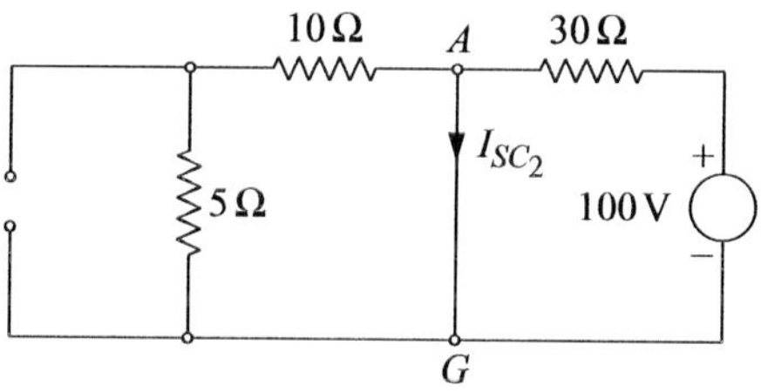

No current flows through 10 and 5 Ω resistor as AG is short circuited.

$$\therefore \quad I_{SC_2} = \frac{100}{30} = \frac{10}{3}\text{ A}$$

$$\therefore \quad I_{SC} = I_{SC_1} + I_{SC_2} = \frac{10}{3}+\frac{10}{3} = \frac{20}{3}\text{ A}$$

$$I_N = I_{SC} = \frac{20}{3}\text{ A.}$$

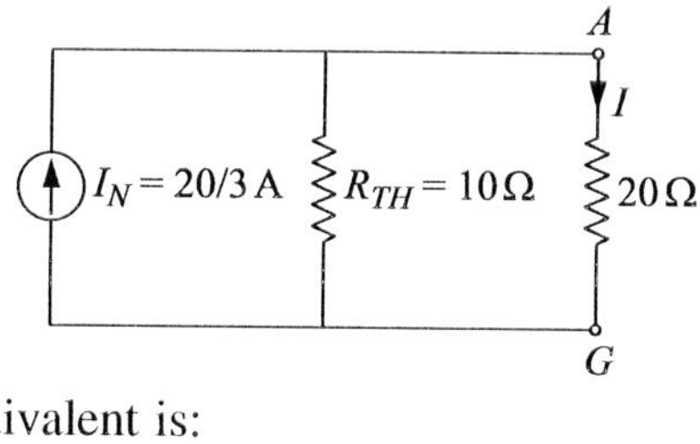

The Norton's equivalent is:

Current, $$I = \frac{10\times 20}{10+20}\times\left(\frac{20}{3}\right)\times\frac{1}{20}$$

$$I = \frac{10 \times 20}{3(10+20)} = \frac{200}{30 \times 3} = \frac{20}{9}$$

or $\qquad I = 2.22$ A.

Example 4.38. Using Norton's theorem, find current which would flow in a 25 Ω resistor connected between points N and O in figure given under:

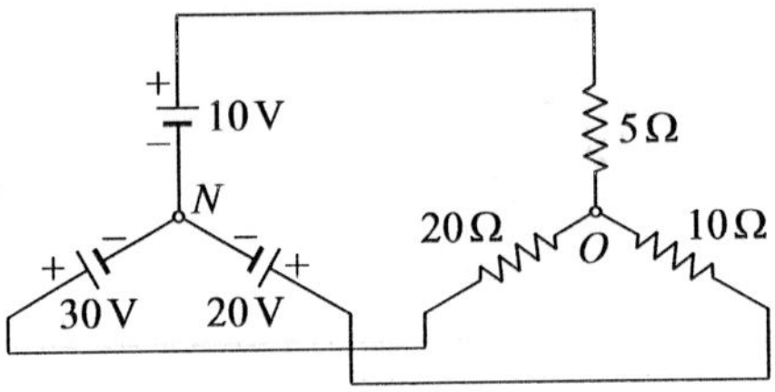

Solution: For G_N:

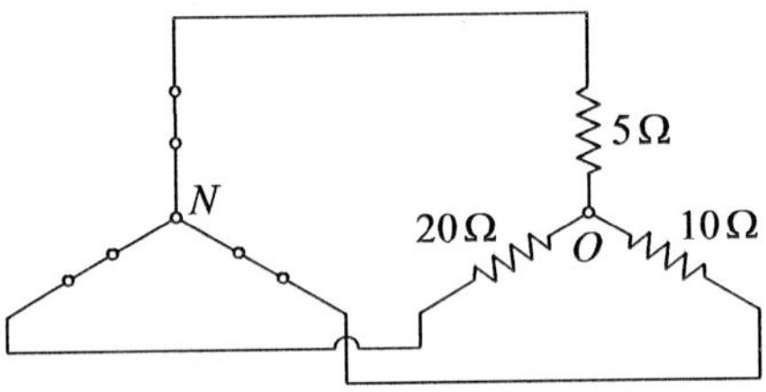

$$G_N = \frac{1}{R_{TH}} \text{ and } R_{TH} = R_{NO}$$

All resistances n parallel with NO, hence,

$$R_{TH}\ 5 \parallel 10 \parallel 20 = \frac{1}{\frac{1}{5} + \frac{1}{10} + \frac{1}{20}}$$

$$= \frac{20}{4+2+1} = \frac{20}{7}\ \Omega.$$

For I_N:

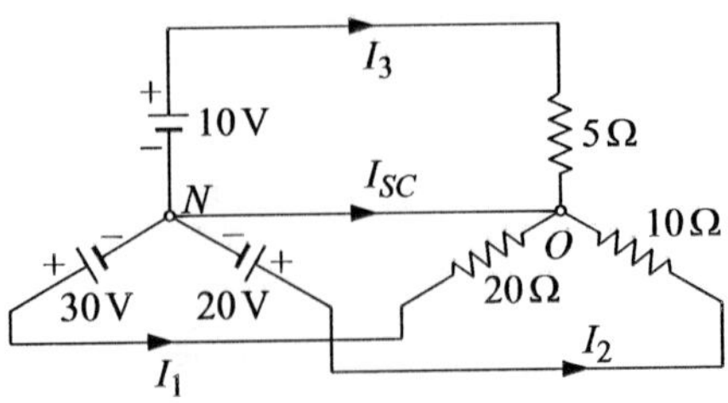

$$I_{SC} = I_1 + I_2 + I_3 = \frac{30}{20} + \frac{20}{10} + \frac{10}{5} = \frac{3}{2} + 2 + 2$$

$$I_{SC} = 5.5 \text{ A}$$

$$\therefore \quad I_N = I_{SC} = 5.5 \text{ A}$$

The Norton's equivalent circuit is:

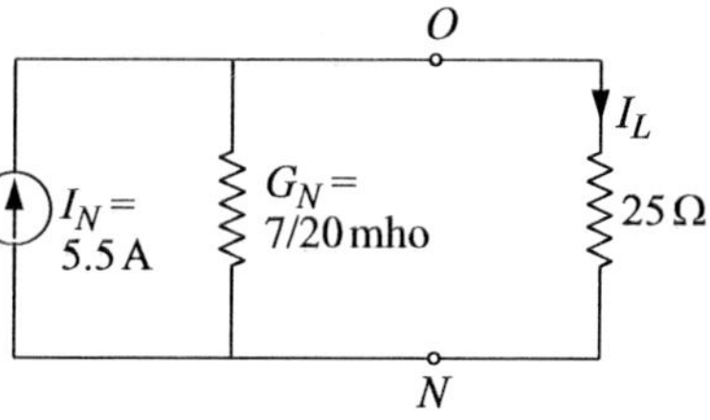

$$I_L = \frac{\frac{20}{7} \times 25}{\frac{20}{7} + 25} \times 5.5 \times \frac{1}{25}$$

$$= \frac{20 \times 25}{20 + 25 \times 7} \times 5.5 \times \frac{1}{25}$$

$$= \frac{20 + 5.5}{20 + 175} = \frac{11}{195}$$

4.8 MAXIMUM POWER TRANSFER THEOREM

4.8.1 DC Networks

Any resistor load connected to a dc network receives maximum power when the load resistance equals to the Thevenin's equivalent resistance of the network as seen from the load terminals.

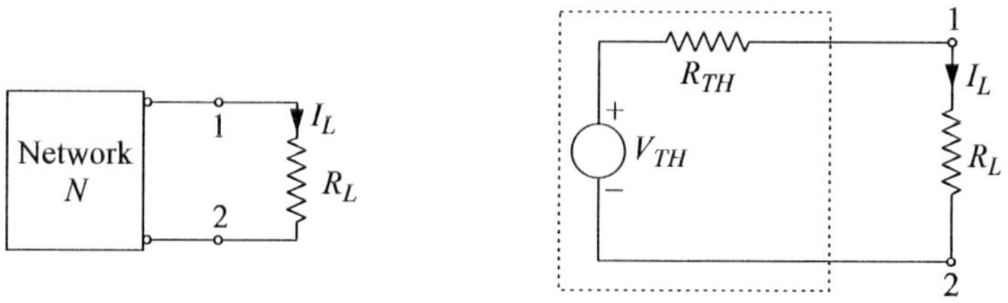

(i) *(ii)*

Fig. 4.14 *Network with load shown in (i) is converted in (ii).*

If the network conversion is done as shown in the Fig 4.14, then for maximum power transfer:

$$R_L = R_{TH}$$

The above can also be proved as follows:

The load current $$I_L = \frac{V_{TH}}{R_{TH} + R_L}$$

The power transferred to R_L in given by:

$$P_L = I_L^2 R_L = \left(\frac{V_{TH}}{R_{TH} + R_L} \right)^2 \times R_L$$

For maximum power transfer to R_L, we should have,

$$\frac{P_L}{d\,R_L} = 0$$

$$\therefore \quad \frac{V_{TH}^2}{(R_{TH}+R_L)^2} + R_L(-2)\frac{V_{TH}^2}{(R_{TH}+R_L)^3} = 0$$

or $$1 - \frac{2R_L}{R_{TH}+R_L} = 0$$

or $$R_{TH} + R_L = 2R_L$$

or $$R_L = R_{TH}$$

Hence, it is prove that for maximum power transfer to load $R_L = R_{TH}$.

$$P_{\max} = \left(\frac{V_{TH}}{R_{TH}+R_L}\right)^2 \times R_{TH} = \frac{V_{TH}^2}{4R_{TH}}$$

Example 4.39. Calculate the value of R which will absorb maximum power from the circuit shown in the following figure. Also compute the value of the maximum power.

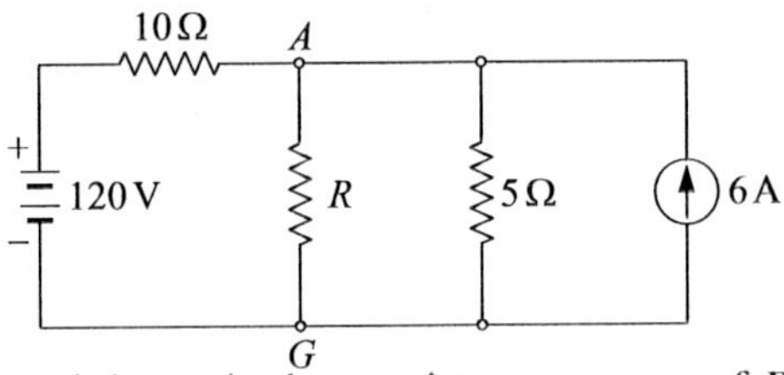

Solution: The Thevenin's equivalent resistance across of R needs to be found out.

For R_{TH}:

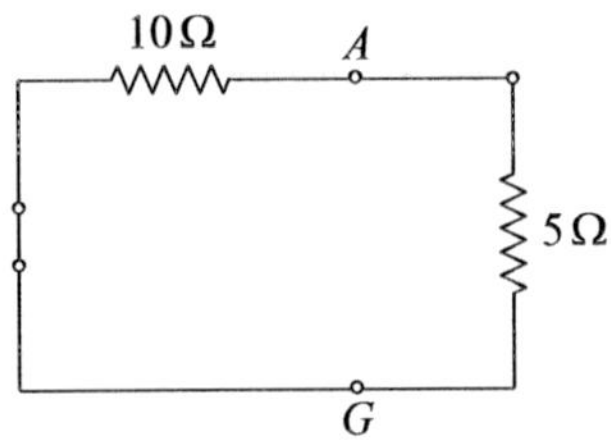

$$R_{AG} = \frac{10\times 5}{10+5} = \frac{50}{15} = \frac{10}{3}\,\Omega$$

$$\therefore \quad R_{TH} = R_{AG} = \frac{10}{3}\,\Omega$$

The V_{TH} can be found out by superposition theorem.

Eliminate Battery:

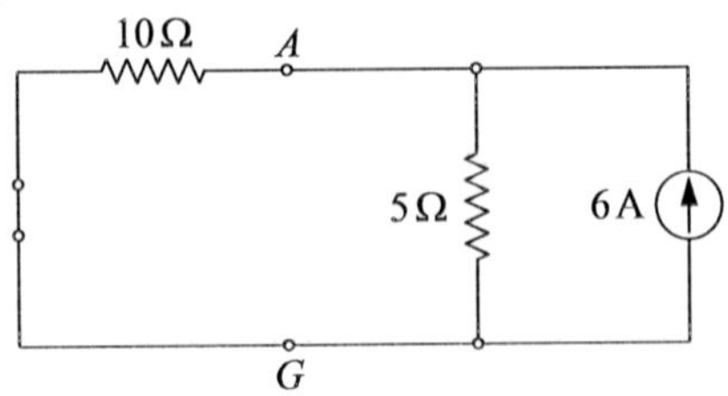

$$V'_{AG} = \frac{10 \times 5}{10+5} \times 6 = \frac{10 \times 5 \times 6}{15} = 20 \text{ V}$$

Eliminate Current Source:

$$V''_{AG} = \frac{120}{10+5} \times 5 = \frac{120 \times 5}{15} = 40 \text{ V}$$

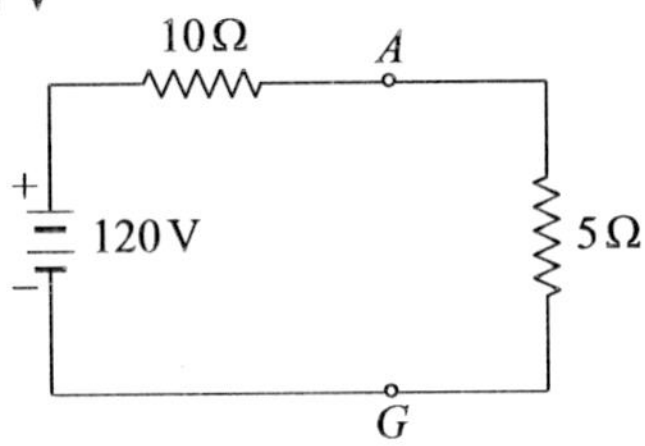

$$\therefore \quad V_{AG} = V'_{AG} + V''_{AG}$$

$$= 20 + 40 = 60 \text{ V}$$

$$V_{TH} = V_{AG} = 60 \text{ V}$$

The Thevenin's equivalent is given as:

$$R_{TH} = \frac{10}{3} \Omega$$

$$R = R_{TH} = \frac{10}{3} \Omega \text{ for maximum power transfer.}$$

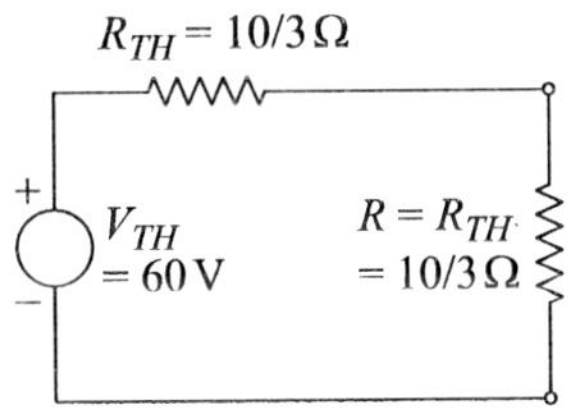

$$P_{max} = \left(\frac{60}{\frac{10}{3} + \frac{10}{3}} \right)^2 \times \frac{10}{3}$$

$$= \left(\frac{60}{20} \times 3 \right)^2 \times \frac{10}{3} = \frac{81 \times 10}{3}$$

or $\quad P_{max} = 270 \ \Omega.$

Example 4.40. Consider the electric circuit shown in the figure under. Determine (*i*) the value of R so that the load of 20 Ω should draw the maximum power, (*ii*) the value of the maximum power drawn by the load. **[UPTQ, 2001]**

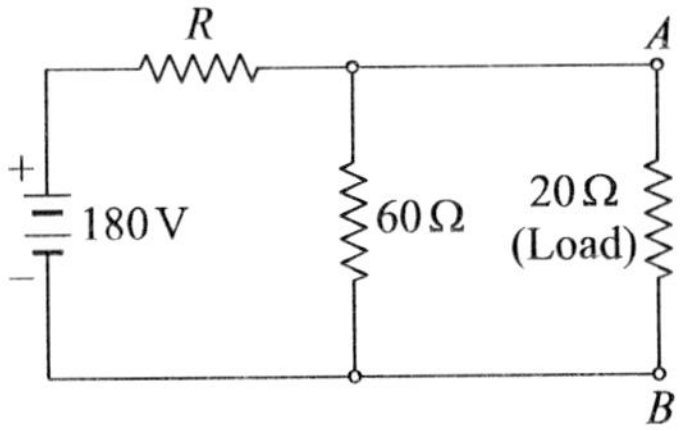

Solution: (*i*) The load should draw the maximum power if

$$R_L = R_{TH} \qquad ...(1)$$

where R_{TH} = Thevenin's equivalent resistance

R_L = Load resistance

For R_{TH}:

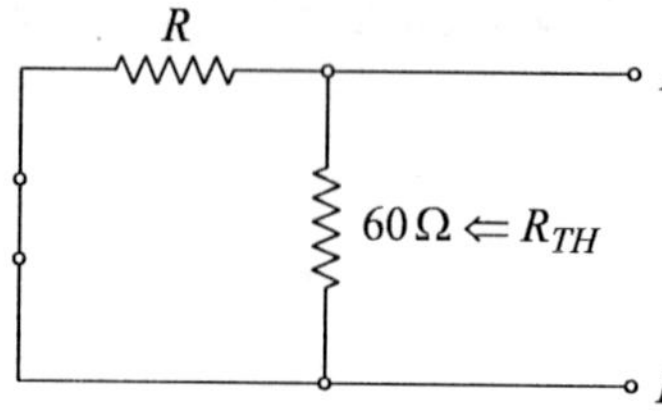

$$R_{TH} = \frac{60R}{60+R}\ \Omega \qquad ...(i)$$

$$R_L = 20\ \Omega \qquad ...(ii)$$

Acording to equation (*i*)

$$\frac{60R}{60+R} = 20$$

$$60R = 1200 + 20R$$

$$40R = 1200$$

$$R = \frac{1200}{40} = 30\ \Omega$$

(*ii*) Value of maximum power draw by the load

$$I_{max} = \frac{(V_{TH})^2}{4R_{TH}}$$

$$V_{TH} = V_{AB}$$

According to the given circuit

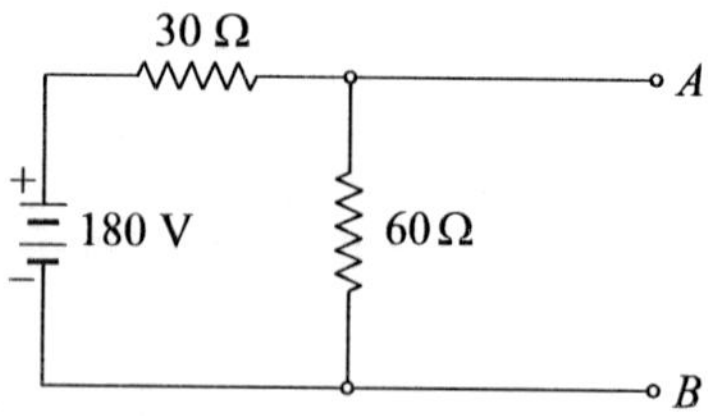

$$V_{AB} = \frac{180}{30+60} \times 60 = \frac{180}{90} \times 60$$

$$V_{AB} = 2 \times 60 = 120\ \text{V}$$

So

$$V_{AB} = 2 \times 60 = 120\ \text{V}$$

$$R_{TH} = R_L = 20\ \Omega$$

For maximum power:

$$P_{max} = \frac{(V_{TH})^2}{4R_{TH}} = \frac{(120)^2}{4 \times 20}$$

$$= \frac{120 \times 120}{40+20} = 30 \times 6 = 180\ \text{W}$$

Example 4.41. Find the value of R in the circuit of figure given below such that the maximum power transfer takes place. **[UPTQ, 2001]**

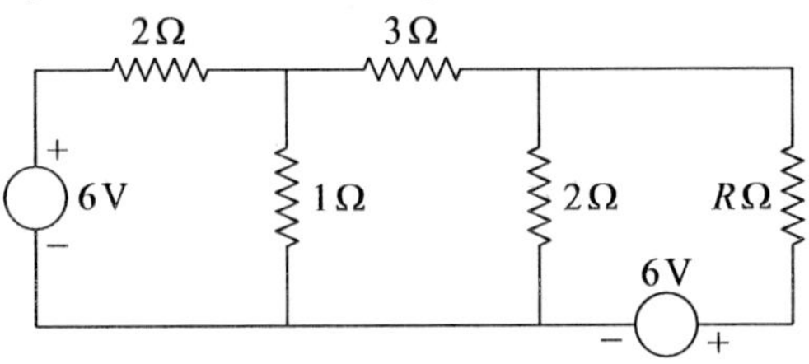

Solution: Thevenin's resistance across R must be equal to R for maximum power transfer to R.

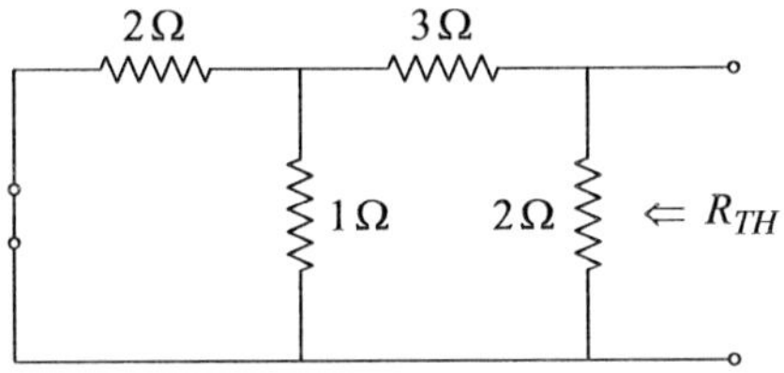

$$R_{TH} = \left(3+\frac{2\times1}{2+1}\right) \| 2 = \left(3+\frac{2}{3}\right) \| 2$$

$$= \frac{\frac{11}{3}\times 2}{\frac{11}{3}+2} = \frac{11\times 2}{11+6} = \frac{22}{17}\ \Omega.$$

$$\therefore \qquad R = R_{TH} = \frac{22}{17}\ \Omega.$$

Example 4.42. In the circuit shown in figure, find the condition for maximum power transfer to R_L. Hence, determine the maximum power transferred.

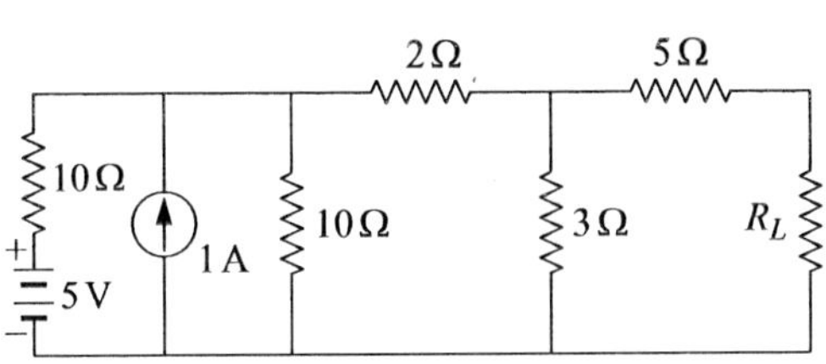

Solution: For maximum power transfer to R_L, Thevenin's resistance is to be found across R_L, hence, the circuit for R_{TH} is:

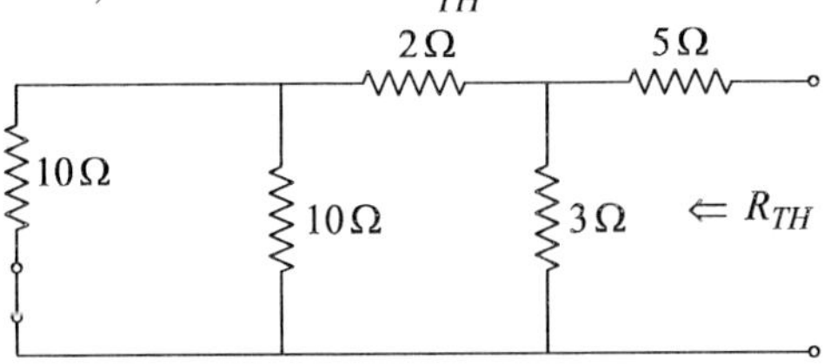

$$R_{TH} = 5 + \left[\left(\frac{10\times10}{10+10} + 2\right) \| 3\right] = 5 + [(5+2) \| 3]$$

$$= 5 + \frac{7\times3}{7+3} = 5 + \frac{21}{10} = 7.1$$

∴ Condition for maximum power transfer to R_L is that $R_L = R_{TH} = 7.1\ \Omega$.

1 A current is parallel with 10 Ω resistance is just like Norton's equivalent circuit. This can be represented as Thevenin's equivalent, hence, the circuit becomes:

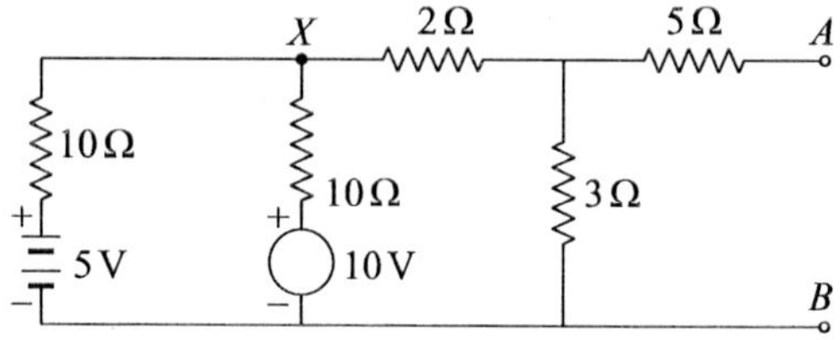

Using nodal analysis:

$$V_X\left[\frac{1}{10}+\frac{1}{10}+\frac{1}{5}\right]-\frac{5}{10}-\frac{10}{10}=0$$

$$V_X\,[1+1+2]-5-10=0$$

$$V_X = \frac{15}{4}\text{ V.}$$

∴ $$V_{AB} = \frac{\left(\frac{15}{4}\right)}{2+3}\times3 = \frac{15}{4\times5}\times3 = \frac{9}{4} = 2.25\text{ V}$$

$$\text{Max. power transferred} = \frac{V_{TH}^2}{4\,R_{TH}} = \frac{(2.25)^2}{4\times7.1} = 0.178\text{ W.}$$

Example 4.43. In the network shown in figure given below find (*i*) the Norton's equivalent circuit at terminals *A* – *B* (*ii*) the maximum power that can be provided to a resistor *R* connected to terminals *A* – *B*. **[UPTU, 2002]**

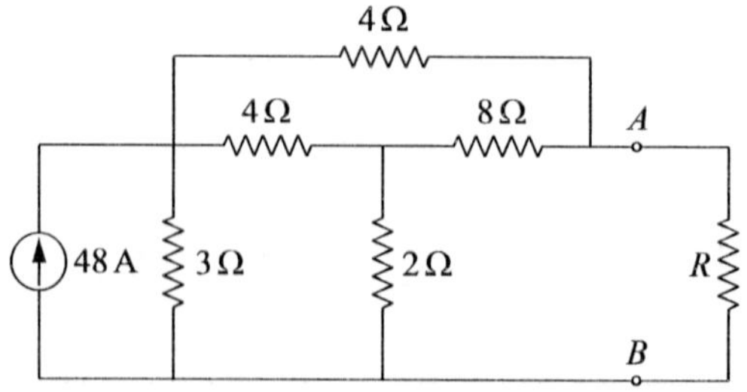

Solution: For R_{TH}:

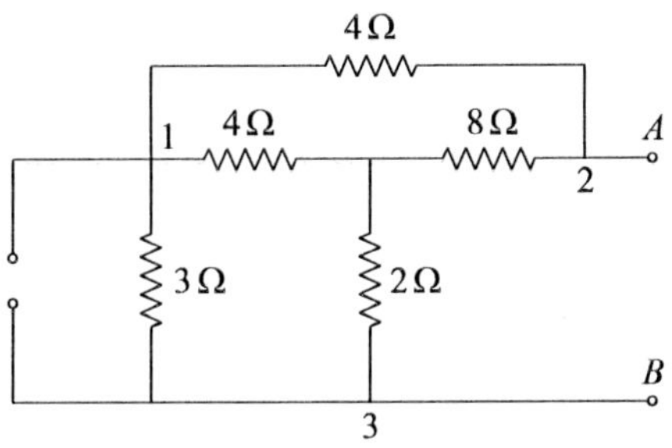

Star-Delta conversion is needed as follows:

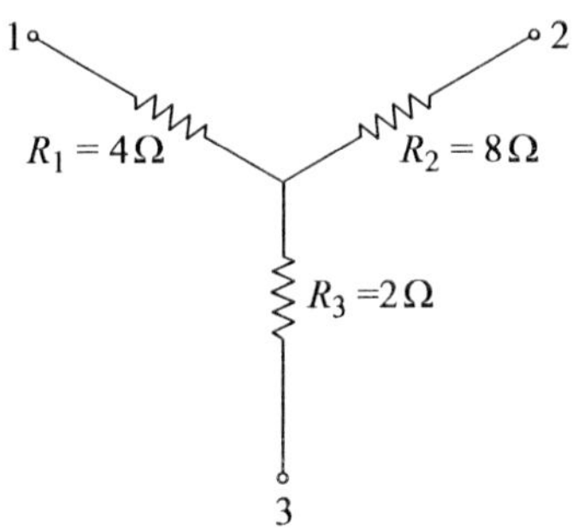

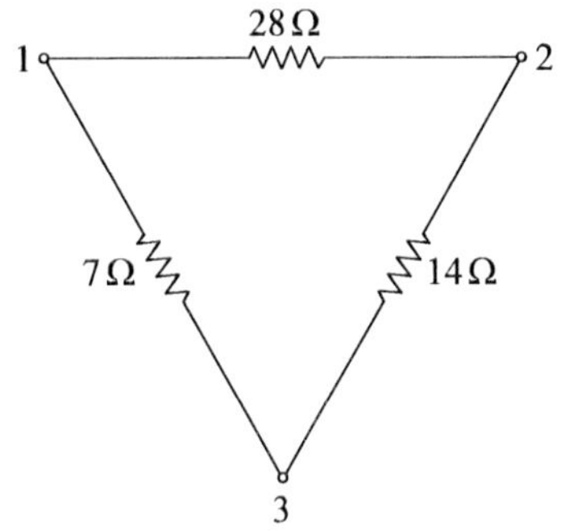

$$R_{12} = R_1 + R_2 + \frac{R_1 R_2}{R_3} = 4 + 8 + \frac{4 \times 8}{2} = 28\ \Omega$$

$$R_{23} = R_2 + R_3 + \frac{R_2 R_3}{R_1} = 8 + 2 + \frac{8 \times 2}{4} = 14\ \Omega$$

$$R_{31} = R_3 + R_1 + \frac{R_3 R_1}{R_2} = 2 + 4 + \frac{2 \times 4}{8} = 7\ \Omega$$

The circuit can be made as follows:

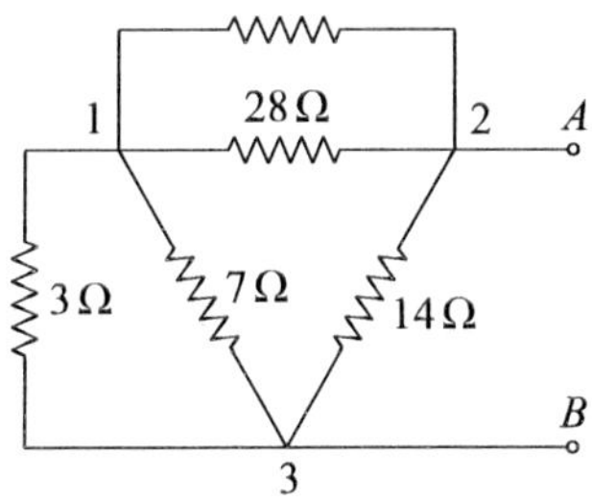

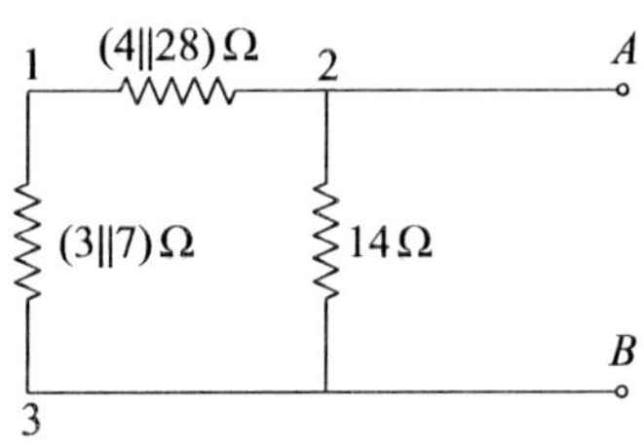

$$4 \parallel 28 = \frac{4 \times 28}{4 + 28} = \frac{4 \times 28}{32} = \frac{7}{2}\ \Omega$$

$$3 \parallel 7 = \frac{3 \times 7}{3 + 7} = \frac{21}{10}\ \Omega$$

∴ The circuit becomes as follows:

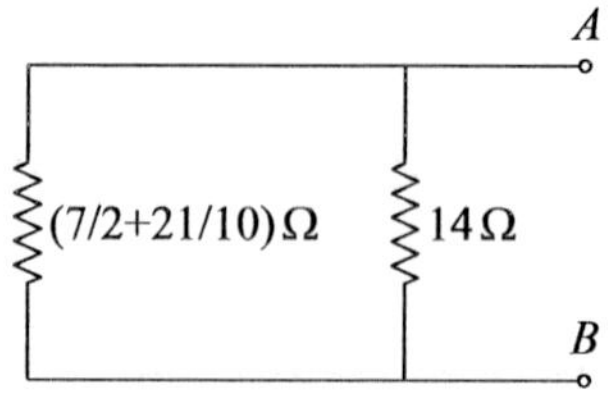

$$= \frac{35+21}{10} = \frac{56}{10} = \frac{28}{5}$$

∴ $$R_{AB} = \frac{\frac{28}{5}\times 14}{\frac{28}{5}+14} = \frac{28\times 14}{28+70} = \frac{28\times 14}{98} = 4\ \Omega$$

$$R_{TH} = R_{AB} = 4\ \Omega$$

For I_N calculation the star-delta connection is to be retained and circuit becomes:

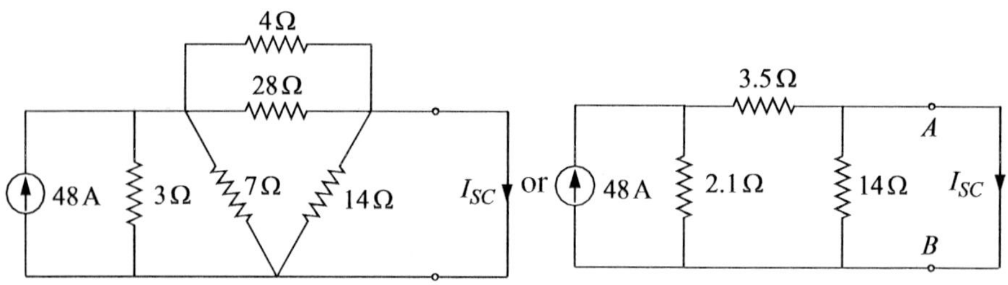

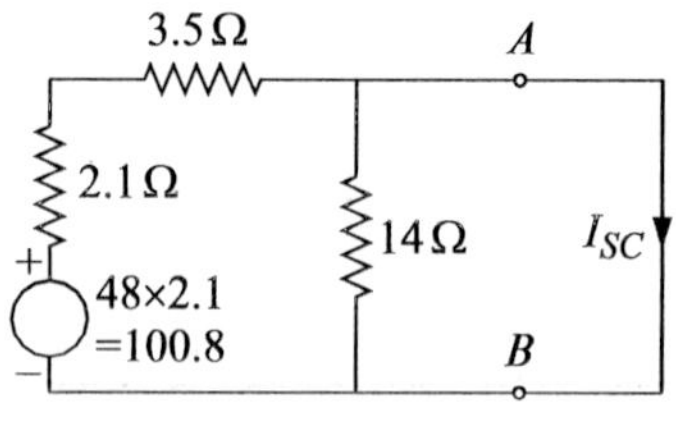

Now current flows through 14 Ω resistor as *A*-*B* is shorted.

∴ $$I_{SC} = \frac{100.8}{2.1+3.5} = 18\text{ A}.$$

∴ $$I_N = I_{SC} = 18\text{ A}$$

(*i*) Hence, Norton's equivalent circuit at terminals *A*–*B*:

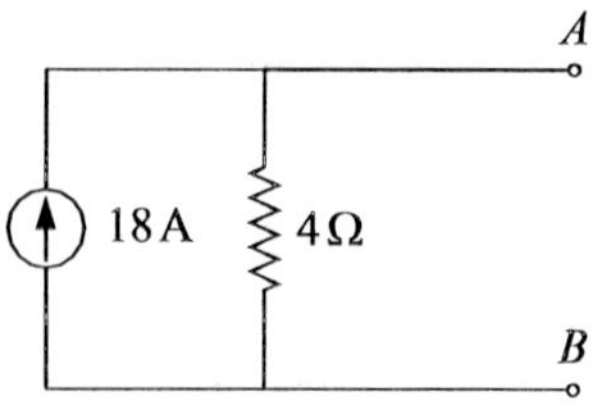

(*ii*) The maximum power transfer to resistor *R* across *A*-*B* will occur if $R = R_{TH} = 4\ \Omega$.

$\therefore$ The maximum power transfer $= \dfrac{V_{TH}^2}{4\,R_{TH}}$

$$= \frac{(18 \times 4)^2}{4 \times 4} = 18 \times 18 = 324 \text{ W}.$$

4.8.2 AC Networks

In any active network, maximum power transfer to the load takes when the load impedance is complex conjugate of an equivalent impedance of the network as viewed from terminals of the load.

A network is shown the Fig. 4.15 (*i*). Suppose Z_{eq} (*see* Fig. 4.15 (*ii*) is the equivalent impedance of the network as viewed from the terminals *ab* and replacing all independent sources by their internal impedance.

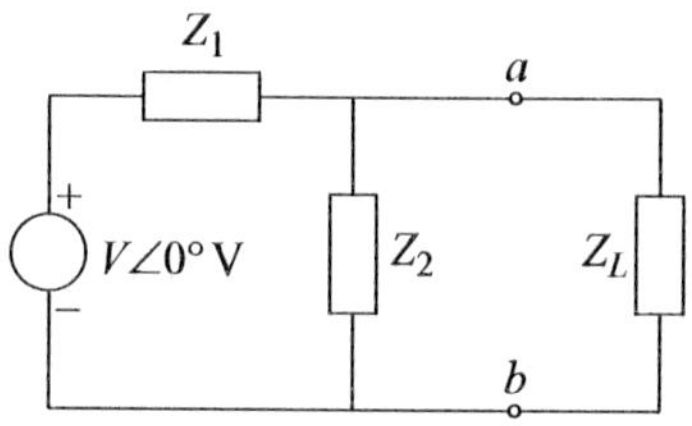

Fig. 4.15 (*i*)

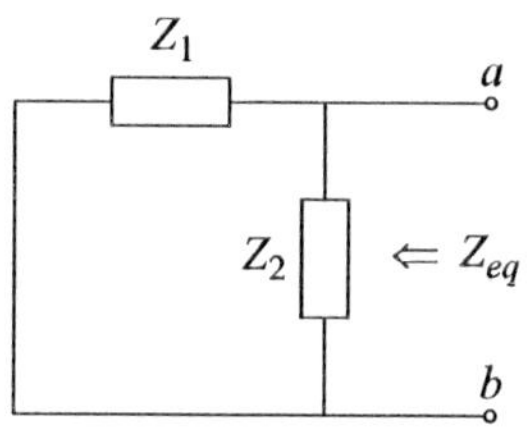

Fig. 4.15 (*ii*)

If $$Z_{eq} = R + jx,$$

then for maximum power transfer to the load, the load impedance should be:

$$Z_L = Z_{eq} = R - jx$$

i.e., If Z_V is inductive, the Z_{eq} should be capacitive or *vice-versa.*

Proof of maximum power transfer theorem

In above theorem can be proved also, consider that the network is converted into Thevenin's equivalent across *ab*, *see* Fig. 4.15 (*iii*).

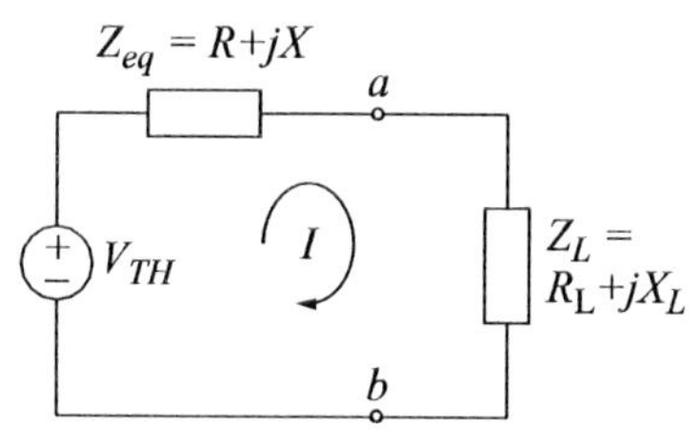

Fig. 4.15 (*iii*)

Current, $$I = \frac{V_{TH}}{Z_{eq} + Z_L} = \frac{V_{TH}}{(R + R_L) + j(X + X_L)}$$

The power transferred to the load;

$$P_L = I^2 R_L = \frac{V_{TH}^2 \ R_L}{(R + R_L)^2 + j(X + X_2)^2}$$

The magnitude of the current is given by

$$I = \frac{V_{TH}}{\sqrt{(R + R_L)^2 + (X + X_L)^2}} R_L \qquad ...(i)$$

The maximum power *w.r.t* X_L is given by:

$$\frac{\partial P_L}{\partial X_2} = 0$$

$$\therefore \qquad \frac{\partial}{\partial X_L}\left[\frac{V_{TH}^2 \ R_L}{(R + R_L)^2 + (X + X_L)^2}\right] = 0$$

or
$$\frac{-2 V_{TH}^2 \ R_L (X + X_L)}{\left[(R + R_L)^2 + (X + X_L)^2\right]^2} = 0$$

or
$$X + X_L = 0$$

i.e.,
$$X_L = -X \qquad ...(ii)$$

Now, taking maximum power transfer condition *w.r.t* R_L, we get

$$\frac{\partial P_L}{\partial R_L} = 0$$

Since we already know one condition

i.e.,
$$X_L = -X$$

Hence,
$$P_L = \frac{V_{TH}^2 \ R_L}{(R + R_L)^2}$$

$$\therefore \qquad \frac{\partial}{\partial R_L}\left[\frac{V_{TH}^2 \ R_L}{(R + R_L)^2}\right] = 0$$

or
$$\frac{V_{TH}^2}{(R + R_L)^2} - \frac{V_{TH}^2 \ R_L . 2(R + R_L)}{(R + R_L)^4} = 0$$

or
$$V_{TH}^2 (R + R_L)^2 - 2R_L (R + R_L) V_{TH}^2 = 0$$

or
$$R - R_L = 0$$

or
$$R_L = R \qquad ...(iii)$$

Then, maximum power transfer,

$$P_{\max} = I^2 R_L = \frac{V_{TH}^2 \ R_L}{(2 R_L)^2} \text{ as } Z_{eq} + Z_L = R_L$$

or $$P_{max} = \frac{V_{TH}^2}{4\,R_L} \qquad \text{...(iv)}$$

If only magnitude of load variable

However, if only magnitude of the load (Z_L) is variable but, its angle cannot be varied, then maximum power transfer takes place if absolute magnitude of the load (Z_L) is made equal to the circuit equivalent impedance Z_{eq}.

In order to prove consider the fig. 4.15 (*iv*). Suppose to beep load angle constant, load is taken as purely resistive *i.e.*, $R_L = |Z_{eq}|$.

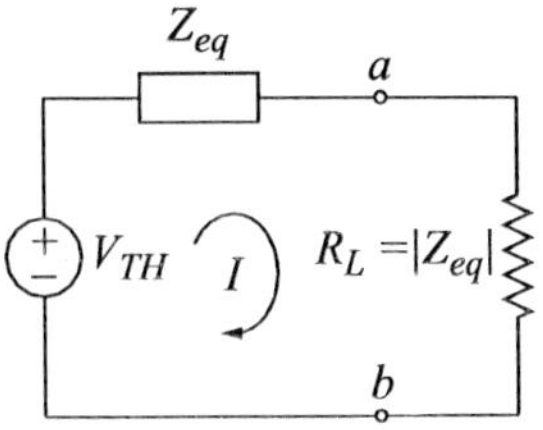

Fig.4.15 (*iv*)

$$I = \frac{V_{TH}}{Z_{eq} + R_L}$$

∴ Maximum power transferred,

$$P_{max} = I^2 R_L = \left(\frac{V_{TH}}{Z_{eq} + R_L}\right)^2 R_L \qquad \text{...(v)}$$

4.9 DELTA-STAR TRANSFORMATION

Delta-star transformation is replacement of any delta (Δ) configuration to star (*Y*) configuration Δ-*Y* transformation is also known as π-*T* transformation.

Similarly, star (*Y*) to delta (Δ) transformation is also done. *Y*-Δ transformation is also known as *T*-π transformation. Such transformations are done in a network to facilitate easy circuit analysis. A delta-star transformation is shown in Fig. 4.16.

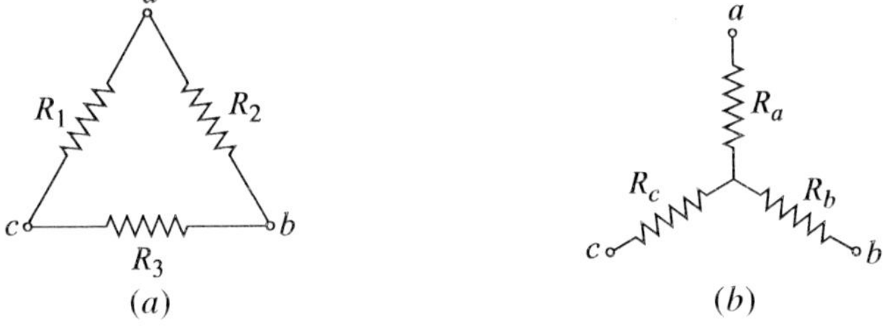

Fig. 4.16. *Delta-star transformation*

The resistance when measure any of the two terminals is same across same terminals in the other case.

From delta system (*a*), we can write for the delta system:

$$R_{ab} = [R_2 \parallel (R_1 + R_3) = \frac{R_2(R_1 + R_3)}{R_1 + R_2 + R_3}$$

In star system (*b*) :

$$R_{ab} = R_a + R_b$$

As both star and delta are equivalent to each other, hence,

$$R_a + R_b = \frac{R_2\,(R_1 + R_3)}{R_1 + R_2 + R_3} \qquad ...(i)$$

Similarly terminals *b-c*, give:

$$R_b + R_c = \frac{R_3(R_1 + R_2)}{R_1 + R_2 + R_3} \qquad ...(ii)$$

For terminals *c-a*:

$$R_c + R_a = \frac{R_1(R_2 + R_3)}{R_1 + R_2 + R_3} \qquad ...(iii)$$

By adding equations (*i*), (*ii*) and (*iii*), we get:

$$2(R_a + R_b + R_c) = \frac{R_2(R_1 + R_3) + R_3(R_1 + R_2) + R_1(R_2 + R_3)}{R_1 + R_2 + R_3}$$

or
$$2(R_a + R_b + R_c) = \frac{R_1R_2 + R_2R_3 + R_1R_3 + R_2R_3 + R_1R_2 + R_1R_3}{R_1 + R_2 + R_3}$$

or
$$2(R_b + R_b + R_c) = \frac{2(R_1R_2 + R_2R_3 + R_1R_3)}{R_1 + R_2 + R_3}$$

or
$$R_a + R_b + R_c = \frac{R_1R_2 + R_2R_3 + R_1R_3}{R_1 + R_2 + R_3} \qquad ...(iv)$$

By substracting equation (*ii*) form equation (*iv*), we get:

$$R_a = \frac{R_1R_2}{R_1 + R_2 + R_3} \qquad ...(v)$$

Similarly by subtracting eqn. (*iii*) and (*i*) from equation (*iv*), we get;

$$R_a = \frac{R_2R_3}{R_1 + R_2 + R_3} \qquad ...(vi)$$

$$R_c = \frac{R_1R_3}{R_1 + R_2 + R_3} \qquad ...(vii)$$

It can be observed that the resistance of any branch of the equivalent star connection is obtained by multiplying the adjacent delta resistances and dividing by the sum of three delta resistances.

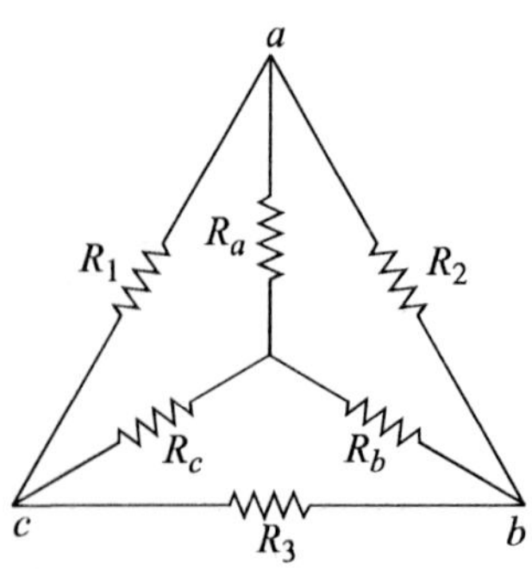

In order to replace a delta by an equivalent star resistance is given by

$$Y\text{-Resistance} = \frac{\text{Product of adjacent } \Delta \text{ resistances}}{\text{Sum of } \Delta \text{ resistances}}$$

4.10 STAR-DELTA TRANSFORMATION

If we multiply equations (*v*) and (*vi*), we get:

$$R_a R_b = \frac{R_1 R_2^2 R_3}{(R_1 + R_2 + R_3)^2} \quad \text{...}(viii)$$

Similarly multiply equations (*vi*) and (*vii*), we get:

$$R_b R_c = \frac{R_2 R_3^2 R_1}{(R_1 + R_2 + R_3)^2} \quad \text{...}(ix)$$

and also by multiply equations (*v*) and (*vii*), we get:

$$R_c R_a = \frac{R_1^2 R_2 R_3}{(R_1 + R_2 + R_3)^2} \quad \text{...}(x)$$

Adding equations (*viii*), (*ix*) and (*x*) we get:

$$R_a R_b + R_b R_c + R_c R_a = \frac{R_1 R_2^2 R_3 + R_1 R_3^2 R_2 + R_1^2 R_2 R_3}{(R_1 + R_2 + R_3)^2}$$

$$= \frac{R_1 R_2 R_3 (R_1 + R_2 + R_3)}{(R_1 + R_2 + R_3)^2} = \frac{R_1 R_2 R_3}{(R_1 + R_2 + R_3)} \quad \text{...}(xi)$$

Now, by dividing equation (*xi*) by equation (*v*), we get

$$R_3 = \frac{R_a R_b + R_b R_c + R_c R_a}{R_a}$$

or

$$R_3 = R_b + R_c + \frac{R_b R_c}{R_a} \quad \text{...}(xii)$$

Similarly by dividing equation (*xi*) by (*vi*) and (*vii*), we get

$$R_1 = R_a + R_c + \frac{R_a R_c}{R_b} \quad \text{...}(xiii)$$

and

$$R_2 = R_a + R_b + \frac{R_a R_b}{R_c} \quad \text{...}(xiv)$$

It can be observed now that the equivalent delta resistances between two terminals is the sum of the two star resistances connected to those terminals plus the product of the same divided by the third star resistance.

Example 4.44. Figure below has three resistances r, $2r$, and $3r$ are connected in delta. Determine the resistances for an equivalent star connection. **[UPTU, 2003]**

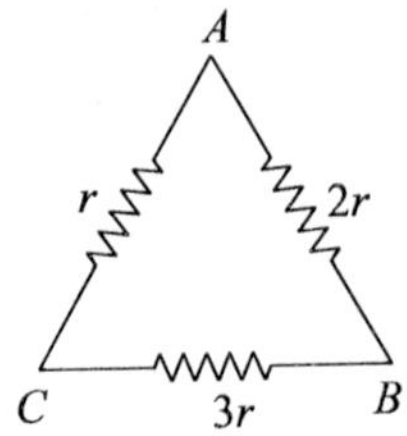

Solution: The equivalent star connected circuit is as under:

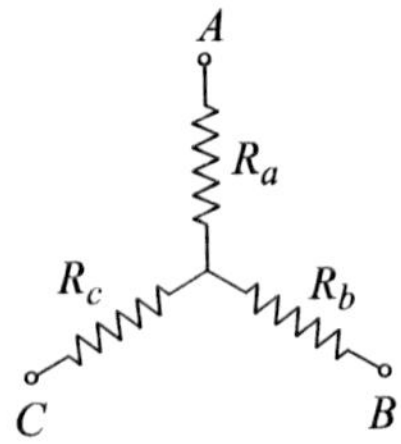

$$R_a = \frac{r \times 2r}{r + 2r + 3r} = \frac{r}{3}$$

$$R_b = \frac{2r \times 3r}{6r} = r$$

$$R_a = \frac{r \times 3r}{6r} = \frac{r}{2}$$

Example 4.45. Find the resistances at the A-B terminals in the electric circuit of figure given below using Δ-Y transformation. **[UPTU, 2001]**

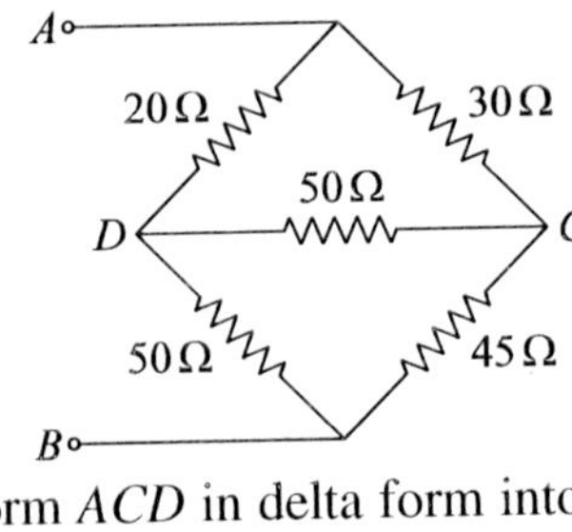

Solution: Let us transform ACD in delta form into star form:

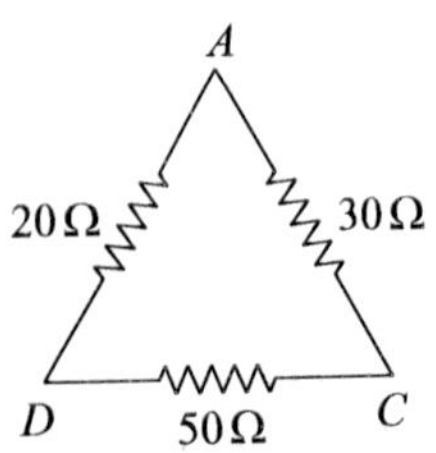

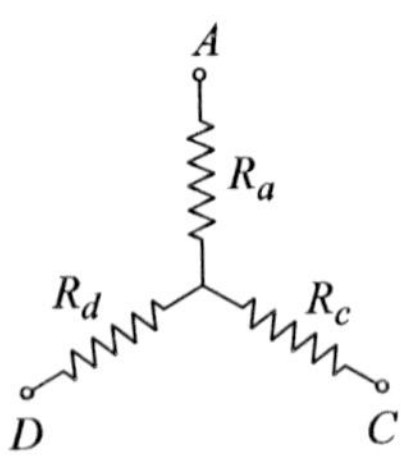

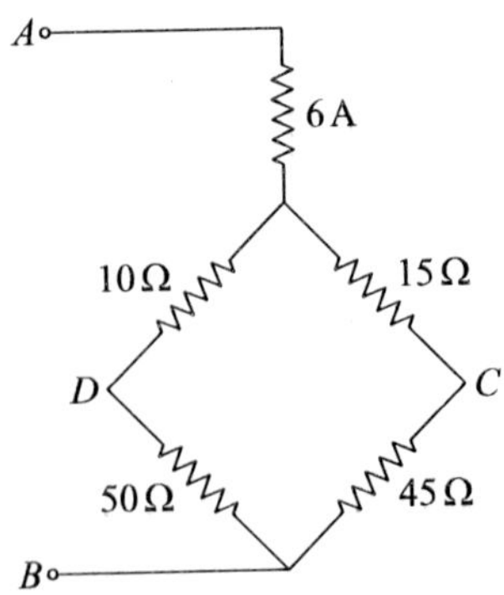

$$R_a = \frac{20 \times 30}{20 + 30 + 50} = 6\ \Omega$$

$$R_c = \frac{30 \times 50}{100} = 15\ \Omega$$

$$R_d = \frac{20 \times 50}{100} = 10\ \Omega$$

The original circuit is converted into following:

$$\therefore \quad R_{AB} = 6 + (10 + 50) \parallel (15 + 45)$$

$$= 6 + \frac{60 \times 60}{60 + 60} = 36\ \Omega.$$

Example 4.46. Find the resistances between terminals *XY* of the bridge circuit shown in figure given under by using delta-star transformations.

[UPTU, 2002]

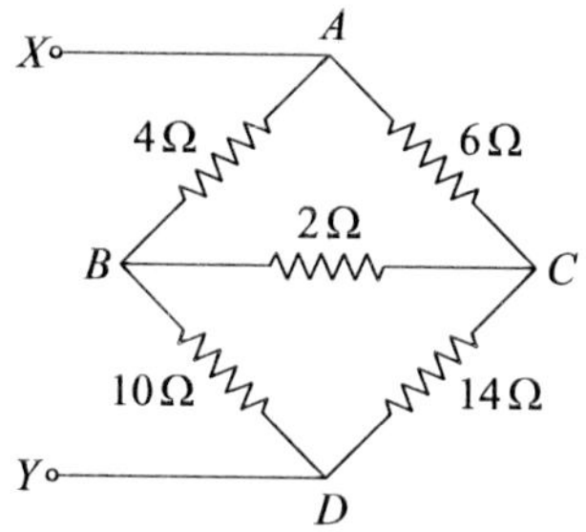

Solution: The portion *ACB* can be converted into star and the new circuit becomes:

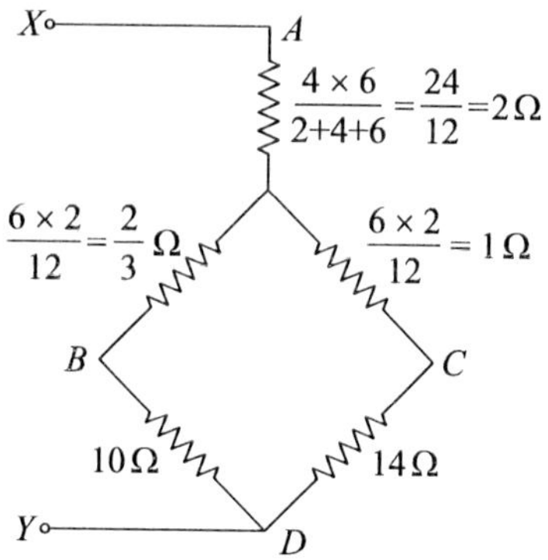

$$\therefore \quad R_{XY} = 2 + \left[\left(10 + \frac{2}{3}\right) \parallel (1 + 14)\right]$$

$$= 2 + \frac{\frac{32}{3} \times 15}{\frac{32}{3} + 15}$$

$$= 2 + \frac{32 \times 15}{32 + 45} = 2 + \frac{480}{77}$$

or $\quad R_{XY} = 8.234\ \Omega.$

Example 4.47. A network of resistance is shown in figure as follows. Computer the equivalent resistances measured between (*i*) *A* and *B*, (*ii*) *B* and *C* and (*iii*) *C* and *A*.

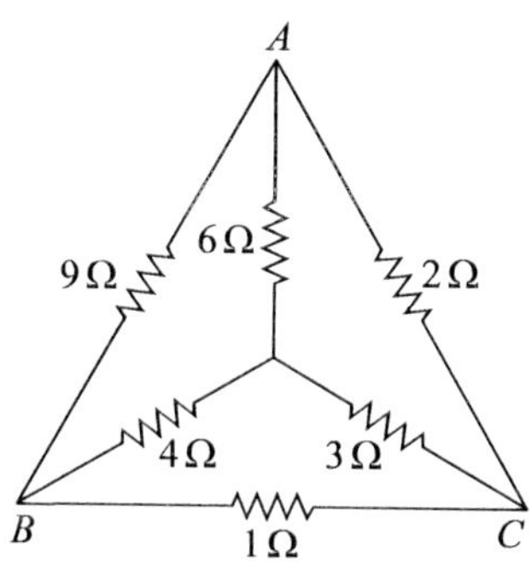

Solution: The internal star network is to be transformed into a delta as follows:

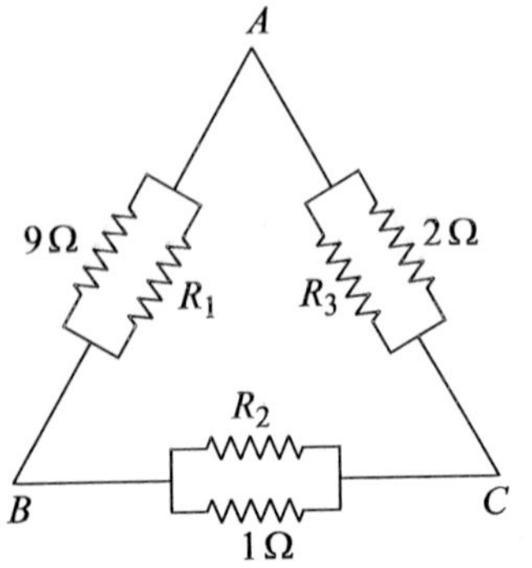

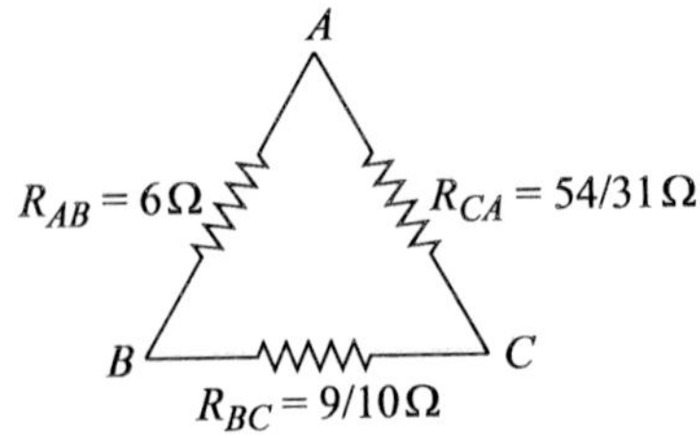

$$R_1 = 4 + 6 + \frac{4 \times 6}{3} = 10 + 8 = 18\ \Omega$$

$$R_2 = 4 + 3 + \frac{4 \times 3}{6} = 7 + 2 = 9\ \Omega$$

$$R_3 = 3 + 6 + \frac{3 \times 6}{4} = 9 + 2.5 = 13.5\ \Omega$$

In the first figure, the net resistance values of the delta branches are:

$$A_{AB} = \frac{9 \times 18}{9 + 18} = \frac{9 \times 18}{27} = 6\ \Omega$$

$$R_{BC} = \frac{1 \times 9}{1 + 9} = \frac{9}{10}\Omega$$

$$R_{CA} = \frac{13.5 \times 2}{13.5 + 2} = \frac{27}{15.5} = \frac{54}{31}\Omega$$

(*i*) Resistance measured between *A* and *B*

$$= 6 \| \left[\frac{54}{31} + \frac{9}{10}\right] = 6 \| \left[\frac{819}{310}\right]$$

$$= \frac{6 \times \frac{819}{310}}{6 + \frac{819}{310}} = \frac{4914}{2679}\Omega.$$

(*ii*) Resistance measured between *B* and *C*

$$= \frac{9}{10} \| \left[6 + \frac{54}{31}\right] = \frac{9}{10} \| \frac{240}{31}$$

$$= \frac{\frac{9}{10} \times \frac{240}{31}}{\frac{9}{10} + \frac{240}{31}} = \frac{2160}{2679}\,\Omega$$

(*iii*) Resistance measured between C and A

$$= \frac{54}{31} \,||\, \left[6 + \frac{9}{10}\right] = \frac{54}{31} \,||\, \frac{69}{10}$$

$$= \frac{\frac{54}{31} \times \frac{69}{10}}{\frac{54}{31} + \frac{69}{10}} = \frac{3726}{2679}\,\Omega.$$

Example 4.48. With the help of star/delta transformation, obtain the value of current supplied by battery in the circuit shown in figure as under.

[UPTU, 2001]

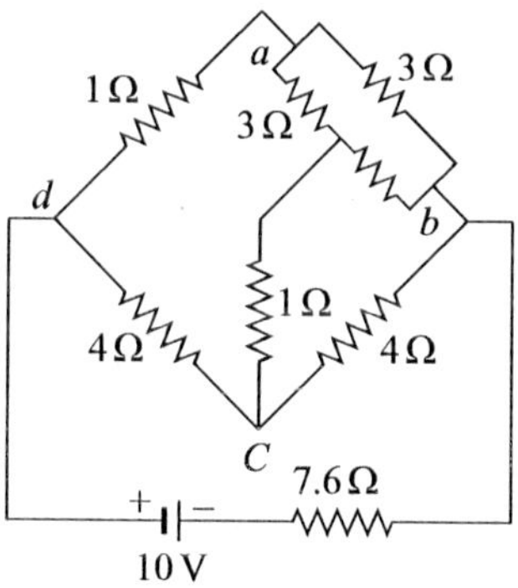

Solution: In the figure a b and c star needs transformation to delta, a, b and c *i.e.*,

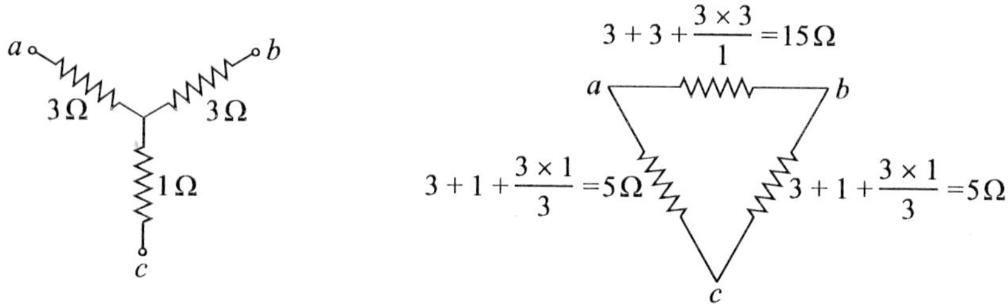

The circuit after star-delta transformation is follows:

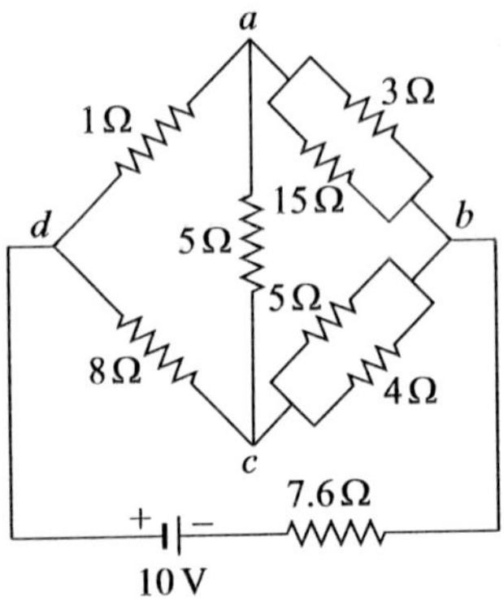

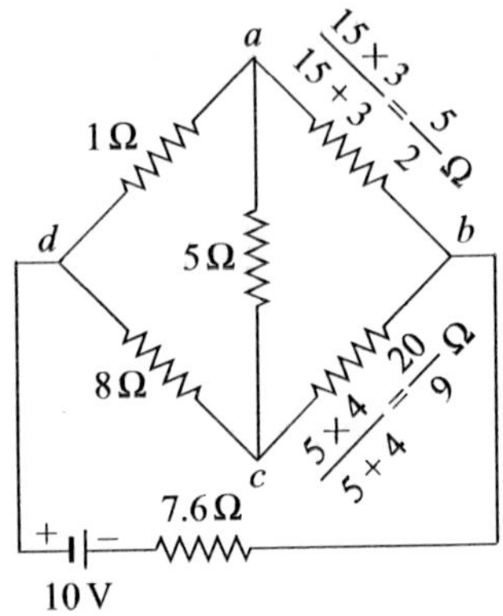

Now, transformation of delta figure a, c and d to be done into star *i.e.*,

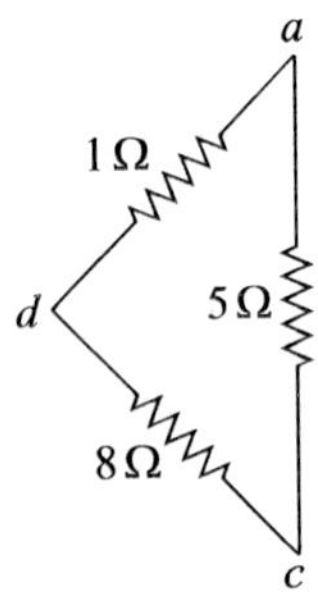

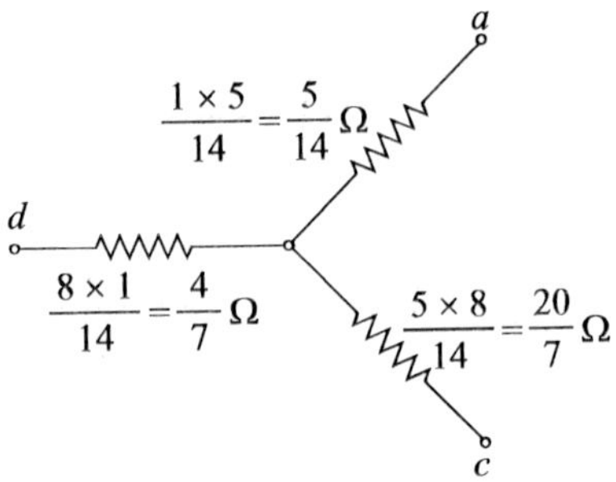

Therefore, the above main circuit becomes:

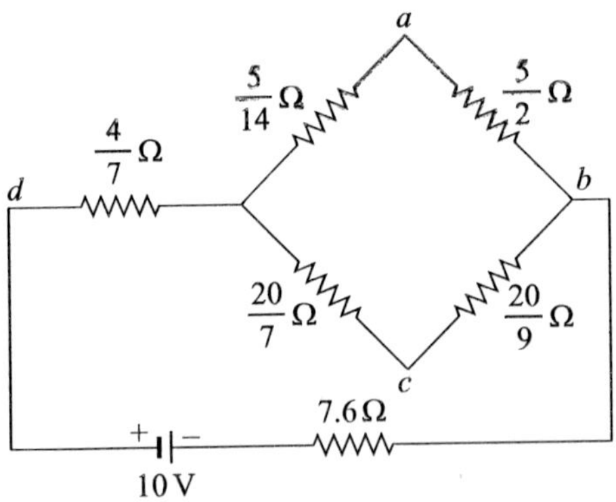

⇒

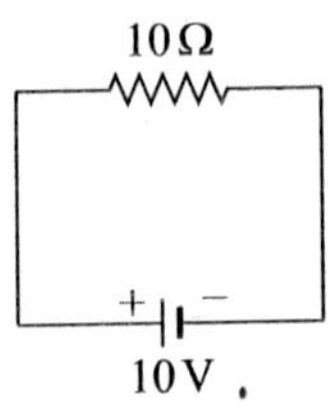

Equivalent resistance across battery, R_e

$$= \left(\frac{4}{7} + 7.6\right) + \left[\left(\frac{5}{14} + 2.5\right) \| \left(\frac{20}{7} + \frac{20}{9}\right)\right]$$

$$= \left(\frac{4}{7} + 7.6\right) + \left[\frac{\frac{20}{7} \times \frac{320}{63}}{\frac{20}{7} + \frac{320}{63}}\right] = 10\ \Omega.$$

Hence, current supported by the battery $= \dfrac{10\text{ V}}{10\,\Omega} = 1$ A.

4.11 RECIPROCITY THEOREM

In any linear network comprised of linear and bilateral impedances and active sources, the ratio of voltage V introduced in one loop to the current I in other loop is same as the ratio obtained if the positions of V and I are interchanged in the network. While calculating the ratio, the sources other than one which is considered to obtain the ratio, must be replaced by their internal impedances.

The above can be explained with the help of circuits. Consider Figs. 4.17 and 4.18.

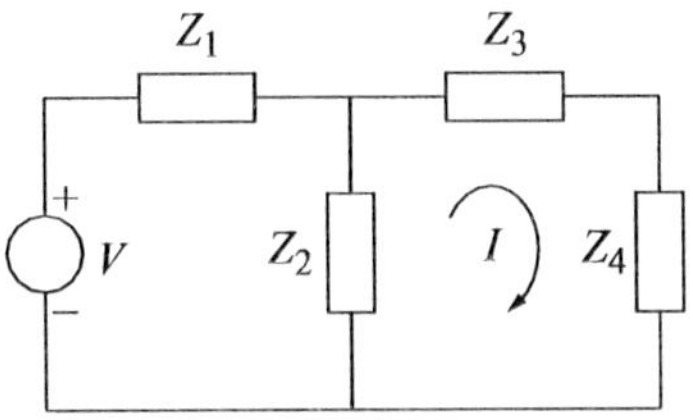

Fig. 4.17

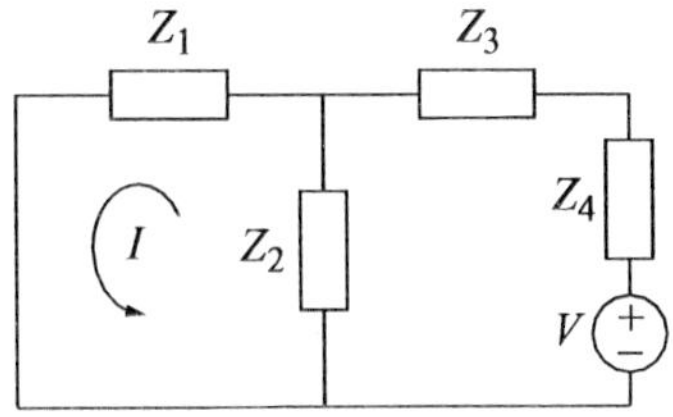

Fig. 4.18

The voltage to current ratio is $\frac{V}{I}$. In the Fig. 4.18, V and I are interchanged as shown, then as per reprocity theorem, $\frac{V}{I}$ should remain same $\frac{V}{I}$ is also known as transfer impedence denoted as Z_{12} where V is the voltage introduced in loop 1 and I is current response due to V in loop 2.

Roof of Reprocity Theorem.

Let the network of Fig. 4.19 be taken for this purpose and the Fig. 4.20 shown interchanged positions of V and I.

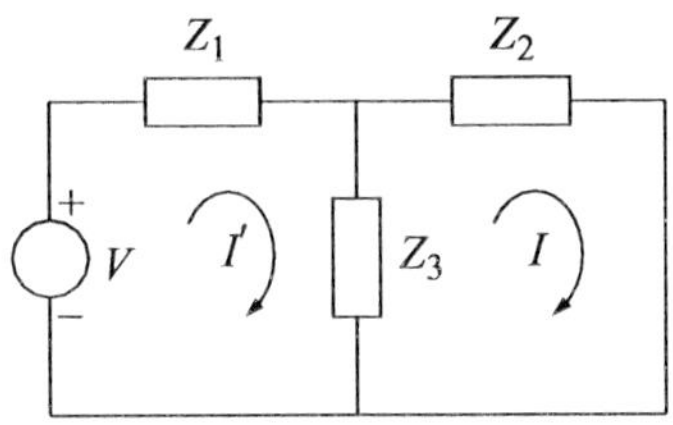

Fig. 4.19 *Network with labelled loop currents*

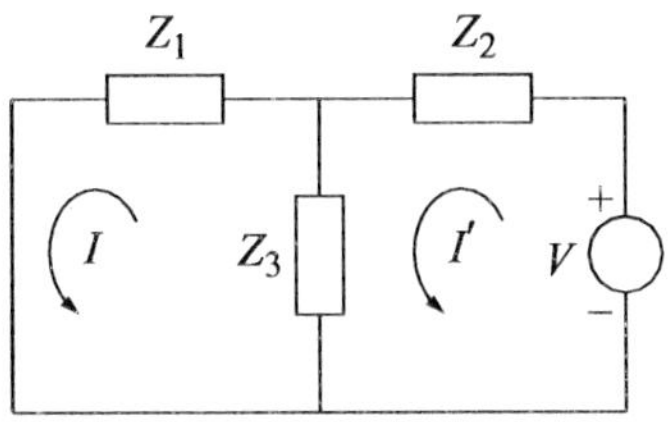

Fig. 4.20 *Interchanged positions of V and I*

By KVL in two loops of Fig. 4.23, we get:

$$-I' Z_1 - I'Z_3 + I Z_3 + V = 0$$

or $\quad I' (Z_1 + Z_3) - I Z_3 = V_2 \quad \ldots (i)$

and $\quad I' Z_3 - I (Z_2 + Z_3) = 0 \quad \ldots (ii)$

By Cramer's rule:

$$I = \frac{\begin{vmatrix} Z_1 + Z_3 & V \\ Z_3 & 0 \end{vmatrix}}{\begin{vmatrix} Z_1 + Z_3 & -Z_3 \\ Z_3 & -(Z_2 + Z_3) \end{vmatrix}}$$

$$= \frac{V\,Z_3}{(Z_1 Z_2 + Z_2 Z_3 + Z_3 Z_1)}$$

or $$\frac{V}{I} = \frac{Z_1 Z_2 + Z_2 Z_3 + Z_3 Z_1}{Z_3} \quad \text{... } (iii)$$

Now, Fig. 4.20 circuit is combined which has interchanged positive of V and I. By applying KVL, we get:

$$I Z_1 - I Z_3 + I' Z_3 = 0$$

or $$I Z_3 - I' (Z_1 + Z_3) = 0 \quad \text{... } (iv)$$

and $$-I Z_2 - I Z_3 + I' Z_3 + V = 0$$

or $$I(Z_2 + Z_3) - I' Z_3 = V \quad \text{... } (v)$$

By applying Cramer's rule, we get

$$I = \frac{\begin{vmatrix} 0 & -(Z_1 + Z_3) \\ V & -Z_3 \end{vmatrix}}{\begin{vmatrix} Z_3 & -(Z_1 + Z_3) \\ Z_2 + Z_3 & -Z_3 \end{vmatrix}} = \frac{Z_3\,V}{Z_1 Z_2 + Z_2 Z_3 + Z_3}$$

or $$\frac{V}{I} = \frac{Z_1 Z_2 + Z_2 Z_3 + Z_3 Z_1}{Z_3} \quad \text{... } (vi)$$

The equations (*iii*) and (*vi*) give same value of $\frac{V}{I}$, hence, reciprocity theorem stands proven.

4.12 COMPENSATION THEOREM

In any linear comprised of linear and bilateral impedances and active sources, if impedance Z of the branch carrying current I increases by SZ, then the increment of voltage or current in each branch of the network is that voltage or current that could be produced by an opposing source of value V_C (= δZI) introduced in the altered branch after replacing original sources by their internal impedances

Consider the circuits of the Figs. 4.21 and 4.22.

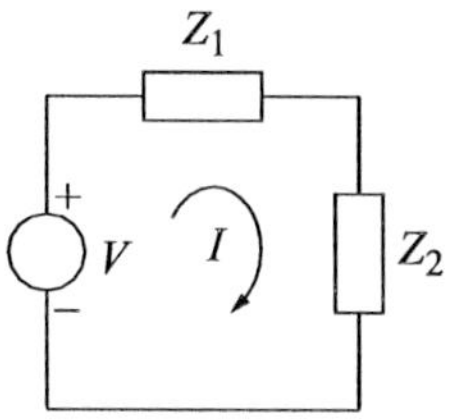

Fig. 4.21 *Network*

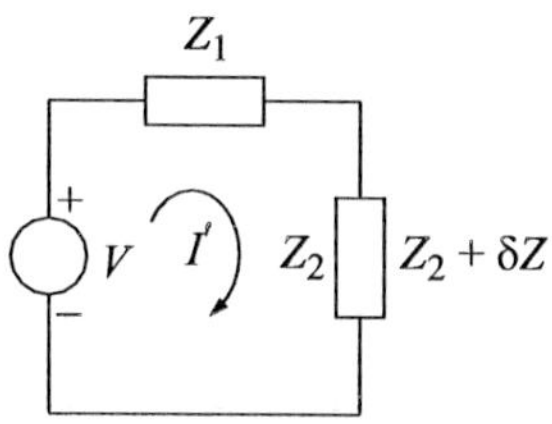

Fig. 4.22 *Network with changed impedence* Z_2

The changed Z_2 to $Z_2 + \delta Z$ changes current I to I' in the circuit of Fig. 4.22

i.e. $$\delta I = I - I'$$

The above current changed can be compensated by introducing a voltage source $V_C = \delta ZI$ with proper purity in the branch having changed value of impedence, *see* Fig. 4.23.

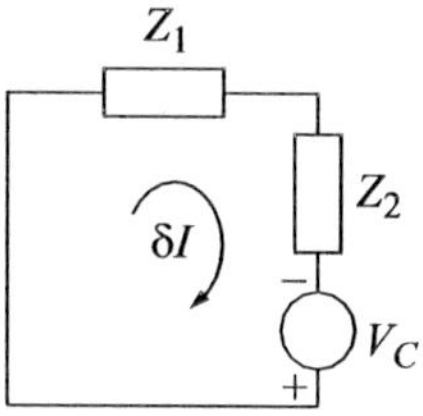

Fig. 4.23 *Compensation source voltage has polarity*

The voltage source introduced in the modified branch is known as compensation source voltage with value $V_C = \delta IZ$ which has polarity to modify current I change.

Proof of compensation theorem

Consider networks of Figs. 4.21 and 4.22. In circuit of Fig. 4.21, we get:

$$I = \frac{V}{Z_1 + Z_2} \quad \text{... } (i)$$

and from circuit Fig. 4.22, we get:

$$I' = \frac{V}{Z_1 + (Z_2 + \delta Z)} \quad \text{... } (ii)$$

Change in current is given by:

$$\delta I = I - I'$$

$$= \frac{V}{Z_1 + Z_2} - \frac{V}{Z_1 + (Z_2 + \delta Z)}$$

or $$\delta I = \frac{V}{(Z_1 + Z_2)} \times \frac{\delta Z}{(Z_1 + Z_2 + \delta Z)}$$

B introducing value of V from equation (i), we get:

$$\delta I = \frac{\delta\, ZI}{(Z_1 + Z_2 + \delta Z)}$$

or $$\delta I = \frac{V_C}{(Z_1 + Z_2 + \delta Z)} \quad \text{... (iii)}$$

Change branch is shown in Fig. 4.28, thus, by applying KVL we get:

$- Z_1 \times \delta I - (Z_2 + \delta Z) \times S_I + V_C = 0$

or $$\delta I = \frac{8\ ZI}{(Z_1 + Z_2 + \delta Z)}$$

or $$\delta I = \frac{V_C}{(Z_1 + Z_2 + \delta Z)} \quad \text{... (iv)}$$

It can be seen that δI values in both cases are same, hence, compensation theorem stands proven.

4.13 MILLMAN'S THEOREM

Voltage sources V_1, V_2, ..., V_n with their internal impedances Z_1, Z_2, ..., Z_n respectively, can be replaced by single voltage source V_m having series of impedances Z_m where V_m and Z_m are given as:

$$V_m = \frac{V_1Y_1 + V_2Y_2 + ... + V_nY_n}{Y_1 + Y_2 + ...,+ Y_n}$$

i.e., $$V_m = \frac{\sum_{j=1}^{n} V_j\, Y_j}{\sum_{j=1}^{n} Y_j}$$

and $$Z_m = \frac{1}{Y_1 + Y_2 + ... + Y_n}$$

or $$Z_m = \frac{1}{\sum_{j=1}^{n} Y_j}$$

where Y_1, Y_2, ..., Y_n are the admittances corresponding to the impedances Z_1, Z_1, ..., Z_n respectively.

Look in the Fig. 4.24 having n voltage sources V_1, V_2, ..., V_n with series impedances of Z_1, Z_2, ..., Z_n connected in parallel.

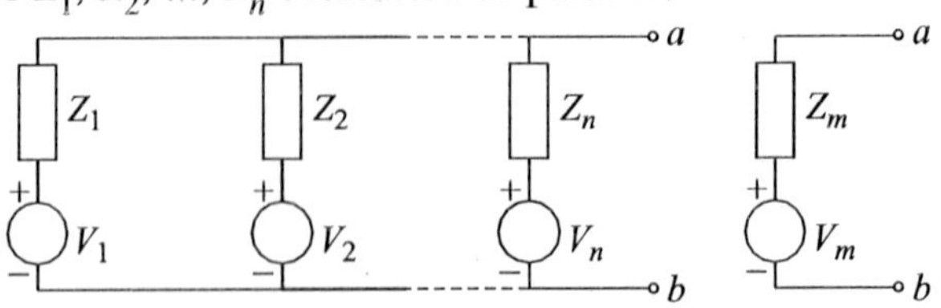

Fig. 4.24 *Network and Millman's equivalent*

The admittances will be given by:

$$Y_1 = \frac{1}{Z_1},\ Y_{12} = \frac{1}{Z_2}, ..., Y_n = \frac{1}{Z_n}$$

If Millman's Theorem is applied, then

$$V_m = \frac{V_1Y_1 + V_2Y_2 + ..., + V_nY_n}{Y_1 + Y_2 + ..., + Y_n}$$

and $$Z_m = \frac{1}{Y_1 + Y_2 + Y_3 + ..., + Y_n}$$

Proof of Millman's Theorem

From Fig. 4.25, the voltage source in series with impedance can be converted into current source in parallel admittance.

$\therefore$ $$I_1 = \frac{V_1}{Z_1}$$

or $$I_1 = V_1Y_1 \text{ and } Y_1 = \frac{1}{Z_1}$$

Similarly, $I_2 = V_2\ Y_2$, $I_3 = V_3Y_3$, ..., $I_n = V_n\ Y_n$ Thus, the network can be written in current source forms as given in Fig. 4.25.

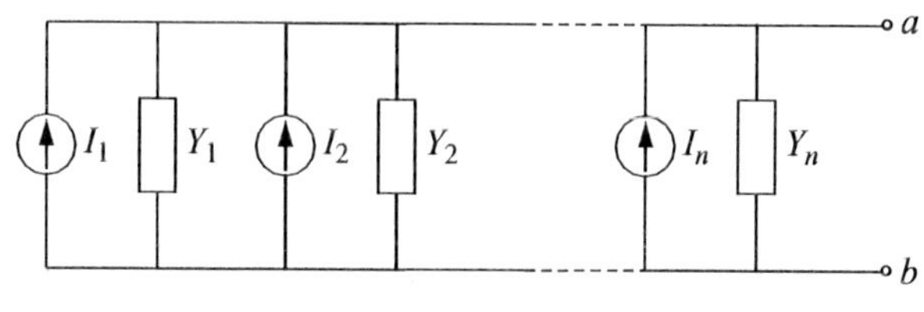

Fig. 4.25 *Network with current sources*

When, the current source and admittances are combined, we get:

$$I_m = I_1 + I_2 + ... + I_n \qquad ... (i)$$

and $$Y_m = Y_1 + Y_2 + ..., + Y_n \qquad ... (ii)$$

Therefore, the equivalent circuit of Fig. 4.25 can be made as shown in Fig 4.26.

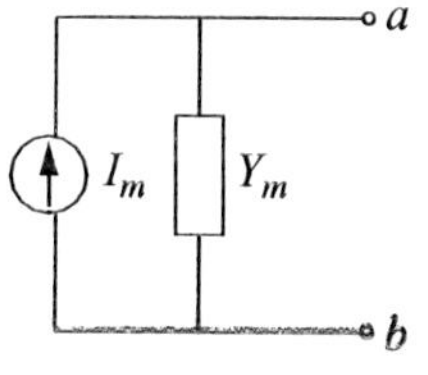

Fig. 4.26 *Equivalently circuit with combined current sources*

The equivalent, circuit with voltage sources can be made as shown in Fig. 4.27.

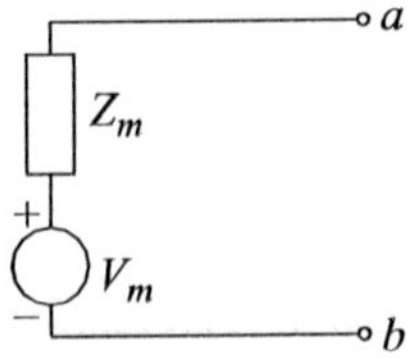

Fig. 4.27 *Equivalent circuit with combined voltage source*

We know that:

$$V_m = \frac{I_m}{Y_m} \text{ as } Z_m = \frac{1}{Y_m} \text{ and } V_m = I_m Z_m$$

By substituting values of V_m and Y_m from equations (*i*) and (*ii*), we get:

$$V_m = (I_1 + I_2 + ..., + I_n) \times \frac{1}{Y_1 + Y_2 + ... + Y_n}$$

and it is known that:

$$I_1 = \frac{V_1}{Z_1} = V_1 Y_1, I_2 = V_2 Y_2 + ..., + V_n Y_n$$

Hence,

$$V_m = \frac{V_1Y_1 + V_2Y_2 + ..., + V_nY_n}{Y_1 + Y_2 + ..., Y_n} = \frac{\sum_{j=1}^{n} V_j V_j}{\sum_{j=1}^{n} Y_j}$$

$$Z_m = \frac{1}{Y_m} = \frac{1}{Y_1 + Y_2 + ..., + Y_n} = \frac{1}{\sum_{j=1}^{n} Y_j}$$

4.14 TELLEGEN'S THEOREM

In any arbitrary lumped network, the algebraic sum of the instantaneous powers in all the branches, at any instant is zero. All the branch currents and the voltages in that network must satisfy Kirchoff's laws. It can also be stated in an easier way *i.e.*, the algebraic sum of the power delivered by all sources is equal to the algebraic sum of the powers absorbed by all the elements.

An arbitrary network with *b* branches and *n* nodes is shown in the Fig. 4.28.

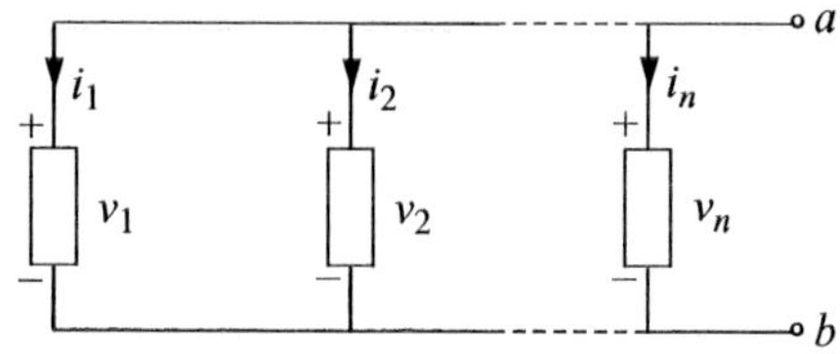

Fig. 4.28 *Network at any instant as Tellegen's Theorem*

$$\sum_{j=1}^{n} V_j(t)\, I_j(t) = 0 \qquad \text{... } (i)$$

Example 4.49. Determine the current in 1Ω resistor of the network shown in figure.

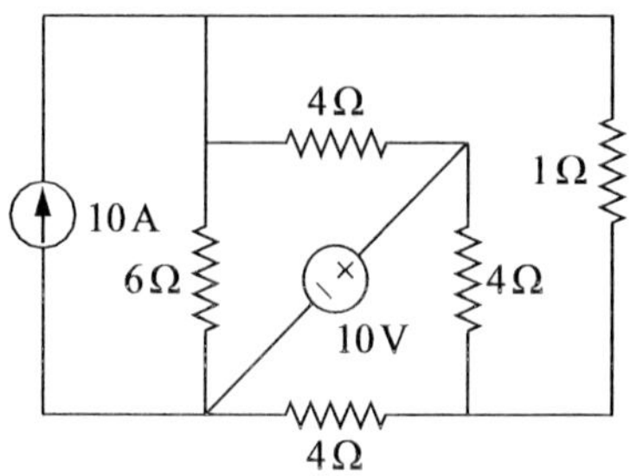

[UPTU, 2003]

Solution: By converting current source of 10 *A* and parallel resistance of 6Ω into voltage source, and redrawing the circuit we get:

writing KVL equations, we get:

$$6I_1 + 4(I_1 - I_2) + 10 = 60$$
$$6I_1 + 1 \times I_2 + 4(I_2 + I_3) = 60$$
$$4I_3 + 4(I_2 + I_3) = 10$$

or
$$10I_1 - 4I_2 = 60$$
$$6I_1 + 5I_2 + 4I_3 = 60$$
$$4I_2 + 6I_3 = 10$$

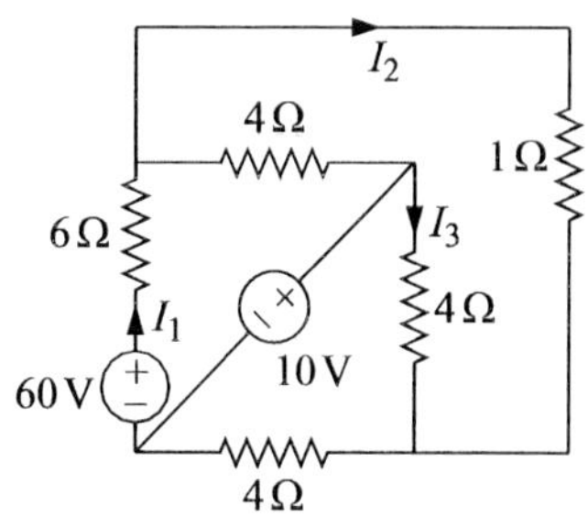

Using Euler's technique, we get:

$$I_2 = \frac{\begin{vmatrix} 10 & 50 & 0 \\ 6 & 60 & 4 \\ 0 & 10 & 8 \end{vmatrix}}{\begin{vmatrix} 10 & -4 & 0 \\ 6 & 5 & 4 \\ 0 & 4 & 8 \end{vmatrix}}$$

$$= \frac{10(60 \times 8 - \times 10) - 50\,(6 \times 8 - 0) + 0}{10(5 \times 8 - 4) + 4(6 \times 8) + 0}$$

$$= \frac{10(480-40)-50\times 48}{10(40-16)+4\times 48}$$

$$= \frac{10\times 440-50\times 48}{10\times 24+4+48}$$

$$= \frac{400\,(11-6)}{48\,(5+4)}$$

$$= \frac{400\times 5}{48\times 9} = \frac{125}{27}$$

or $I_2 = 4.63$ A

Hence, current through resistor 1Ω = 4.63 A

Example 4.50. Determine the necessary values of v and i in the network shown in the figure. **[UPTU, 2005]**

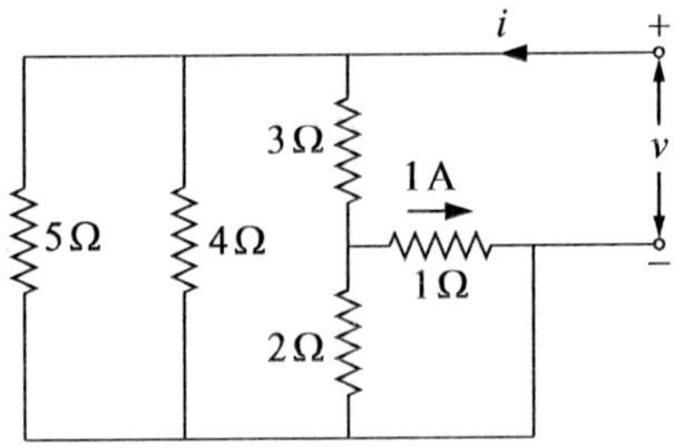

Solution: The circuit can be simplified shown in Fig. (a).

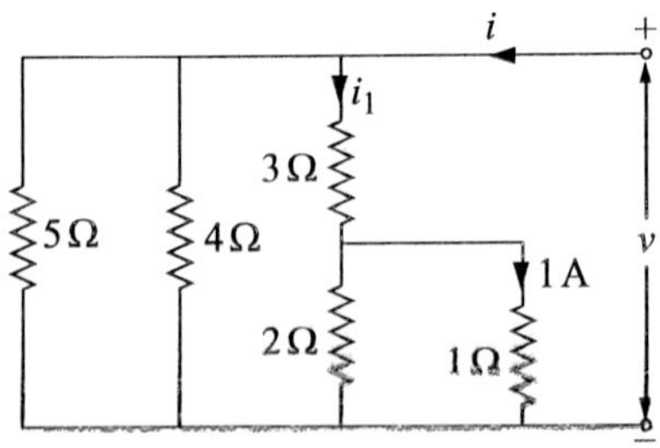

(a) Simplified circuit

Now, if current through 1 Ω resistor is 1A, the current through 2 Ω resistor will be $\frac{1}{2}$ A . Thus 3 current through 3Ω resistor shall be, $i_1 = \left(1+\frac{1}{2}\right) = \frac{3}{2}$A.

The circuit (a) can be reduced to Fig. (b).

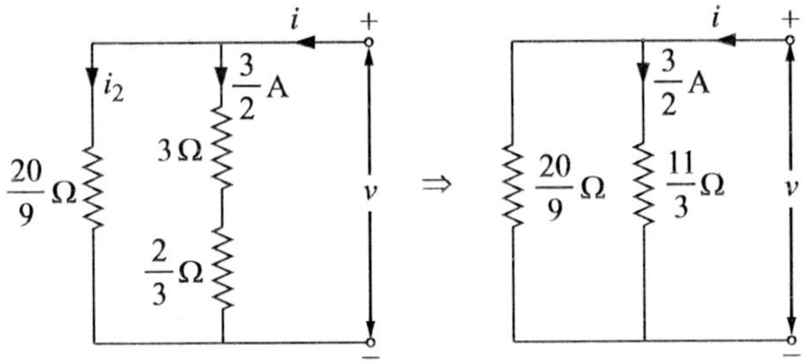

(b) *Further simplified circuit.*

Current through $\frac{20}{9}\Omega$ is given by:

$$i_2 = \left(\frac{11}{3}\times\frac{3}{2}\right)\times\frac{9}{20} = \frac{11}{2}\times\frac{9}{20}$$

or $$i_2 = \frac{99}{40}$$

Thus, $$i = i_1 + i_2 = \frac{3}{2}+\frac{99}{40}$$

$$= \frac{60+99}{40} = \frac{159}{40}\text{A}$$

or $$i = \frac{159}{40}\text{A}$$

Net resistance across can be given as:

$$R_{TH} = \frac{\frac{20}{9}\times\frac{11}{3}}{\frac{20}{9}+\frac{11}{3}} = \frac{20\times 11}{60+99} = \frac{220}{159}$$

∴ $$v = iR_{TH} = \frac{159}{40}\times\frac{220}{159} = \frac{11}{2}$$

or $$v = 5.5 \text{ volt.}$$

Example 4.51. For the circuit shown in figure, find the voltage across 4 Ω resistor by using nodal analysis. **[UPTU, 2005]**

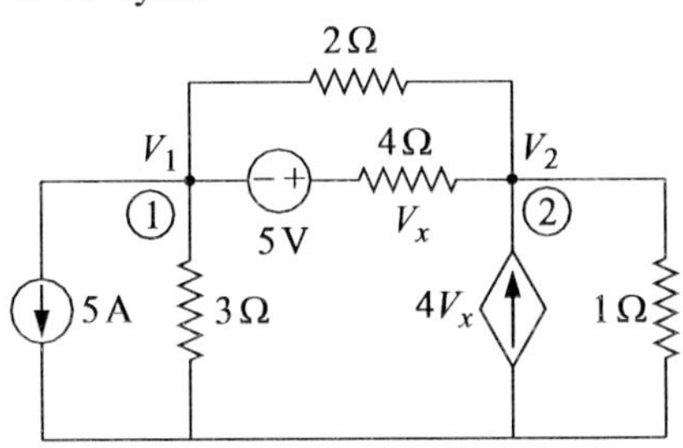

Solution: Nodes (1) and (2) have voltages V_1 and V_2 respectively.

The voltage $V_x = V_1 + 5 = -V_2$ ... (*i*)

By writing KCL equation at node 1, we get

$$\frac{V_2 - V_1}{2} = 5 + \frac{V_1}{3} + \frac{V_1 + 5 - V_2}{4}$$

or $-13V_1 + 9V_2 = 75$ ... (*ii*)

By writing KCL equation at node 2, we get:

$$\frac{V_2 - V_1}{2} + \frac{V_2}{1} = 4V_x + \frac{V_1 + 5 - V_2}{4}$$

or $3V_1 - 7V_2 + 16V_x = -5$ ... (*iii*)

By replacing value of V_x from equation (*i*) is (*iii*), we get:

$3V_1 - 7V_2 + 16(V_1 + 5 - V_2)$

or $19V_1 - 23V_2 = -85$ .... (*iv*)

Using Euler's technique in equation (*ii*) and (*iv*) we get:

$$V_1 = \frac{\begin{vmatrix} 75 & 9 \\ -85 & -23 \end{vmatrix}}{\begin{vmatrix} -13 & 9 \\ 19 & -23 \end{vmatrix}} = \frac{-75 \times 23 + 9 \times 85}{13 \times 23 - 9 \times 9}$$

$$= -\frac{960}{128}$$

or $V_1 = -7.5\ V$

By substituting V_1 in equation (*ii*), we get:

$$(-13 \times -7.5) + 9V_2 = 75$$

or $$V_2 = \frac{75 - 13 \times 7.5}{9}$$

or $V_2 = -2.5$ V

Hence, from equation (*i*) we get:

$$V_x = -7.5 + 5 - (-2.5)$$

i.e., $V_x = 0$

Example 4.52 Determine the current in the capacitor branch by the superposition theorem in the circuit shown in the figure. **[UPTU, 2001]**

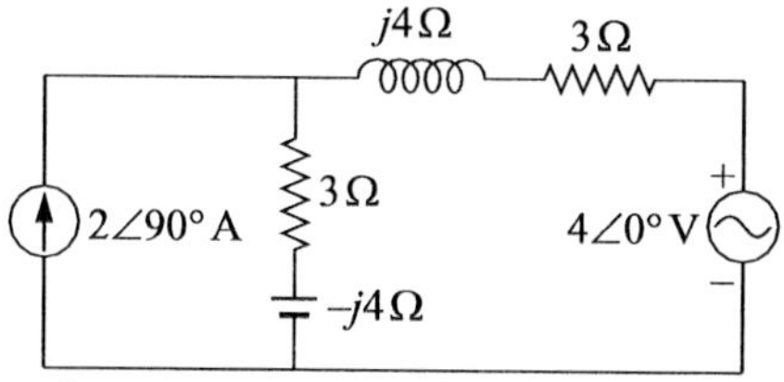

Solution: Consider only current source and in that case circuit becomes as shown in the Fig. (*a*).

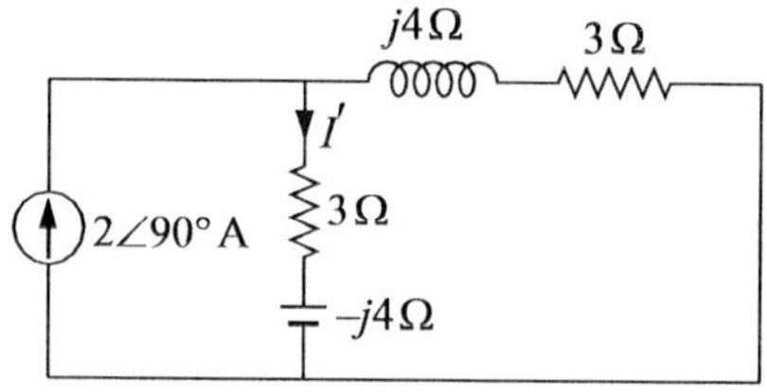

(*a*) *Only current source exists and voltage source becomes just a shorting*

$$I = 2\angle 90^\circ \times \left[\frac{3+j4}{(3+j4)\,(3-j4)}\right]$$

$$= \frac{2\angle 90^\circ \times 5\angle 33.13^\circ}{6\underline{|0^\circ}}$$

$$= 1.6667\ \underline{|123.130}$$

$$= (-\,1.3333 + \mathrm{j}1)\ \mathrm{A}$$

or $$I' = \left(-\frac{4}{3} + j\right)A$$

Now, consider only voltage source and current source becomes open circuit. The circuit obtained in shown in Fig. (*b*).

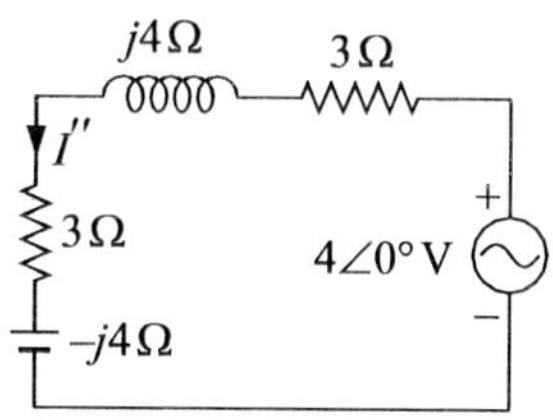

$$I'' = \frac{4\angle 0^\circ}{(3-j4)+(3+j4)} = \frac{4}{6}$$

or $$I'' = \frac{2}{3}\mathrm{A}$$

The net current through capacitor = $I' + I''$

$$= -\frac{4}{3} + j + \frac{2}{3} = -\frac{2}{3} + j = 1.2018\ \angle 123.68^\circ\ \mathrm{A}$$

Example 4.53. Determine current I in figure, using super position theorem.

[UPTU, 2003]

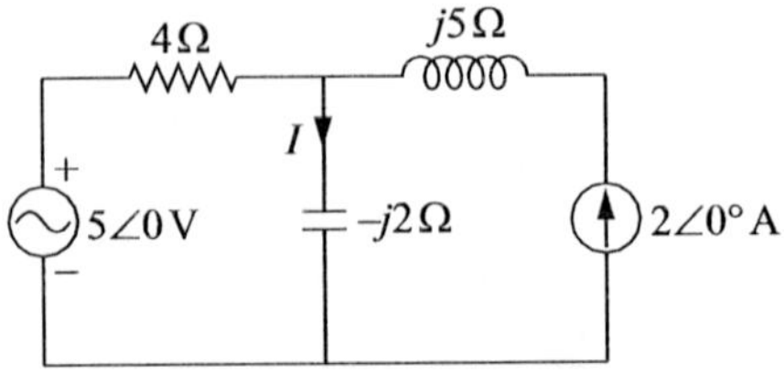

Solution: Consider only current source as shown in circuit of Fig. (*a*) with voltage source taken as short circuit.

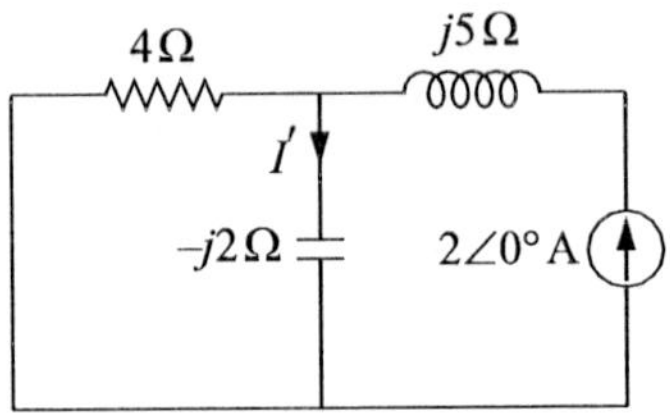

(*a*) *Circuit with only current source*

∴ Current through capacitor,

$$I' = 2\times[4\,||\,(-j2)]\times\frac{1}{(-j2)}$$

$$= 2\times\frac{4\times(-j2)}{4-2j}\times\frac{1}{(-j2)}$$

or $$I' = \frac{4}{2-j}\,\text{A}$$

Now, consider only voltage source as shown in Fig. (*b*) where in current source is taken as open circuited.

∴ Current through capacitor

$$I'' = \frac{5}{4-2j}\,\text{A}$$

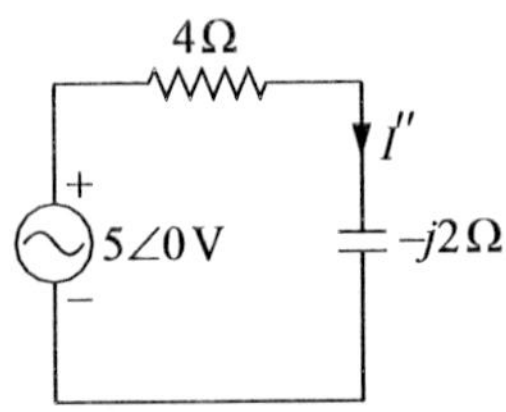

(*b*) *Circuit with only voltage source*

The net current I through capacitor,

$$I` = I' + I''$$

$$= \frac{4}{2-j} + \frac{5}{4-2j}$$

$$= \frac{1}{(2-j)}\left[4+\frac{5}{2}\right]$$

$$= \frac{6.5}{2-j} = \frac{6.5\,(2+j)}{4+1} = 1.3\,(2+j)$$

or $$I = 2.9\,\angle 26.56^\circ\ \text{A}$$

Example 4.54. For the circuit shown in figure, obtain the voltage across each current source, using superposition theorem. **[UPTU, 2004]**

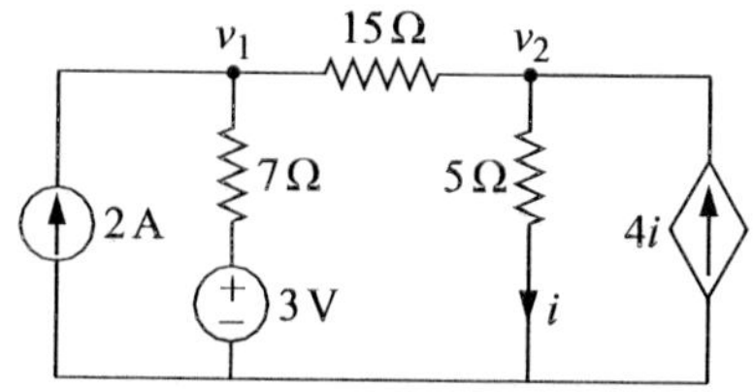

Solution. Consider that only 3V source remains in the circuit, hence, the circuit can be reduced to Fig.(*a*) and dependent current source value becomes $4i_1$.

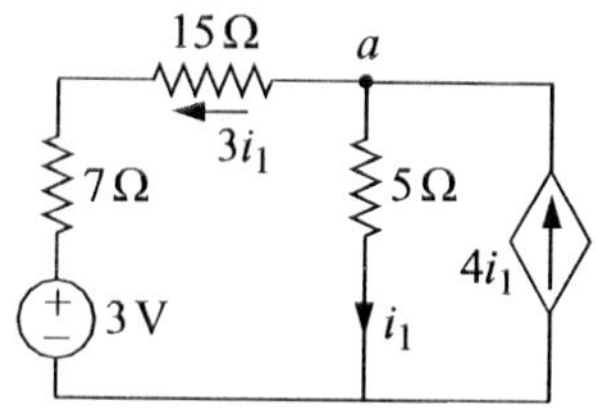

(*a*) *Only 3V source considered.*

Using KVL we get:

$$(15 + 7)(-3i_1) + 5i_1 = 3$$

or
$$i_1 = -\frac{3}{61}\text{ A} \qquad \text{... } (i)$$

Now, consider that only 2A source remains in the circuit, hence, the circuit can be reduced to Fig.(*b*) and dependent current source value becomes $4i_2$

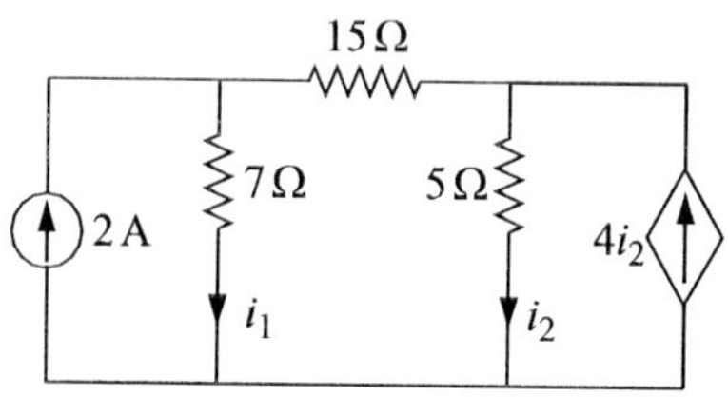

(*b*) *Only current source 2A remains*

By converting 2A current source and parallel resistance 7Ω into Thevenin's equivalent, the circuit becomes as shown in Fig.(*c*).

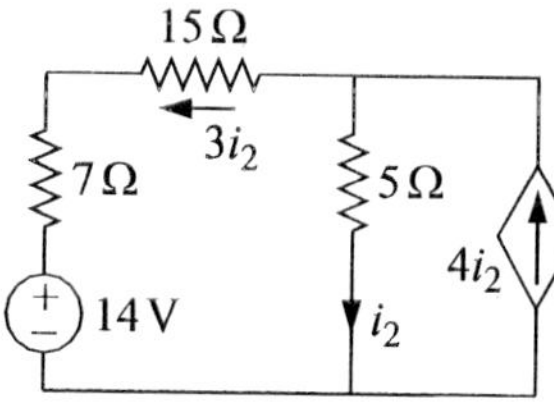

(*c*) *Current source 2A converted into Thevenin's equivalent*

Using KVL equation, we get:

$$(15 + 7)(-3i_2) + 5i_2 = 14$$

or $$i_2 = -\frac{14}{61} \text{ A.} \quad \text{... (ii)}$$

Thus, now current using superposition theorem, $I = i_1 + C_2 = -\frac{3}{61} - \frac{14}{61} = -\frac{17}{61}$ A. The complete circuit is redrawn and labelled as shown in Fig. (*d*).

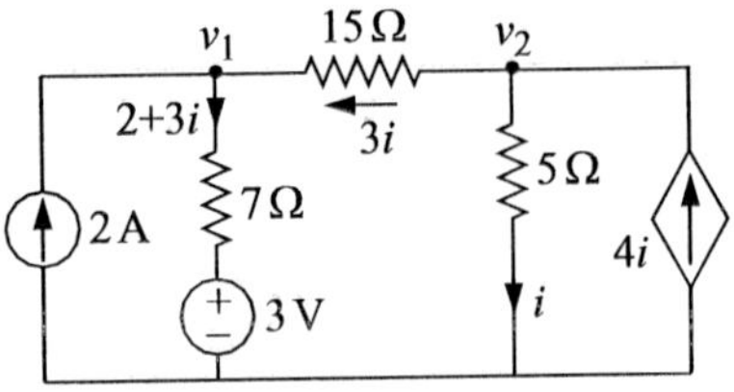

$$\therefore \quad v_1 = 7(2 + 3i) + 3 = 7\left(2 - \frac{17 \times 3}{61}\right) + 3$$

$$= 14 - \frac{21 \times 17}{61} + 3 = 17 - \frac{357}{61}$$

$$= \frac{1037 - 357}{61} = \frac{680}{61}$$

or $$v_1 = 11.15\text{V}$$

and $$v_2 = 5\left(-\frac{17}{61}\right) = -\frac{85}{61}$$

or $$v_2 = -1.39 \text{ V}$$

Example 4.55. Using superposition theorem, calculate the current through $(2 + j3)\ \Omega$ impedence of the circuit shown in the figure. **[UPTU, 2004]**

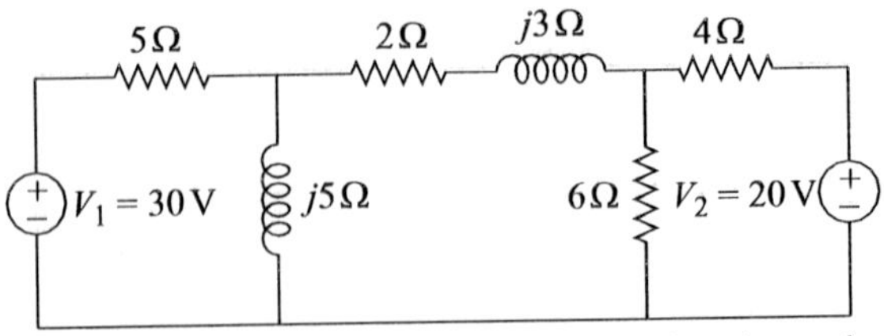

Solution: Consider that source V_2 remains in the circuit as shown in Fig. (*a*).

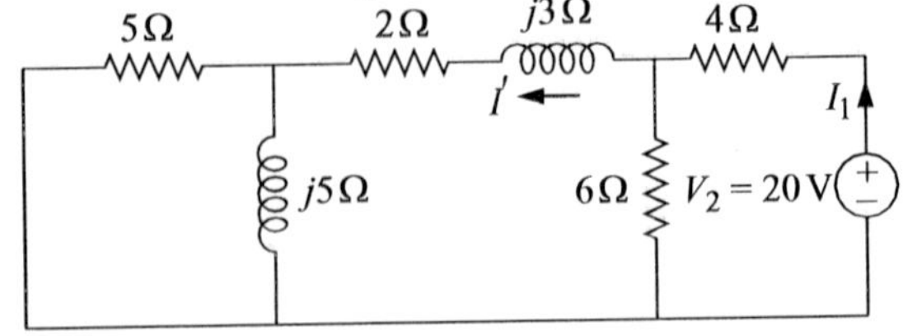

(*a*) Only V_2 source remains

$$I_1 = \frac{20}{4+[6 \| (5 \| j5)+(2+3j)3]}$$

$$= \frac{20}{\left[4+6\|\left(\frac{5j}{1+j}+2+3j\right)\right]} = \frac{20}{4+6\|\left(\frac{-1+10j}{1+j}\right)}$$

$$= \frac{20}{4+6\|(4.5+5.5j)} = \frac{20}{\left(4+\frac{6(4.5+5.5j}{6+4.5+5.5j}\right)}$$

$$= \frac{20(10.5+5.5j)}{42+22j+27+33j} = \frac{20(10.5+5.5j)}{69+55j}$$

$\therefore$ $$I' = \frac{6}{6+4.5+5.5j} \times I_1$$

$$= \frac{6}{(10.5+5.5j)} \times \frac{20(10.5+5.5j)}{69+55j}$$

$$I' = \frac{120}{69+55j} \qquad \text{... (i)}$$

Now, consider that only 30V source remains in the circuit as shown in the Fig. (*b*).

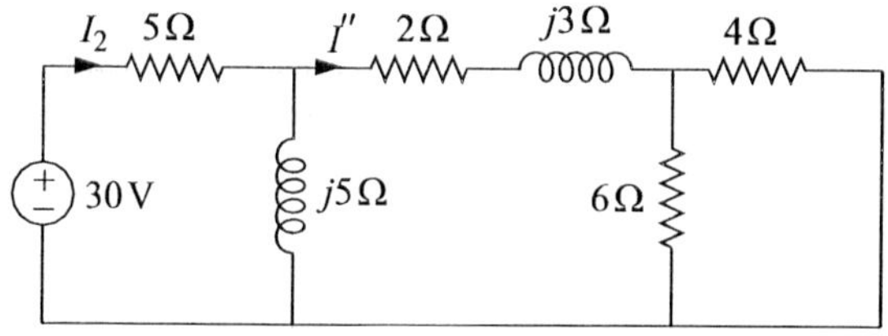

(*b*) *Only 30 V source remains*

$$I_2 = \frac{30}{5+[\{(4\|6)+(2+3j)\}\|5j]}$$

$$= \frac{30}{5+\{(2.4+2+3j)\|5j\}} = \frac{30}{5+\{4.4+3j)\|5j\}}$$

$$= \frac{30}{5+\left(\frac{5j(4.4+3j}{4.4+3j+5j}\right)}$$

or $$I_2 = \frac{30(4.4+8j)}{5(4.4+8j)+5j(4.4+3j)}$$

or $$I'' = \frac{5j}{5j+(2+3j)+2.4} \times I_2$$

$$\text{or} I'' = \frac{5j}{\{5j+(2+3j)+2.4\}} \times \frac{30(4.4+8j)}{5(4.4+8j)+5j(4.4+31)}$$

$$= \frac{30j}{4.4+8j+4.4j-3}$$

or $$I' = \frac{30j}{1.4+12.4j}$$

$\therefore$ $$I = \frac{30j}{1.4+12.4j} - \frac{120}{69+j55}$$

$$= 2.40\angle 6.4^\circ - 1.36\angle -37.6^\circ$$

$$= 2.39 + 0.27j - 1.06 + 0.85j$$

$$= 1.33 + 1.12j$$

or $$I = 1.74\angle 40^\circ \text{ A}$$

Example 4.56. In the network of the figure, verify the substitution theorem by replacing the 6Ω-resistor by a voltage source. **[UPTU, 2005]**

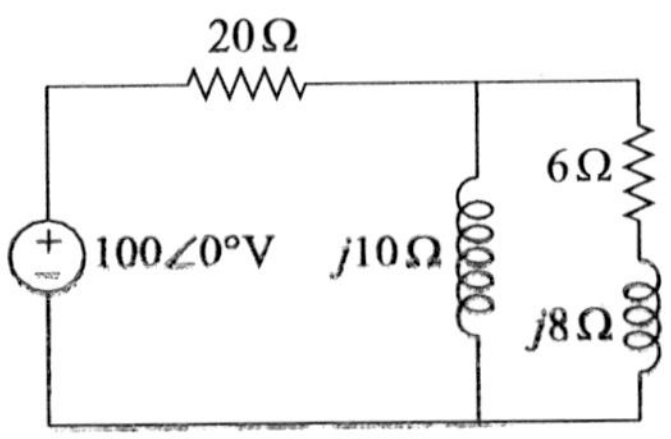

Solution: Net current from the voltage source as per Fig. (*a*)

$$I_s = \frac{100}{20 + \dfrac{j10 \times (6+j8)}{j10+6+j8}} = \frac{10(6+18j)}{(4+42j)}$$

or $$I_s = 4.50\angle -13^\circ \text{ A}$$

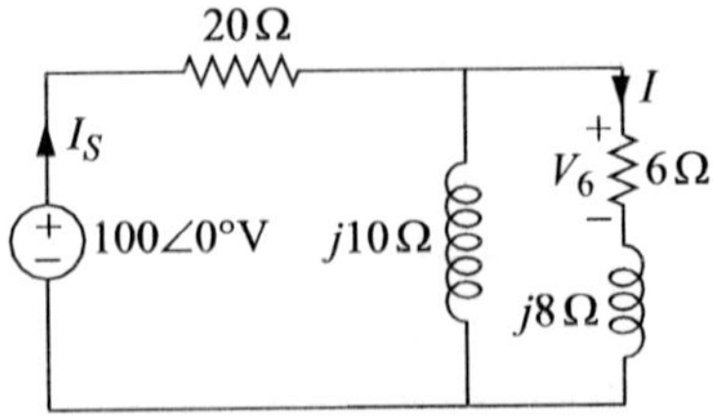

(*a*) *Current in the given circuit*

The current, $$I = I_s \frac{j10}{j10+6+j8}$$

$$= 4.50\angle{-13^\circ} \times \frac{10\angle 90^\circ}{18.97\angle 71.56^\circ}$$

or $$I = 2.37\angle 5.5 \text{ A}$$

The voltage drop across 6Ω resistor,

$$V_6 = 6(2.37\angle 5.5^\circ) = 14.22\angle 5.5^\circ \text{ Volt.}$$

Now, we replace 6 Ω resistor by a voltage source, $V_6 = 14.22\angle 5.5^\circ$ V as shown in Fig. (*b*).

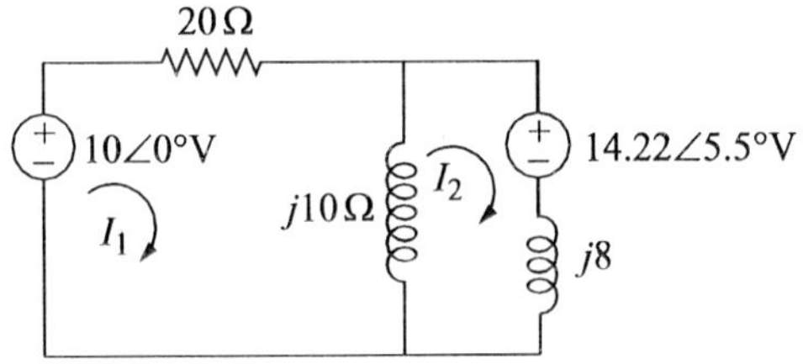

(*b*) *6 Ω resistor replaced a voltage source*

The mesh equations for both the loops:

$$(20 + j10)\, I_1 - j10\, I_2 = 100 \quad \text{... } (i)$$

and $$j8\, I_2 + j10\,(I_2 - I_1) = -\,14.22\angle 5.5^\circ$$

or $$-j10\, I_1 + j18\, I_2 = -\,14.22\angle 5.5^\circ \quad \text{... } (ii)$$

From equations (*i*) and (*ii*), we get:

$$I_1 = \frac{\begin{vmatrix} 100 & -j10 \\ -14.22\angle 5.5^\circ & j18 \end{vmatrix}}{\begin{vmatrix} 20+j10 & -j10 \\ -j10 & j18 \end{vmatrix}} = 4.50\angle{-13^\circ} \text{ A}$$

and $$I_2 = \frac{\begin{vmatrix} 20+j10 & 100 \\ -j10 & -14.22\angle 5.5^\circ \end{vmatrix}}{\begin{vmatrix} 20+j10 & -j10 \\ -j10 & j18 \end{vmatrix}} = 2.37\angle 5.5^\circ \text{ A}$$

We can see that $I_1 = I_s$ and $I_2 = I$, hence, substitution theorem stands verified.

Example 4.57. Using Thevenin's theorem and superposition theorem, convert the active network shown in the figure by a single voltage source in series with equivalent impedance and hence, find the current '*i*' when *A*-*B* short circuited.

[UPTU, 2005]

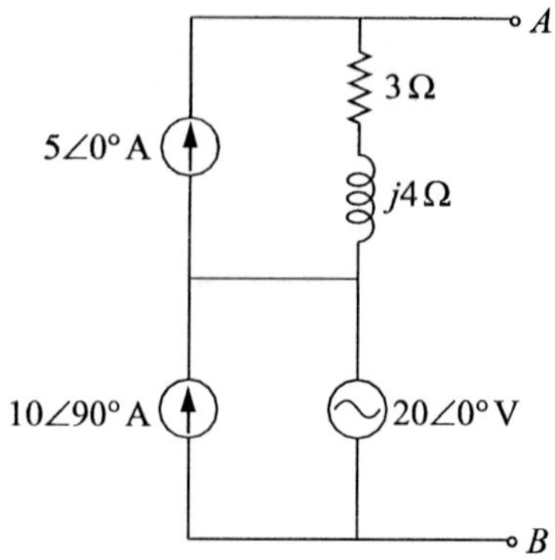

Solution: First let us find out Z_{TH} and for this all the energy sources will be removed. Thus, the resultant circuit is shown in Fig. (*a*).

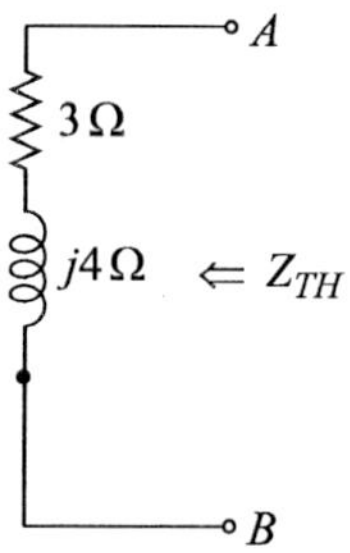

Fig. (*a*) *Circuit for* Z_{TH}

$$\therefore \qquad Z_{TH} = (3 + 4j)\ \Omega$$

The value of V_{TH} can be found out using superposition theorem. Consider that only voltage source is present as shown in Fig. (*b*).

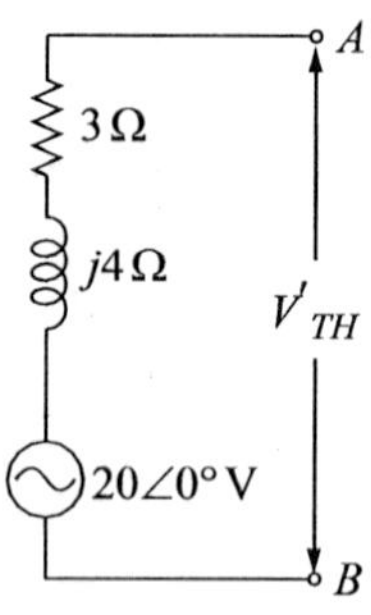

(*b*) *Only voltage source is present*

$$\therefore \qquad V'_{TH} = 20\angle 0^\circ \text{ V} \qquad \text{... (}i\text{)}$$

Now, consider that only $5\angle 0^\circ$ A current source is present and circuit becomes as shown in Fig. (*c*).

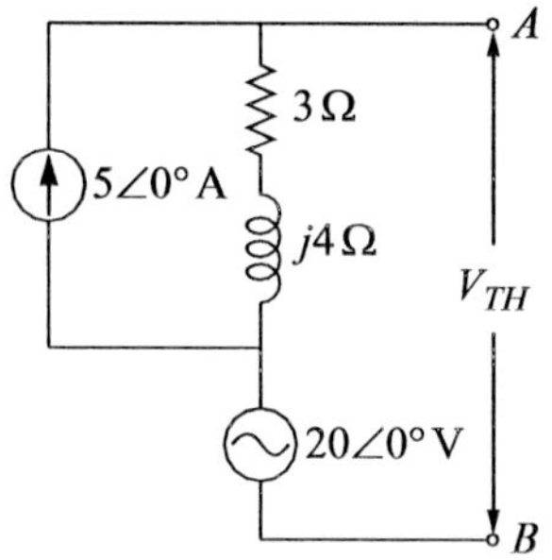

(*c*) *Only* 5∠0° A *current source present*

$$\therefore \qquad V''_{TH} = 5(3 + 4j)\ \text{V}.$$

Finally, consider that only 10∠90° A current source is present as shown in Fig. (*d*).

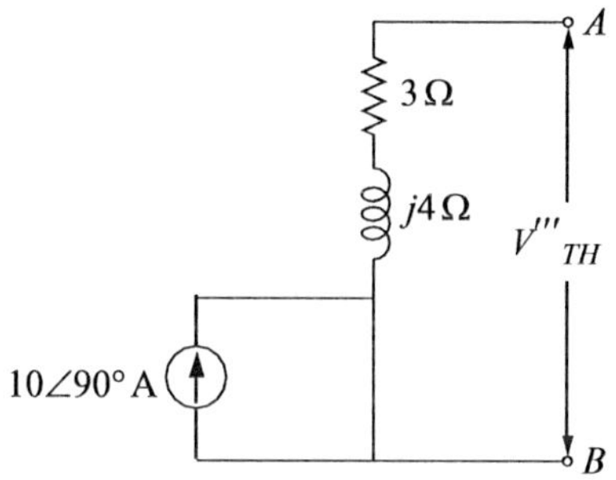

(*d*) *Only* 10|90° A *current source present*

The current source 10∠90° A is short circuited in the resultant circuit, thus,

$$V'''_{TH} = 0$$

Applying superposition theorem, the Thevenin's equivalent voltage source is given by:

$$\begin{aligned} V_{TH} &= V'_{TH} + V''_{TH} + V'''_{TH} \\ &= 20 + 5(3 + 4j) + 0 \\ &= 20 + 15 + 20j = 35 + 20\,j \end{aligned}$$

$$\text{or} \qquad V_{TH} = 40.31\angle 29.74^\circ\ \text{V}$$

Hence, Thevenin's equivalent of the network given is shown in Fig. (*e*).

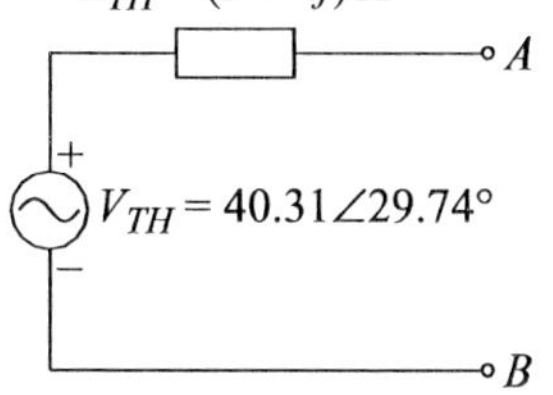

(*e*) *Thevenin's equivalent*

If A–B is shorted, then the shorting current is given by:

$$I_{sc} = \frac{V_{TH}}{Z_{TH}} = \frac{5(7+4j)}{(3+4j)} = \frac{5(7+4j)(3-4j)}{9+16}$$

$$= \frac{5(21+16+12j-28j)}{25}$$

$$= \frac{37-16j}{5}$$

$$= (7.4 - 3.2j)$$

or $$I_{sc} = (7.4 - 3.2j) \text{ A}$$

From the given circuit:

$$5\angle 0^\circ + i = I_{sc}$$

∴ Current, $$i = (I_{sc} - 5)$$

$$= (7.4 - 3.2j) - 5 = 2.4 - 3.2j$$

or $$i = 4\angle -53.13^\circ \text{A}$$

Example 4.58. Find the current through the capacitor and voltage across 4 Ω resistance of the ac network shown in the figure by using superposition theorem.

[UPTU, 2006]

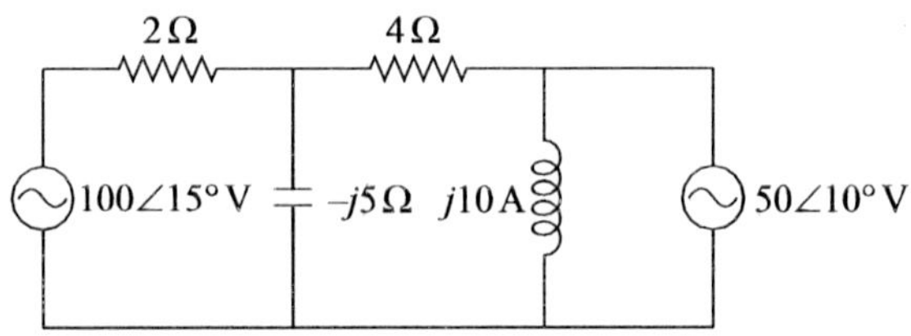

Solution: Consider that only 50 ∠10° V source remains in the circuit as shown in Fig. (*a*).

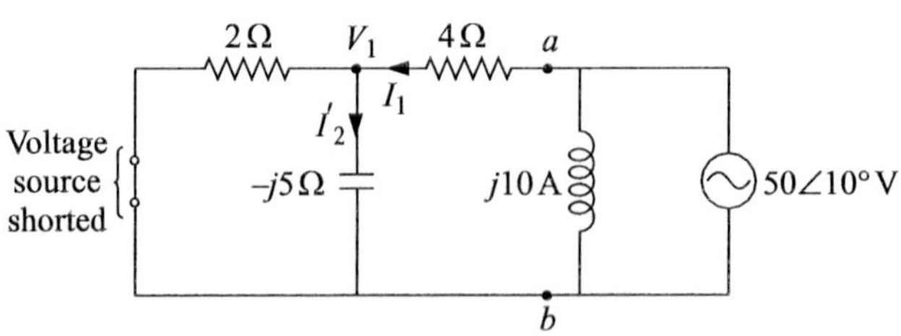

(*a*) Only 50 ∠10° V source present in the circuit

The impedance left of $a - b$ is given by:

$$Z_{ab} = (2 \parallel -j5) + 4 = \frac{8-30j}{2-5j}$$

$$= \frac{31.05\angle{-75.07^\circ}}{5.385\angle{-68.20^\circ}} = 5.766\angle{-6.87^\circ}$$

Current through 4Ω resistor,

$$I_1 = \frac{50\angle 10^\circ}{5.766\angle{-6.87^\circ}} = 8.672\angle 16.87^\circ \text{ A}$$

$$I_2 = \frac{2}{2-j5} \times I_1' = \frac{2\times 8.672\angle 16.87^\circ}{5.385\angle{-68.20^\circ}}$$

or $$I_2' = 3.221\angle 85.07^\circ \text{ A}$$

Voltage across capacitor,

$$V_1 = 4I_1' = 34.69\angle 16.87^\circ \text{ V}$$

Now, consider that only $100\ \underline{|150}$ V source is present as shown in Fig. (*b*).

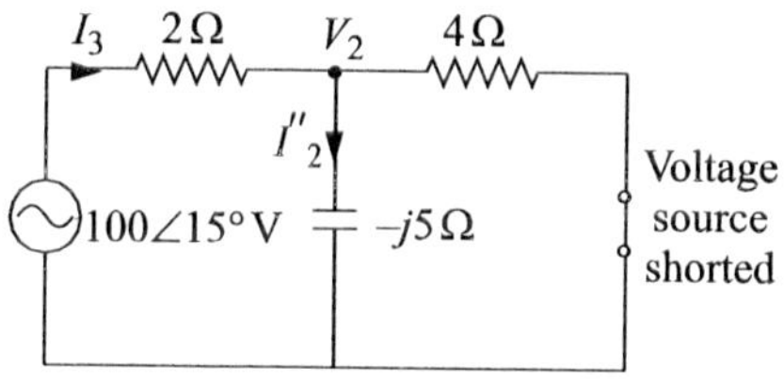

(*b*) Only $100\angle 15^\circ$ V source is converted

Net impedance across the source,

$$Z = 2 + (-j_5 \parallel 4) = 2 + \frac{-5j\times 4}{4-5j} = \frac{8-30j}{4-5j}$$

$$= \frac{31.05\angle{-75.07^\circ}}{6.403\angle{-51.34^\circ}}$$

$$= 4.849\angle{-23.73^\circ}\ \Omega$$

Current $$I_3 = \frac{100\angle 15^\circ}{2} = \frac{100\angle 15^\circ}{4.849\angle{-23-75^\circ}}$$

$$= 20.623\angle 38.73^\circ \text{ A}$$

The current through capacitor,

$$I_2'' = \frac{4}{4-5j} \times I_3 = \frac{4\times 20.623\angle 38.73^\circ}{6.403\angle{-51.34^\circ}}$$

or $$I_2'' = 12.88\angle 90.07^\circ = j12.88 \text{ A}$$

The voltage across 4Ω resistor,

$$V_2 = -j5 \times I_2'' = -j5 \times j12.88 = 64.4 \angle 0 \text{ V}$$

Using superposition theorem, total current through capacitor,

$$I_2 = I_2' + I_2'' = j12.88 + 3.221 \angle 85.0^\circ$$
$$= 0.2768 + j16.089 = 16.09 \angle 89^\circ \text{ A}$$

or $$I_2 = 16.09 \angle 89^\circ \text{ A}$$

and the total voltage across the capacitor,

$$V = V_2 - V_1$$

is complex number and subtracting

$$= 64.4 - 34.69 \angle 16.87$$

$= 31.2 - j10.07$ (by converting in complex number and then solving)

$$= 32.78\angle -17.89 \text{ V}$$

Example 4.59. Find voltage 'v' using superposition theorem of the network shown in the figure. **[UPTU, 2007]**

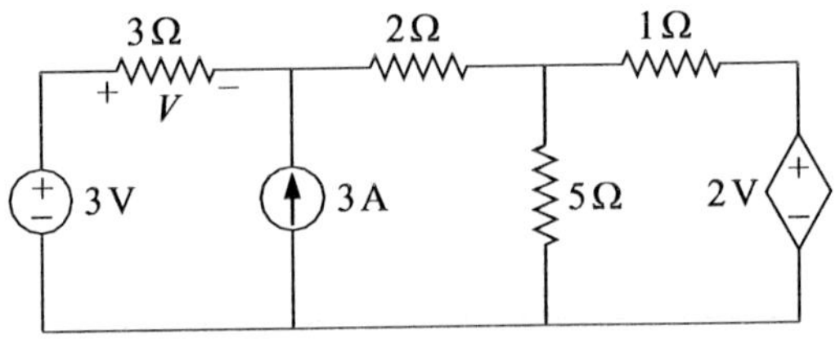

Solution: Consider that only 3 volts source is present as shown in the Fig. (a).

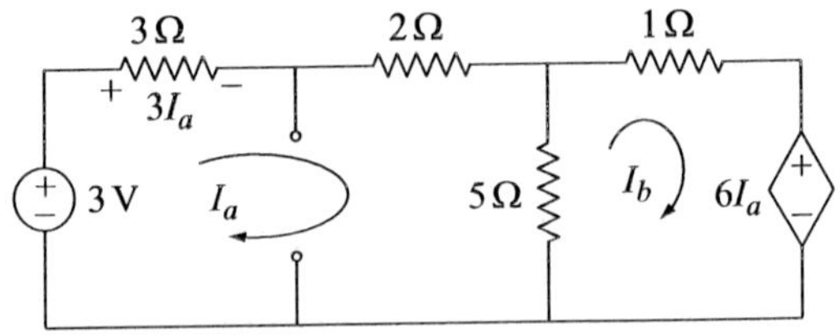

(a) *Only 3 volts source*

From KVL equations for the loops:

$$(3 + 2)\, I_a + 5(I_a - I_b) = 3$$

or $$5I_a + 5I_a - 5I_b = 3$$

or $$10I_a - 5I_b = 3 \quad \text{... }(i)$$

and $$1 \times I_b + 5(I_b - I_a) = -6I_a$$

or $$I_b + 5I_b - 5I_a = -6I_a$$

or $$I_a + 6I_b = 0 \quad \text{... (ii)}$$

From equations (*i*) and (*ii*), we get:

$$I_a = \frac{\begin{vmatrix} 3 & -5 \\ 0 & 6 \end{vmatrix}}{\begin{vmatrix} 10 & -5 \\ 1 & 6 \end{vmatrix}} = \frac{18}{60+5}$$

or $$I_a = \frac{18}{65}\text{ A}$$

Now, consider that only 3A source is present as shown in Fig. (*b*).

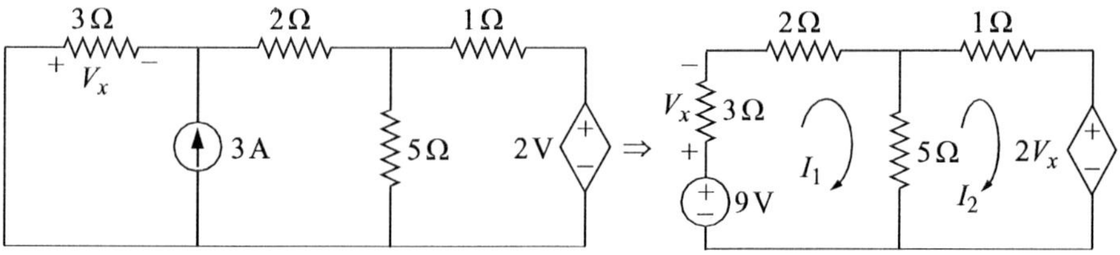

(b) *Only current source is present*

$$(3 + 2)\, I_1 + 5(I_1 - I_2) = 9$$

or $$5I_1 + 5I_1 - 5I_2 = 9$$

or $$10I_1 - 5I_2 = 9 \quad \text{... (iii)}$$

and $$1 \times I_2 + 5(I_2 - I_1) = -\,2(3I_1)$$

or $$I_2 + 5I_2 - 5I_1 = -\,6I_1$$

$$I_1 + 6I_2 = 0 \quad \text{... (iv)}$$

From equations (*iii*) and (*iv*), we get:

$$I_1 = \frac{\begin{vmatrix} 9 & -5 \\ 0 & 6 \end{vmatrix}}{\begin{vmatrix} 10 & -5 \\ 1 & 6 \end{vmatrix}} = \frac{54}{60+5}$$

or $$I_1 = \frac{54}{65}\text{ A}$$

By superposition theorem, we get:

$$V = 3(I_a + I_1) = 3\left(\frac{18}{65} + \frac{54}{65}\right)$$

$$= 3 \times \frac{72}{65} = \frac{216}{65}$$

or $$V = 3.323 \text{ Volt.}$$

Example 4.60. Replace the network to the left of terminals in the given figure by Thevenin's equivalent circuit. Find the current and power delivered to the load $Z_L = (4 + j4)$ *W* **[UPTU, 2002]**

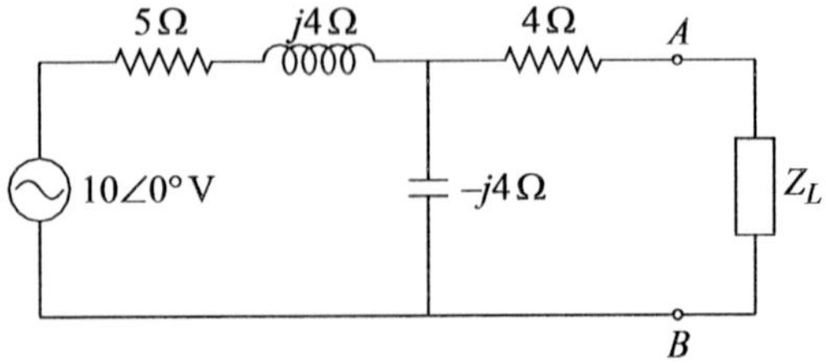

Solution: The value of V_{TH} and Z_{TH} will be found as shown in Figs. (*a*) and (*b*).

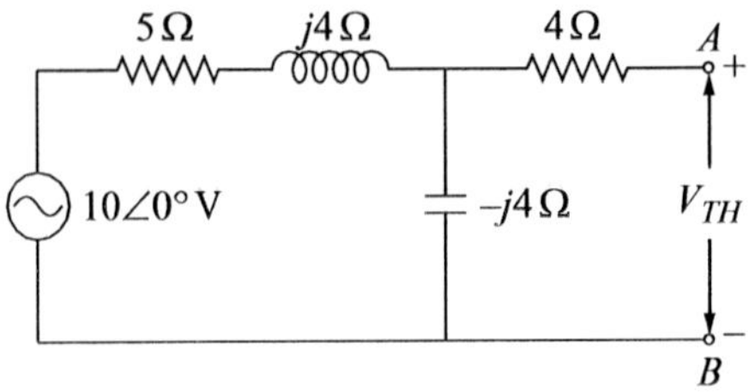

(*a*) *Circuit for* V_{TH}

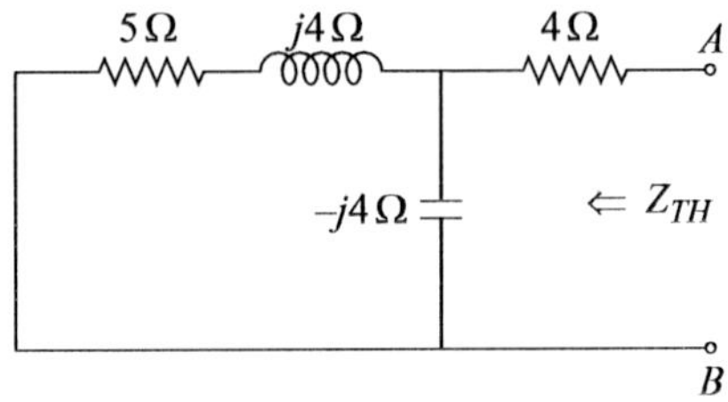

(*b*) *Circuit for* Z_{TH}

From figure (*a*), we get:

$$V_{TH} = 10\angle 0^\circ \times \left[\frac{-j4}{5 + j4 - j4}\right]$$

or $$V_{TH} = -j8 \text{ V}$$

From figure (*b*), we get:

$$Z_{TH} = [(5 + j4) \parallel (-j4)] + 4$$

$$= \frac{(5 + j4)(-j4)}{5 + j4 - j4} + 4 = \frac{16 - 20j}{5} + 4$$

$$= \frac{16 - 20j + 20}{5} = \frac{36 - 20j}{5}$$

or $$Z_{TH} = (7.2 - j4)\ \Omega$$

The Thevenin's equivalent circuit is shown in Fig. (*c*).

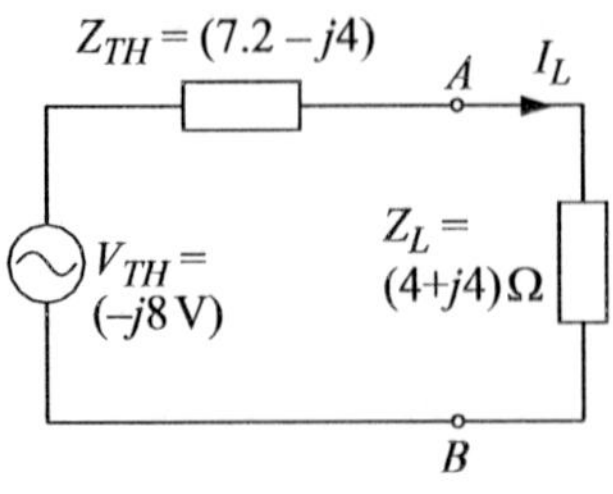

(*c*) *Thevenin's equivalent circuit with the load* Z_L

$\therefore$ Current, $$I_L = \frac{(-j8)}{(7.2 - j4) + (4 + j4)} = \frac{-j8}{11.2} = \frac{-j}{1.4}$$

and power derived, $P_L = I^2_L R_L = \left(-\frac{j}{1.4}\right)^2 \times 4 = 2.04$ W

Example 4.61. Obtain Thevenin's equivalent circuit shown in the figure given.

[UPTU, 2003]

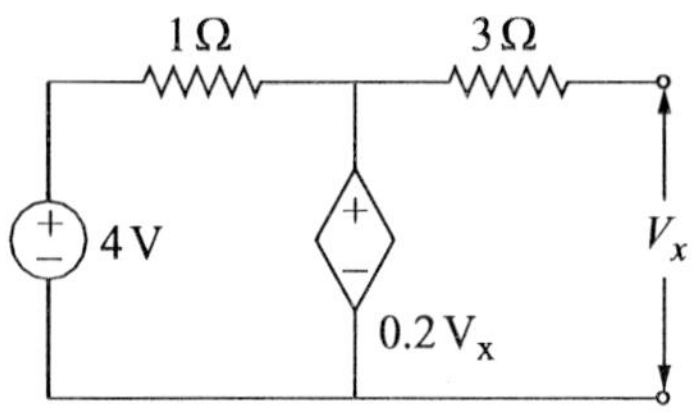

Solution: The current 0.2 V_x will flow through 1 Ω resistor, keeping in mind that terminals across V_x are open as shown in circuit (*a*).

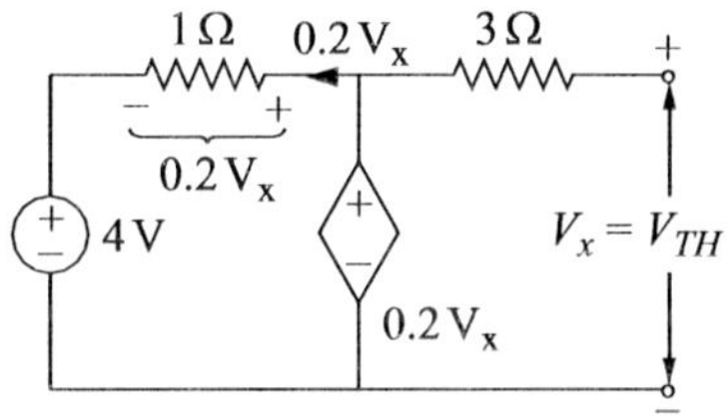

(*a*) Circuit with current 0.2 V_x through 1 Ω resistor.

$$\therefore \quad V_x = 4 + 0.2\ V_x$$

$$\text{or} \quad 0.8\ V_x = 4 \text{ or } V_x = 5\text{ V}$$

as load is not known, therefore, *AB* should be shorted and short circuit current calculated as shown in Fig. (*b*).

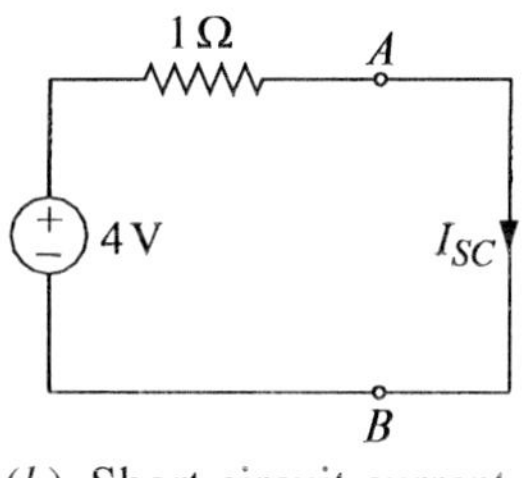

(*b*) Short circuit current.

$$\therefore \quad I_{sc} = \frac{4\text{ V}}{1\,\Omega} = 4\text{ A}.$$

$$\therefore \quad Z_{TH} = \frac{V_{TH}}{I_{sc}} = \frac{5}{4} = 1.25\,\Omega.$$

The Thevenin's equivalent circuit is shown in Fig. (*c*).

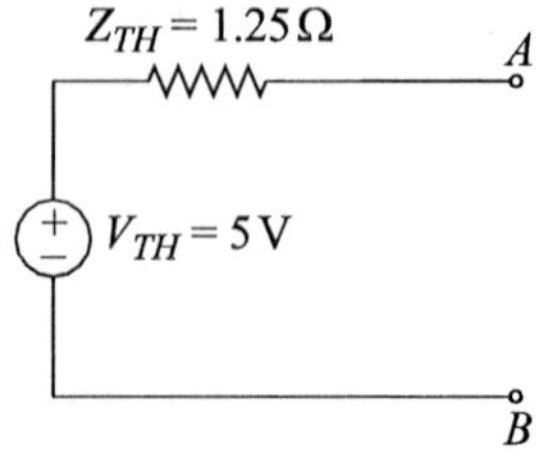

Example 4.62. Find the Thevenin's equivalent for the network shown in the figure at the right of terminals *a*–*b* and hence find the current.

[UPTU, 2004]

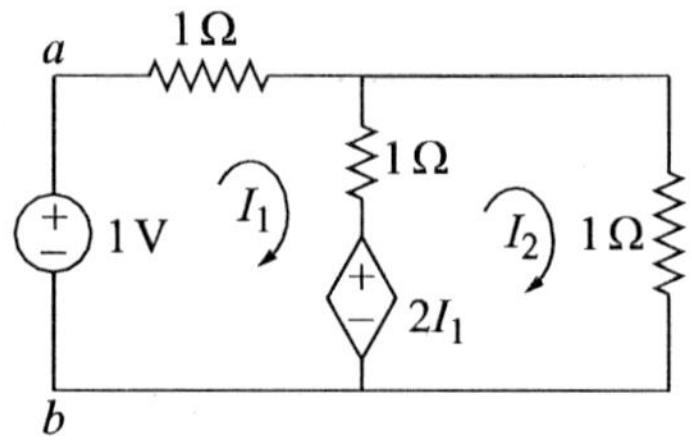

Solution: In view of circuit having dependent source also, therefore, current I_1 is to be found out and then $R_{TH} = \dfrac{V}{I_1}$.

By writing mesh equations, we get:

$$1 \times I_1 + 1 \times (I_1 - I_2) + 2I_1 = V$$

or $$I_1 + I_1 - I_2 + 2I_1 = 1$$

or $$4I_1 - I_2 = V \qquad \text{... (i)}$$

and $$1 \times I_2 - 2I_1 + 1 \times (I_2 - I_1) = 0$$

or $$I_2 - 2I_1 + I_2 - I_1 = 0$$

$$-3I_1 + 2I_2 = 0 \qquad \text{... (ii)}$$

Using Euler's method, we get:

$$I_1 = \frac{\begin{vmatrix} 1 & -1 \\ 0 & 2 \end{vmatrix}}{\begin{vmatrix} 4 & -1 \\ -3 & 2 \end{vmatrix}} = \frac{2}{8-3}$$

$$= \frac{2}{5}\,\text{A}$$

$\therefore$ $$R_{TH} = \frac{1\,\text{V}}{\frac{2}{5}\,\text{A}} = \frac{5}{2}\,\Omega$$

Voltage across a–b is 1V, hence, $V_{TH} = 1\text{V}$.

Thus, Thevenin's equivalent is

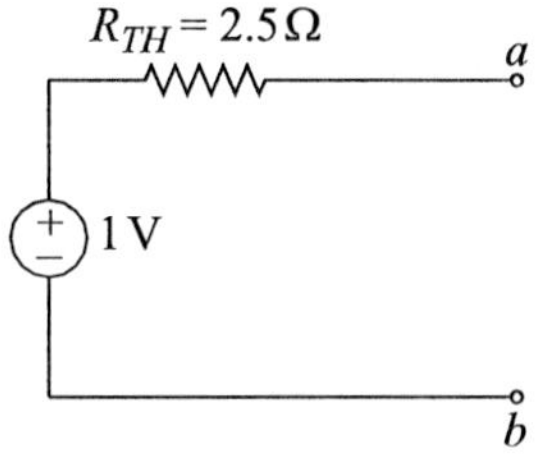

Example 4.63. In the network shown in the figure, find Thevenin's equivalent network across A–B terminals. **[UPTU, 2006]**

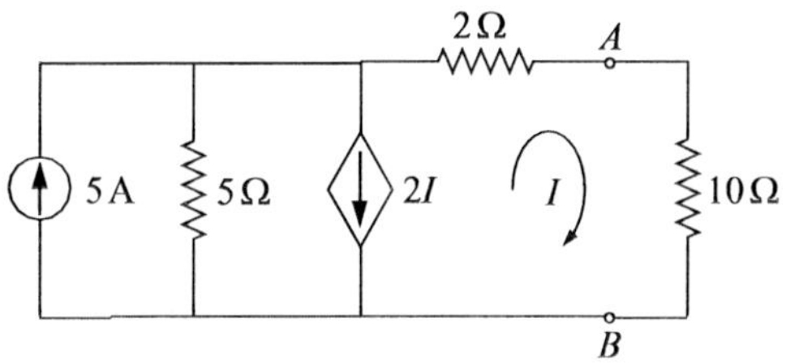

Solution: If terminals A and B are open circuited, I current becomes zero and current source of $2I$ also becomes zero as shown in Fig. (a). Hence, voltage across $A - B$ is:

$$V_{TH} = 5\text{ A} \times 5\ \Omega = 25\text{ V}$$

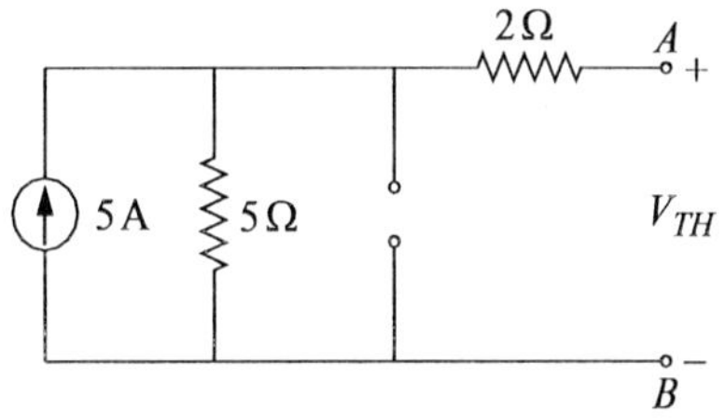

(a) Circuit for VTH

Now, as dependent source exists,

then, $R_{TH} = \dfrac{V_{TH}}{I_{sc}}$, where I_{sc} = current through $A - B$ of A and B are shorted.

The circuit (b) is shown in above conditions:

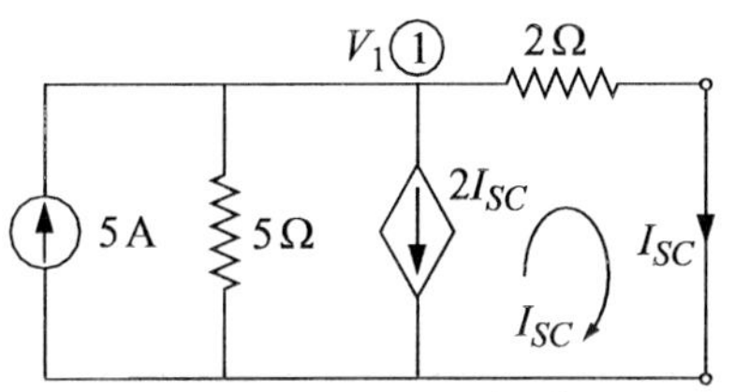

(b) Circuit for R_{TH} calculation

KCL at node (1) gives:

$$\frac{V_1}{5} = 2I_{sc} + I_{sc} = 5$$

and $$2 \times I_{sc} = V_1$$

∴ $$\frac{2}{5} I_{sc} + 3I_{sc} = 5$$

or $$I_{sc} = \frac{25}{17} \text{ A}$$

Hence, $$R_{TH} = \frac{V_{TH}}{I_{sc}} = \frac{25}{\frac{25}{17}} = 17 \ \Omega$$

Example 4.64. Find the Norton's equivalent of the network shown in the figure.

[UPTU, 2001]

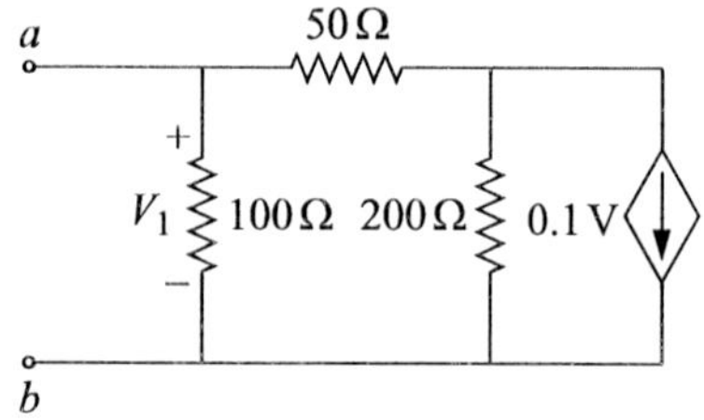

Solution: Since there is no independent energy source in the circuit, therefore, Norton equivalent source $I_N = 0$.

In order to find value of R_N, consider that V is applied across *ab* as shown in Fig. (*a*).

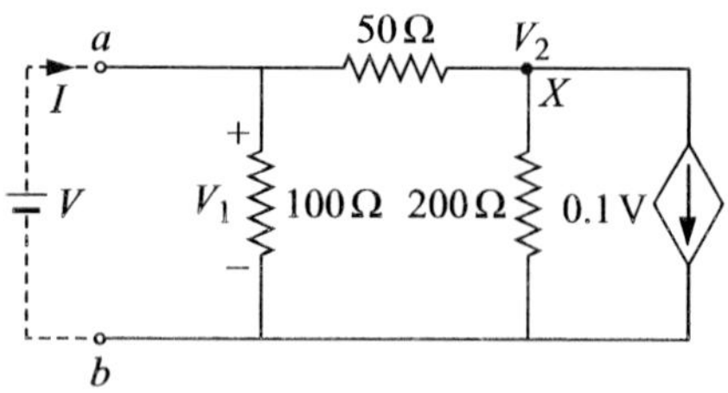

(*a*) *Fictitious voltage source applied across ab*

Let voltage at node × be V_2.

Thus, using KCL at node × we get:

$$I = \frac{V_1}{100} + \frac{V_1 - V_2}{50} \qquad \text{... (i)}$$

and $$\frac{V_1 - V_2}{50} = \frac{V_2}{200} + 0.01\,V_1 \qquad \text{... (ii)}$$

Simplifying equation (i), we get:

$$100I = V_1 + 2V_1 - 2V_2$$

or $$100\, I = 3V_1 - 2V_2 \qquad \text{... }(iii)$$

and simplifying equation (*ii*), we get:

$$4V_1 - 4V_2 = V_2 + 20V_1$$

or $$0 = 5V_2 + 16V_1$$

or $$V_2 = -\frac{16}{5}V_1 \qquad \text{... }(iv)$$

By substituting value of V_2 from equation (*iv*) into equation (*iii*), we get:

$$100I = 3V_1 - 2\times\left(-\frac{16}{5}\right)V_1$$

$$= 3V_1 + \frac{32}{5}V_1$$

or $$100\, I = \frac{47}{5}V_1$$

∴ $$\frac{V_1}{I} = \frac{500}{47}$$

Thus, $$R_N = \frac{V_1}{I} = \frac{500}{47}\,\Omega$$

or $$R_N = 10.63\ \Omega$$

Example 4.65. Find Norton's equivalent to the left side of terminals $a - b$.

[UPTU, 2004]

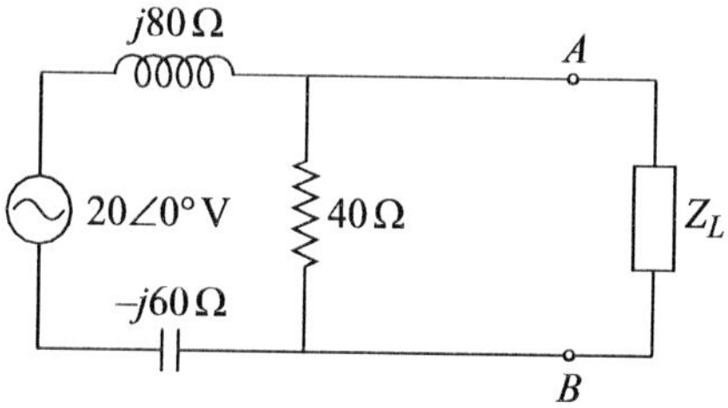

Solution: Remove voltage source and Z_L to find Z_N as shown in Fig. (*a*).

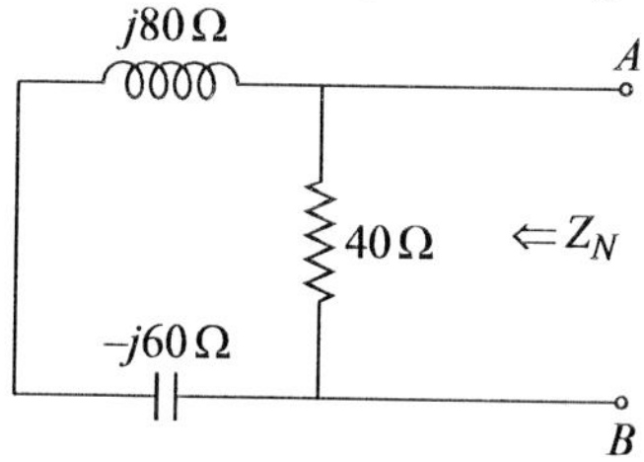

(*a*) Circuit for R_N

$$Z_N = (j80 - j60) \parallel 40 = (j20) \parallel 40$$

$$= \frac{j20 \times 40}{40 \times j20} = \frac{80\,j(40 - j20)}{(40)^2 + (20)^2}$$

or $$Z_N = (8 + j16)\ \Omega$$

Remove Z_L and put a shorting in its place to I_N as shown in Fig. (*b*).

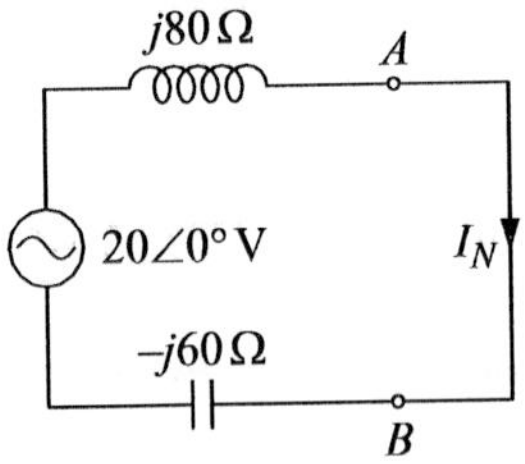

(*b*) Circuit for I_N

$$I_N = \frac{20}{(j80 - j60)} = \frac{20}{j20}$$

or $$I_N = -jA = 1\angle -90^\circ\ \text{A}$$

Thus, Norton's equivalent is shown in Fig. (*c*).

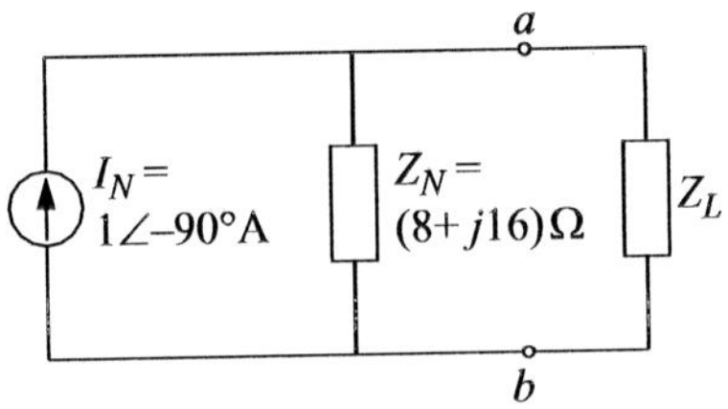

Fig. (*c*) Norton's equivalent.

Example 4.66. Verify the reciprocity theorem for the network shown in the figure in which voltage source of 5V cause a current I in the 2A resistor. Find the value of I. **[UPTU, 2005]**

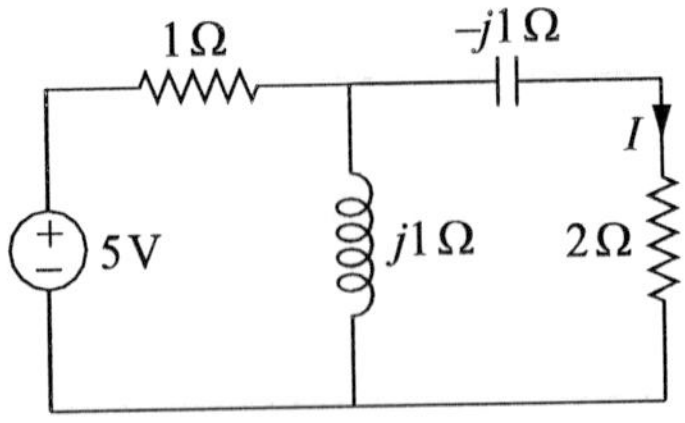

Solution: The current in 2 Ω resistor from the given figure:

$$I = \frac{5}{1 + j \parallel (2 - j)} \times \frac{j}{j \parallel (2 - j)}$$

$$= \frac{5}{1 + \left(\frac{j(2-j)}{j+2-j}\right)} \times \frac{j}{\left(\frac{j(2-j)}{j+2-j}\right)}$$

or $$I = \frac{5j}{3+2j} \text{ A} \qquad \text{... } (i)$$

Now, 5V source is placed in second loop as shown in Fig. (*a*).

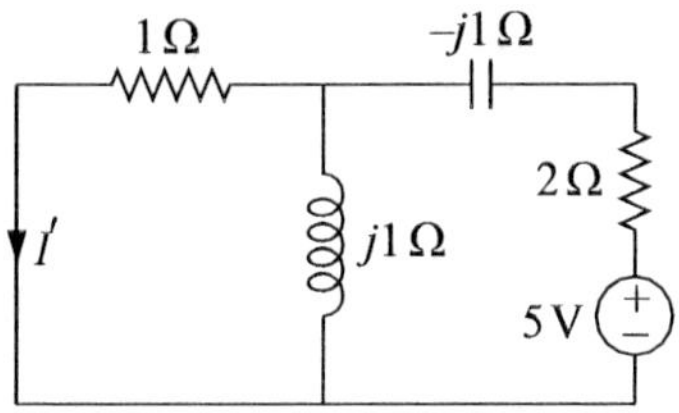

∴ The current in the first loop is:

$$I' = \frac{5}{(2-j) + (1 \| j)} \cdot \frac{j}{1+j}$$

or $$I' = \frac{5j}{3+2j} \text{ A} \qquad \text{... } (ii)$$

From equations (*i*) and (*ii*), we can see the current values are same, hence, reciprocity theorem stands proven.

Example 4.67. In the network two voltages, sources act on a bad impedance connected to terminals *a* and *b* as shown in the figure. If the load is variable both in resistance and reactance, what load Z_L will receive maximum power?

[UPTU, 2001]

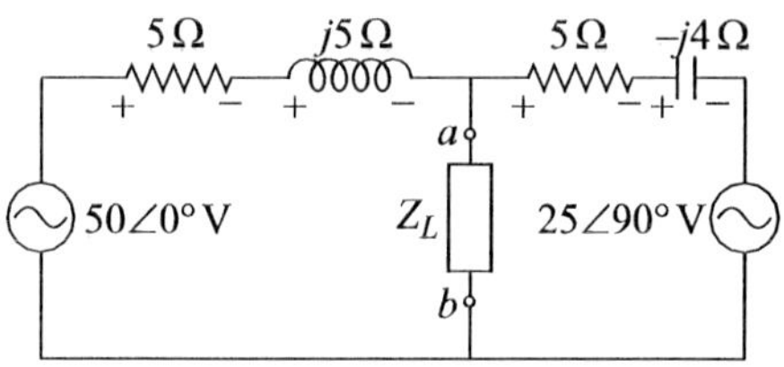

Solution: First we need to find Thevenin's equivalent impedance across load Z_L. For this, let us find Z_{TH} by removing both voltage sources which become just shortings, and Z_L is removed, as shown in Fig. (*a*).

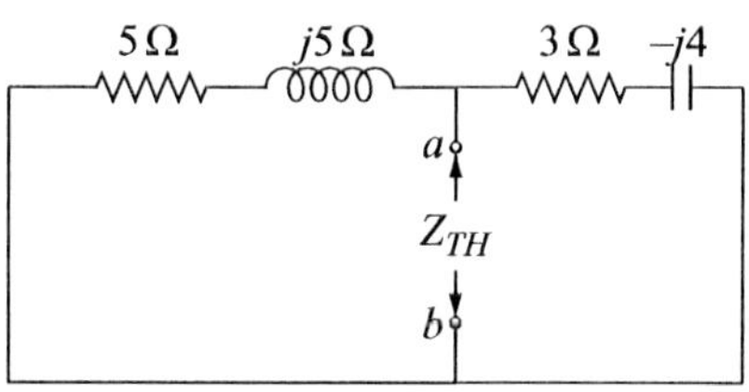

$$Z_{TH} = (5 + j5) \parallel (3 - j4)$$

$$= \frac{(5 + j5)(3 - j4)}{5 + j5 + 3 - j4}$$

or $$Z_{TH} = \frac{5(1 + j)(3 - j4)}{8 + j}$$

$$= \frac{5(3 + 4 + 3j - 4j)(8 - j)}{8^2 + 1}$$

$$= \frac{5(7 - j)(8 - j)}{65}$$

$$= \frac{5(56 + 1 - 8j - 7j)}{65}$$

$$= \frac{57 - 15j}{13}$$

or $$Z_{TH} = \frac{57}{13} - \frac{15}{13} j$$

For maximum power transfer, the Z_L must be complex conjugate of Z_{TH} and circuit becomes as shown in Fig. (*c*).

$\therefore$ $$Z_L = Z_{TH} = \frac{57}{13} + \frac{15}{13} j$$

Example 4.68. For the circuit shown in figure, find the value of Z_{TH} at will receive maximum power. Also determine this power. **[UPTU, 2005]**

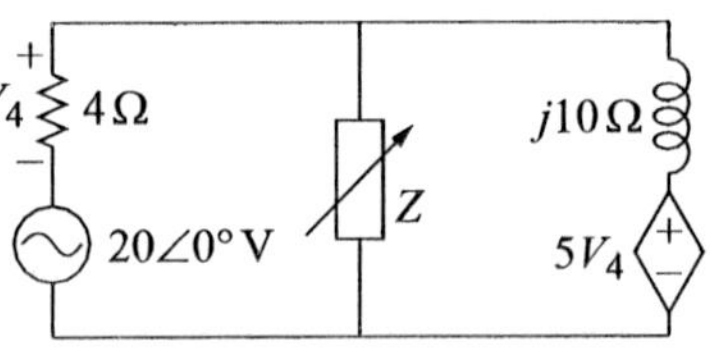

Solution: As there is dependent source also, hence, Thevenin impedence with the formula:

$$Z_{TH} = \frac{V_{TH}}{I_{sc}}$$

where I_{sc} is the current when Z is removed and shorted, and V_{TH} is the voltage across Z when it is removed.

For, V_{TH} take the circuit of Fig. (*a*).

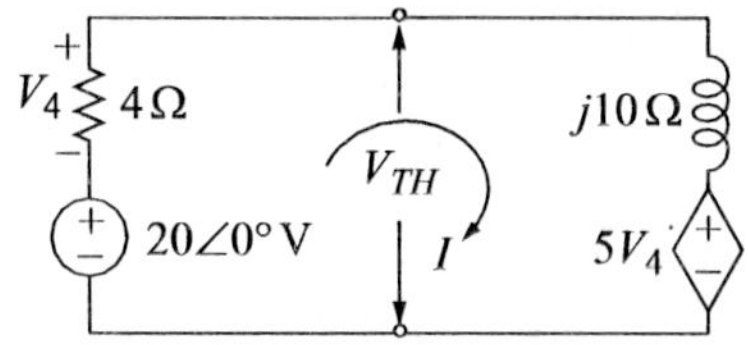

(*a*) Circuit for V_{TH}

By writing KVL equation in the loop, we get:

$$I + j10I + 5V_4 = 20$$

and $$V_4 = -4I$$

∴ $$4I + j10I + 5(-4I) = 20$$

or $$4I + j10\,I - 20\,I = 20$$

or $$I = \frac{20}{-16 + j10} = \frac{20\angle 0^\circ}{18.87\angle 148^\circ}$$

$$= 1.06\angle{-148^\circ}\text{ A}$$

∴ $$V_{TH} = 20 - 4(1.06\angle{-148^\circ})$$

$$V_{TH} = 20 + 3.59 + 2.25j = 23.59 + 2.25j$$

$$= 23.7\angle 5.45^\circ\text{ V}$$

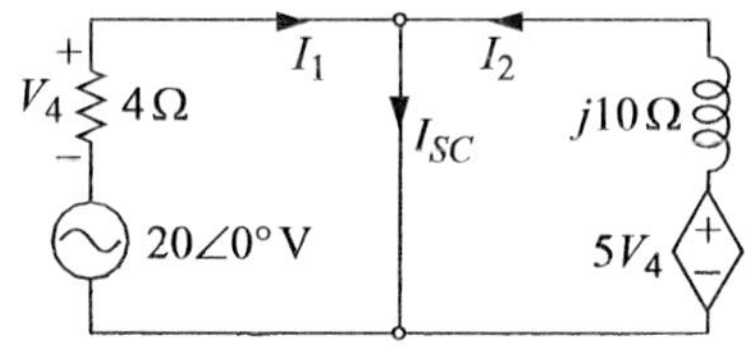

(*b*) Load removed and short circuited.

∴ $$I_{sc} = I_1 + I_2$$

$$= \frac{20}{4} + \frac{5V_4}{10j}$$

From circuit (*b*), we have $V_4 = -\frac{20}{4}$

∴ $$I_{sc} = 5 + \left(\frac{5}{10j} \times -20\right)$$

or $$I_{sc} = (5 + 10j)\text{ A} = 11.18\angle 63.4^\circ\text{ A}$$

$$Z_{TH} = \frac{V_{TH}}{I_{sc}} = \frac{23.7\angle 5.45^\circ}{11.18\angle 63.4^\circ}$$

or $$Z_{TH} = 2.12\angle{-57.95^\circ}\ \Omega$$

For maximum power transfer load Z:

$$Z = Z^* = 2.12\angle 57.95°$$

$$P_{max} = \frac{V_{TH}^2}{4\,R_L} = \frac{(23.7)^2}{4 \times 1.125} = 124.8 \text{ W}$$

or $$P_{max} = 124.8 \text{ W}$$

Example 4.69. In the network shown in the figure, find the value of R_L which results in maximum power transfer. Also find maximum power transfer.

[UPTU, 2006]

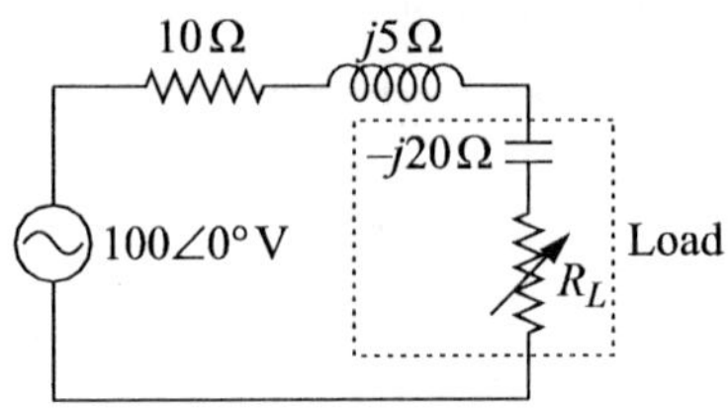

Solution: In this case, X_L is fixed and only R_L is variable, hence, for maximum power transfer to the load:

$$R_L = \sqrt{R^2 + (X + X_L)^2}$$

$$= \sqrt{(10)^2 + (5j - 2j)^2}$$

$$= \sqrt{100 + 225} = \sqrt{325}$$

∴ $$R_L = 18.03 \text{ W}$$

The maximum power transferred to the load,

$$P_{max} = \frac{V_{TH}^2\,R_L}{(R + R_L)^2 + (X + X_L)^2}$$

$$= \frac{(100)^2 \times (18.03)}{(10 + 18.03)^2 + (5 - 20)^2}$$

or $$P_{max} = 178.4 \text{ W}$$

Example 4.70. Calculate the load current I in the circuit of the figure by Millman's theorem.

[UPTU, 2001]

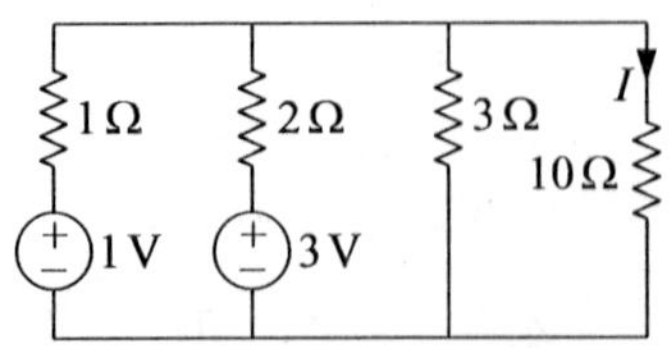

Solution: In order to apply, the circuit is labelled as shown in figure (*a*).

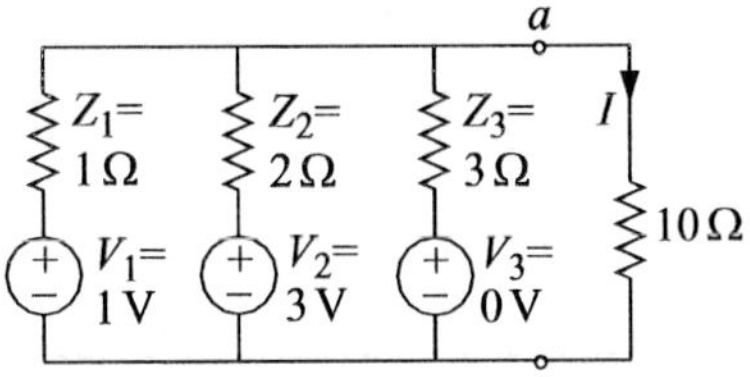

(*a*) Circuit with labelling

The admittances of the impedances in series with voltage sources are:

$$Y_1 = \frac{1}{Z_1} = \frac{1}{1} = 1 \text{ ℧}$$

$$Y_2 = \frac{1}{Z_1} = \frac{1}{2} \text{ ℧}$$

$$Y_3 = \frac{1}{Z_3} = \frac{1}{3} \text{ ℧}$$

$\therefore$ $$V_m = \frac{V_1Y_1 + V_2Y_2 + V_3Y_3}{Y_1 + Y_2 + Y_3} = \frac{1\times 1 + 3\times\frac{1}{2} + 0}{1 + \frac{1}{2} + \frac{1}{3}}$$

or $$\frac{1 + \frac{3}{2}}{\left(\frac{11}{6}\right)} = \frac{5}{2}\times\frac{6}{11}$$

or $$V_m = \frac{15}{11} \text{ V}$$

and $$Z_m = \frac{1}{Y_1 + Y_2 + Y_3} = \frac{1}{1 + \frac{1}{2} + \frac{1}{3}}$$

or $$Z_m = \frac{6}{11}\,\Omega$$

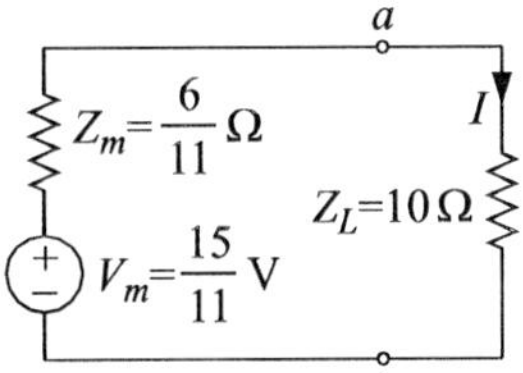

The circuit of Fig. (*a*) can be replaced by Fig. (*b*).

$\therefore$ $$I = \frac{V_m}{Z_m + Z_L} = \frac{15}{11}\times\frac{1}{\left(\frac{6}{11} + 10\right)}$$

$$I = \frac{15}{11} \times \frac{1}{\left(\frac{6}{11} + 10\right)}$$

$$= \frac{15}{11} \times \left(\frac{1}{\frac{6+10}{11}}\right) = \frac{15}{11} \times \frac{11}{116} = \frac{15}{116}$$

or $$I = 0.1293 \text{ A}$$

Example 4.71. Verify Tellegen's theorem in the network shown in the figure.

[UPTU, 2002]

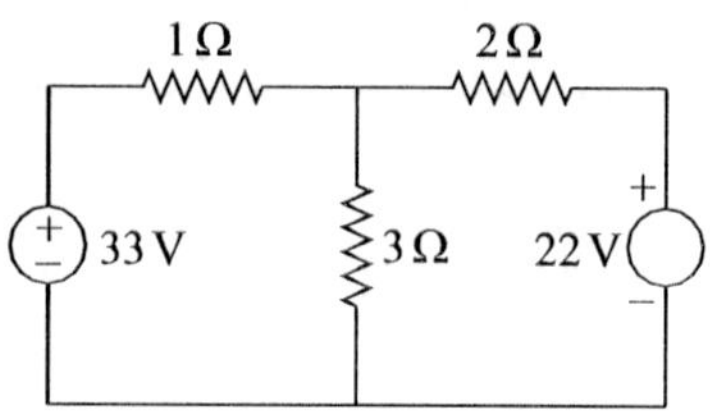

Solution: The current and voltages are marked as shown in the Fig. (*a*).

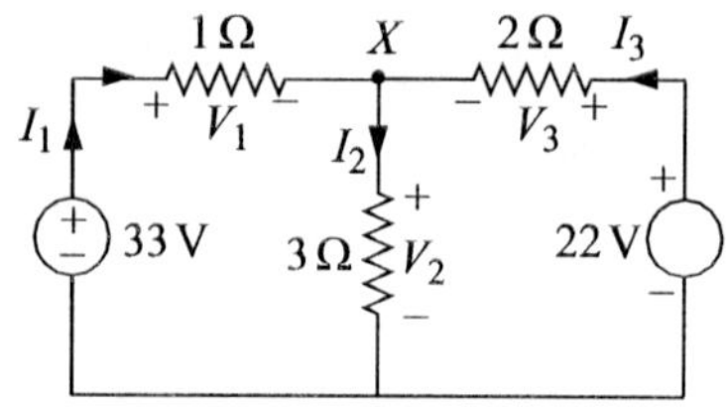

By writing KVL equations, we get:

$$33 = 1 \times I_1 + 3 \times I_2$$

or $$33 = I_1 + 3\,I_2 \quad \text{... } (i)$$

and $$22 = 2 \times I_3 + 3 \times I_2$$

or $$22 = 2I_3 + 3I_2 \quad \text{... } (ii)$$

By applying KCL at node *X*, we get:

From equations (*i*), (*ii*) and (*iii*) we get:

$$I_1 = 9\text{A}$$

$$I_2 = 8\text{A}$$

and $$I_3 = -1\text{A}$$

$$\text{Power delivered} = 33I_1 + 22I_3$$

$$= 33 \times 9 + 22 \times -1$$

$$= 11[27 - 2] = 11 \times 25 = 275 \text{ W}$$

$$\text{Power absorbed} = I_1^2 \times 1 + I_2^2 \times 3 + I_3^2 \times 2$$

$$= (9)^2 \times 1 + (8)^2 \times 3 + (-1)^2 \times 2$$

$$= 81 + 64 \times 3 + 2 = 275 \text{ W}$$

It can be seen that power delivered is equal to the power absorbed in the circuit, hence, Tellegen's theorem stands proven.

Example 4.72. Illustrate Tellegen's theorem for the network shown in the figure. **[UPTU, 2006]**

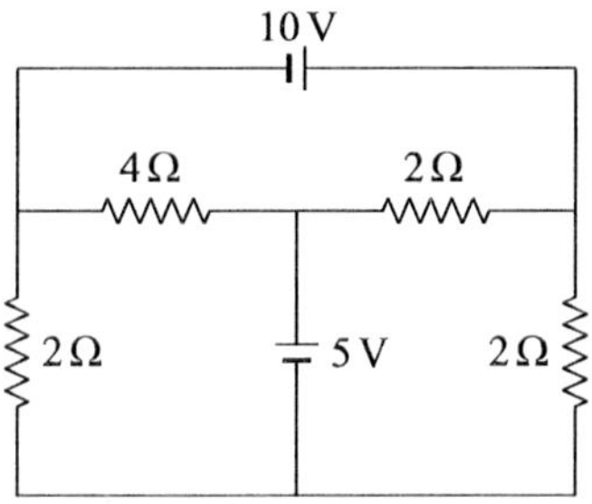

Solution: Consider the circuit of Fig. (*a*) with mesh currents.

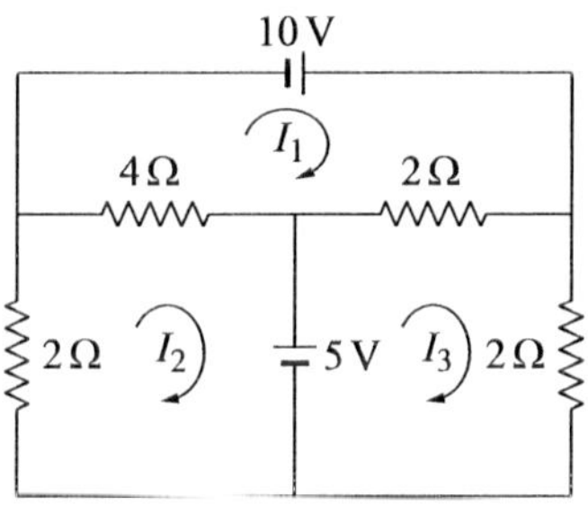

(*a*) *Circuit with mesh currents*

KVL equations for the mesh give:

$$6I_1 - 2I_2 - 4I_3 = 10$$

$$3I_1 - I_2 - 2I_3 = 5 \quad \text{... }(i)$$

$$-2I_1 + 4I_2 = 5 \quad \text{... }(ii)$$

$$4I_1 - 6I_3 = 5 \quad \text{... }(iii)$$

Using Euler's technique, we get from above three equations;

$$I_1 = \frac{\begin{vmatrix} 5 & -1 & -2 \\ 5 & 4 & 0 \\ 5 & 0 & -6 \end{vmatrix}}{\begin{vmatrix} 3 & -1 & -2 \\ -2 & 4 & 0 \\ 4 & 0 & -6 \end{vmatrix}} = \frac{-110}{-28}$$

or $$I = \frac{55}{14}\ \text{A}$$

$$I_2 = \frac{\begin{vmatrix} 3 & 5 & -2 \\ -2 & 5 & 0 \\ 4 & 5 & -6 \end{vmatrix}}{-28} = \frac{-90}{-28}$$

or $$I_2 = \frac{45}{14}\ \text{A}$$

and $$I_3 = \frac{\begin{vmatrix} 3 & -1 & 5 \\ -2 & 4 & 5 \\ 4 & 0 & 5 \end{vmatrix}}{-28} = \frac{-50}{-28}$$

$$I_3 = \frac{25}{14}\ \text{A}$$

Power delivered = $10\, I_1 + 5(I_2 - I_3)$

$$= 10 \times \frac{55}{14} + 5 \times \left(\frac{45}{14} - \frac{25}{14}\right)$$

$$= \frac{325}{7}\ \text{W}$$

and power absorbed = $(I_1 - I_3)^2 \times 4 + (I_1 - I_2)^2 \times 2 + I_2^2 \times 2 + I_3^2 \times 2$

$$= \left(\frac{55}{14} - \frac{25}{14}\right)^2 \times 4 + \left(\frac{55}{14} - \frac{45}{14}\right)^2 \times 2 + \left(\frac{45}{14}\right)^2 \times 2 + \left(\frac{25}{14}\right)^2 \times 2 = \frac{325}{7}\ \text{W}$$

Thus, Power delivered = Power absorbed.

Hence, Tellegen's theorem is illustrated.

Example 4.73. State the Tellegen's theorem and verify it for the network shown in the figure. **[UPTU, 2006]**

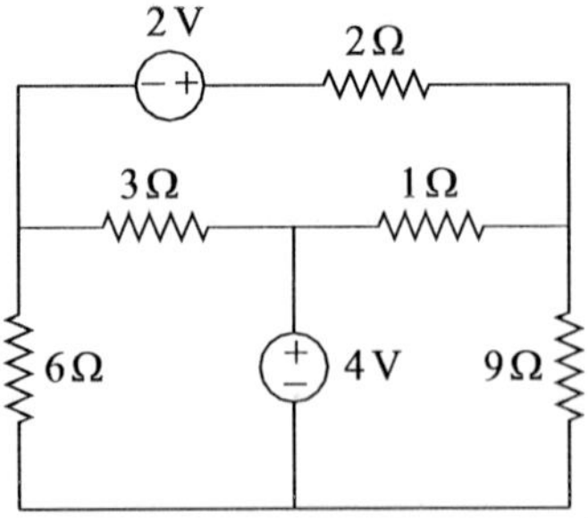

Solution: Fig. (*a*) shows the loop current and voltage drops across each element.

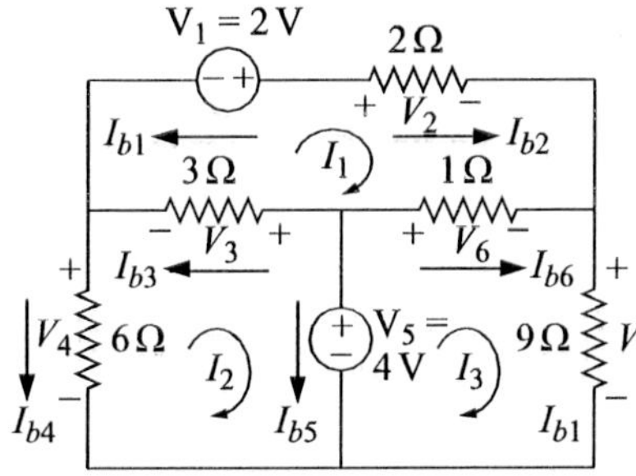

(*a*) Loop currents and Branch currents.

By KVL equations in the loops, we get:

$$2I_1 + 1 \times (I_1 - I_3) + 3(I_1 - I_2) = 2$$

or $$6I_1 - 3I_2 - I_3 = 2 \quad \text{... } (i)$$

$$1 \times (I_3 - I_1) + 9I_3 = 4$$

or $$-I_1 + 10\, I_3 = 4 \quad \text{... } (ii)$$

$$6I_2 + 3(I_2 - I_1) = -4$$

or $$-3I_1 + 9I_2 = -4 \quad \text{... } (iii)$$

By solving equations (*i*), (*ii*) and (*iii*) for the values of I_1, I_2 and I_3, we get:

$$I_1 = \frac{96}{441}\text{ A, } I_2 = \frac{-164}{441}\text{ A and } I_3 = \frac{186}{441}\text{ A}$$

The currents through various elements:

$$I_{b_1} = -I_1 = -\frac{96}{441}\text{ A and } V_1 = 2\text{V}$$

$$I_{b_2} = I_1 = \frac{96}{441}\text{ A, } V_2 = \frac{192}{441}\text{ V}$$

$$I_{b_3} = I_1 - I_2 = \frac{260}{441}\text{ A, } V_3 = \frac{780}{441}\text{ V}$$

$$I_{b_4} = -I_2 = \frac{164}{441}\text{ A, } V_4 = \frac{984}{441}$$

$$I_{b_5} = I_2 - I_3 = -\frac{350}{441}\text{ A, } V_5 = 4\text{V}$$

$$I_{b_6} = I_3 - I_1 = \frac{90}{441}\text{ A, } V_6 = \frac{90}{441}\text{ V}$$

$$I_{b_7} = I_3 = \frac{186}{441}\text{ A, } V_7 = \frac{1674}{441}\text{ V}$$

The Tellegen's theorem:

$$\sum_{j=1}^{7} V_j I_j = \left(\frac{1}{441}\right)^2 [-(2)(441)(96) + (192)(96) + (780)(260) + (984)(164)$$

$$-(4)(441)(350) + (90)(90) + (1674)(186)]$$

i.e., $\sum_{j=1}^{7} V_j I_j = 0$, as the algebraic sum of power delivered and absorbed are zero, hence, Tellegen's theorem is verified.

SUMMARY

1. **Basic Network Analysis:** Method is by using Ohm's law, Kirchhoff's current law and Kirchhoff's voltage law. The supply sources can be voltage source or current source. These supply sources can be independent or dependent sources. It is possible to convert voltage source into current source or *vice-versa*. The dependent source circuit analysis need caution. Matrix technique of solving simultaneous equations is a very handy tool.
2. **Maxwell's Mesh Current:** Method is deduced to: mesh current × sum of resistances of the mesh + Algebraic sum of multiplication of common other mesh currents and respective resistances = sum of voltage rise in the mesh. A mesh is a loop which does not contain any other loop.
3. **Nodal Voltage Method is Deduced to:** Node voltage × Sum of conductances connected to node – Node voltages to other and × Conductance of the branch = 0.
   The above is valid for independent sources only. If network contains dependent sources, the values of node voltages using KCL method is obtained.
   A node is a junction of two more circuit elements.
4. **Superposition Theorem:** According to this theorem states that the current through or the voltage across any element can be found out by the algebraic addition of the individual currents or voltages due to each independent voltage or current sources acting alone. The non-acting independent sources are replaced by their internal resistances or short circuits in the case of voltage sources and by their internal conductances or open circuits for current sources.
5. **Thevenin's Theorem:** States that a pair of terminals of any linear network can be replaced by an equivalent circuit composed of a voltage V_{TH} in series with a resistance R_{TH}. The V_{TH} is the voltage across the open circuited terminals considered and R_{TH} is the equivalent resistance of the network as seen from those terminals with all independent sources removed.

6. **Norton's Theorem:** This theorem states that a pair of terminals of any linear network containing resistors and energy sources can be replaced by an equivalent circuit consisting of a current source I_N in parallel with a conductance G_N. The current I_N can be found out by shorting the terminals, whereas, G_N is the ratio of short circuit current I_N to the open circuit voltage across the pair of terminals.
   Norton's equivalent can be found from the Thevenin's equivalent circuit or *vice-versa.*

$$I_N = G_N V_{TH} \text{ and } G_N = \frac{1}{R_{TH}}$$

7. **Maximum Power Transfer Theorem:** It states that any resistive load connected to a dc network receives maximum power when the load resistance equals the Thevenin's equivalent resistance R_{TH} of the network as seen from the load terminals.

$$P_{max} = \frac{V_{TH}^2}{4 R_{TH}}$$

8. **Delta (Δ) – Star (Y) Transformation:** It states that the resistances of any branch of the equivalent star connection is obtained by multiplying the adjacent delta resistances and dividing by the sum of three delta resistances. Δ-Y transformation π-T transformation.
9. **Star (Y) – Delta (Δ): Transformation** states that the equivalent delta resistances between two terminals is the sum of the two star resistances connected to these terminals plus the product of the same divided by the third star resistance.

EXERCISES

4.1 What do you understand by KCL and KVL? Explain in detail.

4.2 State and explain superposition theorem.

4.3 State Thevenin's theorem. What are V_{TH} and R_{TH}?

4.4 State Norton's theorem. How to convert Thevenin's equivalent circuit to Norton's equivalent circuit or *vice-versa.*

4.5 State and prove maximum power transfer theorem. **[UPTU, 2003]**

4.6 Show that in pure resistive network, the load will draw maximum power from the source if the load resistance is equal to the source resistance.

4.7 Explain maximum power transfer theorem and prove according to this theorem that R_L will abstract maximum power the network when $R_L = R_{TH}$.

4.8 What do you understand by star/delta transformation. If three resistances R_1, R_2, R_3 are connected in star-connection, then calculate the equivalent values of these resistances for delta connection.

4.9 Write loop equations for the circuit given below, then with the help of Cramer's rule, solve for the current I_3 flowing through the 3 Ω resistance. Use the specified loop currents.

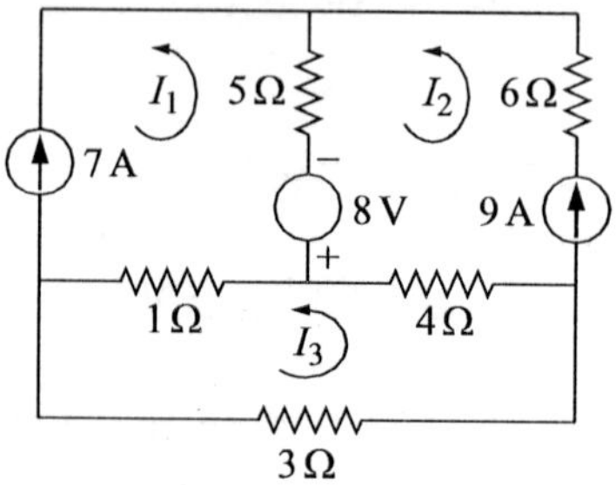

4.10. Determine the loop currents in the network given under by mesh analysis method. (**Ans.** 5A, 8A)

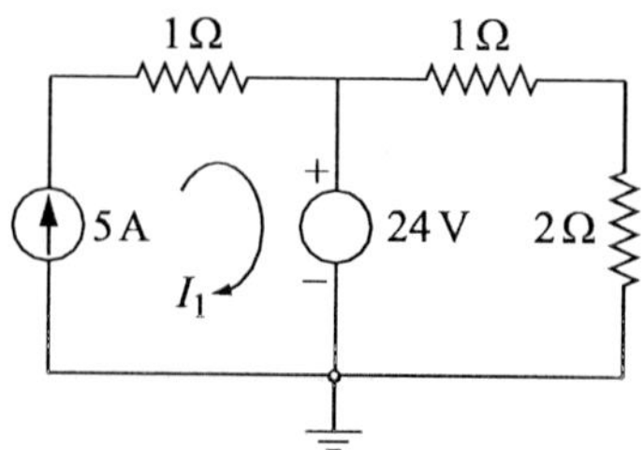

4.11. Find the various branch currents in the passive elements of the network as shown in the figure below:

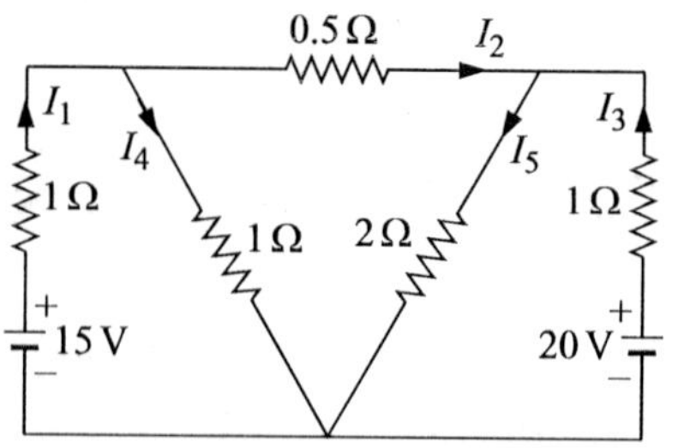

(**Ans.** $I_1 = 5.75$ A, $I_2 = -3.5$ A, $I_3 = 9$ A, $I_4 = 9.25$ A, $I_5 = 5.5$ A)

4.12. Solve for current I of the figure given below by applying the superposition theorem. (**Ans.** – 1.68 A)

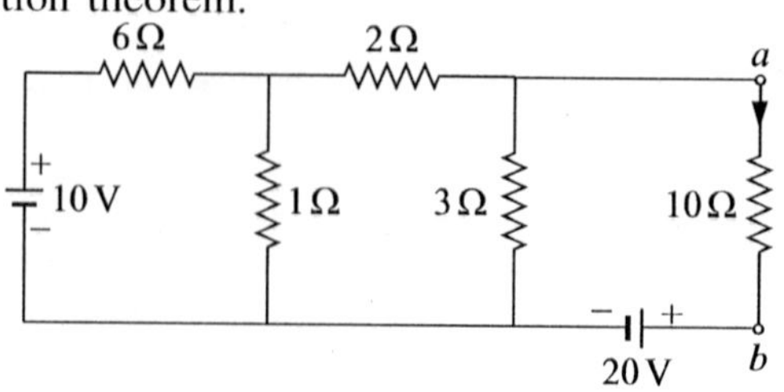

4.13. Replace the network of above figure to the left of terminals *ab* by its Thevenin equivalent circuit, hence determine *I*. (**Ans.** 1.68 A)

4.14. Determine *I* of the figure of Q 4.12 by mesh analysis. (**Ans.** – 168. A)

4.15. Evaluate *I* of the figure of Q. 4.12 by nodal analysis. (**Ans.** 1.68 A)

4.16. A resistive network is shown in figure given below. Determine the value of the equivalent resistance at the terminals *AB*.

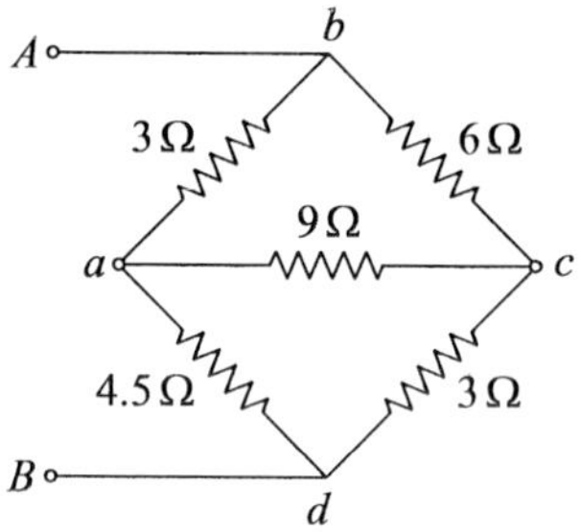

4.17. In the network of the figure given below: Determine: (*a*) the value of the load resistance to give maximum power transfer and (*b*) the power delivered to the load. (**Ans.** 5 Ω, 1.25 W)

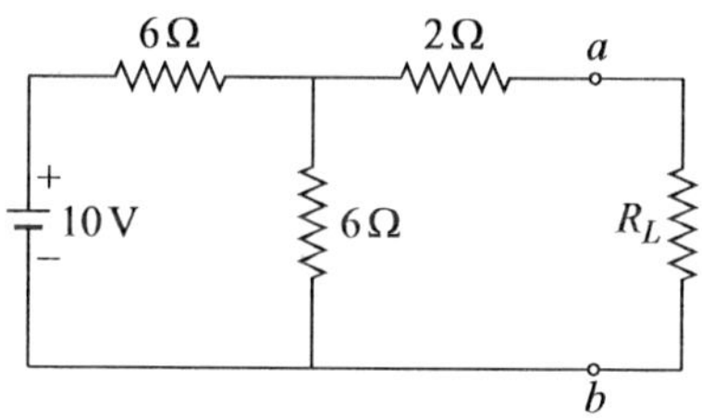

4.18. Use Thevenin's theorem to determine the voltage across the 25 Ω-resistor given in the figure under. Also determine Thevenin's equivalent resistance across 25 Ω resistor. (**Ans.** 10 V, 15 Ω)

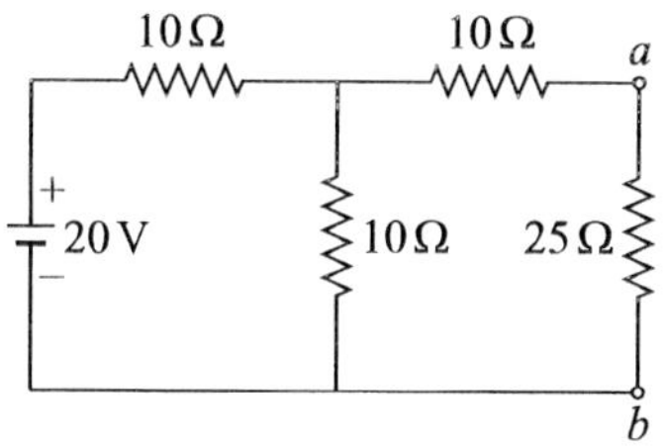

4.19. What is a network? What are active and passive elements? Explain with examples.

4.20. What are linear and non-linear networks? How does it affect the network analysis ?

4.21. Explain unilateral and bilateral elements with examples.

4.22. Find the voltages across the impedances in the circuit shown in figure. Then transform the voltage source and $10\angle 30°\Omega$ impedance to an equivalent current source and again find the voltages. Compare results.

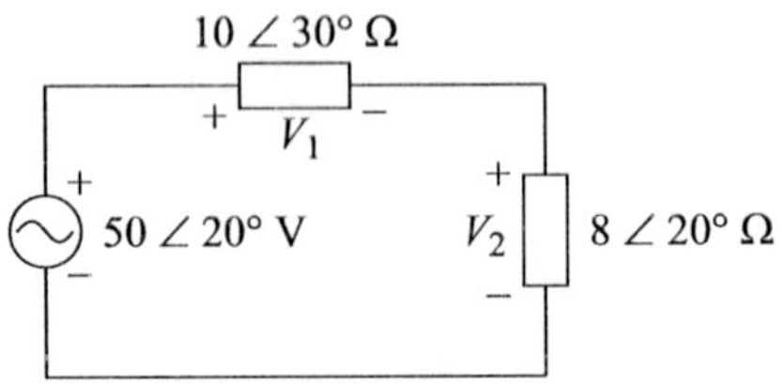

4.23. Calculate the current in 3Ω resistor for the circuit shown in figure.

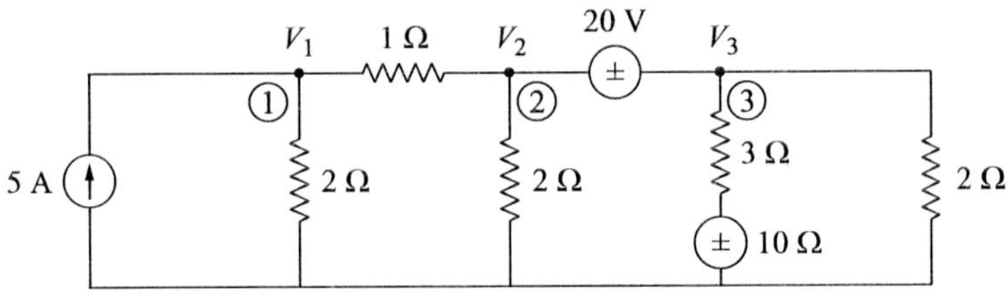

4.24. Calculate the current in 3Ω resistor for the circuit shown in figure.

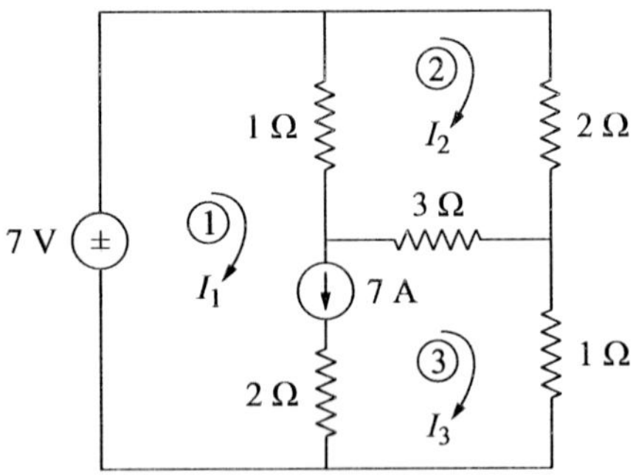

4.25. Calculate V in the circuit of figure.

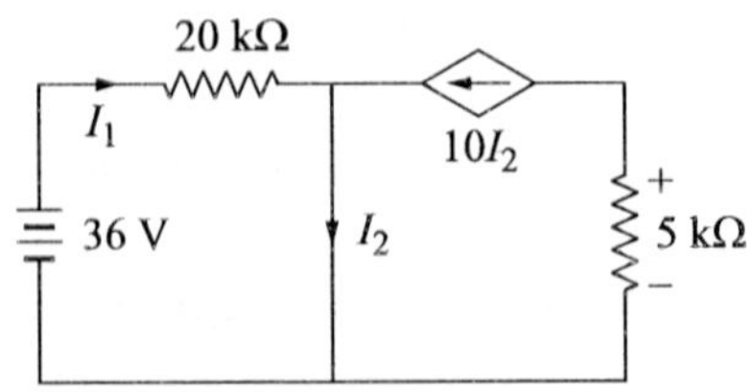

4.26. Determine I in the circuit of figure.

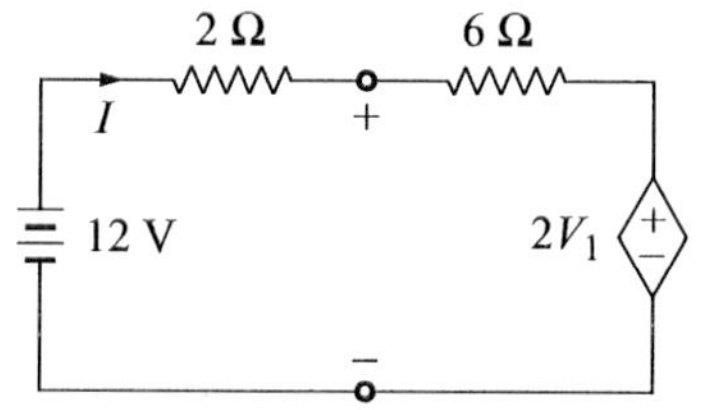

4.27. For the circuit shown in figure, determine the current I through the 10Ω resistance by

(*i*) Nodal analysis (application of KCL), and

(*ii*) Mesh analysis (application of KVL).

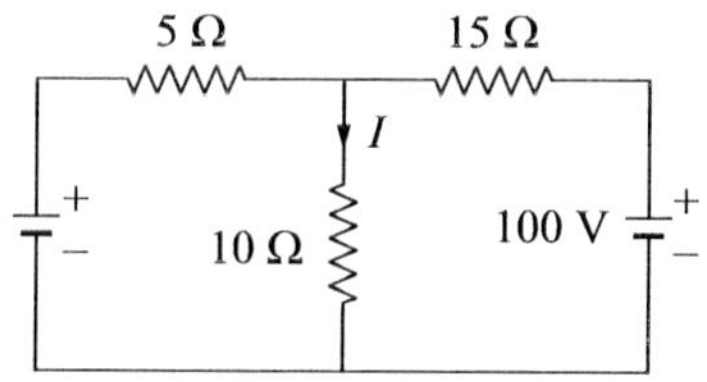

4.28. Verify reciprocity theorem for the *T*-network shown in figure.

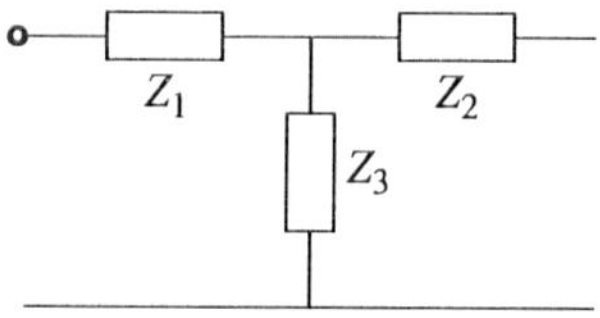

4.29. Verify the reciprocity theorem for the circuit shown in figure.

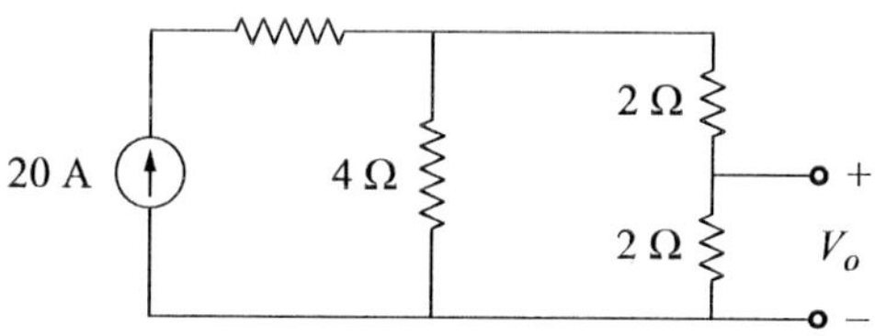

4.30. For the circuit shown in figure, determine the current I-through the 10 Ω resistance by superposition theorem, by Thevenin's theorem and by Norton's theorem.

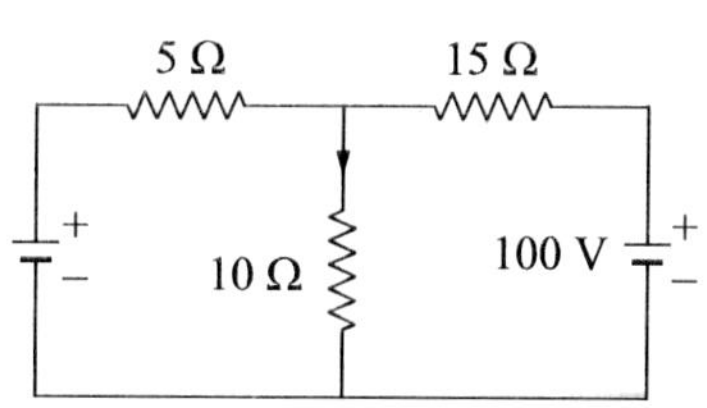

4.31. By superposition theorem, calculate current I in the circuit shown in figure.

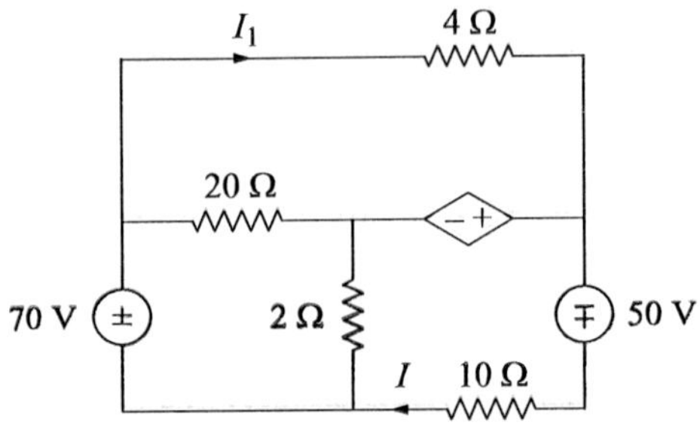

4.32. Verify reciprocity theorem for circuit as shown in figure.

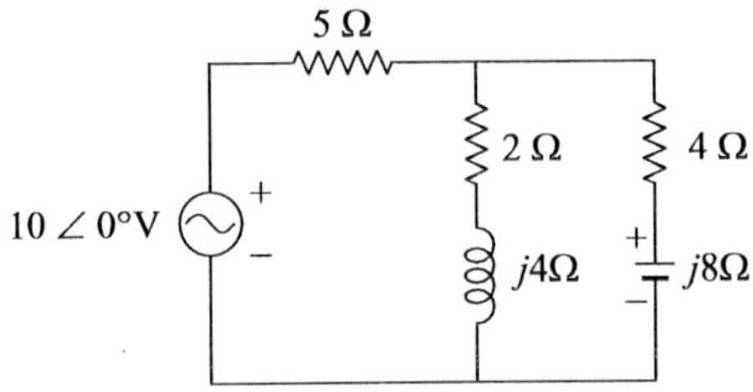

4.33. Obtain the Thevenin's equivalent parameters of the circuit shown in figure at terminals AB. Then obtain Norton's equivalent parameter.

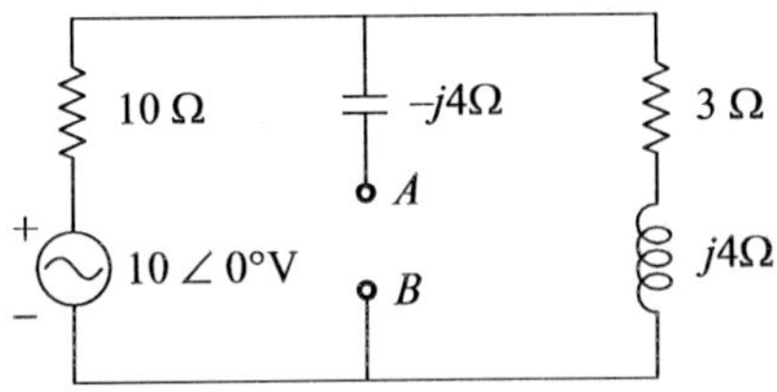

4.34. Find the Thevenin's equivalent circuit at terminals AB of the circuit given in figure.

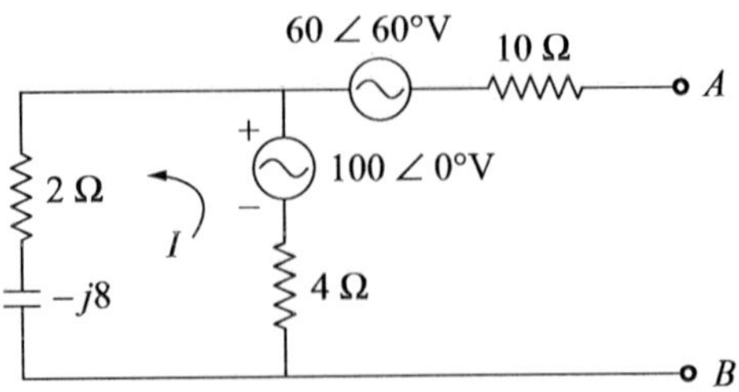

4.35. Find the Thevenin's equivalent circuit at terminals AB for the circuit shown in figure.

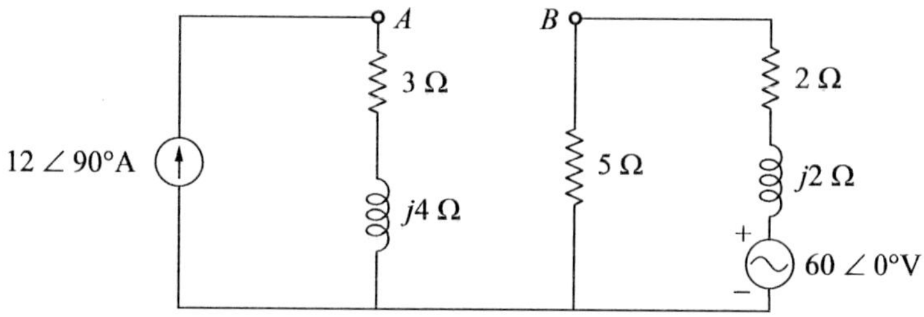

4.36. Obtain the Thevenin's and Norton's equivalent parameters at terminals AB of the circuit in figure.

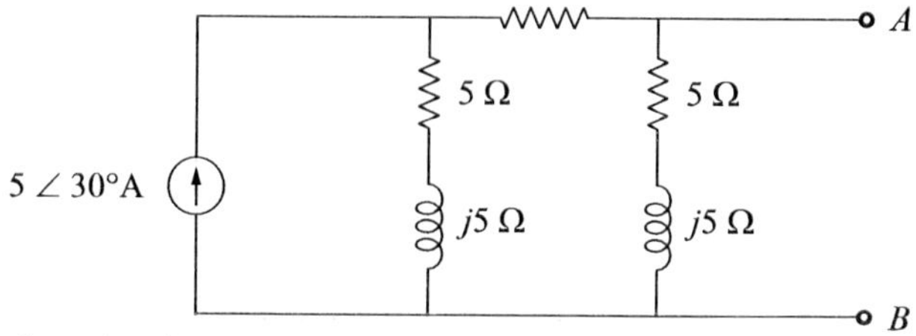

4.37. For the circuit shown in figure. Using Millmen's theorem find the current in the load impedance, $Z_L = (2 + j4)\Omega$

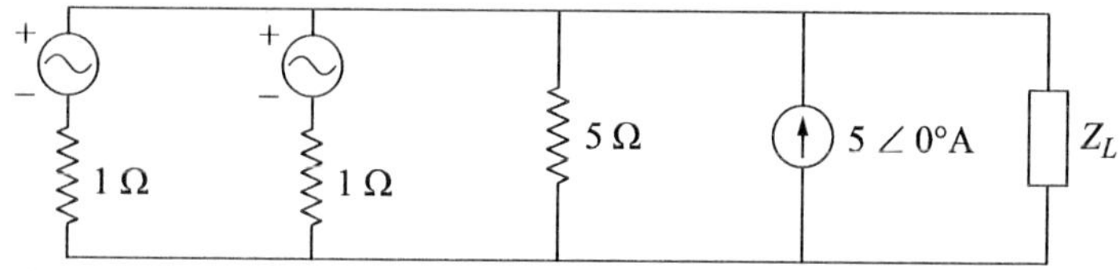

4.38. Find the Thevenin's equivalent circuit at terminals AB for the circuit shown in figure.

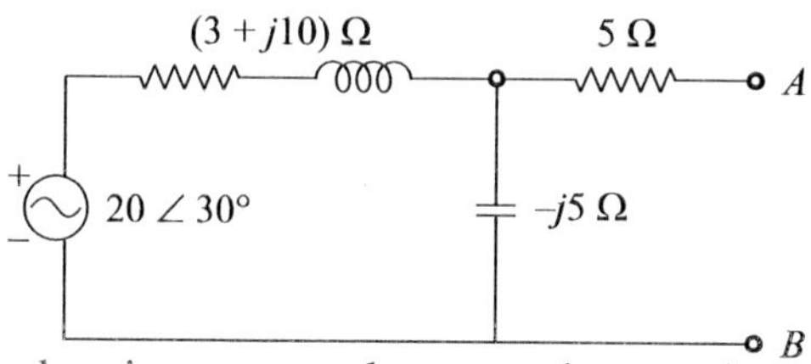

4.39. A loudspeaker is connected across the terminals A and B of the circuit shown in figure. What should be the impedance of the speaker to obtain the maximum power transferred to it and what is the maximum power.

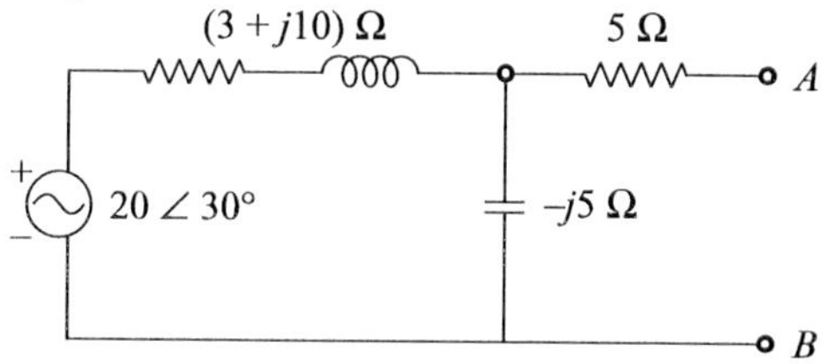

4.40. Determine the impedance to be connected across AB of figure for maximum power transfer.

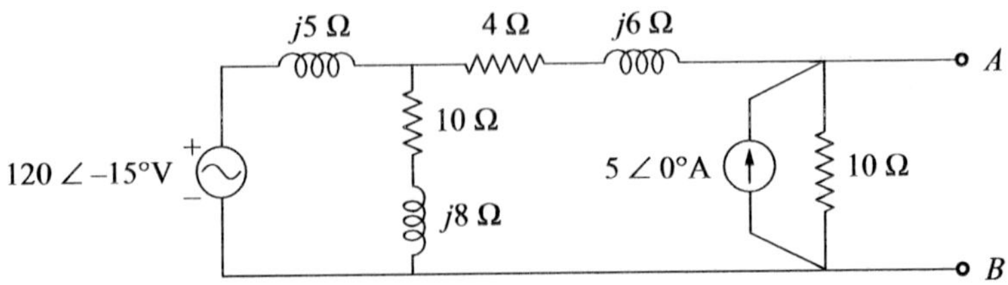

4.41. In the circuit shown in figure, determine the value of R and C that maximum power is absorbed by R. Determine the power for these value. The signal frequency is $\omega = 200$ rad/sec, $V_S = 200 \angle 0°$V.

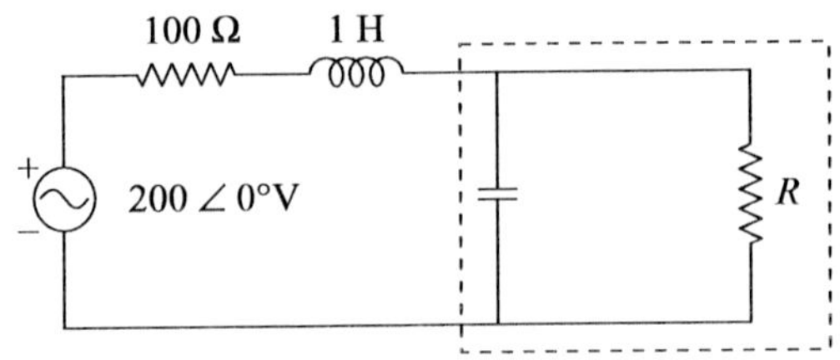

4.42. In the circuit shown in figure, determine Z_L so that the power absorbed by it is maximum and the value of the power absorbed.

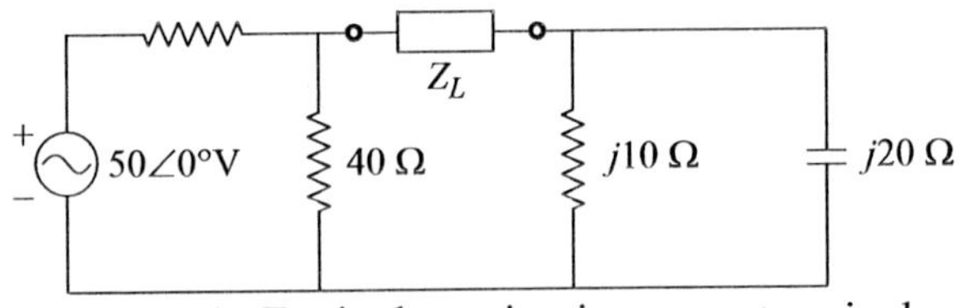

4.43. Find the Nortan's Equivalent circuit across terminals AB of the circuit shown in figure.

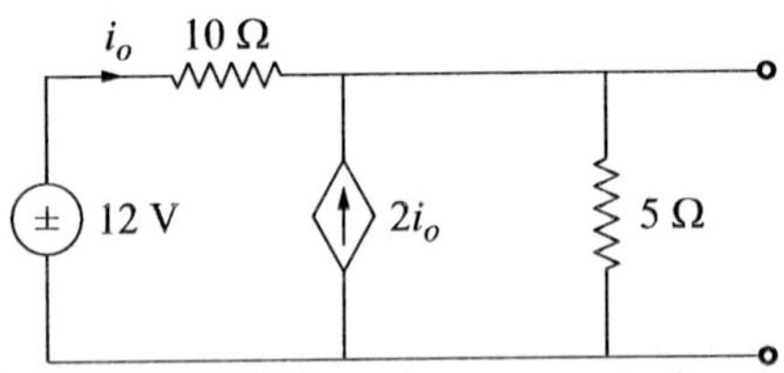

4.44. Determine the value of Z_L to be connected across AB in figure for maximum power transfer. Given $V_S = 25$ V, obtain the power, absorbed by Z_L.

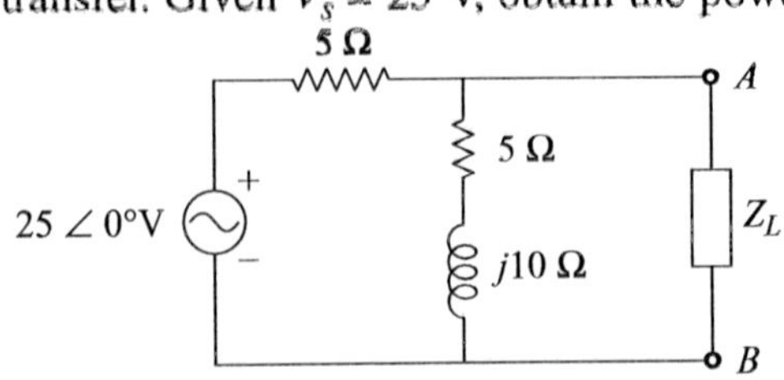

4.45. Determine the value of R_L to be connected across AB in figure for maximum power transfer. Also calculate the maximum power absorbed by the R_L.

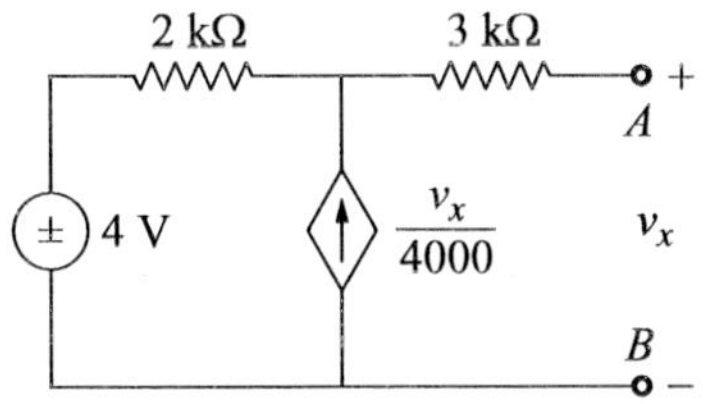

4.46. Calculate the current in the 6Ω resistor of the circuit of figure by:

(*i*) Thevenin's theorem

(*ii*) Superposition theorem.

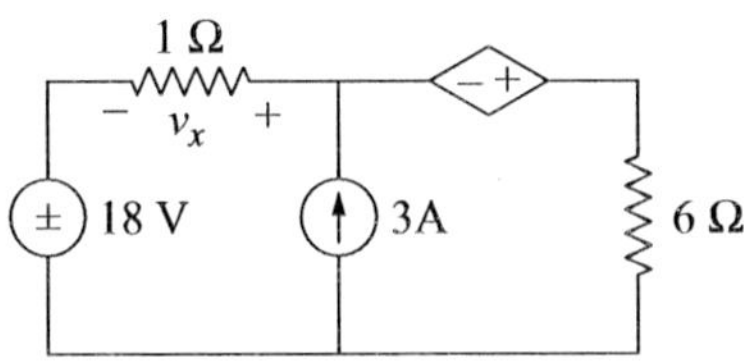

4.47. Find the current in 14Ω resistor of the circuit of figure by Thevenin's theorem.

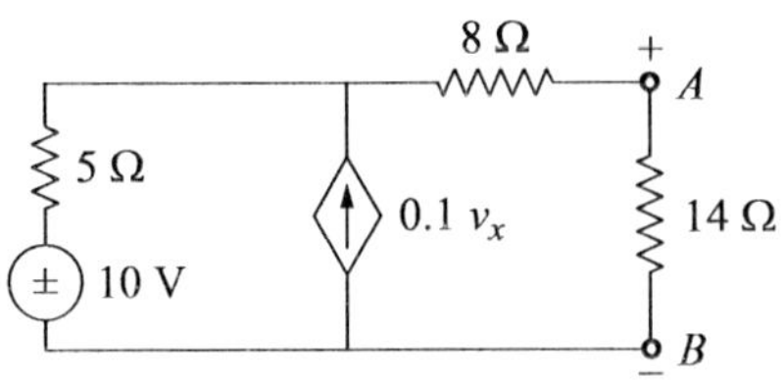

4.48. Verify Tellegen's theorem for the circuit of figure given.

$$v_1 = 4\text{V},\ v_2 = 2\text{V},\ v_4 = 3\text{V}$$
$$i_1 = 2\text{A},\ i_2 = 2\text{A},\ i_5 = -6\text{A}$$

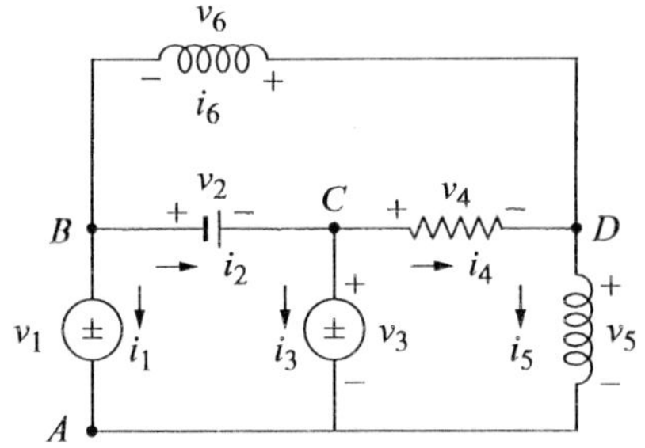

4.49. Consider the figure verify Tellegen's theorem if

$$v_1 = 5 \text{ V}, v_4 = 2 \text{ V, and } v_2 = 0 \text{ and } i_1 = 2 \text{ A}, i_3 = 1 \text{ A}.$$

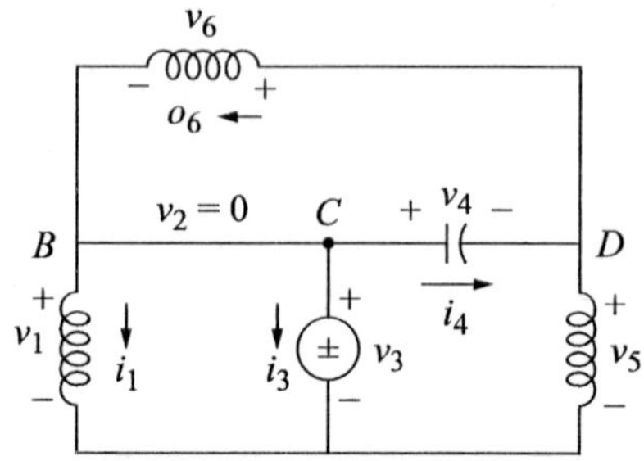

4.50. Verify Tellegen's theorem for the circuit shown in figure, where the branch voltages and currents have the following values:

$$v_0 = 40 \text{ V}, v_1 = 32 \text{ V}, v_2 = 4 \text{ V}$$
$$v_3 = 8\text{V}, v_4 = 28 \text{ V}, v_5 = 12 \text{ V}$$
$$i_0 = -32\text{A}, i = 24 \text{ A}, i_2 = 4 \text{ A}$$
$$i_3 = 28\text{A}, i_4 = 8 \text{ A}, i_5 = 4 \text{ A}$$

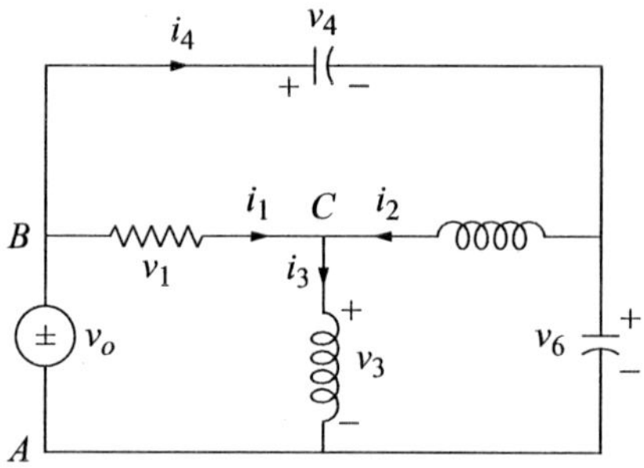

4.51. Determine the value of I_1 and I_2 in given figure using mesh analysis

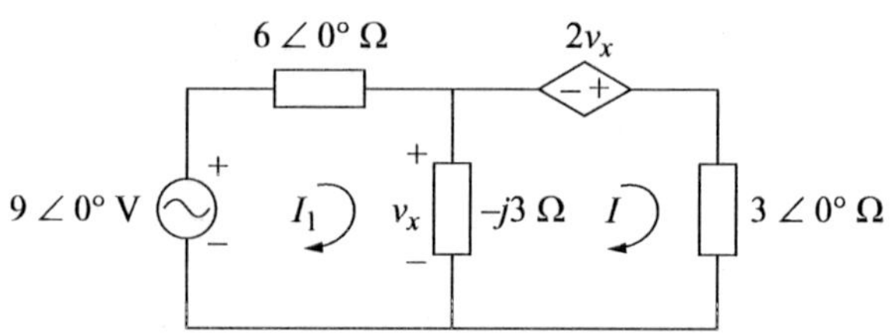

4.52. Using Thevenin theorem find the power in $(4 + j6)\Omega$ impedance connected across terminals AB in given figure:

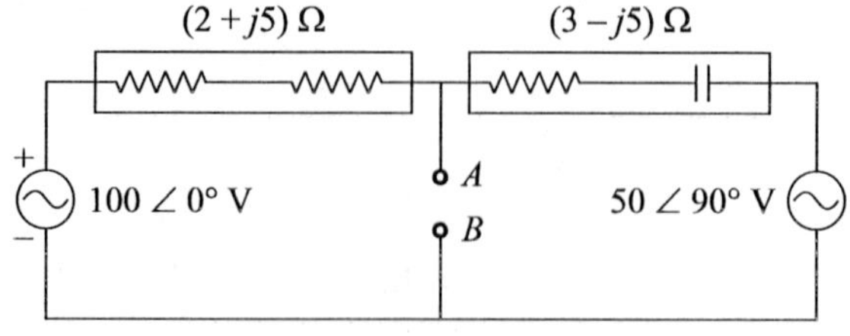

❑❑❑

5

Network Functions

5.1 INTRODUCTION

Solution of network problems involve a set of linear differential equations. The classical method of solving complex differential equations is tedious. One of the easier method is transforming the problem into an imaginary s-domain from real t-domain. All the analysis is done in s-domain and solution are obtained in t-domain and relations are obtained in t-domain by inverse-transformation. All the networks are designed to produce a particular output, when the input is applied to it. The network parameters perform some operations on the applied 000input and produce the required output. One such network is shown in Fig. 5.1.

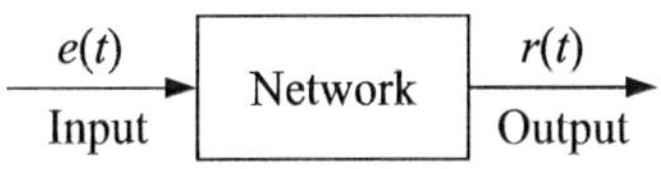

Fig. 5.1 *A network*

$e(t)$ = input or excitation

$r(t)$ = output or response

The ratio of output to the applied input is nothing, but the effect of the network perimeters. This effect is mathematically expressed in the s-domain *i.e.*, the Laplace domain, called "network function". The network function is the characteristics of the network and not dependent on input or output.

Thus, the network function is defined as the ratio of Laplace transform of output differential to the Laplace transform of input with all initial conditions being zero. A network function is shown in Fig. 5.2.

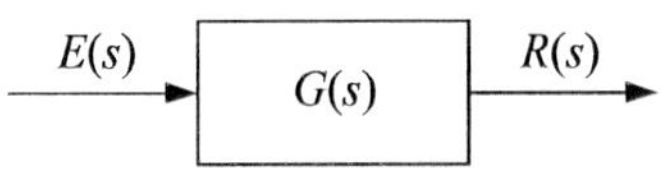

Fig. 5.2 *A network function*

$$\text{Network function} = G(s) = \frac{R(s)}{E(s)}$$

where $G(s)$= Network function

$E(s)$ = Laplace transform of the input

R(s) = Laplace transform of the output

5.2 LAPLACE TRANSFORM

5.2.1 Laplace Transform and Inverse Laplace Transform

The Laplace transform is one of the mathematical bases for the solution of linear ordinary differential equation. In comparison with the classical method of solving linear differential equation, the Laplace transform method has the following two attractive features:

(*i*) The homogeneous equation and the particular integral of the solution are obtained in one operation.

(*ii*) The Laplace transform converts the differential equation into an algebraic equation. Then simple algebraic rules are used to manipulate and solve in the *s*-doman. The final solution is obtained by taking the inverse Laplace transform.

The Laplace transform is a good method for analysing process control systems. It is applied for the analytical solution of differential equations in this section.

The Laplace Transformation is defined as

$$£(f(t) = F(s) = \int_0^{\infty} f(t)e^{-st}dt \qquad ...(5.1)$$

The important properties of Laplace Transform are:

(*i*) The function $f(t)$ is considered for $t < 0$.

(*ii*) Laplace transform may not exist for all functions.

(*iii*) Laplace transform converts a function in the time domain to a function in the *s*-domain where-ins is a complex variable *i.e.*

$$s = \sigma + j\omega \qquad ...(5.2)$$

$$\sigma = \text{Real part}$$

$$j\omega = \text{Imaginary part.}$$

(*iv*) The Laplace transform of a function $f(t)$ will be designated as F(s).

(*v*) The Laplace transform is a linear operator as it satisfies the following equation:

$$£\,[af_1(t) + bf_2(t)] + b\,£\,[f_2(t)] \qquad ...(5.3)$$

(*vi*) Tables of Laplace transform are available for easy analysis. Similarly inverse Laplace transform tables are also available as follows:

$$£^{-1}[F(s)] = f(t) \text{ for } t \geq 0$$

or $$£(t) = £^{-1}[F(s)] = \frac{1}{2\pi j}\int_{c-j\infty}^{c+j\infty} F(s)\, e^{st} ds \qquad (5.4)$$

The Laplace transforms define only for single valued functions, therefore, the transform and its inverse are unique.

Laplace transform of some of the important functions are given in Table 5.1 and some are derived here for understanding follows:

(*a*) **Constant:** $f(t) = k$

$$£\,[k = \int_0^{\infty} K\, e^{-st}\, dt = \frac{-K}{s}\, e^{-st}\Big|_0^{\infty} = \frac{K}{s}$$

(*b*) Step: K at $t = 0$

$$f(t) = K\,u(t) \text{ for } u(t) = \begin{cases} 0 \text{ at } t = 0^+ \\ 1 \text{ for } t > 0^+ \end{cases}$$

$$£[K\,u(t)] = K\,£[u(t)] = K\left[\int_0^{\infty} e^{-st}\, dt\right] = \frac{K}{s}$$

Table 5.1 Laplace Transforms

S.No.	$f(t)$	$F(s) = \int_0^{\infty} f(t)\, e^{-st} dt$
1.	$\delta(t)$, unit impulse	1
2.	$u(t)$, unit step	$1/s$
3.	(t)	$1/s^2$
4.	t^n	$\frac{\lfloor n}{s^{n+1}}$
5.	$00e^{\pm at}$	$\frac{1}{s \mp a}$
6.	$te^{\pm at}$	$\frac{1}{(s \mp a)^2}$
7.	$\sin \omega t$	$\frac{\omega}{s^2 \mp \omega^2}$
8.	$\cos \omega t$	$\frac{S}{s^2 + \omega^2}$
9.	$e^{-at} \sin \omega t$	$\frac{\omega}{(s^2 + a^2) + \omega^2}$
10.	$e^{-at} \cos \omega t$	$\frac{S + a}{(S + a)^2 + \omega^2}$

11.	sin h at	$\frac{a}{s^2 - a^2}$
12.	cos h at	$\frac{s}{s^2 - a^2}$
13.	$e^{\pm at} f(t)$	$F(s \mp a)$
14.	$f(t \pm t_0)$	$e^{\pm to\, s} F(s)$

(*c*) Exponential: $f(t) = e^{-at}$

$$£[e^{at}] = \int_0^\infty e^{at} e^{-st}\, dt$$

$$= \frac{1}{a-s} e^{-(s-a)} \Big|_0^\infty = \frac{1}{s-a}$$

(*d*) Since: $$f(t) = \sin \omega t = \frac{e^{j\omega t} - e^{-j\omega t}}{2j}$$

$$£[\sin \omega t] = \int_0^\infty \sin \omega t\, e^{-st}\, dt$$

$$= \int_0^\infty \left(\frac{e^{j\omega t} - e^{-j\omega t}}{2j} \right) e^{-st} dt$$

$$= \int_0^\infty \left(\frac{e^{-(s-j\omega)} - e^{-(s+j\omega)t}}{2j} \right) dt$$

$$= \frac{1}{2j} \left[\frac{e^{-(s-j\omega)t}}{s-j\omega} + \frac{e^{-(s-j\omega)t}}{s+j\omega} \right]_0^\infty$$

$$= \frac{s}{s^2 + \omega^2}$$

(*e*) Derivative of a function

$$£\left[\frac{d\, f(t)}{dt}\right] = \int_0^\infty \frac{d\, f(t)}{dt} e^{-St}\, dt$$

$$f(t) = u, \quad \text{So} \left(\frac{d\, f(t)}{dt}\right) dt = du$$

and $e^{-St}\, dt = dv$, So $v = -\frac{e^{-S}t}{S}$

$$£\left[\frac{df\ (t)}{dt}\right] = \int_0^\infty u dv = uv \int_0^\infty - \int_0^\infty u dv$$

$$= f(t)\left[-\frac{e^{-st}}{s}\right]_0^{\infty} = \int_0^{\infty}\left(\frac{-e^{-st}}{s}\right)\left(\frac{df(t)}{dt}\right) dt$$

$$= \frac{f(0^+)}{s} + \frac{1}{s}£\left[\frac{df(t)}{dt}\right]$$

Hence, $$£\left[\frac{df(t)}{dt}\right] = S\,F(s) - f(0^+)$$

(*f*) Integration of a function:

$$£\left[\int_0^t f(t)\,dt\right] = \int_0^{\infty}\left[\int_0^t f(t)\,dt\right]e^{-st}\,dt$$

Let $$u = \int_0^t f(t)\,dt \text{ so } du = £(t)\,dt$$

and $$dv = e^{-st}\,dt, \qquad \text{So} \quad v = \frac{e^{-st}}{s}$$

$\therefore$ $$£\left[\int_0^t f(t)\,dt\right] = \int_0^{\infty} udv = uv\int_0^{\infty} - \int_0^{\infty} udv$$

$$= \frac{-e^{st}}{s}\int_0^t f(t)\,dt\Bigg|_0^{\infty} + \frac{1}{s}\int_0^{\infty} f(t)\,e^{st}\,dt$$

$$= 0 - 0 + \frac{1}{s}£[f(t)] = \frac{F(s)}{s}$$

(*g*) Time Translation or Dead Time:

$$£[f(t-\theta)] = \int_0^{\infty} f(t-\theta)\,e^{-st}\,dt$$

$$= e^{-\theta s}\int_0^{\infty} F(t')\,e^{-st'}\,dt'$$

$$= e^{-\theta s}\int_0^{\infty} F(t')\,e^{-st'}\,dt' = e^{-\theta s}F(s)$$

where θ is dead time in time units and $t' = t - \theta$

Some Important Relationships:

(*h*) Final Value Theorem:

$$F(\infty) = \lim_{t\to\infty} f(t) = \lim_{S\to 0} f(t)\,[sF(s)]$$

(*i*) Initial Value Theorem

$$F(0+) = \lim_{t\to 0} f(t) = \lim_{S\to\infty} [sF(s)]$$

(*j*) Convolution Integral

Given $£[f_1(t)] = F_1(s)$

and $£[f_2(t)] = F_2(s)$

then $£^{-1}[F_1(s) \cdot F_2(s)] = f_1(t) \cdot f_2(t)$

(*k*) **Sealing:**

$$£[f(at)] = \frac{1}{a} F\left(\frac{s}{a}\right)$$

(*l*) Theorem for Periodic Function:

If $f(t)$ is the periodic function *i.e.*, wave with period T,

then $$£[f(t)] = \frac{1}{(1-e^{-Ts})} F_1(s)$$

where $f(t)$ is function which repeats at each interval of T.

(*m*) Differential with respect to S

$$£[t\, f(t)] = -\frac{d\, F(s)}{ds}$$

(*n*) Multiplication by a Constant

$$£[k\, f(t)] = K\, F(s)$$

(*o*) Sum and Difference

$$£[f_1(t) \pm f_2(t)] = F_1(s) \pm F_2(s)$$

(*p*) Shifting Theorem

$$£[f(t-a) \,.\, u(t-a)] = e^{-as}\, F(s)$$

(g) Integration by s

$$£\left[\frac{f(t)}{dt}\right] = \int_0^\infty F(s)\, ds$$

Inverse Laplace Transform by Partial-Fraction Expression

In majority of problems in central system, the evaluation of the inverse Laplace transform does not rely on the use of the inverse integral of equation (5.4) Rather, the inverse Laplace transform operation involving rational function can be carried out using Laplace transform table and partial fraction expression. The looping up of the inverse transform in the transform table can be conducted by computer programs. Partial Fraction-Expansion method is discussed here.

When the Laplace transform solution of differential equation is a rational function in s, it can be written as:

$$G(s) = \frac{Q(s)}{P(s)}$$

where $P(s)$ and $Q(s)$ and polynomial of (s). The order of $P(s)$ may be written as:

$$P(s) = s^n + a_{n-1}s_{n-1} + ... + q_1 s + q_0$$

where a_0, a_1, ..., a_{n-1} are real coefficients. This method of partial-fraction expansion will now be given for the cases of simple poles, multiple-order places, and complex conjugate poles of $G(s)$

$G(s)$ With Simple Poles

If all the poles of $G(s)$ are simple and real, equation (9.24) can be written as:

$$G(s) = \frac{Q(s)}{P(s)} = \frac{Q(s)}{(s+s_1)(s+s_2)...(s+s_n)}$$

where $s_1 \neq s_2 ... \neq s_n$. Applying the partial fraction expansion equation (5.25) is written as:

The coefficient K_{si} $(i = 1, 2, n)$ is determined by multiplying both side of equation 5.25 or 5.26 by the factor $(s + s_i)$ and setting s equal to $-s_i$. To find the coefficient K_{s_1}, for instance, we multiply both sides-of equation 5.25 by $(s + s_1)$ and let $s = s - 1$, therefore, we get

$$K_{s1} = \left[(s+s_1)\frac{Q(s)}{P(s)}\right]\Bigg|_{s=s_1}$$

$$= \frac{Q(-s_1)}{(s_2 - s_1)(s_3 - s_1)...(s_n - s_1)}$$

Consider the function:

$$G(s) = \frac{5s+3}{(s+1)(s+2)(s+3)}$$

which is written in the partial fraction expanded form as:

$$G(s) = \frac{K_{-1}}{s+1} + \frac{K_{-2}}{s+2} + \frac{K_{-3}}{s+3}$$

The coefficients K_{-1}, K_{-2} and K_{-3} are determined as:

$$K_{-1} = [(s+1)\,G(s)]|_{s=-1} = \frac{5(-1)+3}{(2-1)(3-1)} = -1$$

$$K_{-2} = [(s+2)\,G(s)]|_{s=-2} = \frac{5(-2)+3}{(1-2)(2-3)} = 7$$

$$K_{-3} = [(s+3)\,G(s)]|_{s=-3} = \frac{5(-3)+3}{(1-3)(2-3)} = -6$$

Hence equation (5.28) can be written as:

$$G(s) = \frac{-1}{s+1} + \frac{7}{s+2} - \frac{6}{s+3}$$

G(s) with Multiple-order Poles:

If r of the poles of $G(s)$ are identical or we say that the pole at $s = -s_i$ is of the multiplicity r, $G(s)$ can be written as:

$$G(s) = \frac{Q(s)}{P(s)} = \frac{Q(s)}{(s+s_1)(s+s_2)\ldots(s+s_{n-r})(s+s_i)^r}$$

$(i \neq 1, 2, \ldots, n-r)$, then $G(s)$ can be expanded as:

$$G(s) = \frac{Ks_1}{s+s_1} + \frac{K_{s2}}{s+s_2} + \ldots + \frac{K_{s(n-r)}}{s+s_{n-r}}$$

$\leftarrow n-r$ terms of simple poles $\rightarrow$

$$+ \frac{A}{s+s_1} + \frac{A_2}{(s-s_i)^2} + \ldots + \frac{A_r}{(s+s_i)^r} \quad \ldots(5.35)$$

$\leftarrow r$ terms of repeated poles $\rightarrow$

The $(n-r)$ coefficients, $K_{s_1}, K_{s_2}, \ldots K_{s(n-r)}$ which correspond to simple poles, may be evaluated by the method described by equation (5.27). The determination of the coefficient that correspond to the multiple order poles is described as follows:

$$A_G = [(s+s_i)' s_1(s)]|_{s=-s_i} \quad \ldots(5.36)$$

$$A_{r-1} = \frac{d}{ds}[(s+s_i)^r G(s)]|_{s=-s_i} \quad \ldots(5.37)$$

$$A_{n-2} = \frac{1}{\underline{|2}} \frac{d^2}{ds^2}[(s+s_i)^r G(s)]|_{s=-s_i} \quad \ldots(5.38)$$

$$A_1 = \frac{1}{\underline{|r-1}} \frac{d^r-1}{ds^{r-1}}[(s+s_i)^r G(s)].|_{s=s_i} \quad \ldots(.5.39)$$

Consider the function

$$G(s) = \frac{1}{s(s+1)^3(s+2)} \quad \ldots(5.40)$$

By using the approach of equation (5.35), $G(s)$ can be written as:

$$G(s) = \frac{K_0}{s} + \frac{K_{-2}}{s+2} + \frac{A_1}{s+1} + \frac{A_2}{(s+1)^2} + \frac{A_3}{(s+1)^3} \quad \ldots(5.41)$$

The coefficient corresponding to the samples are:

$$K_0 = [s\,G(s)]|_{\,s=0} = \frac{1}{2} \qquad \text{...(5.42)}$$

$$K_{-2} = [(s+2)\,G(s)]|_{\,s=-2} = \frac{1}{2} \qquad \text{...(5.43)}$$

and those of the third-order pole are:

$$A_3 = [(s+1)^3\,G(s)]\,|_{S=-1} = -1$$

$$A_2 = \frac{d}{ds}[(s+1)^2\,G(s)]\Big|_{s=-1}$$

$$= \frac{d}{ds}\left[\frac{1}{s(s+2)}\right]\Bigg|_{s=-1} = 0 \qquad \text{...(5.44)}$$

$$A_3 = \frac{1}{\underline{|2}}\frac{d^2}{ds^2}\left[(s+2)\,3a(s)\right]\Big|_{s=-1}$$

$$= \frac{1}{2}\frac{d^2}{ds^2}\left[\frac{1}{s(s+2)}\right]\Bigg|_{s=-1} = -1 \qquad \text{...(5.45)}$$

The completed partial fraction expression is:

$$G(s) = \frac{1}{2s} + \frac{1}{2(s+2)} - \frac{1}{s+1} - \frac{1}{2(s+1)^3} \qquad \text{...(5.46)}$$

$G(s)$ has simple complex conjugate poles

The partial-fraction expansion of equation (5.23) is also valid for simple complex-conjugate poles. Since complex conjugate poles are more difficult to handle and are of special interest in control system studies the decrease special treatment.

Suppose that $G(s)$ of equation (5.25) contains a pair of complex poles:

$$s = -d + j\omega \quad \text{and} \quad s = -d - j\omega$$

The corresponding coefficient of these poles are:

$$K - d + j\omega = (s + d - j\omega)\,G(s)\big|_{s=-dij\omega} \qquad \text{...(5.47)}$$

$$s{-}d + j\omega = (s + d + j\omega)\,G(s)\big|_{s=-d-j\omega} \qquad \text{...(5.48)}$$

Consider the function

$$G(s) = \frac{\omega^2 n}{s^2 + 2s\omega_n s + \omega_n^2} \qquad \text{...(5.49)}$$

Let us presume that the value of 5 and ω_n are such that the poles of $G(s)$ are complex.

The $G(s)$ is expanded as follows:

$$G(s) = \frac{K - a + j\omega}{s + d - j\omega} + \frac{K - a - j\omega}{s + d + j\omega} \quad ...(5.50)$$

where $$a = 5\omega_n \quad ...(5.51)$$

and $$\omega = \omega_n\sqrt{1 - 5^2} \quad ...(5.52)$$

The coefficient in equation (5.50) are determined as follows:

$$K - d + j\omega = (s + d - j\omega)\, G(s)|_{s = -d + j\omega} = \frac{\omega^2 n}{2\, j\omega} \quad ...(5.53)$$

$$K - d - j\omega = (s - d - j\omega)\, Gs\,|_{s = -d - j\omega} = \frac{\omega^2 n}{2\, j\omega} \quad ...(5.54)$$

The complete partial fraction expansion of equation (5.50) is given by

$$G(s) = \frac{\omega^2 n}{2\, j\omega}\left[\frac{1}{s + d - j\omega} - \frac{1}{s + d + j\omega}\right] \quad ...(5.55)$$

Taking the inverse Laplace transform on bath sides of equation (5.55), we get:

$$g(t) = \frac{\omega^2 n}{2\, j\omega} e^{-at}\left(e^{j\omega t} - e^{-j\omega t}\right)$$

$$= \frac{\omega_n}{\sqrt{1 - 5^2}} e^{-5\omega_n t} \sin \omega_n \sqrt{1 - 5^2 t}$$

for $t \geq 0$

5.2.2 Routh-Hurwitz Criterion

This criterion is an algebraic method which provides information as the absolute stability of a linear time-invariant system that has a characteristic equation with constant coefficient *i.e.*, the right half of *s*-plane. The number of roots which lie on the $j\omega$ axis and in right half plane is also indicated. The Routh-Hurwitz criterion in short represents a method of determining the location of zeros of a polynomial with constant real coefficient with respect to the left half of the *S*-plane, without actually solving for the zero. The characteristic equation of a linear system is defined as the equation obtained by setting the denominator polynomial of the transfer function to zero. Consider that the characteristic equation of a linear time invariant single output system is given by:

$$F(s) = a_n s^n + a_{n-1} s^{n-1} + a_1 s + a_0 = 0 \quad ...(5.56)$$

Where all the coefficients are real in order that equation (5.62) not have roots with positive real parts it is necessary and sufficient the following conditions are met:

(1) All the coefficients of the equation have the same sign.

(2) None of the coefficients vanishes.

The conditions of above case based on the laws of algebra which relate the coefficient of equation (5.56) as follows:

$$\frac{a_{n-1}}{a_n} = -\Sigma \text{ all roots} \qquad ...(5.57)$$

$$\frac{a_{n-2}}{a_n} = \Sigma \text{ product of the roots taken two at a time} \qquad ...(5.58)$$

$$\frac{a_{n-3}}{a_n} = -\Sigma \text{ product of the roots taken three at a time} \qquad ...(5.59)$$

$$\frac{a_0}{a_n} = (-1)4\ s \text{ product of all the roots.}$$

Hence, all these ratios must be positive and zero under at least one of the roots has a positive real part.

The two necessary conditions for equation (5.62) to have no roots in the right half suppose can easily be checked by inspection of the equation. However, these conditions are not sufficient, as it is quite possible that an equation with all its coefficient non-zero and of the same sign still may not have all the roots in the left half of the s-plane.

5.2.1.1 Hurwitz Criterion

The Routh Hurwitz criterion is based on the Hurwitz Criterion, which is the necessary and sufficient conditions that all roots of equation (5.18) lie in the left half of the s-plane is that the equation's. Hurwitz determinant, D_K, $K = 1, 2, 3 ..., n$ must all be positive

The Hurwitz determinants of equation (5.18) is given by

$$D_1 = a_{n-1}$$

$$D_2 = \begin{vmatrix} a_{n-1} & a_{n-3} \\ a_n & a_{n-2} \end{vmatrix}$$

$$D_3 = \begin{vmatrix} a_{n-1} & a_{n-3} & a_{n-5} \\ a_n & a_{n-2} & a_{n-4} \\ 0 & a_{n-1} & a_{n-3} \end{vmatrix}$$

$$D_n = \begin{vmatrix} a_{n-1} & a_{n-3} & a_{n-5} & \cdots & 0 \\ a_n & a_{n-2} & a_{n-4} & \cdots & 0 \\ 0 & a_{n-1} & a_{n-3} & \cdots & 0 \\ \vdots & \vdots & \vdots & \cdots & \vdots \\ 0 & 0 & 0 & \cdots & 0 \end{vmatrix}$$

where the coefficients with indices larger than or negative should be replaced with zero. At the first glance, the application of the Huriwit determinants may seem to be formidable for high order equations due to the amount of work involved in evaluations and determinants in the equation 5.26.

5.2.1.2 Routh's Tabulation

The first step in the simplification of the Hurwitz Criterion is known as the Routh-Hurwitz Criterion, is to arrange coefficients of the equation (5.62) into the rows. The first row consists of the first, third, and fifth ... coefficients and the second row consists of the second, fourth sixth ... coefficients all accounting from the highest order item, as shown in the following tabulation:

$$\begin{array}{cccc} a_n & a_{n-2} & a_{n-4} & a_{n-6} \\ a_{n7} & a_{n-7} & a_{n-5} & a_{n-7} \end{array}$$

The next step is to form the following array of the numbers of the indicated operations, illustrated here for a sixth-order equation

$$a_6 s^6 + a_5 s^5 + a_4 s^4 + a_3 s^3 + a_2 s^2 + a_1 s_1 + a_0 = 0$$

s^6	a_6	a_4	a_2	a_6
s^5	a_5	a_3	a_1	a
s^4	$A = \dfrac{a_5 a_4 - a_6 a_3}{a_5}$	$B = \dfrac{a_5 a_2 - a_6 a_1}{a_5}$	$Q_0 = \dfrac{a_5 a_0 - a_6 \times 0}{a_5}$	0
s^3	$C = \dfrac{A a_3 - a_5 B}{A}$	$D = \dfrac{A a_1 - a_5 a_0}{A}$	$0 = \dfrac{A \times 0 - a_5 \times 0}{A}$	0
s^2	$F = \dfrac{BC - AD}{C}$	$a_0 = \dfrac{C a_0 - A \times 0}{C}$	$0 = \dfrac{C \times 0 - A \times 0}{C}$	0
s^1	$\dfrac{ED - D a_0}{E}$	$0\,0$	0	
s^0	$a_0 = \dfrac{F a_b - E \times 0}{F}$	0	0	0

The array above is called the Routh tabulation or Routh's array. The column of 3's on the left side is used for identification purposes there reference column keep tract of the calculations and the last row of the Routh's tabulation should always by the S raw.

The roots of the equation are all in the left half of the s-plane if all the elements of the first column of the Routh's tabulation are of the same sign. The number of changes of signs in the elements of the first column equals the number of roots with positive real parts or in the right half s-plane.

The reason for the foregoing conclusion is simple, based on the requirements of the Hertz determinants. The relations between the elements in the first column of the Routh's tabulation and the Hurwitz determinants are:

s_6	$a_5 = a_6$
s_5	$a_5 = D_1$
s_4	$A = D_2/D_1$
s_3	$C = D_3/D_2$
s_2	$E = D_4/D_3$
s_1	$F = D_5/D_4$
s_0	$a_0 = \dfrac{D_6}{D_5}$

Hence, if all the Hurwitz determinants are positive, elements in the first column of the Routh's tabulation would be of the same sign.

Consider the equation:

$$(s - 2)\,(s + 1)\,(s - 3) = s^3 - 4s^2 + s + 6 = 0$$

which has one negative coefficient. Hence, from the necessary condition, we know without applying Routh's test that not all the rats of the equations one in the left half of s-plane. In fact, from the factored form of the equation, we know that there are two roots in the right half of s-plane, at $s = 2$, and $s = 3$. For the purpose of illustrating the Routh-Hurwitz criterion, Routh tabulation is made as follows:

s^3	1	1
Sign change		
s^2	–4	6
Sign change		
s^1	$\dfrac{(-4)\,(-1) - (6)\,(1)}{-4} = 2.5$	0
s^0	$\dfrac{(2.5)\,(6) - (-4)\,(0)}{2.5} = 6$	0

Since these are two sign changes in the first column of the tabulation, the equation has two roots located in the right half of s-plane. This is conformity with known result.

Consider the equation:

$$2s^4 + s^3 + 3s^2 + 5s + 10 = 0$$

Since the equation has no missing terms and the coefficient are all of the same sign it statistics the necessary conditions for not having roots in the right half or on the imaginary axis, of the s-plane. However, the sufficient condition must still be checked. Routh's tabulation is as follows:

s^4	2	3	10
s^3	1	5	0
s^2	$\dfrac{(1)(3)-(2)(5)}{1} = -7$	10	0
s^1	$\dfrac{(-7)(5)-(1)(10)}{-7} = 6.43$	0	0
s^1	10	0	0

Since there are two changes in sign in the first column of the tabulation, the equation has two roots in the right half of the s-plane. Hence, the system is unstable.

5.2.1.3 Special Cases when Routh's Tabulation terminates prematurely

Suppose (*i*) The first element in any one row of the Routh's tabulation is zero, but the others are not; or (*ii*) the element in one row of Routh'

σ tabulation: In the first, if a zero appears in the first element of a row, the element in the next row will all become infinite and Routh's tabulation cannot continue. To remedy the situation, we replace the zero element in the first column by an arbitrary small positive number ε, and then proceed with Routh's tabulation.

In the second special case, when all the elements in one row of Routh's tabulation are zeros before the tabulation is properly terminated, it indicates that one or more of the following may exist:

(*i*) The equation has at least one pair of real roots with equal magnitude but opposite signs.

(*ii*) The equation has one or more pairs of imaginary roots.

(*iii*) The equation has pairs of complex-conjugate roots forming symmetry about the region of the s-plane.

The situation with the entire row of zeros can be remedied by using auxiliary equation $A(S) = 0$ which is formed from the coefficient of the row just above the row of zeros in Routh's tabulation. The auxiliary equation is always an even polynomial *i.e.*, only one power of S appear. To continue with Routh's tabulation when a row of zeros appears. The following steps are conducted:

(*i*) From the auxiliary equation $A(s) = 0$ by use of the coefficients from the just proceeding the row of zeros.

(*ii*) Take the derivative of the auxiliary equation with respect to s, this gives

$$\frac{dA(s)}{ds} = 0$$

(*iii*) Replace the row of zeros with the coefficients of $\frac{dA(s)}{d(s)} = 0$

(*iv*) Continue with Routh's tabulation in the usual manner with the newly formed row of coefficient replacing the row of zeros.

(*v*) Interpret the change of signs, if any, of the coefficient in the first column of Routh's tabulation in the usual manner.

5.3 CONCEPT OF COMPLEX FREQUENCY

The solution of differential equation gives rise to terms such as:

$$I = I_0 e^{st}$$

Where $s = \sigma + j\omega$ which is known as complex frequency. σ is real part and $j\omega$ is imaginary part with ω as angular frequency can be understood in the following way:

$$\begin{aligned} I &= I_0 e^{st} = I_0\, e^{(\sigma + j\omega)t} \\ &= I_0\, e^{\sigma t} \cdot e^{j\omega t} \\ &= I_0\, e^{\sigma t} (\cos\omega t + j \sin \omega t) \end{aligned}$$

σ effect the overall quantity. The effects of σ can be seen in the following case:

(*i*) If $s = 0$, then the signal is unaffected and behaviour is shown in the Fig. 5.3. The signal is unaffected.

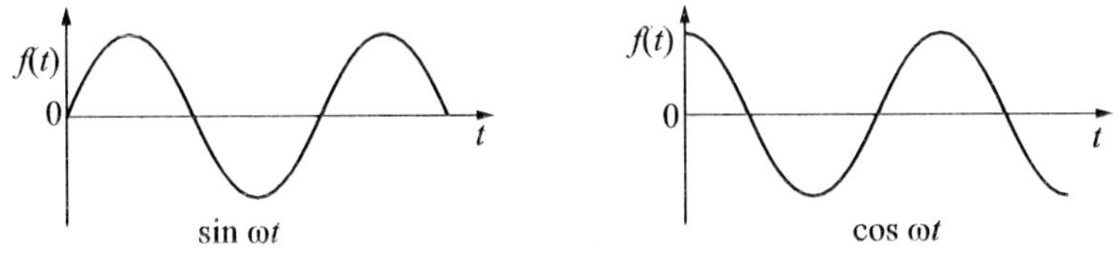

Fig. 5.3 *Real part $\sigma = 0$*

(*ii*) If $\sigma < 0$, then the signal reduces exponentially and the behaviour is shown in the Fig. 5.4

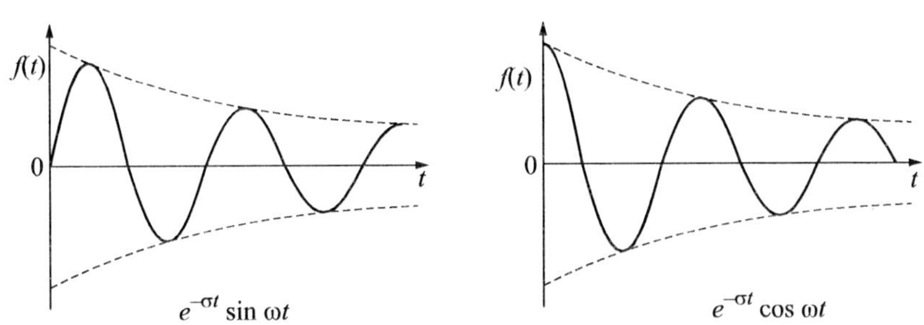

Fig. 5.4 *Real part $\sigma < 0$*

(*iii*) If $\sigma > 0$, then the signal increase exponentially as shown in the Fig. 5.5

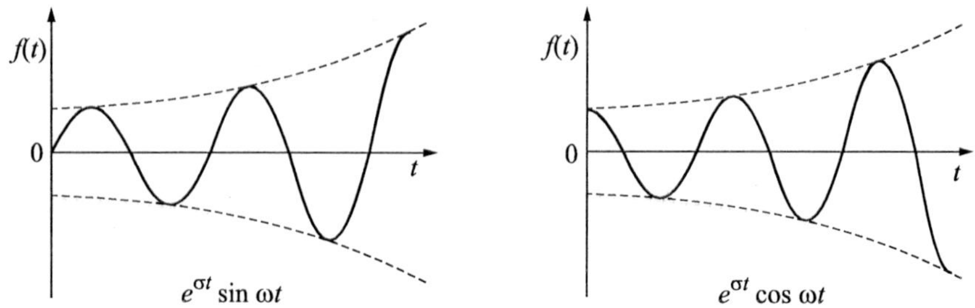

Fig. 5.5 *Real part* $\sigma > 0$

It can be seen that *σ decides the magnitude and ω decides the number of oscillations per second of the function.* The role of two kinds of frequencies is same even though the consequences are different. The two frequencies when combined, become complex frequency.

Network Analysis using Laplace Transform

Impedance and admittance are important parameters of a network analysis. These are given a common name of Immittance.

Network can be transformed in *s*-domain and their representation can be using Laplace transformed elements.

(*i*) Register in *s*-domain: A register R coming a $i(t)$ having a $v(t)$ a voltage across it is considered.

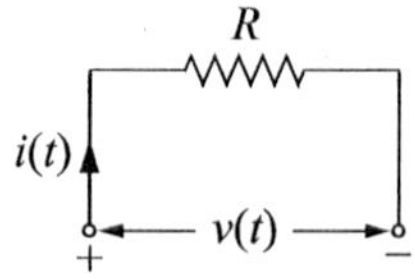

As per Ohm's Law:

$$v(t) = i(t) \times R$$

By taking Laplace transform of both the sides of the equation, we get:

$$V(s) = I(s)\, R$$

Thus, the Laplace transformed circuit becomes:

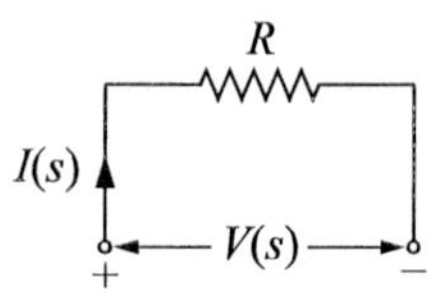

Transformed impedance,

$$Z(s) = \frac{V(s)}{I(s)} = R$$

and transformed admittance:

$$Y(s) = \frac{I(s)}{V(s)} = \frac{I}{R} = G = \text{Conductance.}$$

(*ii*) Inductor in *s*-domain: An inductor L carrying a current $i(t)$ having a voltage $V_L(t)$ across it is considered.

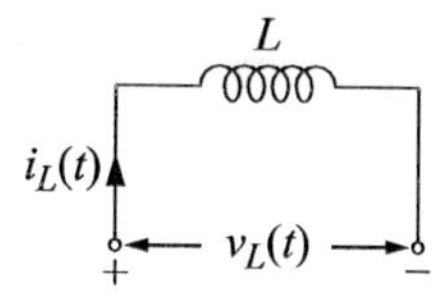

The voltage $V_L(t)$ is given by:

$$V_L(t) = L\frac{di_L(t)}{dt}$$

By taking Laplace transform of both the sides of the equation, we get:

$$V_L(s) = L[s\,I_L(s) - i_2(0^-)]$$

or $\quad V_L(s) = I_L(s) \times (sL)$, if initial current through the inductor is zero.

Thus, the Laplace transformed circuit becomes:

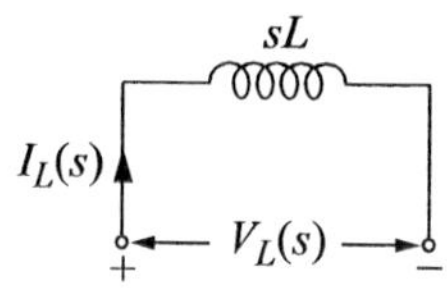

Transformed impedance,

$$Z(s) = \frac{V_L(s)}{I_L(s)} = sL$$

and transformed admittance:

$$Y(s) = \frac{I_L(s)}{V_L(s)} = \frac{1}{sL}$$

(*iii*) Capacitor in *s*-domain: A capacitor C carrying a current $i_c(t)$ having a voltage $v_c(t)$ across it is considered.

The voltage $v_c(t)$ is given by:

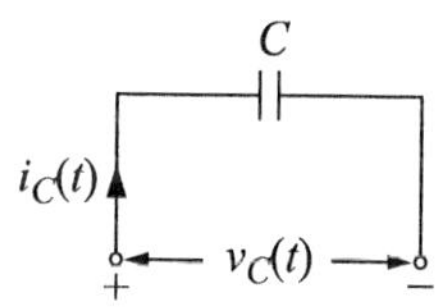

$$v_c(t) = \frac{1}{C}\int_{-\infty}^{t} i_c(t)\,dt$$

$$= \frac{1}{C}\int_{-\infty}^{0^-} i_c(t)\,dt + \frac{1}{C}\int_{0}^{t} i_c(t)\,dt$$

$$= v_c(0^-) + \frac{1}{C}\int_{0}^{-\infty} i_c(t)\,dt$$

By taking Laplace transform of both sides of the equation, we get:

$$V_c(s) = \frac{V_c(0^-)}{s} + \frac{I_c(s)}{sc}$$

or $\quad V_c(s) = \dfrac{I_c(s)}{sc}$, if initial voltage across the capacitor is zero.

Thus, the Laplace transformed circuit becomes:

Transformed impedance,

$$Z(s) = \frac{V_c(s)}{I_c(s)} = \frac{1}{sc}$$

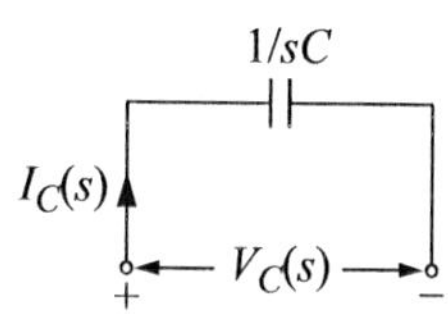

and transformed admittance:

$$y(s) = \frac{I_c(s)}{V_c(s)} = sc$$

***(iv)* Network in *s*-domain:** We have seen transformations of *R*, *L*, *C* elements based circuits. It can be observed that the *R*, *L*, *C* elements in *s*-domain are given by:

Element	*Value in s-domain*
R	*R*
L	*sL*
C	$\frac{1}{sc}$

The networks with above replacements for respective elements, give the transformed network in *s*-domain. A such impedances can be combined and network can be simplified using series combinations, parallel combinations and techniques like star-delta or delta-star conversions. Ohm's law is applicable *i.e.*, $V(s) = I(s) \times Z(s)$.

Example 5.1 Find the driving point impedance function $z(s)$ and driving point admittance function $y(s)$ for the one-part network shown in the figure.

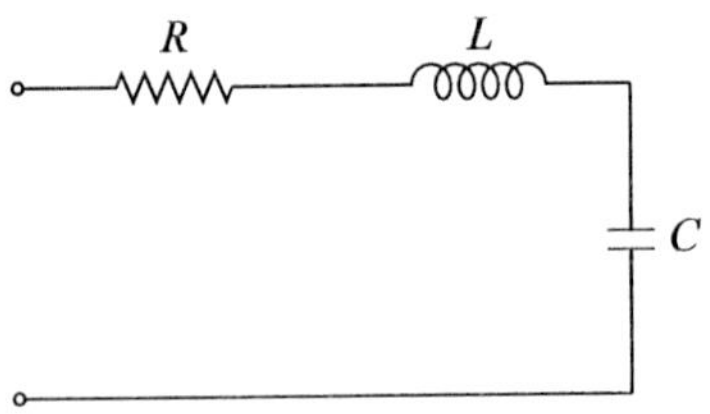

Solution: The network in *s*-domain is shown in Fig. (*a*) as shown below.

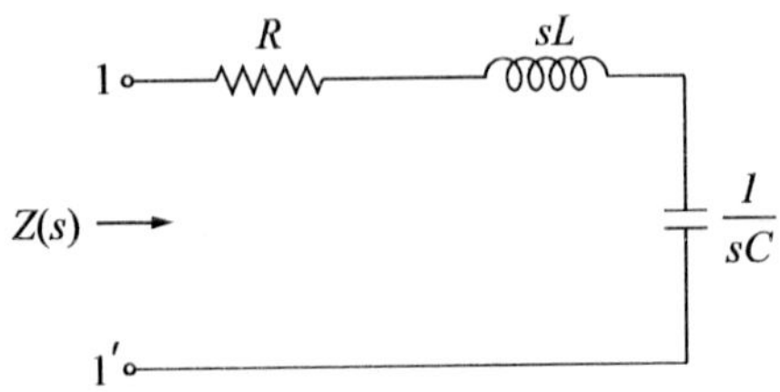

The driving point impedance function,

$$Z(s) = R + sL + \frac{1}{sc} = \frac{scR + s^2\ Lc + 1}{sc}$$

or
$$Z(s) = \frac{s^2\ Lc + sRc + 1}{sc}$$

The driving point admittance function,

$$y(s) = \frac{sc}{s^2 Lc + sRc + 1}$$

Example 5.2 For the given one-part network as shown in the figure, find driving point admittance function.

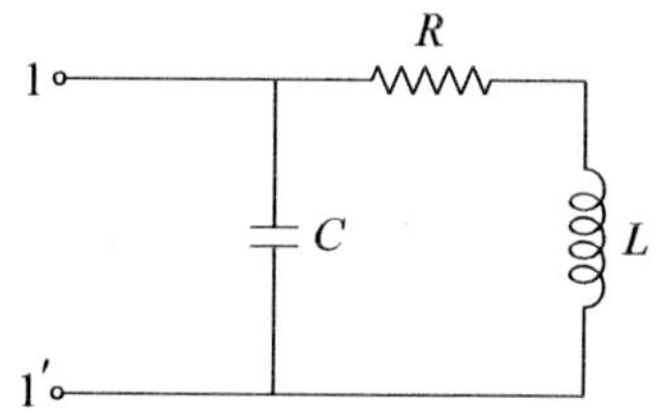

Solution: The network in s-domain is given in Fig. (a) as shown below:

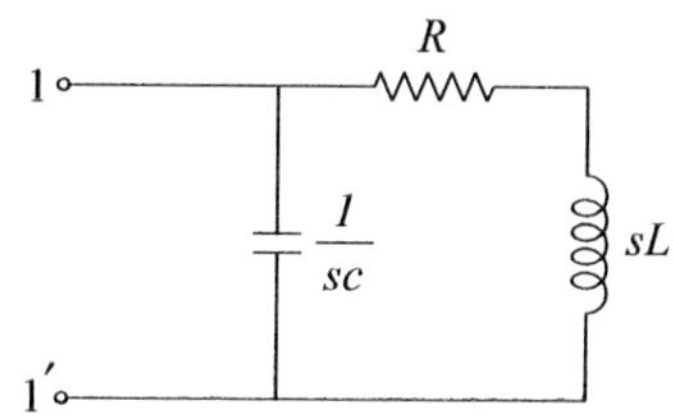

(a) *Network in s-domain*

The driving point admittance,

$$y(s) = y_1(s) + y_2(s)$$

where, $$y_1(s) = sc$$

and $$y_2(s) = \frac{1}{R + sL}$$

$\therefore$ $$y(s) = y_1(s) + y_2(s) = sc + \frac{1}{R + sL}$$

or $$y(s) = \frac{s^2 Lc + sRc + 1}{R + sL}$$

5.4 ONE OR TWO PORT NETWORKS

The network is treated as a black box. A pair of terminals are connected to the black box at which signal may enter the network or leave the network. Such black box is known as one port network as shown in Fig. 5.6.

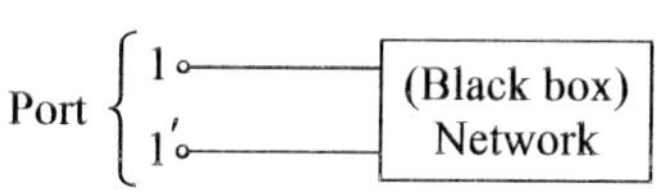

Fig. 5.6 *One port network*

When the black box has one pair of terminals to enter the signals and another pair of terminals to leave the network, then it is known as two port network as shown in the Fig. 5.7.

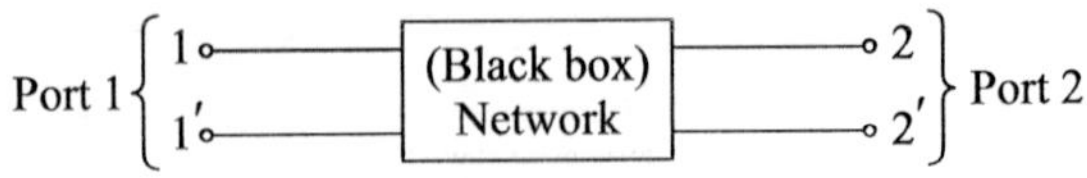

Fig. 5.7 *Two port network*

5.4.1 Network Function of One Port Network

In the case of one port network, there are two variables namely voltage. $v(t)$ and current $i(t)$ variables w.r.t time. When the network functions are related to the two variables and measured at the same part, then it is known as *driving point function*. One port network and the two variables are shown in the Fig. 5.8.

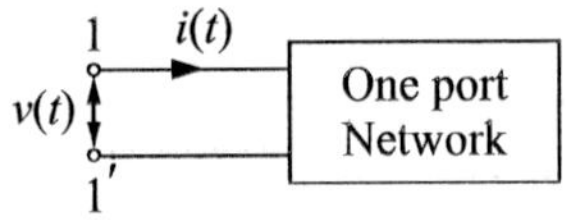

Fig. 5.8 *One port network with two variables i.e., voltage and current w.r.t time*

(*i*) **Driving Point Impedance:** The ratio of the Laplace transform of the voltage and the current at the port of the single port network is known as driving point impedance function. *In such* case, the initial conditions are neglected.

The driving point impedance is defined as:

$$Z(s) = \frac{V(s)}{I(s)} \qquad ...(5.60)$$

where $V(s)$ = Laplace transform of $v(t)$

$I(s)$ = Laplace transform of $i(t)$.

(*ii*) **Driving Point Admittance:** The ratio of transform of current $i(t)$ to voltage $v(t)$ is known as driving point admittance.

$$\therefore \qquad Y(s) = \frac{I(s)}{V(s)} \qquad ...(5.61)$$

where $I(s)$ = Laplace transform of the current

$V(s)$ = Laplace transform of the voltage.

5.4.2 Network Functions of Two Port Network

Two port network is shown in the Fig. 5.9. The variables measured at port 1 are $v_1(t)$ and $i_1(t)$, while the variables measured at port 2 are $v_2(t)$ and $i_2(t)$,

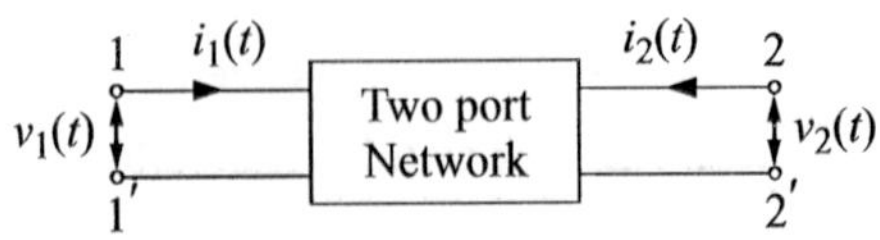

Fig. 5.9 *Two port network with voltage and current variables at each port.*

The ratio of the variables of same port is known as point function.

1. **Driving Point Impedance Functions:** Ratio of Laplace transform of $v_1(t)$ and $i_1(t)$ is known as driving point impedance of port 1.

$$Z_{11}(s) = \frac{V_1(s)}{I_1(s)} \qquad ...(5.62)$$

Further, ratio of Laplace transform of $v_2(t)$ and $i_2(t)$ for port 2 is known as driving point impedance of port 2.

$$Z_{22} = \frac{V_2(s)}{I_2(s)} \qquad ...(5.63)$$

2. **Driving Point Admittance Functions:** Reciprocal of driving point impedance of port 1 is known as driving point admittance of port 1.

$$Y_{11}(s) = \frac{I_1(s)}{V_1(s)} \qquad ...(5.64)$$

Similarly, reciprocal of driving point impedance of port 2 is known as driving point admittance of port 2.

$$Y_{22}(s) = \frac{I_2(s)}{V_2(s)} \qquad ...(5.65)$$

3. **Transfer Functions of Two Port Network:** Ratio of Laplace transforms of two variables such that one variable is defined at one port and the other variable is defined at the second port, is known as transfer function. As such, these are four transfer functions of two port networks.

(*i*) **Voltages Ratio Transfer Function:** Ratio of Laplace transforms of voltage at one port to voltage at the other port is known as voltage ratio transfer function.

$$G(s) = \frac{V_2(s)}{V_1(s)} \qquad ...(5.66)$$

(*ii*) **Current Ratio Transfer Function:** Ratio of Laplace transforms of current at one port to current at other port is known as current ratio transfer function.

$$\alpha(s) = \frac{I_2(s)}{I_1(s)} \qquad ...(5.67)$$

(*iii*) **Transfer Impedance Function:** Ratio of Laplace transforms of voltage at one port to current at the other port is known as transfer impedance function.

$$Z_{21} = \frac{V_2(s)}{I_1(s)} \qquad ...(5.68)$$

$$\text{and} \qquad Z_{12}(s) = \frac{V_1(s)}{I_2(s)} \qquad \text{...(5.69)}$$

(*iv*) **Transfer Admittance Function:** Ratio of Laplace transforms of current at one port to voltage at the other port is known as transfer admittance function.

$$Y_{21}(s) = \frac{I_2(s)}{V_1(s)} \qquad \text{...(5.70)}$$

$$\text{and} \qquad Y_{12}(s) = \frac{I_1(s)}{V_2(s)} \qquad \text{...(5.71)}$$

(*v*) **Ladder Networks:** A ladder network shown in the figure has its arms marked with impedances and admittances.

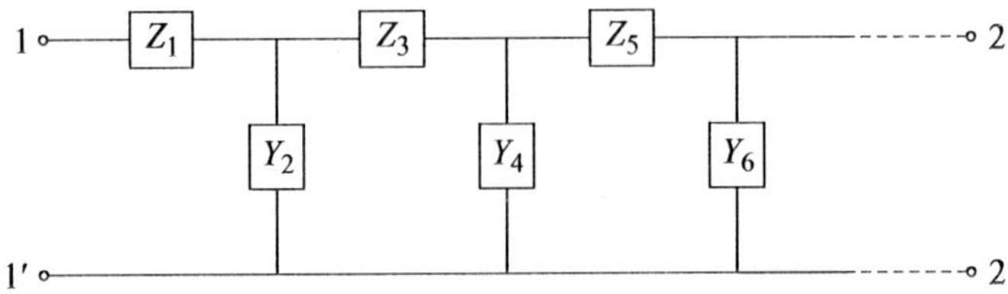

Fig. 5.10 *Ladder diagram*

The driving point impedance at port 1 is given by continued fraction:

$$Z = z_1 + \cfrac{1}{y_2 + \cfrac{1}{z_3 + \cfrac{1}{y_4 + \cfrac{1}{z_5 + \cfrac{1}{y_6 + \cdots}}}}}$$

5.5 IMPULSE RESPONSE AND NETWORK FUNCTION

A system is shown in Fig. 5.11 where in it has $e(t)$ as input or output and $r(t)$ as output or response of the network.

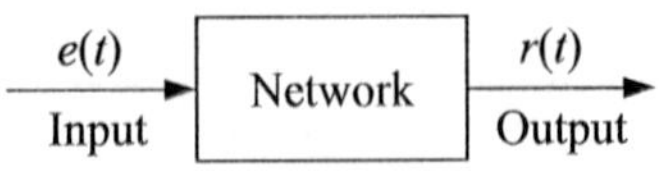

Fig. 5.11 *Network with input and output.*

If Laplace transforms of input, network equations and output are taken, then the Laplace transform representation will be as shown in Fig. 5.11.

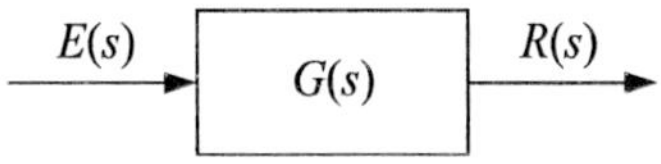

Fig. 5.12 *Laplace transform Representation of network with input and output*

Thus, $$G(s) = \frac{R(s)}{E(s)} \quad ...(5.72)$$

i.e. $$R(s) = G(s) \cdot E(s)$$

The input is unit impulse $\delta(t)$,

then $$e(t) = \delta(t)$$

and $$E(s) = £[\delta(t)] = 1$$

$\therefore$ $$R(s) = G(s) \cdot 1$$

or $$G(s) = R(s) \quad ...(5.73)$$

This implies that the Laplace transform of the response is same as the system function, for unit impulse input. In terms of time function, the impulse response is given by:

$$g(t) = £^{-1}\,[R(s)] = £^{-1}[G(s)]$$

or $$g(t) = £^{-1}[G(s)] \quad ...(5.74)$$

Example 5.3 A transformed network is shown below. Find $z_{11}(s)$ and $G_{12}(s)$.

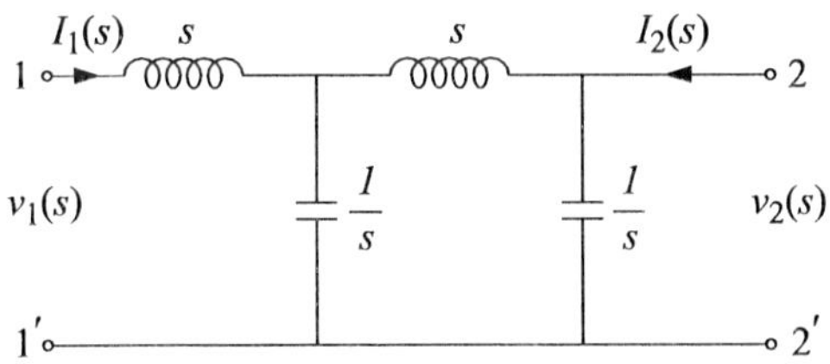

Solution: From the given figure, we get:

$$Z_{11}(s) = s + \frac{t \times \left(s + \frac{1}{s}\right)}{\frac{1}{s} + s + \frac{1}{s}}$$

$$= s + \frac{\left(1 + \frac{1}{s^2}\right)}{\left(\frac{2}{s} + s\right)} = s + \frac{s^2 + 1}{s^3 + 2s}$$

$$= \frac{s^4 + 2s^2 + s^2 + 1}{s^3 + 2s}$$

or $$Z_{11}(s) = \frac{s^4 + 3s^2 + 1}{s^3 + 2s}$$

$$I_1(s) = \frac{V_1(s)}{\left(\dfrac{s^4+3s^2+1}{s^3+2s}\right)} = v_1(s)\left(\frac{s^3+2s}{s^4+3s^2+1}\right)$$

$$I_2(s) = I_1(s)\times\frac{\frac{1}{s}\left(1+\frac{1}{s}\right)}{\frac{1}{s}+s+\frac{1}{s}}\times\frac{1}{\left(s+\frac{1}{s}\right)}$$

or
$$I_2(s) = I_1(s)\times\frac{1}{s}\times\frac{1}{\left(s+\frac{2}{s}\right)}$$

$$V_2(s) = I_2(s)\times\frac{1}{s}$$

$\therefore$
$$V_2(s) = \frac{1}{s^2}\times\frac{1}{\left(s+\frac{2}{s}\right)}\times\left(\frac{s^3+2s}{s^4+3s^2+1}\right)V_1(s)$$

$$= \frac{1}{\left(s^3+2s\right)}\times\frac{(s^3+2s)}{(s^4+3s^2+1)}V_1(s)$$

$$G_{12} = \frac{V_2(s)}{V_1(s)} = \frac{1}{(s^4+3s^2+1)}$$

Example 5.4 For the network shown in the figure find the expressions for the network functions:

(*i*) Voltage ratio transfer function $\left(\dfrac{V_2}{V_1}\right)$.

(*ii*) Current ratio transfer function $\left(\dfrac{I_2}{I_1}\right)$.

(*iii*) Transfer impedance function $\left(\dfrac{V_2}{I_1}\right)$.

(*iv*) Transfer admittance function $\left(\dfrac{I_2}{V_1}\right)$.

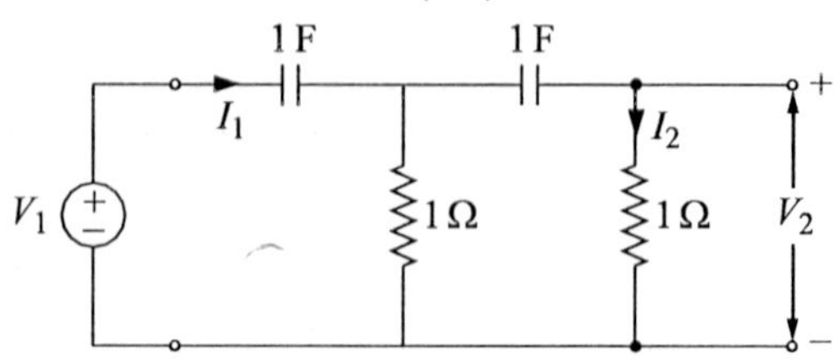

Solution: The network in *s*-domain is as given below:

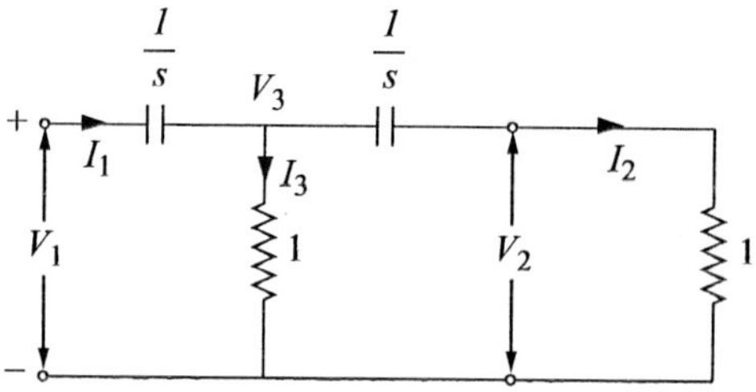

Current $$I_2 = \frac{V_3 - V_2}{\left(\frac{1}{s}\right)} \neq$$

or $$I_2 = s(V_3 - V_2) \quad ...(i)$$

Voltage $$V_2 = I_2 \times 1$$

or $$V_2 = I_2 \quad ...(ii)$$

From equations (*i*) and (*ii*), we get:

$$V_2 = s(V_3 - V_2)$$

or $$V_3 = \frac{s+1}{s} V_2 \quad ...(iii)$$

By applying KCL at node (3), we get

$$I_1 = I_3 + I_2$$

or $$I_1 = \frac{V_3}{1} + \frac{V_2}{1}$$

or $$I_1 = (V_3 + V_2)$$

By putting value of V_3 from equation (*iii*), we get:

$$I_1 = \left(\frac{s+1}{s}\right) V_2 + V_2 = \left(\frac{s+1}{s} + 1\right) V_2$$

or $$I_1 = \left(\frac{2s+1}{s}\right) V_2 \quad ...(iv)$$

We also know that:

$$I_1 = \frac{V_1 - V_3}{\frac{1}{s}} = s(V_1 - V_3) \quad ...(v)$$

From equations (*iv*) and (*v*), we get:

$$\left(\frac{2s+1}{s}\right) V_2 = s(V_1 - V_3)$$

$$\therefore \qquad \left(\frac{2s+1}{s}\right) V_2 = sV_1 - sV_3$$

By substituting value of V_3 from equation (*iii*), we get:

$$\left(\frac{2s+1}{s}\right) V_2 = sV_1 - s\left(\frac{s+1}{s}\right) V_2$$

or $$\left(\frac{2s+1}{s}\right)V_2 + (s+1)V_2 = sV_1$$

or $$\frac{2s+1+s^2+s}{s} V_2 = sV_1$$

or $$V_1 = \frac{s^2+3s+1}{s^2} V_2 \qquad ..(vi)$$

or $$\frac{V_2}{V_1} = \frac{s^2}{s^2+3s+1} = \text{Voltage ratio transfer function.}$$

By putting $V_2 = I_2$ in equation (*iv*), we get:

$$\frac{I_1}{I_2} = \frac{2s+1}{s}$$

or $$\frac{I_2}{I_1} = \frac{s}{2s+1} = \text{Current ratio transfer function.}$$

From equation (*iv*), we can write:

$$\frac{V_2}{I_1} = \frac{s}{2s+1} = \text{Transfer impedance function.}$$

From equation (*vi*) by putting $V_2 = I_2$, we gets

$$\frac{I_2}{V_1} = \frac{s^2}{s^2+3s+1} = \text{Transfer admittance function.}$$

5.6 CONCEPT OF POLES AND ZEROS

System function is the ratio of Laplace transforms of output to input. It can be expressed as the ratio of two polynomials:

$$G(s) = \frac{Q(s)}{P(s)} \qquad ...(5.75)$$

when $Q(s)$ = numerator polynomial in 's' having a degree 'm'.

$P(s)$ = denominator polynomial in 's' having a degree 'n'.

$$\therefore \qquad G(s) = \frac{b_m s^m + a_{m-1} s^{m-1} + b_{m-2} s^{m-2} + \cdots + b_0}{a_n s^n + a_{n-1} s^{n-1} + a_{n-2} s^{n-1} + \cdots + a_0} \qquad ...(5.76)$$

The numerator has m factorization and denominator has n factorizations.

$$\therefore \qquad G(s) = \frac{k(s - z_1)(s - z_2)\cdots(s - z_m)}{(s - p_1)(s - p_2)\cdots(s - p_n)} \qquad ...(5.77)$$

where $z_1, z_2, ..., z_m$ are the roots of the polynomial $Q(s) = 0$ while $p_1, p_2, ..., p_n$ are the roots of the polynomial $P(s) = 0$.

k = system gain factor or scale factor

(*i*) **Poles:** The values of s which make the system function infinite are known as poles of the system. The poles are the roots of the equation $P(s) = 0$. If the denominator of the system function has roots $p_1, p_2, ...,$ p_n, then $p_1, p_2, ... p_n$ are *known* as poles of the system function. In other words, poles are the roots of the equation obtained by equating denominator polynomial to zero. Such an equation is known as characteristic equation of a system. If the pole are real, then it is simple pole. It the pole values repeat then it is repeated pole. A pair of pole with complex conjugate values are known as conjugate poles.

(*ii*) **Zeros:** The values of s which make the system function zero are known as zeros of the system function. The zeros are the roots of the equation $Q(s) = 0$. If the numerator of the system function has roots $z_1, z_2, ..., z_m$, then $z_1, z_2, ..., z_m$ the zeros of the system function.

Again, similar to poles, zero also can be simple zeros, repeated zeros, or complex conjugate zeros.

5.6.1 Restrictions on Poles and Zeros of Driving Point Immittance Functions

The necessary conditions for driving point immittance functions with given

$G(s) = \dfrac{Q(s)}{P(s)}$ are given below:

(*i*) The coefficients in the polynomials $P(s)$ and $Q(s)$ of the function must be real and positive.

(*ii*) The poles and zeros of $G(s)$ must be conjugate if imaginary or complex.

(*iii*) The real part of all poles and zeros must be negative or zero.

(*iv*) If the real part of any pole and zero is zero, then that pole or zero must be simple.

(*v*) The polynomials $P(s)$ and $Q(s)$ may not have missing terms between those of highest and lowest degree, unless all even or odd terms are missing.

(*vi*) The degree of $P(s)$ and $Q(s)$ may differ by either zero or one only.

(*vii*) The terms of lowest degree in $P(s)$ and $Q(s)$ may differ in degree by one at the most.

5.6.2 Restrictions on Poles and Zeros of Transfer Functions

The necessary conditions for transfer functions (z_{12}, y_{12}, a_{12} and G_{12}) being listed. Again consider that transfer function have $Q(s)$ and $P(s)$ polynomials numerator and denominator respectively *i.e.* $N(s) = \dfrac{Q(s)}{P(s)}$. The necessary conditions are given below:

(*i*) The coefficients in the polynomials $P(s)$ and $Q(s)$ of the transfer function $N(s)$ must be real, and those of $P(s)$ must be positive.

(*ii*) Poles and **zeros must** be conjugate if imaginary or complex.

(*iii*) Real part of poles must be negative or zero.

(*iv*) If the real part of the pole is zero, then that pole must be simple which includes the origin.

(*v*) The polynomials $P(s)$ may not have any missing terms between those of highest and lowest degree, unless all even or all odd terms are missing.

(*vi*) The polynomials $Q(s)$ may have terms missing between the terms of lowest and highest degree, and some of the coefficients may be negative.

(*vii*) The degree of $Q(s)$ may be as small as zero independent of the degree of $P(s)$.

(*viii*) For α_{12} and G_{12}, the maximum-degree of $Q(s)$ is the degree of $P(s)$.

(*ix*) For z_{12} and y_{12}, the maximum degree of $Q(s)$ is degree of $P(s)$ plus one.

5.6.3 Pole Zero Plot of Network Function

A complex plane or s-plane with x-axis indicating real axis denoted as σ axis and y-axis indicating imaginary axis denoted as $j\omega$-axis is shown in the figure.

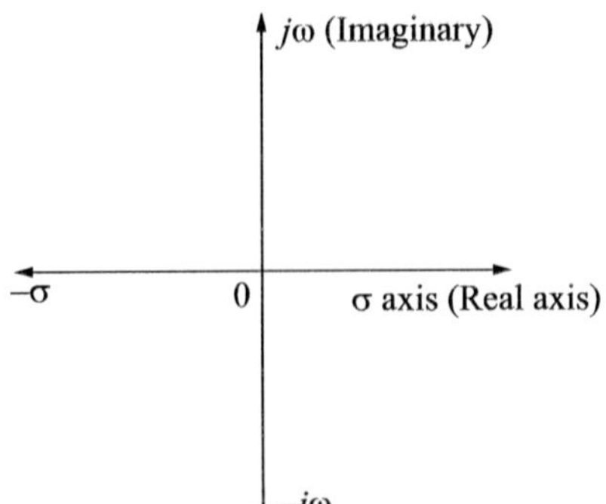

Fig. 5.13 *Complex plane(s-plane)*

All the poles and zeros are basically points on the s-plane. The poles are marked by cross (x) and zeros by zero (0) on the s-plane. In the case of repeated poles or zeros, the marks equal to the number of repeated poles or zeros must be shown.

Example 5.5 For the *Rc* network shown in the figure, find the driving point input impedance z_{11}. Plot the pole-zero plot of this network.

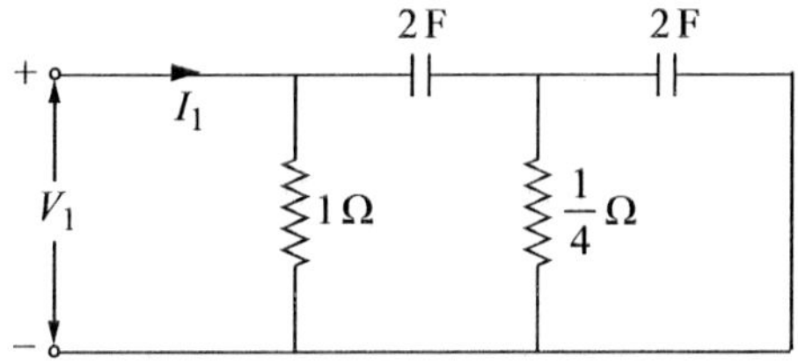

Solution: The network in s-domain is shown Fig. (a).

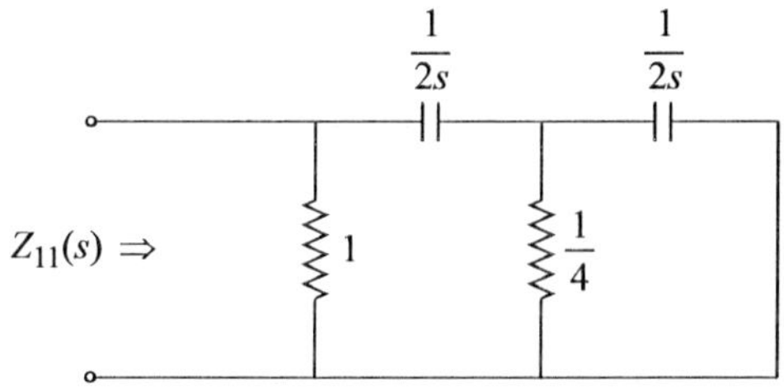

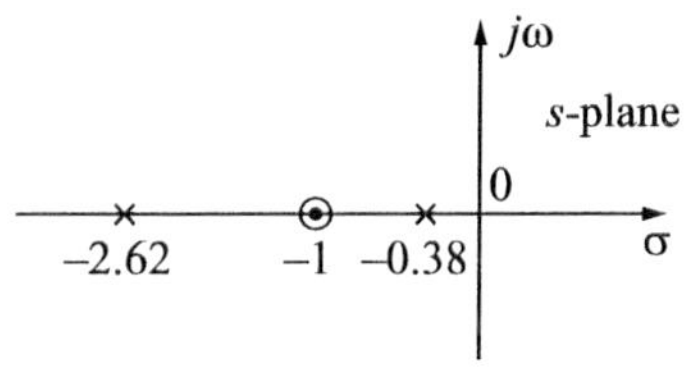

(a) *Network in s-domain* (b) *Pole-zero plot*

The driving point input impedance,

$$Z_{11} = \left[1 \,||\, \left\{\frac{1}{2s} + \left(\frac{1}{4} \,||\, \frac{1}{2s}\right)\right\}\right]$$

$$= 1 \,||\, \left\{\frac{1}{2s} + \frac{\frac{1}{4} \times \frac{1}{2s}}{\frac{1}{4} \times \frac{1}{2s}}\right\} = \left[1 \,||\, \left\{\frac{1}{2s} + \frac{1}{2(s+2)}\right\}\right]$$

$$= \left[1 \,||\, \left(\frac{s+2+s}{2s\,(s+2)}\right)\right]$$

$$= \left[1 \,||\, \frac{s+1}{s(s+2)}\right] = \frac{1 \times \frac{s+1}{s(s+2)}}{1 + \frac{s+1}{s(s+2)}}$$

$$= \frac{s+1}{s^2 + 3s + 1}$$

or $$Z_{11} = \frac{(s+1)}{s^2+3s+1}$$

The roots of the denominator are:

$$s^2 + 3s + 1 = 0 \text{ or } s = \frac{-3 \pm \sqrt{9-4}}{2}$$

or $$s = -1.5 + 1.12$$

∴ $$p_1 = -1.5 \pm 1.12 = -0.38$$

and $p_2 = -1.5 - 1.12 = -2.62$

Thus, $$Z_{11} = \frac{(s+1)}{(s-z_1)(s-z_2)}$$

Hence, zero at one point = –1 and poles are two

one at point = –0.38

and other at point = –2.62

The plot of pole-zero is given in Fig. (*b*).

5.7 TIME RESPONSE FROM POLE-ZERO PLOT

Time response of a network can be obtained from the pole-zero plot. The residues of partial-fraction expansion are obtained by the graphical methods.

Consider a current transform of a network as:

$$I(s) = \frac{Q(s)}{P(s)}$$

which is represented in terms of poles and zeros as follows:

$$I(s) = \frac{k(s-z_1)(s-z_2)\cdots(s-z_m)}{(s-p_1)(s-p_2\cdots(s-p_n)}$$

The above can be written as partial-fraction expansion:

$$I(s) = \frac{k_1}{(s-p_1)} + \frac{k_2}{(s-p_2)} \cdots + \frac{k_n}{(s-p_n)}$$

A particular coefficient k_i can be found as:

$$k_i = \left[(s-p_i)\,I(s)\right]_{s=p_i} \quad \text{for } i = 1, 2, \cdots, n$$

or $$k_i = \frac{k(p_i-z_1)(p_i-z_2)\cdots(p_i-z_m)}{(p_i-p_1)(p_i-p_2)\cdots(p_i-p_n)}$$

where p_i, z_m and p_n are complex numbers.

The difference of complex numbers *i.e.* $(p_i - z_m)$ and $(p_i - p_n)$ are also complex numbers.

A plot of pole-zero is shown in the figure. Suppose residue is to be found out corresponding to pole p_i *i.e.* k_i is to be obtained from the pole-zero plot.

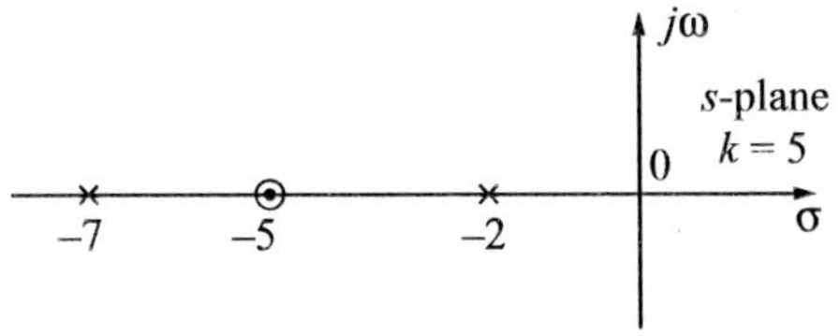

Pole-zero plot for obtaining k_i

From pole-zero plot k_i is given by:

$$k_i = k \frac{\text{Product of the vectors from the zeros to the pole } p_i}{\text{Product of the vectors from the other poles to the pole } p_i}$$

Note that the residues of the conjugate poles are also conjugate in nature.

Thus,

$$k_i = k \frac{r_{1i}\, r_{2i} \cdots r_{mi}}{R_{1i}\, R_{2i} \cdots R_{ni}} \angle(\alpha_{1i} + \alpha_{2i} + \cdots + \alpha_{mi}) - (\beta_{1i} + \beta_{2i} + \cdots + \beta_{ni})$$

where

$$\begin{aligned} p_i - z_1 &= r_{1i} \angle\alpha_{1i} \\ p_i - z_2 &= r_{2i} \angle\alpha_{2i} \\ &\vdots \\ p_i - z_m &= r_{mi} \angle\alpha_{mi} \end{aligned}$$

and

$$\begin{aligned} p_i - p_1 &= R_{1i} \angle\beta_{1i} \\ p_i - p_2 &= R_{2i} \angle\beta_{2i} \\ &\vdots \\ p_i - p_n &= R_{ni} \angle\beta_{ni} \end{aligned}$$

The current in time-domain can be obtained as:

$$i(t) = £^{-1}\left[I(s)\right] = £^{-1}\left[\frac{k_1}{s - p_1} + \frac{k_2}{s - p_2} + \cdots + \frac{k_n}{s - p_n}\right]$$

Example 5.6 For the given network function, draw pole-zero plot and obtain time domain response voltage.

$$V(s) = \frac{5(s+5)}{(s+2)\,(s+7)}$$

Solution: The pole-zero plot is as given in Fig. (*a*).

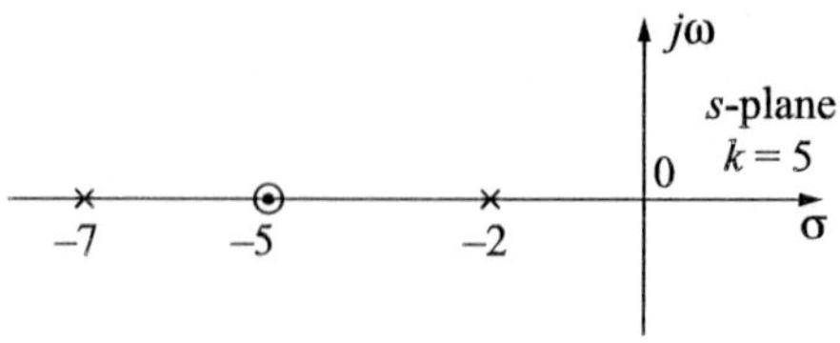

(*a*) *Pole-zero plot*

The partial-fraction expansion is given as:

$$V(s) = \frac{k_1}{s+2} + \frac{k_2}{s+7}$$

The values of k_1 from the pole-zero plot is obtained as given in Fig. (*b*).

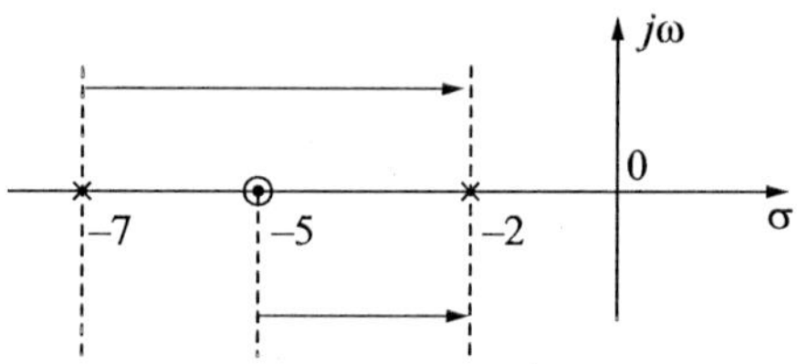

$$k_1 = 5\left[\frac{3\angle 0^\circ}{5\angle 0^\circ}\right] = 3$$

The value of k_2 from the pole-zero plot is obtained as given in Fig. (*c*).

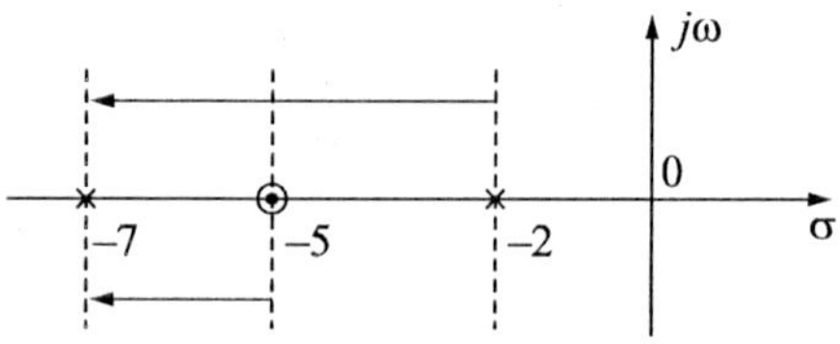

$$k_2 = 5\left[\frac{2\angle 180^\circ}{5\angle 180^\circ}\right] = 2$$

∴ $$V(s) = \frac{3}{s+2} + \frac{2}{s+7}$$

By taking inverse Laplace transform of above equation, we get:

$$V(t) = 3e^{-2t} + 2e^{-7t} \text{ Volts.}$$

5.8 STABILITY ANALYSIS FROM POLE-ZERO PLOT

A system is said to be stable if it remains in the same state when there is no disturbance. When the system is found to be stable, then it is necessary to know the strength of its stability or degree of stability.

The stability analysis is based on the properties of the transfer function. In analysis the characteristic equation is of importance to describe the transient response of the system. From the roots of characteristic equation, some of the conclusions drawn may be summarised as follows:

(*i*) When all the roots of the characteristic equation lie in the left half of the s-plane, the system response due to initial condition will decrease to zero at time $t = \infty$. Thus, the system will be termed as stable.

(*ii*) When one or more pair of roots lie on the imaginary axis and there are no roots on the right half of the s-plane, the response will be oscillatory without damping. Thus, the system will be termed as unstable.

(*iii*) When one or more roots lie in the right half of the s-plane, the response will exponentially increase in magnitude, there by system will be unstable.

Example 5.7 Network transfer functions given by:

(*i*) $F(s) = \dfrac{k}{s+\alpha}$

(*ii*) $F(s) = \dfrac{k}{(s+\alpha)^r}$

(*iii*) $F(s) = \dfrac{k}{s+\alpha-j\beta} + \dfrac{k}{s+\alpha+j\beta}$

Draw the pole-zero plot for each transfer function, find its time response and also plot the same. Comment on stability of the function.

Solution: (*i*)Pole-zero plots of $\dfrac{k}{s+\alpha}$ is shown in Fig. (*a*)

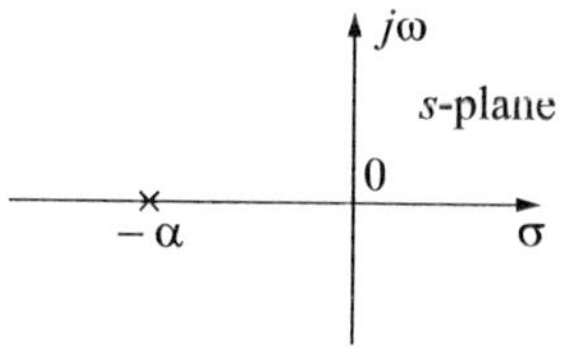

(*a*)

By taking inverse Laplace transform of the transfer function, we get:

$$f(t) = £^{-1}\left(\frac{k}{s+\alpha}\right)$$

or $$f(t) = k\,e^{-at}$$

The time-domain plot of the function is shown in Fig. (*b*)

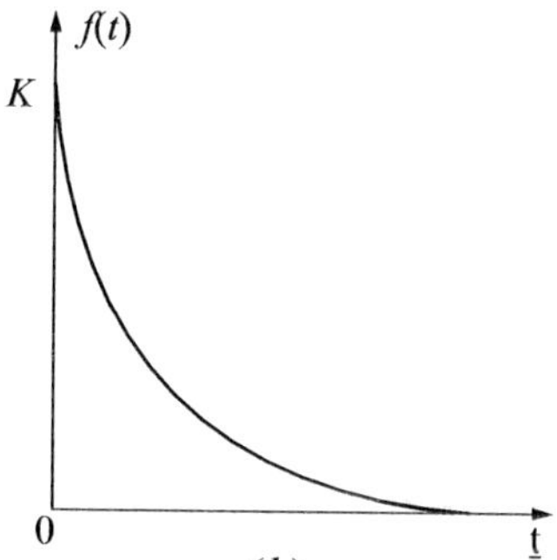

(*b*)

It can be noted that the pole zero plot in Fig. (*a*) has roots in left-half of *s*-plane, hence, system is stable as seen in the Fig. (*b*)

(*ii*) Pole zero plot of $\dfrac{k}{(s+\alpha)^r}$ is shown in Fig. (*c*).

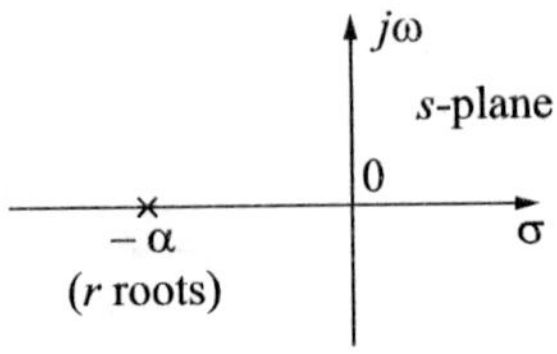

(*c*)

The time response is found by taking inverse Laplace transform of the function:

$$f(t) = k\, t^r\, e^{-at}$$

As $\lim\limits_{t\to\infty} f(t) = \lim\limits_{t\to\infty} k\, t^r\, e^{-at} = 0$, therefore, multiple pole in left-half of *s*-plane gives stable system.

(*iii*) The pole zero-plot of $\dfrac{k}{s+\alpha-j\beta}\,\dfrac{k}{s+\alpha+j\beta}$ is shown in Fig. (*d*).

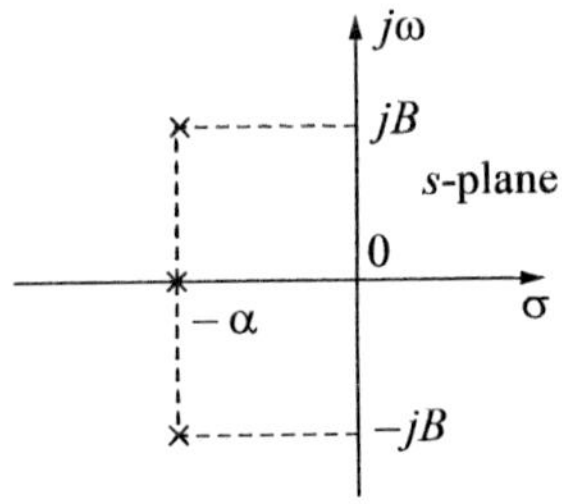

By taking inverse Laplace transform of the function, we get:

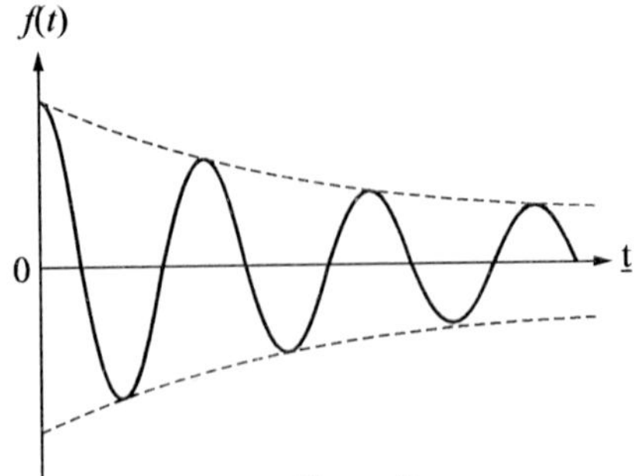

$$f(t) = 2k\, e^{-at} \cos \beta t$$

As the time tends to ∞, the response of the functions reduces as shown in Fig. (*e*). Thus, complex conjugate poles in left-half of *s*-plane of give stable system.

Example 5.8 Network transfer functions are given by:

(i) $F(s) = \dfrac{k}{s-\alpha}$

(ii) $F(s) = \dfrac{k}{s-\alpha-j\beta} + \dfrac{k}{s-\alpha+j\beta}$

Draw the pole zero plot of each transfer function, find time response and also plot the same. Comment on stability of the function.

Solution: (i) The pole - zero plot of the function $\dfrac{k}{s-\alpha}$ is shown in Fig. (a).

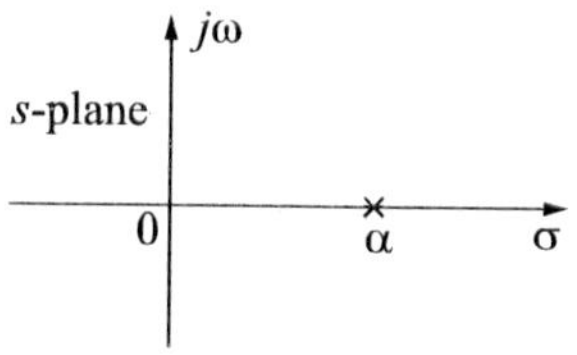

(a) Pole-zero plot

The time response is found by taking inverse Laplace transform of the function.

$$f(t) = £'\left(\frac{k}{s\alpha}\right) = k\,e^{at}$$

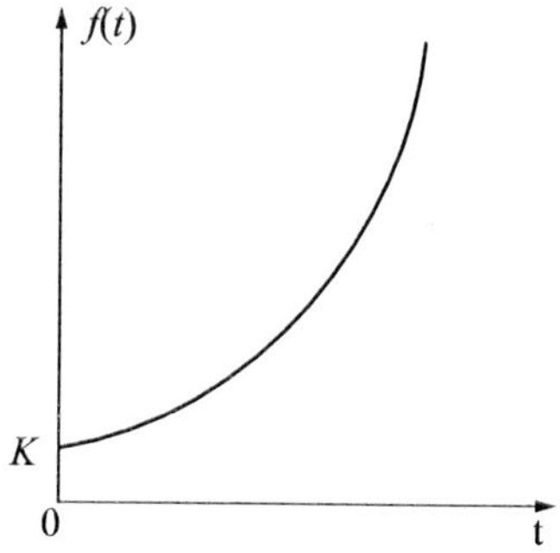

(b) Time response

The function value increases as time lapses as shown in Fig. (b) hence, a pole in right-half of s-plane gives unstable system.

(ii) The pole-zero plot of the function $\dfrac{k}{s-\alpha-j\beta} + \dfrac{k}{s-a+j\beta}$ is shown in figure (c).

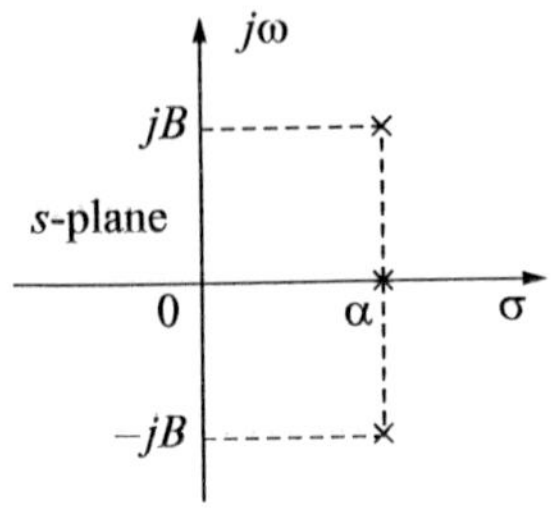

The time response can be found by taking inverse Laplace transform of the function.

Thus, $f(t) = 2k\, e^{at} \cos\beta t$.

The time response is plotted in Fig. (*d*)

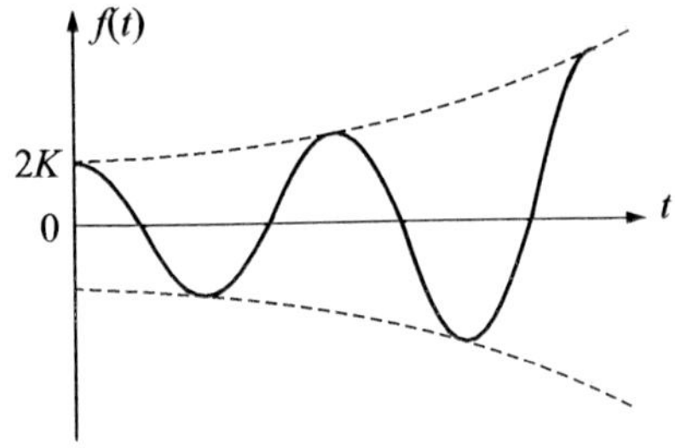

(*d*) *Time-response*

Hence, complex conjugate pole in right-half of *s*-plane gives unstable system.

Example 5.9 A system function shows pole at origin and another system function shows complex conjugate poles at imaginary axis. Draw the time-response and current in stability of the system.

Solution:

(*i*) Pole at origin as shown in Fig. (*a*)

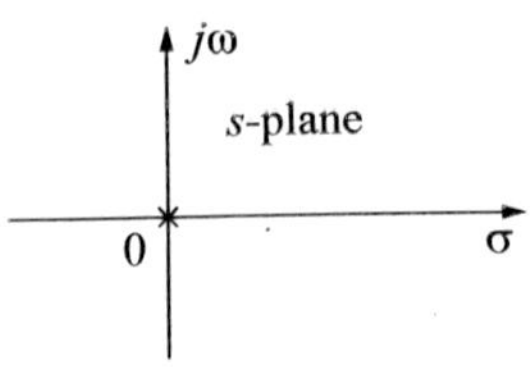

(a) *Pole at origin*

The time function can be obtained by taking inverse Laplace transform.

$$f(t) = \pounds^{1}\left(\frac{k}{s}\right) = k$$

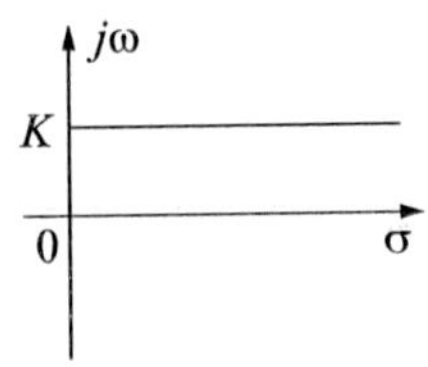

(*b*) *Time response*

The plot is shown Fig. (*b*).

Thus, the pole at origin, gives constant output time response, hence, it gives a stable system, (*ii*) Pole an $j\omega$ axis in complex conjugate form is shown in Fig. (*c*).

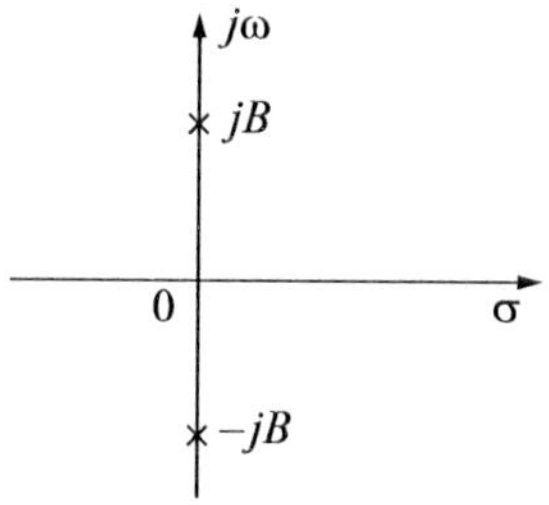

(c) *Pole-zero plot*

$$F(s) = \frac{k}{s + j\beta} + \frac{k}{s - j\beta}$$

By taking Laplace inverse on both sides, we get:

$$f(t) = 2k \cos \beta t$$

The time response is sinusoidal and the same is plotted in Fig. (*d*)

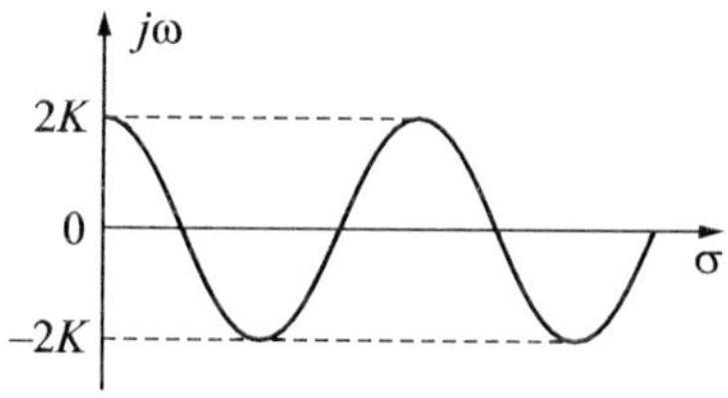

(*d*) *Time response*

Thus, complex conjugate poles on imaginary axis *s*-plane give oscillatory response hence, it is imaginally stable.

5.9 STABILITY ANALYSIS WITHOUT ROOT SOLVING

There are methods of determiner stability without involving root solving.

5.10 FREQUENCY DOMAIN ANALYSIS

The performance of a central system is measured more realistically by its time domain analysis as the performance of most central systems is based on the time response due to certain test signals. The time response of central system is usually more difficult to determine analytically, especially for a high order system. In design problems there are no unified methods of arriving at a designed system which meets the time domain performance, specifications, such as maximum over shoot, rise time, delay time, settling time and so-on. However in the frequencies domain lot of graphical methods are available which are not limited to low order systems. One should realize that these are correlating relations between the frequency and time domain performances in a linear system, such that the time domain properties of the system can be predicted based on frequency domain characteristics. The frequency domain is also more convenient for measurements of system sensitivity to noise and parameter variations.

Most of the signals to be processed are either sinusoidal or composed of sinusoidal components. The starting point of frequency domain analysis is its transfer functions. When the input to a linear time variable system is sinusoidal with amplitude R and frequency ω_0 *i.e.*,

$$r(t) = R \sin \omega_0 t$$

The steady state output of the system $y(t)$ will also be sinusoidal with same frequency ω_0, but possibly with different amplitude and phase *i.e.*,

$$y(t) = y \sin (\omega_0 t + \phi)$$

Where y is the amplitude of the output sine wave and ϕ is the phase shift in degree or radians.

Let the transfer function of a linear signal input signal output to $M(S)$. The Laplace transform of the input and output are related as:

$$Y(s) = M(s)\, R(s)$$

For sinusoidal state Analysis, s is replaced by $j\omega$, and equation 5.78 becomes

$$Y(j\omega) = M(j\omega)\, R(j\omega) \qquad ...(5.78)$$

We can write the function $Y(j\omega)$ as

$$Y(j\omega) = |Y(j\omega)|\, \underline{|Y(j\omega)}$$

we have

Magnitude $\quad |Y(j\omega)| = |M(j\omega)||R(j\omega)|$

Phase, $\quad \underline{|Y(j\omega)} = \underline{|M(j\omega)} + \underline{|R(j\omega)}$

Hence, the value of amplitude and phase characteristics of a closed and phase characteristics of a closed loop system can be used to predict both time domain transient and steady state performances.

5.10.1 Frequency Response of a Closed-loop System

The closed loop transfer function of a single loop central system is:

$$M(s) = \frac{Y(s)}{R(s)} = \frac{G(s)}{1 + G(s)\, H(s)} \qquad ...(5.79)$$

under the sinusoidal steady state $s = j\omega$

$$M(j\omega) = \frac{Y(j\omega)}{R(j\omega)} = \frac{G(j\omega)}{1 + G(j\omega)\, H(j\omega)} \qquad ...(5.80)$$

The magnitude and phase angle shall be

$$M(j\omega) = |M(j\omega)|\, \underline{|M(j\omega)} \qquad ...(5.81)$$

In terms of its real and imaginary points:

$$M(j\omega) = \text{Re}[M(j\omega) + jI_m[M(j\omega)] \tag{5.82}$$

The magnitude of $M(j\omega)$ is

$$|M(j\omega)| = \left|\frac{G(j\omega)}{1 + G(j\omega)\,H(j\omega)}\right| \quad ...(5.83)$$

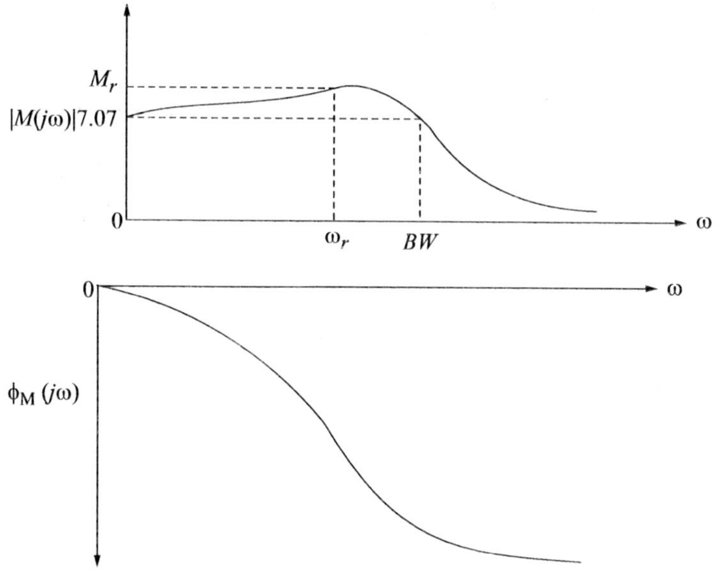

Gain and phase characteristics of feedback central system.

and the phase of $M(j\omega)$ is

$$\angle M(j\omega) = \phi_m(j\omega)$$

$$\angle G(j\omega) - \angle[1 + G(j\omega)\,H(j\omega)]$$

Above figure shows gain and phase characteristics of a typical central, system. The gain and phase of closed loop system can be determined from the equations (9.81) 9.82. In normal practice, the frequency response of $G(S)$ and $H(S)$ can be determined by applying sine wave inputs to the system and sweeping the frequency from 0 to a values beyond the frequency range of the system.

M_r = magnitude of resonant frequency

ω_r = resonant frequency at which peaks Mr occurs

BW = Bandwidth BW is the frequency at which $|M(j\omega)|$ drops to 70.7% or 3dB form its zero frequency value.

5.11 STABILITY ANALYSIS WITH THE BODE PLOT

In a central system having transfer function of equation (5.79), loop transfer function is defined as:

$$L(S) = G(S)\, H(S)$$

Bode plot is a plot of the magnitude *i.e.*, 20 log $|\underline{|(-j\omega)}|$ decibels versus ω (or $\log_{10}$ ω) and phase angle versus a or (log10 w) on semi log.

Bode diagram is illustrated in Fig. 5.14 for a typical loop transfunction $L(S)$.

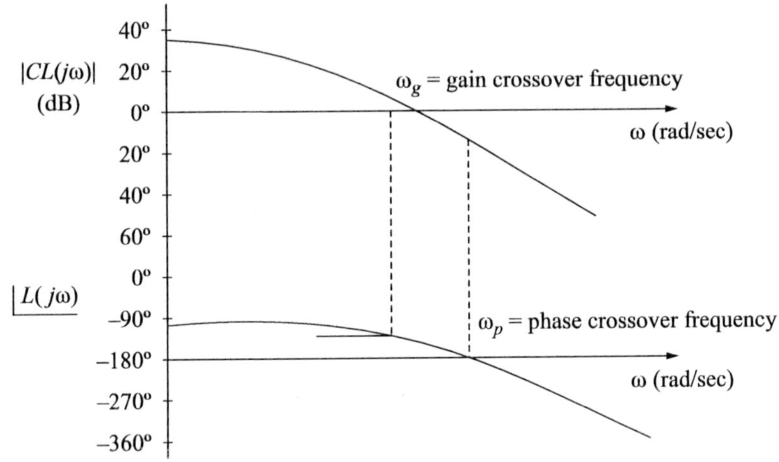

Fig. 5.14 *Bode plot for determination of gain margin and phase margin*

The following observations are to be made from the Bode plot:

(i) The gain margin is posture and the system is stable if the magnitude of $\underline{|(j\omega)}$ at phase cross over is negative in dB *i.e.*, the gain margin is below the 0 dB axis. The gain margin is above the 0-dB axis, the gain margin is negative and the system is unstable.

(ii) The phase margin is positive and the system is unstable if the phase of $\underline{|(j\omega)}$ is greater than –180° at the gain cross over in dB *i.e.*, The phase margin is above –180° axis. If the phase margin is below the –180° axis, the phase margin is negative and the system is unstable.

Gain Crossover Point: The gain crossover point on the frequency domain plot of $\underline{|(j\omega)}$ is a point at which $|\underline{|(j\omega)}| = 1$ or $|\underline{|(j\omega)}|_{dB} = 0$ dB. The frequency at the gain crossover point is known as the gain crossover frequency ω_g.

Phase Crossover Point: The phase crossover point on the frequency domain plot of $\underline{|(j\omega)}$ is a point at which $\underline{|(j\omega)} = 180°$. The frequency at the phase crossover point is known as the phase crossover point ω_p.

5.12 FEEDBACK CONTROL AND EXAMPLES OF STABILITY ANALYSIS

5.12.1 Block Diagram of the Feedback Loop Control System

A block diagram of transfer function to concisely summarize all the informations in the central loop and analyze the complete process central. The block diagram of a process with feedback central loop is shown in Fig. 5.15 (*a*). This method of analysis basically helps in the analytical solution of simultaneous line is differential equatioins with constant coefficients. A constant system can involve several processes and central calculations which must be considered simultaneously. The overall behaviour of a complex system is modelled considering only input and output variables by the use of transfer functions and block diagrams.

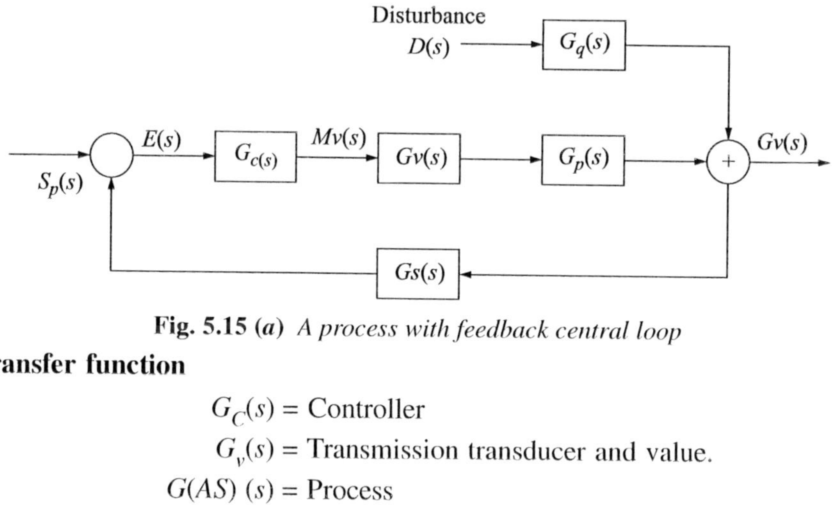

Fig. 5.15 (*a*) *A process with feedback central loop*

Transfer function

$G_C(s)$ = Controller

$G_v(s)$ = Transmission transducer and value.

$G(AS)\ (s)$ = Process

$G(S),\ (s)$ = Sensors, Transducer and Transmission

$G(d)\ (s)$ = Disturbance

Variables

$C_V(s)$ = Controlled variable

$M_V(s)$ = Manipulated variables = $X(S)$

$S_P(s)$ = Set point, desired value

$E(s)$ = The difference between the set point and measured variables

$D(s)$ = Disturbance, an input that changes due to external conditions affects the controlled variables.

Closed-loop transfer function can be easily derived as:

$$\frac{C_v(s)}{D(s)} = \frac{G_d(s)}{1 + G_p(s)\, G_V(s)\, G_C(s)\, G_s(b)}$$

$$\frac{G_v(s)}{S_p(s)} = \frac{G_P(s)}{1 + G_p(s)\, G_V(s)\, G_C(s)\, G_s(b)}$$

The block diagram analysis gives several useful results:

(*i*) The block diagram gives a visual pictures of the equation showing the feedback.

(*ii*) The general closed loop transfer function model is applicable to any specific system. Substitution of the transfer function model for loop elements are to be done for use of the technique.

(*iii*) The overall transfer function eliminator entries demonstrate that only the element in the feedback effect the system stability.

5.12.2 Series Chemical Reactors

The stability of the series chemical reacting shown in the following Fig. 5.15(*b*) is to be analyzed and determined.

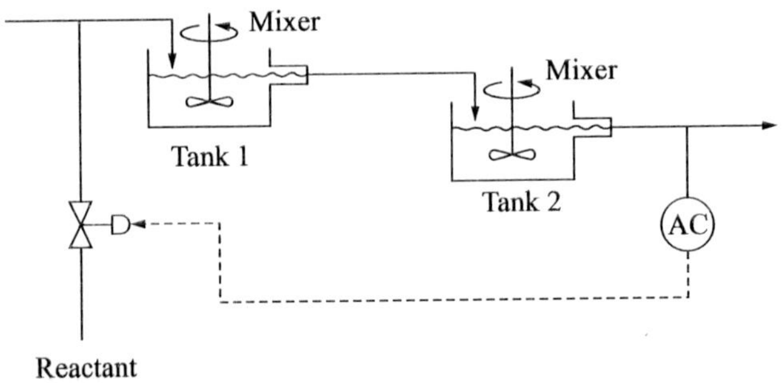

Fig. 5.15 (*b*)

The liquid and reactant flow into tank wherein it is mixed well. This flows into tank wherein again it is mixed subsequently component 'A' is controlled by feedbacks.

$$G_P(s) = \frac{K_p}{(T_S + 1)\,(T_S + 1)}$$

$$G_C(s) = K_C\left(1 + \frac{1}{T_1 s}\right)$$

Process with: $K_P = 0.10 \left(\dfrac{\text{mole/m}^3}{\%}\right)$; $T = 0.5$ min

PT controllable: $K_C = 15$ (% open) | (mole/m^3): $T_1 = 1.0$ min

The individual transfer functions can be combined to give the closed loop transfer function for a set point change, which includes the characteristic equation.

$$\frac{Cv(s)}{Sp(s)} = \frac{G_1(s)\,G_V(s)\,G_C(s)}{1+G_P(s)\,G_V(s)\,G_C(s)\,G_S(s)}$$

$$= \frac{15\left(1+\dfrac{1}{105}\right)\dfrac{0.10}{(0.55+1)^2}}{1+15\left(1+\dfrac{1}{1.05}\right)\left(\dfrac{0.10}{0.55+13^2}\right)}$$

The characteristic equation is

$$0 = 1+15\left(1+\frac{1}{5}\right)\left(\frac{0.1}{(0.55+1)^2}\right)$$

$$0 = 0.2533 + 1.052 + 2.55 + 1.5$$

The solution to this equation gives the exponents in the time domain solution and the values are:

$$\alpha_{12} = 1.60 \pm 2.21j;\ a_3 = -0.81$$

As all the roots have negative real parts, this system is stable.

5.12.3 Three-Tank Mixing Process

The stability of the three tank mixing process shown in Fig. 5.16 is to be evaluated under feedback central with a proportional only controller.

$$\frac{C_V(s)}{D(s)} = \frac{G_d(s)}{1+G_P(s)\,G_V(s)\,G_C(s)\,G_s(s)}$$

All tanks are well mixed. The dynamic of the value and sensor are negligible. No transportation deloy *i.e.*, dead time exists. A linear relationship exists between the value opening and the flow of component A. Densities of components are equal. As the variable to be controlled is the concentration leaving the last tank. Component material balances on the mixing point.

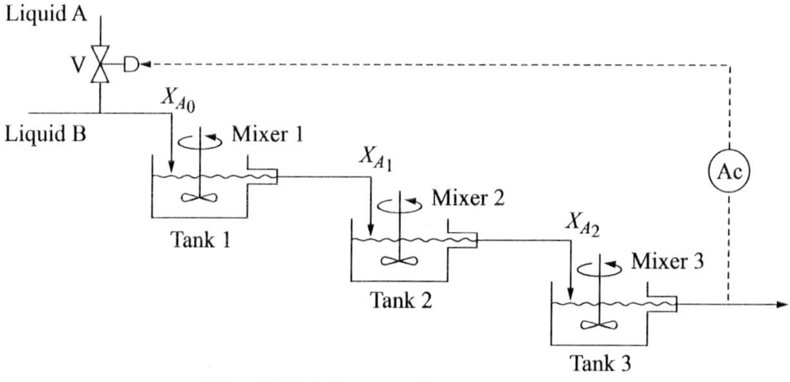

Fig. 5.16. *Three tank mixing process*

The distance transfer function is given by

$$\text{Disturbance} \quad \frac{xA_3(s)}{xA_B(s)} = G_d(s) \simeq \frac{1.0}{(T_s + 1)}$$

For $T = 5.0$ min

It has been found that

$$G_S(s)\, G_V(s)\, G_P(s) = \frac{0.039}{(55+1)}$$

and $$G_C(s) = K_C$$

$$\frac{CV(s)}{D(s)} = \frac{\dfrac{1}{(5s+1)^3}}{1 + K_C \dfrac{0.039}{(5s+1)^3}}$$

The characteristic equation is given by:

$125\, s^3 + 75\, s^2 + 15\, s + (1 + 0.039\, K_c) = 0$

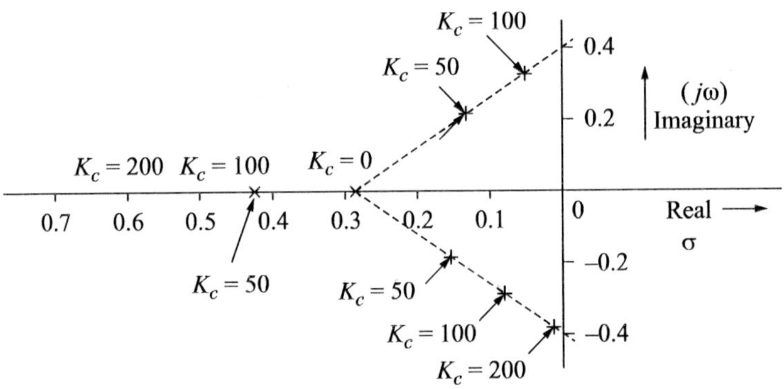

5.12.4 Single-Tank Mixing Process Using Bode Plot

The single-tank mixing process with proportional central system is shown in Fig. 5.17. This process is the same as the three-tank mixer with the last two tanks removed. The process transfer function which includes a censor and fast final element dynamics is given by :

$$G_C(s) = K_C;\ G_P(s)\, G_V(s)\, G_s(s)$$

or $$G_C(s) = \frac{0.039}{5s+1}$$

with time in minutes process is table without central since it has one pale at (–2, 0) *n* the real imaginary plane, so that it satifies the criteria for the bode-method. The stability is to be determined by Bode method.

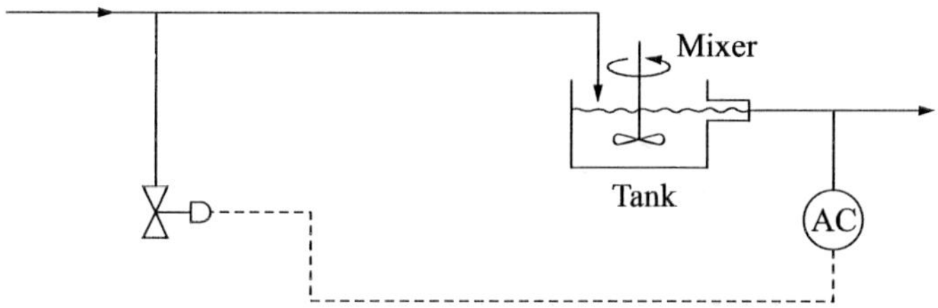

Fig. 5.17 *Single tank mixing process*

The $G_{02}(s)$ is the product of the value, process, senser, controller transfer functions. The $G_{02}(s)$ with proportional only central can be written as:

$$G_{0S}(s) = \frac{0.039\ K_C}{5L+1}$$

The magnitude and phase angle of the $G_0(t)$ can be calculated from $G_{02}(j\omega)$:

$$AR = |G_{0L}(j\omega)|$$

$$= \left|(0.039\ K_C)\left(\frac{1}{1+5\ j\omega}\right)\left(\frac{1-5\ j\omega}{1-5\ j\omega}\right)\right| = \frac{0.039\ K_C}{\sqrt{1+25\ \omega^2}}$$

$$\phi = \underline{|G_{0L}(j\omega)} = \underline{|(0.039\ K_C)} + \underline{\left|\frac{1}{5\ j\omega+1}\right.}$$

$$= \underline{|0} + \tan^{-1}(-5\omega) = \tan^{-1}(-5\omega)$$

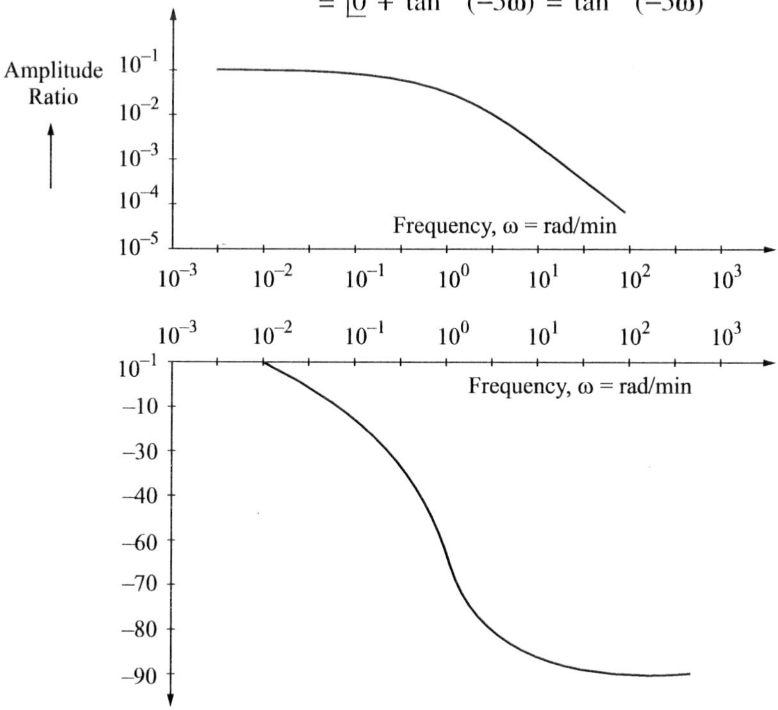

Fig. 5.18 *Single tank mixing process bode plot with $K_C = 1.0$.*

These expressions are presented in Bode plots in Fig. 5.17(*b*) for $K_C = 1$. As the phase angle for this first order system does not decrease below –90° for any controller gain, the phase angle never reaches –180° and the feedback signal cannot create oscillations in the centre drop.

5.12.5 Single-Tank-Mixing Process with Realistic Modelling of Value and Senser Dynamics

The mixing process of Fig. 5.17 is again considered with modification that the value and senser dynamics are more realistically modelled according to the following first-order transfer function with short time constant:

$$G_C(s) = K_C;\ G_P(s) = \frac{0.039}{5s+1}$$

$$G_V(s) = \frac{1}{0.033S+1};\quad G_S(s) = \frac{1}{0.25S+1}$$

The $G_{02}(s)$ can be written as:

$$G_{OL}(s) = G_P(s)\, G_V(s)\, G_C(s)\, G_s(s)$$

$$= \left(\frac{0.039}{5s+1}\right)\left(\frac{1}{0.033s+1}\right)\left(\frac{1}{0.25s+1}\right)$$

$$AR = |G_{OL}(j\omega)|$$

$$= \frac{0.039\,K_C}{\sqrt{1+25\omega^2}}\,\frac{1}{\sqrt{1+0.0625\,\omega^2}}\,\frac{1}{\sqrt{1+0.0011\,\omega^2}}$$

$$\phi = \angle G_{OL}\,(j\omega)$$

$$= \tan^{-1}(-5\omega) + \tan^{-1}(-0.25\omega) + \tan^{-1}(-0.033\omega) + \angle(0.039\,K_C)$$

The amplitude ratio and phase angle and plotted in Fig. 5.19 for a controller gain of 1.0.

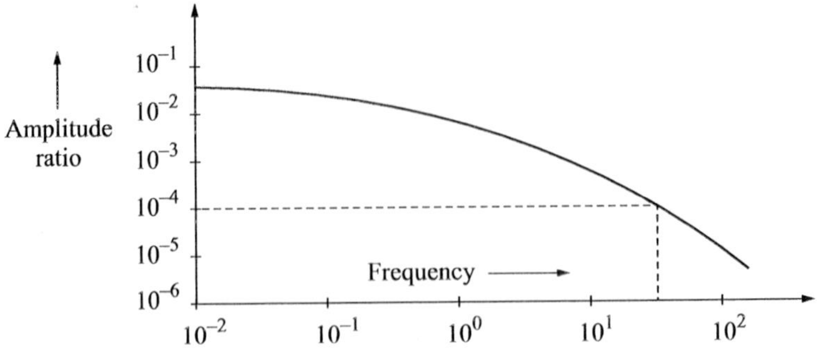

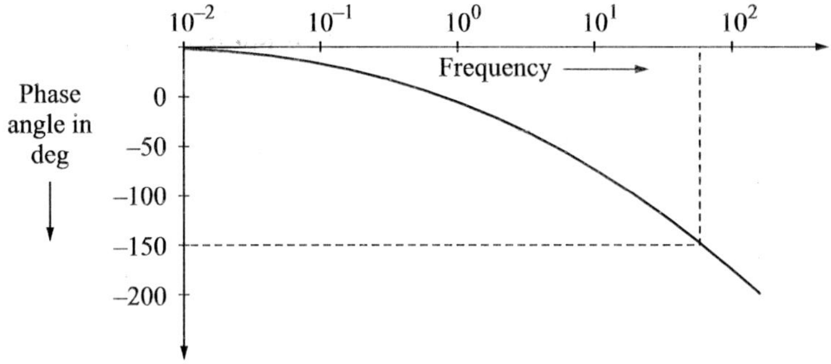

Fig. 5.19 *Bode plot of single-tank mixing process with realistic modelling and K_C = 1.0.*

Due to the added dynamic elements in $G_{OL}(s)$, the phase angle exceeds –180°. The critical frequency is ω_C = 11.6 read/min and the amplitude ratio as follows:

$$K_C = 1.0;\ |G_{OL}(j\omega)| = 0.0002 < 1.0 \text{ stable}$$

$$K_C = 500;\ |G_{OL}(j\omega)| = 0.10 < 1.0 \text{ stable}$$

$$K_C = 6000;\ |G_{OL}(j\omega)| = 1.2 < .10 \text{ unstable}$$

It can be seen by applying the Bode stability criterion, the system is stable for controller gain values of 1.0 and 800 because the amplitude ratios at the critical frequency are less than 1.0. The system is unstable for a controller gain of 6000, which has an amplitude ratio greater then 1.0 at the critical frequency.

It is important to note that all real process central system can be made unstable simply by increasing the magnitude of feedback controller gain.

5.12.6 Bode Plot

Bode-plot is used very widely as an analysis in transfer functions. It is plotted on a semi-log graph paper. The semi-log graph paper may be of 3 cycles, 4 cycles, 5 cycles respectively. Each cycle has 10 distinct divisions. Each of these 10 distinct divisions is further subdivided into another 10 distinct divisions. For frequency response analysis, in the transfer function s is replaced by $j\omega$.

Consider that the transfer function is represented by:

$$G(j\omega) = |G(j\omega)| \angle G(j\omega)$$

or $$l_n\, G(j\omega) = l_n\, |G(j\omega)| + j\phi(\omega)$$

where $$\phi(\omega) = \angle G(j\omega)$$

The real part is the natural logarithm of magnitude in basic unit 'neper' and the imaginary part is a phase characteristic.

On similar lines, we can also write:

$$\log G(j\omega) = \log |G(j\omega)| + \log e^{j\phi(\omega)}$$

$$= \log |G(j\omega)| + 0.434\, j\, \phi(\omega)$$

Thus, plot of 20 log $|G(j\omega)|$ versus ω and phase angle $\phi(\omega)$ versus w is done on a semi-log paper, is known as Bode-plot.

5.12.7 Construction of Bode-Plot

1. Magnitude Plot

(*i*) Use formula 20 log $|k|$ – 20 log $|0.1|$, where k is the dc part of the transfer function, and establish the originating point. The originating point in the first cycle is taken as 0.1.

However, if the originating point in the first cycle is 1, the formula used in 20 log $|k|$ – 20 log $|1|$.

(*ii*) Use formula $\omega T = 1$ *i.e.* $\omega = \frac{1}{T}$ to establish corner frequency.

(*iii*) s in denominator denotes –20 dB/decade.

s^2 in denominator denotes –40 dB/decade.

s in numerator denotes +20 dB/decade.

s^2 in numerator denoted + 40 dB/decade

(*iv*) Step (*ii*) gives first corner frequency which is located on the semi-log paper. This corner frequency point is connected with the preceding point by a straight line. In case the slope is downward (descending), there is a decrease by 20 dB. However, if the profile has a upward slope (ascending), there is an increase by 20 dB.

(*v*) Subsequently, the next corner frequency is obtained and the process is repeated.

2. Phase Plot

The phase angles are calculated for the corner frequencies and the same is plotted on the semi-log graph paper just below the magnitude plot.

3. Specifications from Bode Plot

(*i*) Gain cross-over frequency (ω_{gc}): It is the frequency at which the magnitude of system function is unity *i.e.* o *dB*.

(*ii*) Phase crossover frequency (ω_{pc}) phase: It is the frequency at which the angle of the system function is –180°.

(*iii*) Gain Margin (*GM*): It is the margin in gain allowable by which gain can be increased till the system reaches on the verge of instability. Positive gain means that the system is stable and negative gain means that the system is unstable.

$$GM = \frac{1}{|G(j\omega)|_{\omega=\omega_{pc}}}$$

or $$GM\ (dB) = 20 \log \frac{1}{|G(j\omega)|_{\omega=\omega_{pc}}}$$

$$dB = -20 \log |1+(j\omega)|_{\omega=\omega_{pc}}$$

(*iv*) Phase Margin (*PM*): It is the amount of additional phase log which can be introduced in the system till system reaches on the voice of instability.

$$PM = 180^\circ + \angle |G(j\omega)|_{\omega=\omega_{pc}}$$

Example 5.10 Draw the Bode plot for the system having transfer functions $G(s) = \dfrac{80}{s(s+2)(s+20)}$. Determine *GM*, *PM*, ω_{gc} and ω_{pc}. Comment on stability.

Solution: By arranging *G*(*s*) in the time constant form, we get:

$$G(s) = \frac{2}{s\left(1+\dfrac{s}{2}\right)\left(1+\dfrac{s}{20}\right)}$$

Important Factors

(*i*) $k = 2$

(*ii*) 1 Pole at origin

(*iii*) Simple pole $\dfrac{1}{\left(1+\dfrac{s}{2}\right)}$ with $T_1 = \dfrac{1}{2}$ *i.e.* $\omega_{c_1} = \dfrac{1}{T_1} = 2$

(*iv*) Simple pole $\dfrac{1}{\left(1+\dfrac{s}{20}\right)}$ with $T_2 = \dfrac{1}{20}$ *i.e.* $\omega_{c_2} = \dfrac{1}{T_2} = 20$

Magnitude Plot

(*i*) For $k = 2$, the magnitude is $20 \log k = 20 \log 2 = 6$ d*B*

(*ii*) For one pole at origin, there is a straight line of slope –20*dB*/decade passing through intersection point of $\omega = 1$ on 0 *dB*.

(*iii*) Shift intersection point of $\omega = 1$ and 0 *dB* on 20 log *k* line and draw parallel to –20 *dB*/decade line drawn. and this will continue as a resultant of *k* and $\dfrac{1}{s}$ till first corner frequency *i.e.*, $\omega_{c_1} = 2$.

(*iv*) At $\omega_{c_1} = 2$, it is simple pole, hence, it will contribute the rate of –20 *dB* /decade. Thus, resultant slope after $\omega_{c_1} = 2$ becomes –20 – 20 = –40 *dB*/decade. This is in addition to k, $\frac{1}{s}$ and $\frac{1}{\left(1+\frac{s}{2}\right)}$, and will continue till it intersects next corner to frequency line *i.e.* $\omega_{c_2} = 20$.

At $\omega_{c_2} = 20$, there is simple pole contributing –20dB/decade, therefore, resultant slope after $\omega_{c_2} = 20$ becomes –40 – 20 = –60 dB/decade.

Phase Angle Plot

$$G(j\omega) = \frac{2}{j\omega\left(1+\frac{j\omega}{2}\right)\left(1+\frac{j\omega}{20}\right)}$$

$$\underline{|G(j\omega)} = \frac{\underline{|2+jo}}{\underline{|j\omega}\,\underline{|1+\frac{j\omega}{2}}\,\underline{|1+\frac{j\omega}{20}}}$$

where $$\underline{|2+jo} = 0^\circ$$

$$\frac{1}{\underline{|j\omega}} = -90^\circ$$

$$\frac{1}{\underline{|1+\frac{j\omega}{2}}} = -\tan^{-1}\left(\frac{\omega}{2}\right)$$

$$\frac{1}{\underline{|1+\frac{j\omega}{20}}} = -\tan^{-1}\left(\frac{\omega}{20}\right)$$

Based on above, phase angle table is as follows:

S. No	ω	$\frac{1}{j\omega}$	$-\tan^{-1}\frac{\omega}{2}$	$-\tan^{-1}\frac{\omega}{20}$	Resultant
1	0.2	–90°	–5.7°	–0.57°	–96.27°
2	2	–90°	–4.5°	–5.7°	–140.7°
3	8	–90°	–75.96°	–218°	–187.76°
4	10	–90°	–78.69°	–26.56°	–195.29°
5	20	–90°	–84.28°	–45°	–219.28°
6	40	–90°	–87.13°	–63.43°	–240.58°
7	∞	–90°	–90°	–90°	–270°

First draw point lines of slopes –20, –40, –60 *dB*/decade form intersection point of ω = 1 and 0 *dB* line and subsequently just draw parallel to them from respective points.

Finally the Bode plot *i.e.* Magnitude versus ω and phase versus ω is drawn in the following figure.

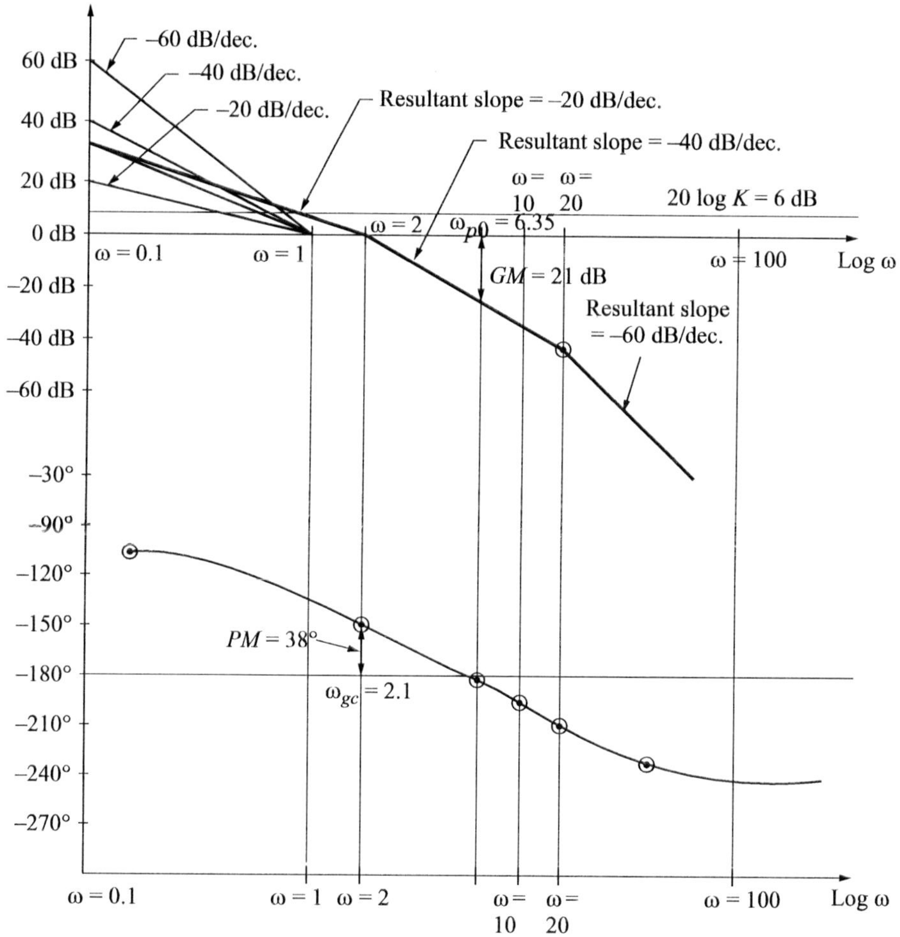

Fig. 5.20 *Bode plot of the transfer function*

From the above Bode plot we get:

Gain Margin, $GM = 21$ dB

Phase Margin, $PP = 38°$

We find that Gain Margin and Phase Margin are both positive, hence, the system is stable.

Example 5.11 Find open circuit transfer impedance V_2/I_1 and open circuit voltage ratio V_2/V_1 for the ladder network shown in the figure.

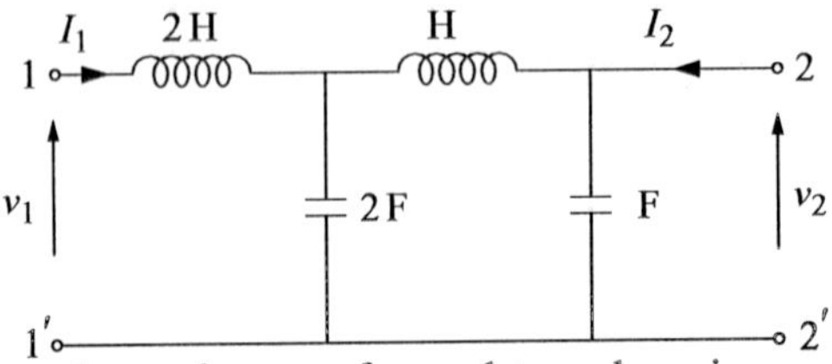

Solution: The network can be transformed to s-domain as given here:

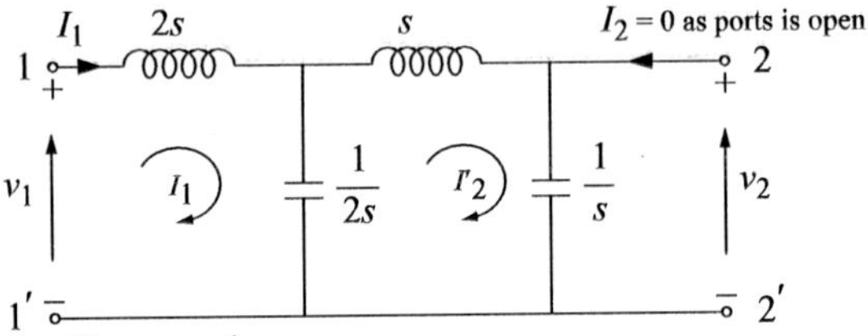

By KVL in the first mesh

$$2s \times I_1 + \frac{1}{2s}(I_1 - I_2') = V_1 \qquad ...(i)$$

and also, $$\frac{1}{s} \times I_2' = V_2 \qquad ...(ii)$$

Further, Current I_2' is:

$$I_2' = \frac{\left(\frac{1}{2s}\right)\left(s + \frac{1}{s}\right)}{\frac{1}{2s} + \left(s + \frac{1}{s}\right)} \times \frac{1}{\left(s + \frac{1}{s}\right)} \times \frac{1}{s} \times I_1$$

or $$I_2' = \frac{\frac{1}{2s}}{\frac{1}{2s} + s + \frac{1}{s}} \times I_1$$

or $$I_2' = \frac{I_1}{2s^2 + 3} \qquad ...(iii)$$

From equations (*ii*) and (*iii*), we get:

$$\frac{V_2}{I_1} = \frac{1}{s(2s^2 + 3)}$$

From equations (*i*) and (*iii*), we get:

$$V_1 = \left[\left(2s + \frac{1}{2s}\right)(2s^2 + 3) - \frac{1}{2s}\right] I_2'$$

$$= \left(\frac{4s^4 + 7s^2 + 1}{s}\right) I_2' \qquad ...(iv)$$

By substituting value of I_2' from equation (*ii*) in the above equation, we get:

$$V_1 = \left(\frac{4s^4 + 7s^2 + 1}{s}\right) \times V_2 \times s$$

$$\therefore \qquad \frac{V_2}{V_1} = \frac{1}{4s^4 + 7s^2 + 1}$$

Example 5.12. Determine driving point impedance function of the network shown in the figure given below:

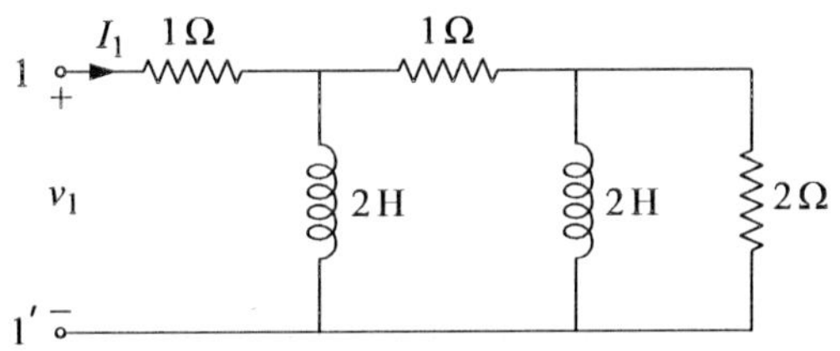

Solution: Network in *s*-domain is given as:

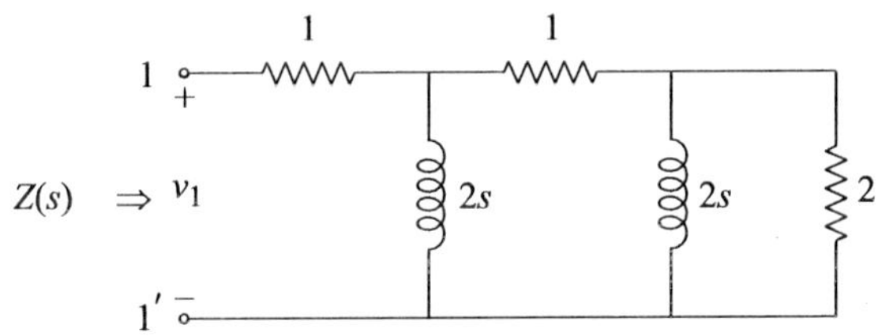

The driving point impedance of the network is,

$$Z(s) = \left[1 + \{2s \parallel (1 + 2s \parallel 2)\}\right]$$

$$= 1 + \left\{2s \parallel \left(1 + \frac{2s \times 2}{2s + 2}\right)\right\}\Bigg]$$

$$= 1 + \left\{2s \parallel \left(1 + \frac{2s}{s + 2}\right)\right\}\Bigg]$$

$$= \left[1 + \left\{2s \parallel \frac{3s + 1}{s + 1}\right\}\right] = \left[1 + \frac{\frac{2s\,(3s + 1)}{(s + 1)}}{2s + \frac{3s + 1}{s + 1}}\right]$$

$$= \left[1 + \frac{2s\,(3s + 1)}{2s^2 + 2s + 3s + 1}\right] = 1 + \frac{2s(3s + 1)}{2s^2 + 5s + 1}\Bigg]$$

$$= \frac{2s^2 + 5s + 1 + 6s^2 + 2s}{2s^2 + 5s + 1}$$

or $$Z(s) = \frac{8s^2 + 7s + 1}{2s^2 + 5s + 1}$$

Consider that:

$$I(s) = \frac{s^2 + 4s + 3}{s^2 + 2s} = 1 + \frac{2s + 3}{s^2 + 2s}$$

$$= 1 + 2\left[\frac{s + \frac{3}{2}}{s(s+2)}\right] = 1 + 2 \times I'$$

where $$I' = \frac{s + \frac{3}{2}}{s(s+2)}$$

Let $$I' = \frac{k_1}{s} + \frac{k_2}{s+2}$$

The pole-zero plot of the I' is given in figure (b).

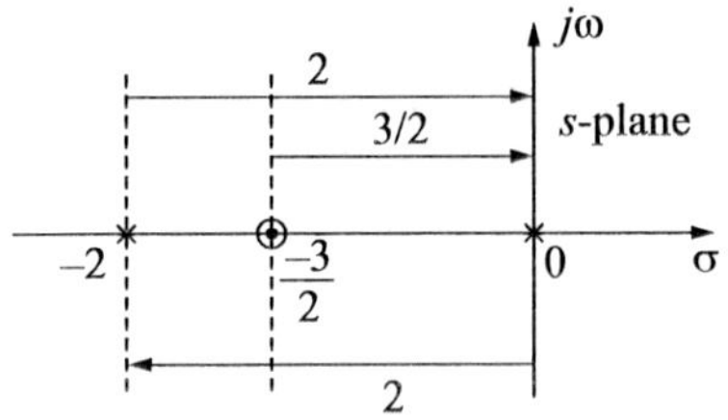

Fig. (b) *Pole-zero plot of I'*

From figure (b), the value of k_1 is:

$$k_1 = \frac{\frac{3}{2}\angle 0^\circ}{2\angle 0^\circ}$$, vectorial lengths upto origin

or $$k_1 = \frac{3}{4}$$

Similarly, $$k_2 = \frac{\frac{1}{2}\angle 180^\circ}{2\angle 180^\circ}$$, vectorial lengths upto pole –2

or $$k_2 = \frac{1}{4}$$

Thus, $$I(s) = 1 + \frac{\frac{3}{2}}{s} + \frac{\frac{1}{2}}{s+2}$$

By taking inverse Laplace transform of above, we get:

$$i(t) = \left[\delta(t) + \frac{3}{2} + \frac{1}{2}e^{-2t}\right] A$$

Example 5.13. Find open circuit transfer impedance $\frac{V_2}{I_1}$ and open circuit voltage ratio $\frac{V_2}{I_1}$ for the ladder network shown in the figure. **[UPTU, 2003]**

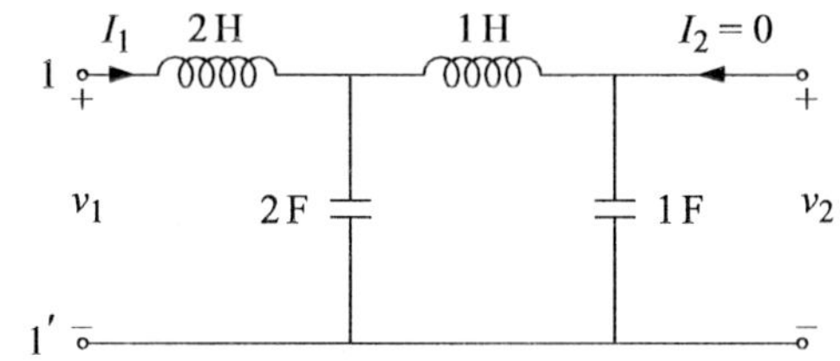

Solution: Laplace transformed network is given below:

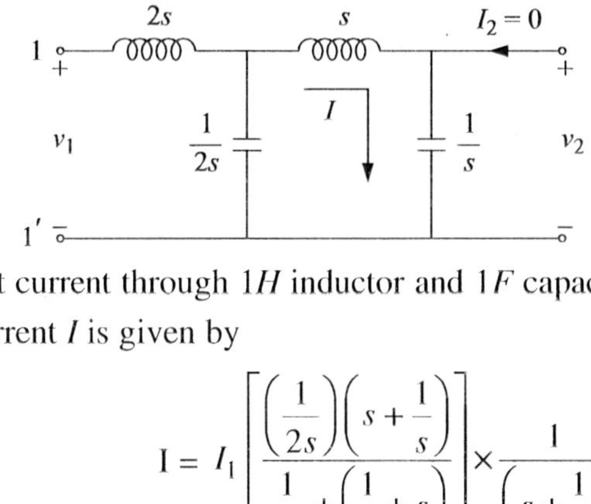

Consider that current through $1H$ inductor and $1F$ capacitor is I.

Then, the current I is given by

$$\text{I} = I_1 \left[\frac{\left(\frac{1}{2s}\right)\left(s + \frac{1}{s}\right)}{\frac{1}{2s} + \left(\frac{1}{s} + s\right)}\right] \times \frac{1}{\left(s + \frac{1}{s}\right)}$$

or $$I = \left[\frac{1}{2s^2 + 3}\right] I_1 \qquad ...(i)$$

Voltage across $2 - 2'$ is given as,

$$V_2 = I\left(\frac{1}{s}\right) = \left(\frac{1}{2s^2 + 3}\right) I_1 \times \frac{1}{s}$$

$\therefore$ $$\frac{V^2}{I^1} = \frac{1}{s(2s^2 + 3)} \qquad ...(ii)$$

By applying *KCL* through the loop containing elements 1, 2*H*, 1*H*, 1*F* and 1′, we get:

$$2\,(2s)\,I_1 + (s)\,I + \left(\frac{1}{s}\right)I = V_1$$

or
$$V_1 = 2sI_1 + \left(s + \frac{1}{s}\right)I$$

$$= 2sI_1 + \left(\frac{s^2+1}{s}\right)\left(\frac{1}{2s^2+3}\right)I_1$$

$$= I_1\left[2s + \frac{s^2+1}{s(2s^2+3)}\right]$$

or
$$V_1 = \frac{4s^4+7s^2+1}{s(2s^2+3)}I_1 \qquad ...(iii)$$

By substituting value of I_1 from equation (*ii*), we get:

$$V_1 = \frac{4s^4+7s^2+1}{s(2s^2+3)} \times \frac{s(2s^2+3)}{1} \times V_2$$

∴
$$\frac{V_2}{V_1} = \frac{1}{4s^4+7s^2+1}$$

Example 5.14. For the network of figure and the element values specified, find the current ratio transfer function given by $\alpha = (I_2/I_1)$. **[IP, 2001]**

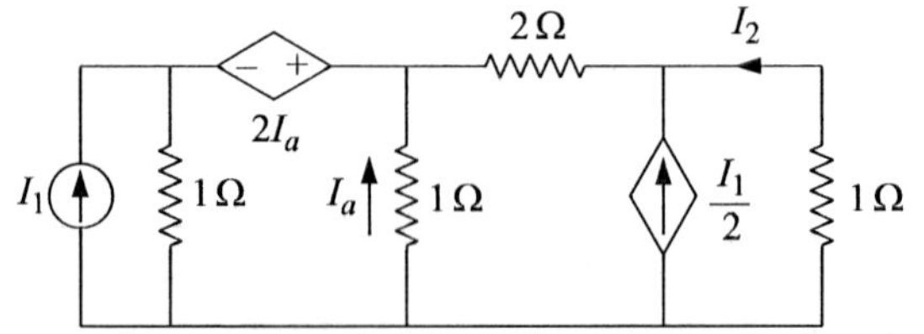

Solution: By converting current I_1 into voltage source, the circuit becomes:

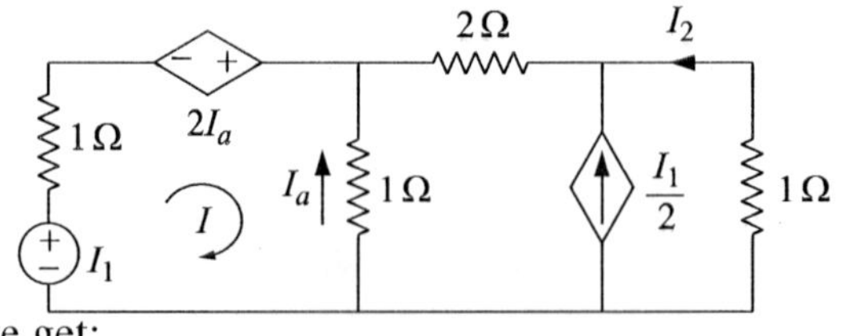

By KVL, we get:

$$I_1 = I - 2I_a - I_a, \text{ where } I = -\left(I_a + \frac{I_1}{2} + I_2\right)$$

In above if I is eliminated, we get:

$$I_1 = -\left(I_a + \frac{I_1}{2} + I_2\right) - 2I_a - I_a$$

or $$I_a = -\frac{1}{4}\left(\frac{3I_1}{2} + I_2\right) \quad ...(i)$$

By KVL in the loop having resistors 1Ω, 2Ω, and 1Ω, we get:

$$1 \times I_2 + 2\left(I_2 + \frac{I_1}{2}\right) = 1 \times I_a$$

or $$I_1 + 3I_2 = I_a \quad ...(ii)$$

From equations (*i*) and (*ii*), we get:

$$I_1 + 3I_2 = -\frac{1}{4}\left(\frac{3I_1}{2} + I_2\right)$$

or $$I_1 + 3I_2 + \frac{3}{8}I_1 + \frac{1}{4}I_2 = 0$$

or $$\frac{11}{8}I_1 + \frac{13}{4}I_2 = 0$$

∴ $$\alpha = \frac{I_2}{I_1} = -\frac{11}{8} \times \frac{4}{13}$$

or $$\alpha = -\frac{11}{26}$$

Example 5.15. Determine the driving point admittance and transfer admittance for the bridged T-network shown in the figure, with a 2Ω-load resistor connected across part 2. **[UPTU, 2003]**

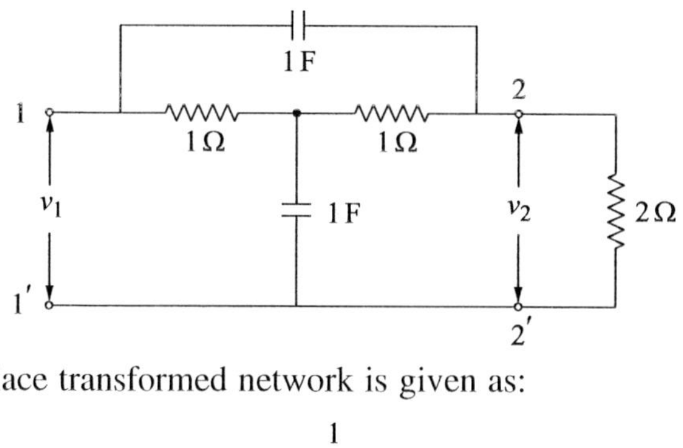

Solution: Laplace transformed network is given as:

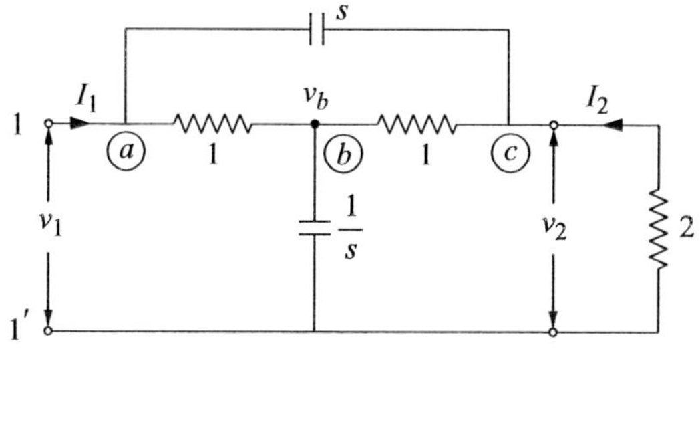

By KCL at node (a), we get:

$$\frac{V_1 - V_b}{1} + \frac{V_1 - V_2}{\frac{1}{s}} = I_1$$

or $\quad (s + 1)\, V_1 - sV_2 - V_b = I_1 \qquad ...(i)$

By KCL at node (b), we get:

$$\frac{V_b}{1} + \frac{V_b}{1/s} = \frac{V_1 - V_b}{1}$$

or $\quad V_1 + V_2 = (s + 2)\, V_b \qquad ...(ii)$

Now by *KCL* at node (c), we get:

$$\frac{V_1 - V_2}{1} + \frac{V_1 - V_2}{1/s} = \frac{V_2}{2}$$

or $\quad -sV_1 + \left(s + \frac{3}{2}\right) V_2 = V_b \qquad ...(iii)$

By substituting value of V_b from equation (ii) into equation (iii), we get:

$$(s + 2)\left\{-sV_1 + \left(s + \frac{3}{2}\right) V_2\right\} = V_1 + V_2$$

or $\quad \left(s^2 + \frac{7}{2}s + 2\right) V_2 = (s^2 + 2s + 1)\, V_1 \qquad ...(iv)$

By substituting value of V_b from equation (iii) in the equation (i), we get:

$$I_1 = (s + 1)\, V_1 - sV_2 + sV_1 - \left(s + \frac{3}{2}\right) V_2$$

Now, by substituting value of V_2 from equation (iv), we get:

$$I_1 = (2s + 1)V_1 - \left(2s + \frac{3}{2}\right)\left(\frac{s^2 + 2s + 1}{s^2 + \frac{7}{2}s + 2}\right) V_1$$

or $$I_1 = \frac{(2s + 1)\left(s^2 + \frac{7}{2}s + 2\right) - \left(2s + \frac{3}{2}\right)(s^2 + 2s + 1)}{\left(s^2 + \frac{7}{2}s + 2\right)} V_1$$

The driving point admittance

$$Y_{11} = \frac{I_1}{V_1} = \frac{1}{2}\left(\frac{5s^2 + 5s + 1}{s^2 + \frac{7}{2}s + 2}\right)$$

At port 2, we get:

$$V_2 = -2 \times I_2$$

or
$$V_2 = -2I_2 \qquad ...(v)$$

From equation (*iv*) and equation (*v*), we get:

$$\left(s^2 + \frac{I}{2}s + 2\right)(-2I_2) = (s^2 + 2s + 1)\, V_1$$

∴ Transfer admittance,

$$Y_{21} = \frac{I_2}{V_1} = -\frac{1}{2}\left[\frac{s^2 + 2s + 1}{s^2 + \frac{1}{2}s + 2}\right]$$

Example 5.16. In the network of the figure, find $\frac{V_c(s)}{V(s)}$. Also find the pole-zero locations and draw pole-zero plot of it. **[UPTU, 2007]**

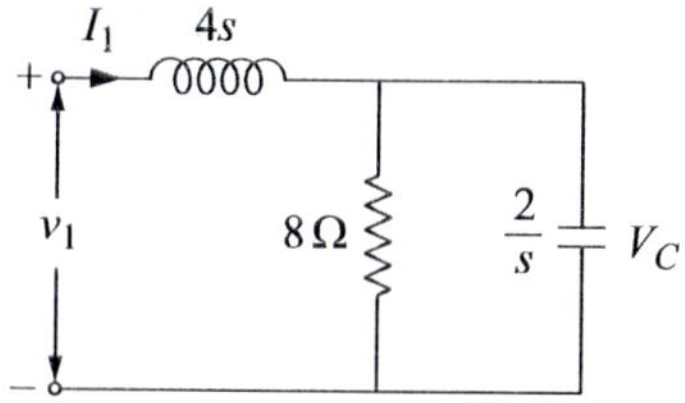

Solution:

$$V_c(s) = \left(\frac{8 \times \frac{2}{s}}{8 + \frac{2}{s}}\right) I = \frac{16}{8s + 2} = \frac{8}{4s + 1}$$

or
$$V_c(s) = \left(\frac{8}{4s + 1}\right) I \qquad ...(i)$$

and
$$V(s) = \left[4s + \frac{8}{4s + 1}\right] I$$

or
$$V(s) = \left(\frac{-16s^2 + 4s + 8}{4s + 1}\right) I \qquad ...(ii)$$

Thus, form equations (i) and (*ii*), we get:

$$\frac{V_c(s)}{V(s)} = \frac{2}{4s^2 + s + 2}$$

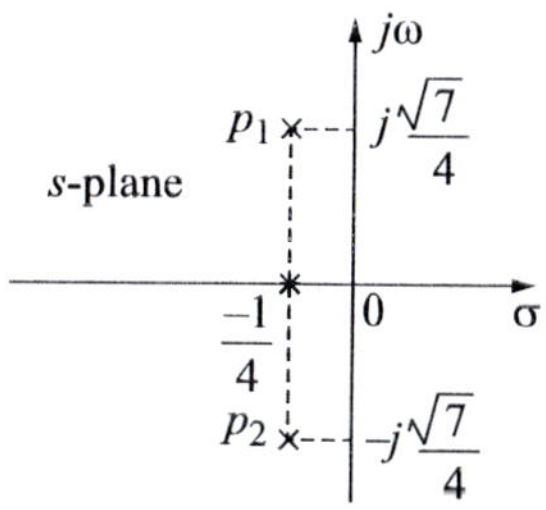

The characteric equation for poles is:

$$4s^2 + s + 2 = 0$$

$$\therefore \qquad p_{1,2} = -\frac{1}{4} \pm j\frac{\sqrt{7}}{4}$$

Thus, the pole-zero plot is as drawn.

Example 5.17. For the given network function, draw the pole-zero diagram and hence obtain time domain response. **[UPTU, 2003]**

$$I(s) = \frac{s^2 + 4s + 3}{s^2 + 2s}$$

Solution: $I(s)$ can be written as:

$$I(s) = 1 + \frac{2s + 3}{s(s+2)}$$

or
$$I(s) = 1 + \frac{2\left(s + \frac{3}{2}\right)}{s(s+2)}$$

or
$$I(s) = 1 + 2\,I'(s)$$

$$I'(s) = \frac{\left(s + \frac{3}{2}\right)}{s(s+2)} = \frac{k_1}{s} + \frac{k_2}{s+2}$$

i.e., zero at $-\frac{3}{2}$ and poles at 0, –2

The pole-zero plot is as follows:

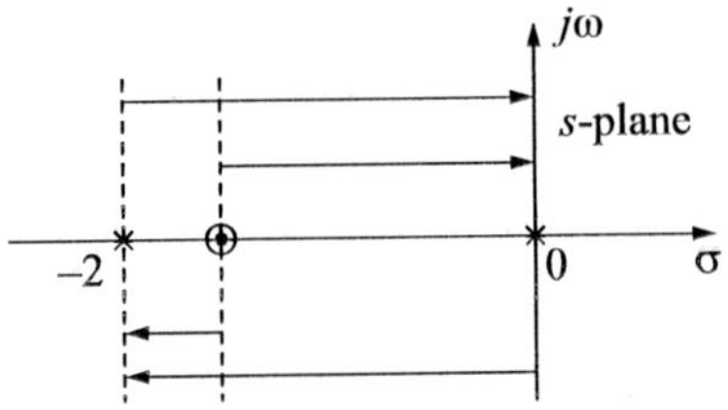

The values of k_1 and k_2 are:

$$k_1 = \frac{\frac{3}{2}}{2} = \frac{3}{4}$$

and
$$k_2 = \frac{\frac{1}{2}}{2} = \frac{1}{4}$$

$$\therefore \qquad I'(s) = \frac{\left(\frac{3}{4}\right)}{s} + \frac{\left(\frac{1}{4}\right)}{s+2}$$

$$\therefore \qquad I(s) = 1 + \left(\frac{3}{2}\right)\frac{1}{s} + \left(\frac{1}{2}\right)\frac{1}{(s+2)}$$

By inverse Laplace transform, we get:

$$i(t) = \left(\delta(t) + \frac{3}{2} + \frac{1}{2}e^{-2t}\right)A$$

Example 5.18. The output voltage $V(s)$ of a network is given by $\frac{4s}{(s+2)(s^2+2s+2)}$. Plot its pole-zero diagram and hence obtain $v(t)$.

[UPTU, 2005]

Solution: $V(s)$ can be factorised as given here:

$$V(s) = \frac{4s}{(s+2)(s+1+j)(s+1-j)}$$

Thus, pole zero plot is:

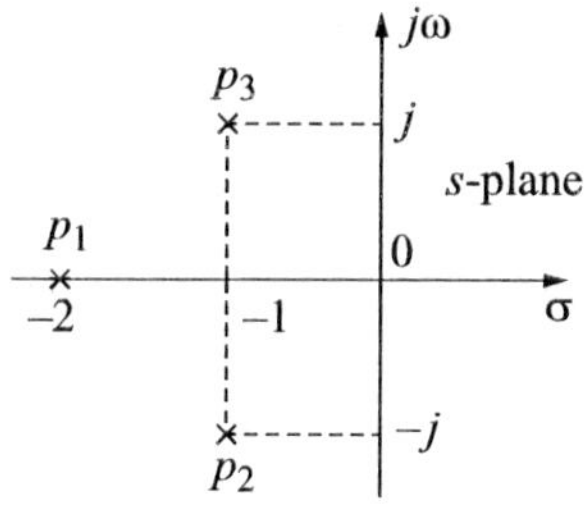

$$\text{Let:} \qquad V(s) = \frac{k_1}{(s+2)} + \frac{k_2}{s+1+j} + \frac{k_3}{s+1-j}$$

$$\text{Thus,} \quad k_1 = 4\times\frac{(p_1 - z)}{(p_1-p_2)(p_1-p_3)} = 4\times\frac{(-2-0)}{(-2+1+j)(-2+1-j)}$$

$$\text{or} \quad k_1 = -4$$

$$k_2 = 4\times\frac{(-1-j-0)}{(-1-j+2)(-1-j+1-j)}$$

$$\text{or} \quad k_2 = 2$$

$$\text{and} \quad k_3 = 4\times\frac{(-1+j-0)}{(-1+j+2)(-1+j+1+j)}$$

$$\text{or} \quad k_3 = 2$$

$$\text{Thus,} \quad V(s) = -\frac{4}{s+2} + \frac{2}{s+1+j} + \frac{2}{s+1-j}$$

By taking inverse Laplace transform on both sides of the equation, we get:

$$v(t) = -4e^{-2t} + 2e^{-(1+j)t} + 2e^{-(1-j)t}$$

$$= -4e^{-2t} + 2e^{-t}(e^{-jt} + e^{jt})$$

$$= -4e^{-2t} + 2e^{-t} \times 2\left(\frac{e^{jt} + e^{-jt}}{2}\right)$$

$$= -4e^{-2t} + 2e^{-t}(2\cos t)$$

or $$v(t) = -4e^{-t}(e^{-t} - \cos t)$$

Example 5.19. The transform current $I(s)$ in a network is given by $I(s) = \dfrac{3s(s+2)}{(s+1)(s+4)}$. Plot the poles and zeros in the s-plane and hence obtain the time domain response. **[UPTU, 2006]**

Solution: $$I(s) = \frac{3s(s+2)}{(s+1)(s+4)}$$

$$= 3 - \frac{(9s+12)}{(s+1)(s+4)}$$

or $$I(s) = 3 - 9I'(s)$$

The poles and zeros of $I'(s)$ are:

$$z = -\frac{4}{3}$$

$$p_1 = -1 \quad p_2 = -4$$

Thus, pole-zero plot is as shown below:

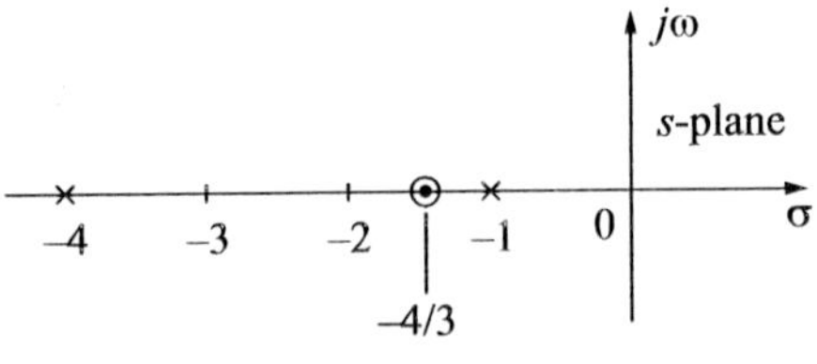

Consider that: $I'(s) = \dfrac{k_1}{s+1} + \dfrac{k_2}{s+4}$. The values of k_1 and k_2 are calculated from the following pole-zero diagram marking.

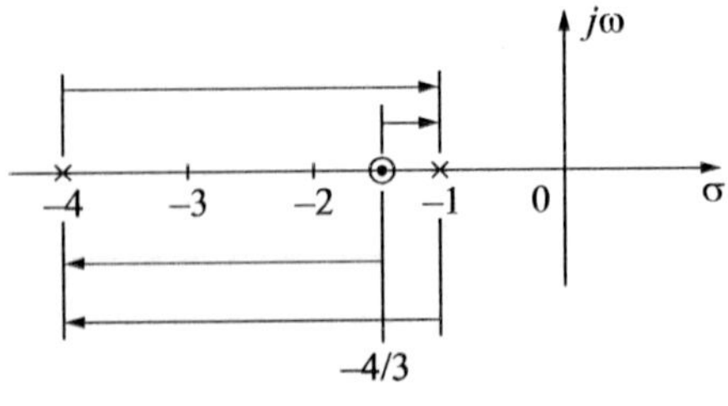

$$k_1 = \frac{\left(\frac{4}{3}-1\right)}{(4-1)} = \frac{\frac{1}{3}}{3} = \frac{1}{9}$$

and $$k_2 = \frac{\left(4-\frac{4}{3}\right)}{(4-1)} = \frac{\frac{8}{3}}{3} = \frac{8}{9}$$

$\therefore$ $$I'(s) = \left(\frac{1}{9}\right)\frac{1}{(s+1)} + \left(\frac{8}{9}\right)\frac{1}{(s+4)}$$

$\therefore$ $$I(s) = 3 - 9 \times \frac{1}{9}\left[\frac{1}{s+1} + \frac{8}{s+4}\right]$$

or $$I(s) = 3 - \frac{1}{s+1} - \frac{8}{s+4}$$

By taking Laplace inverse on both sides, we get:

$$i(t) = 3\,\delta(t) - e^{-t} - 8e^{-4t}$$

Example 5.20. Find the transfer function $\left(\frac{V_2}{V_1}\right)$ of the network given in the figure. Also sketch pole-zero configuration. **[UPTU, 2006]**

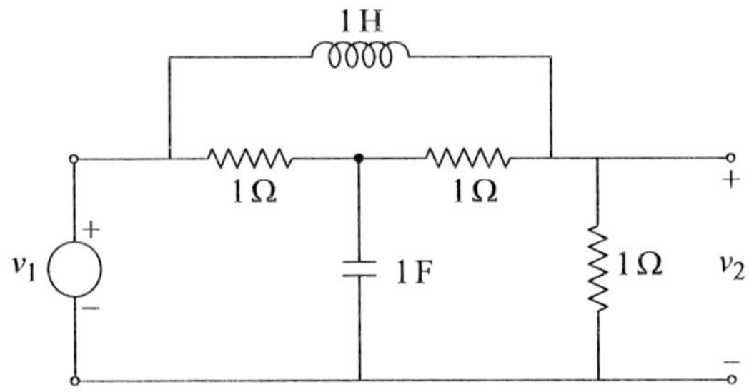

Solution: The transformed network in s-domain is given by:

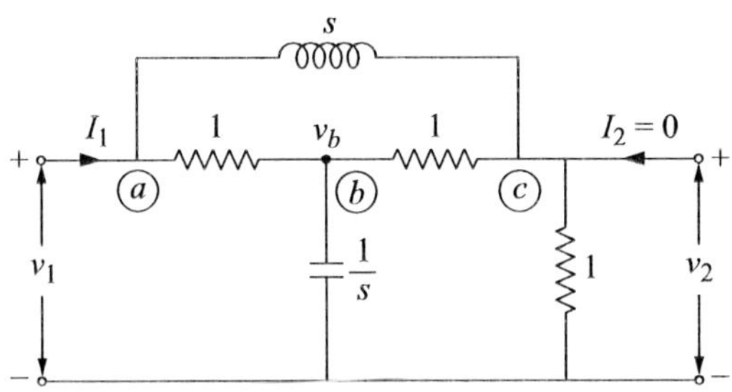

By KCL at node (a), we get:

$$I_1 = \frac{V_1 - V_b}{1} + \frac{V_1 - V_2}{s}$$

or $\left(1+\frac{1}{s}\right)V_1-\frac{V_2}{s}-V_b=I_1$ ...(i)

By KCL at node (b), we get:

$$sV_b=\frac{V_1-V_b}{1}+\frac{V_2-V_b}{1}$$

or $$V_b=\frac{1}{(s+2)}V_1+\frac{1}{(s+2)}V_2 \quad ...(ii)$$

By KCL at node (c), we get:

$$\frac{V_2-V_1}{s}+\frac{V_2-V_b}{1}+\frac{V_2}{1}=0$$

or $$(1+2s)V_2-V_1-sV_b=0 \quad ...(iii)$$

From equations (ii) and (iii), we get:

$$(1+2s)V_2-V_1-\frac{s}{s+2}V_1-\frac{s}{s+2}V_2=0$$

or $$\left[(1+2s)-\frac{s}{s+2}\right]V_2=\left[1+\frac{s}{s+2}\right]V_1$$

or $$\frac{2s^2+5s+2-s}{s+2}V_2=\frac{2s+2}{s+2}V_1$$

or $$\frac{V_2}{V_1}=\frac{2s+2}{2s^2+4s+2}$$

$$=\frac{(s+1)}{(s^2+2s+1)}=\frac{(s+1)}{(s+1)^2}$$

or $$\frac{V_2}{V_1}=\frac{1}{s+1}$$

Pole at $s=-1$

Thus, pole-zero plot is:

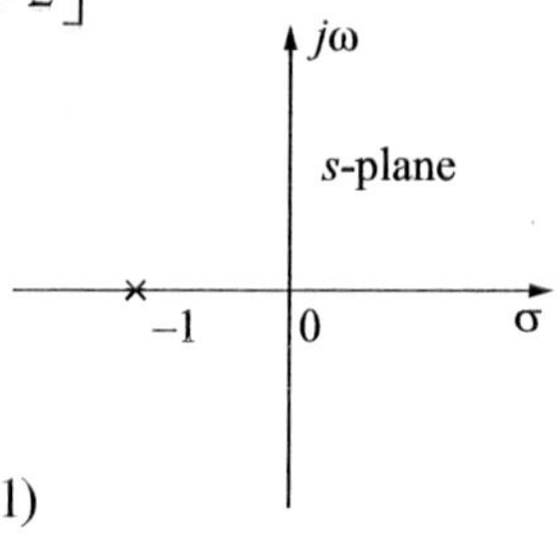

Example 5.21. Find the inverse Laplace of $F_1(s)\cdot F_2(s)$ by convolution integral for the following. $F_1(s)=\frac{1}{s+a}$ and $F_2(s)=\frac{1}{(s+b)(s+c)}$ **[UPTU, 2001]**

Solution: $$F_1(s)=\frac{1}{s+a}$$

and $$F_2(s)=\frac{1}{(s+b)(s+c)}$$

$$= \left(\frac{1}{c-b}\right)\left\{\frac{1}{s+b} - \frac{1}{s+c}\right\}$$

$$f_1(t) = £^{-1}\left[F_1(s)\right] = e^{-at}$$

and $$f_2(t) = £^{-1}\left[F_2(s)\right] = £^{-1}\left[\left(\frac{1}{c-b}\right)\left\{\frac{1}{s+b} - \frac{1}{s+c}\right\}\right]$$

or $$f_2(t) = \frac{1}{c-b}(e^{-bt} - e^{-ct})\,]$$

∴ $$f(t) = £^{-1}\left[F(s)\right] = £^{-1}\left[F_1(s)\cdot F_2(s)\right] = f_1(t)\cdot f_2(t)$$

$$= \int_0^t f_1(t-\tau)\, f_2(\tau)\, d\tau$$

Hence, $$f(t) = \int_0^t e^{-a(t-\tau)} \cdot \left(\frac{e^{-b\tau} - e^{-c\tau}}{c-b}\right) d\tau$$

$$= \frac{e^{-at}}{c-b}\left[\int_0^t e^{a\tau}\left(e^{-b\tau} - e^{-c\tau}\right) d\tau\right]$$

$$= \frac{e^{-at}}{c-b}\left[\int_0^t e^{(a-b)\tau} d\tau - \int_0^t e^{(a-c)\tau} d\tau\right]$$

$$= \frac{e^{-at}}{c-b}\left[\left.\frac{e^{(a-b)\tau}}{(a-b)}\right|_0^t \left.\frac{e^{(a-c)\tau}}{(a-c)}\right|_0^t\right]$$

$$= \frac{e^{-at}}{c-b}\left[\frac{1}{(a-b)}\left(e^{(a-b)t} - 1\right) - \frac{1}{(a-c)}\left(e^{(a-c)t} - 1\right)\right]$$

$$= \frac{e^{-at}}{c-b}\left[\frac{e^{(a-b)t}}{(a-b)} - \frac{e^{(a-c)t}}{(a-c)} - \left\{\frac{1}{a-b} - \frac{1}{a-c}\right\}\right]$$

$$= \frac{e^{-bt}}{(c-b)(a-b)} - \frac{e^{-ct}}{(c-b)(a-c)} - \frac{e^{-at}(a-c-a+b)}{(c-b)(a-b)(a-c)}$$

$$= \frac{e^{-bt}}{(c-b)(a-b)} + \frac{e^{-ct}}{(a-c)(b-c)} + \frac{e^{-at}}{(a-b)-(a-c)}$$

or $$f(t) = \frac{e^{-at}}{(a-b)(a-c)} + \frac{e^{-bt}}{(b-c)(b-a)} + \frac{e^{-ct}}{(c-a)-(c-b)}$$

Example 5.22 Find the inverse Laplace transform of the given function,

$$F(s) = \frac{1}{s^2\,(s+2)}$$, using convolution integral. **[UPTU, 2002]**

Solution: Given: $$F(s) = \frac{1}{s^2(s+2)}$$

Suppose $$F(s) = F_1(s)\cdot F_2(s) = \frac{1}{s^2(s+2)}$$

or $$F_1(s)\,F_2(s) = \left(\frac{1}{s^2}\right)\left(\frac{1}{s+2}\right)$$

$\therefore$ $$F_1(s) = \frac{1}{s^2}$$

and $$F_2(s) = \frac{1}{s+2}$$

By taking inverse Laplace transform of both functions:
$f(t) = t$ and $f_2(t) = e^{-2t}$
By convolution theorem:

$$f(t) = \pounds^{-1}[F(s)] = \pounds^{-1}[F_1(s)\cdot F_2(s)]$$
$$= f_1(t)\cdot f_2(t)$$

$\therefore$ $$f(t) = \int_0^t f_1(t-\tau)\,f_2(\tau)\,d\tau = \int_0^t (t-\tau)\,e^{-2\tau}\,d\tau$$

$$= t\int_0^t e^{-2\tau}\,d\tau - \int_0^t \tau\,e^{-2\tau}\,d\tau$$

$$= \left[t\cdot\left(\frac{e^{-2\tau}}{-2}\right)\Bigg|_0^t\right] - \left[\tau\cdot\frac{e^{-2\tau}}{-2}\Bigg|_0^t - \int_0^t \frac{e^{-2\tau}}{-2}\,d\tau\right]$$

$$= \frac{t}{2}(1-e^{-2t}) + \frac{t}{2}(e^{-2t}) - \frac{1}{2}\frac{e^{-2\tau}}{-2}\Bigg|_0^t$$

$$= \frac{t}{2}(1-e^{-2t}) + \frac{t}{2}e^{-2t} + \frac{1}{4}(e^{-2t}-1)$$

or $$f(t) = -\frac{1}{4}(1-2t-e^{-2t})$$

Example 5.23. Determine Inverse Laplace Transform of the following function using convolution integral: $F(s) = F_1(s)\cdot F_2(s) = \dfrac{s+1}{s(s^2+4)}$ **[UPTU, 2003]**

Solution: Consider: $F_1(s) = \frac{1}{s}$ and $F_2(s) = \frac{s+1}{s^2+4}$

$$\therefore \quad f_1(t) = u(t) \text{ and } f_2(t) = \cos 2t + \frac{1}{2}\sin 2t$$

$$\therefore \quad f(t) = \int_0^t f_1(t-\tau)\, f_2(\tau)\, d\tau = \int_0^t u(t-\tau)\left[\cos 2\tau + \frac{1}{2}\sin 2\tau\right] d\tau$$

$$= \int_0^t \cos 2\tau\, d\tau + \frac{1}{2}\int_0^t \sin 2\tau\, d\tau$$

$$= \left.\frac{\sin 2\tau}{2}\right|_0^t + \frac{1}{2}\left.\frac{(-\cos 2\tau)}{2}\right|_0^t = \frac{1}{2}\sin 2t - \frac{1}{4}(\cos 2t - 1)$$

$$= \frac{1}{4} + \frac{1}{2}\sin 2t - \frac{1}{4}\cos 2t$$

or $$f(t) = \frac{1}{4}(1 + 2\sin 2t - \cos 2t)$$

SUMMARY

1. **Network Functions:** All the networks are designed to produce a particular output, when the input is applied to it. The network function is the characteristics of the network and not dependent on input or output.
2. **Laplace Transform:** The Laplace transform method has the following two attractive features:
   (*i*) The homogeneous equation and the particular integral of the solution are obtained in one operation.
   (*ii*) The Laplace transform converts the differential equation into an algebraic equation. Then simple algebraic rules are used to manipulate and solve in the s-doman. The final solution is obtained by taking the inverse Laplace transform.

   The Laplace Transformation is defined as:

$$\pounds(f(t) = F(s) = \int_0^\infty f(t)e^{-st}\,dt$$

   Where-ins is a complex variable *i.e.*:

$$s = \sigma + j\omega$$
$$\sigma = \text{Real part}$$
$$j\omega = \text{Imaginary part.}$$

   Tables of Laplace transform are available for easy analysis. Similarly inverse Laplace transform tables are also available as follows:

$$\pounds^{-1}[F(s)] = f(t) \text{ for } t \geq 0$$

3. **Routh-Hurwitz Criterion:** This criterion is an algebraic method which provides information as the absolute stability of a linear time-invariant system that has a characteristic equation with constant coefficient *i.e.*, the right half of s-plane. The number of roots which lie on the $j\omega$ axis and in right half plane is also indicated.
4. **Complex Frequency:** Where $s = \sigma + j\omega$ which is known as complex frequency. σ is real part and $j\omega$ is imaginary part with ω as angular frequency.
5. **One or Two Port Networks:** A pair of terminals are connected to the black box at which signal may enter the network or leave the network. Such black box is known as one port network. When the black box has one pair of terminals to enter the signals and another pair of terminals to leave the network, then it is known as two port network.

EXERCISES

5.1. Obtain the driving point impedance functions or transform impedance $Z(s)$ for the networks shown in Figs. (a) to (c), in which transformed impedance marked for each element.

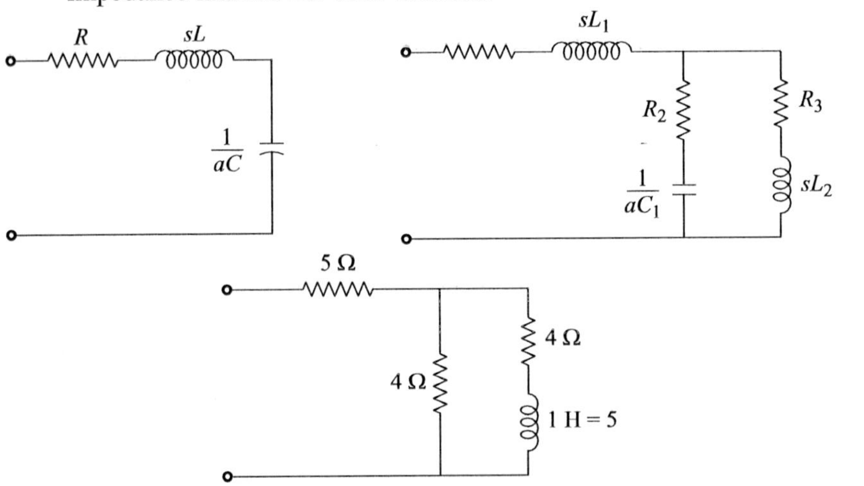

5.2. Determine the driving point admittance functions or Transform admittances $Y_{(S)}$ for the network shown in Fig. (a) to (c).

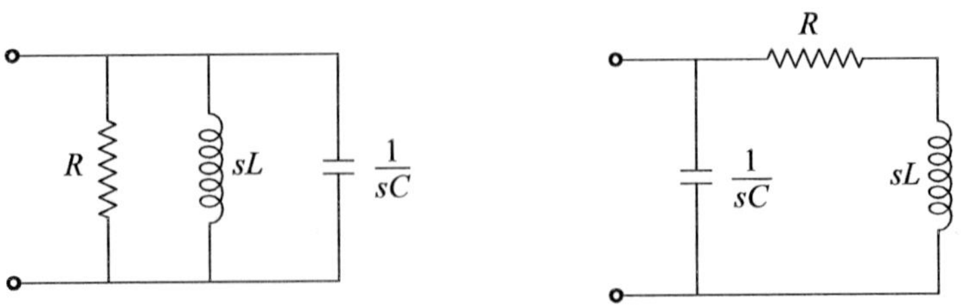

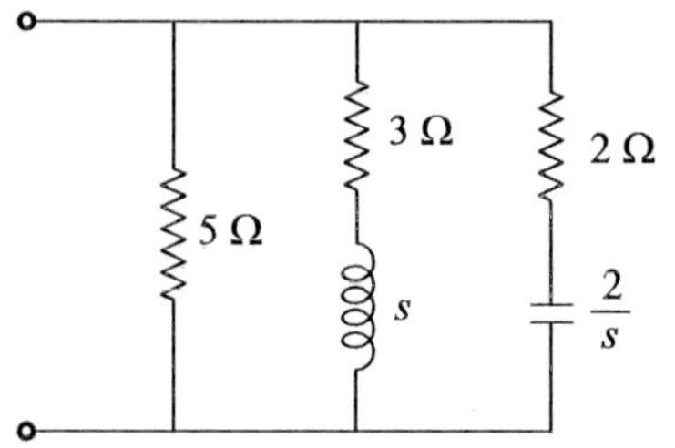

5.3. For the circuit shown in figure. Determine the network transfer function.

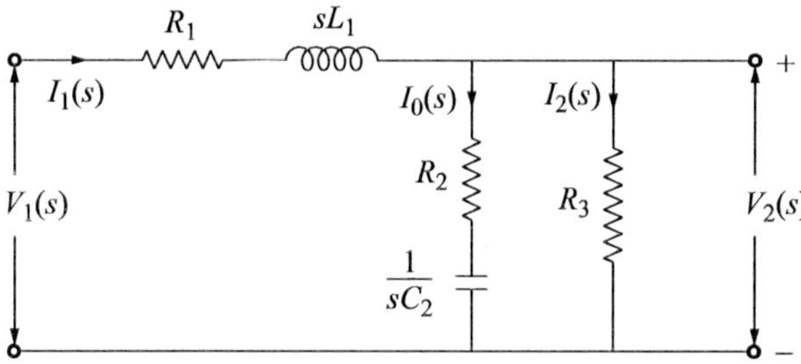

5.4. Find the voltage transfer functions of the networks shown in figures.

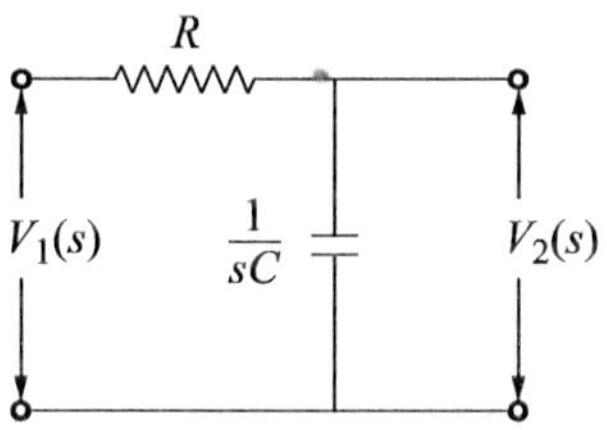

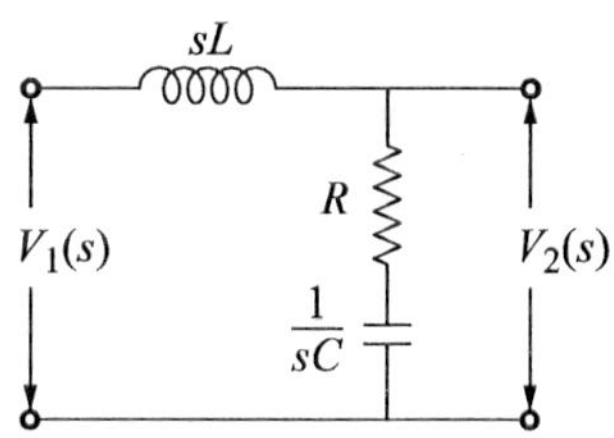

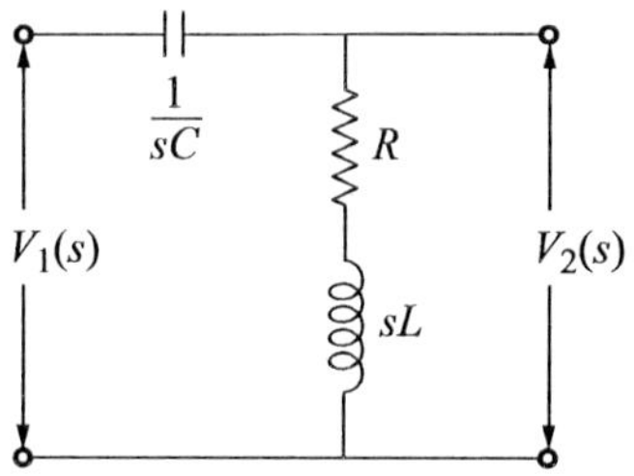

5.5. Find the expression of voltage transfer function for the network shown in figure.

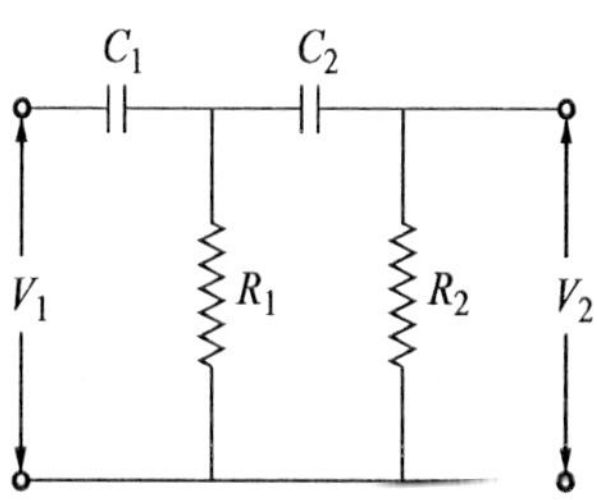

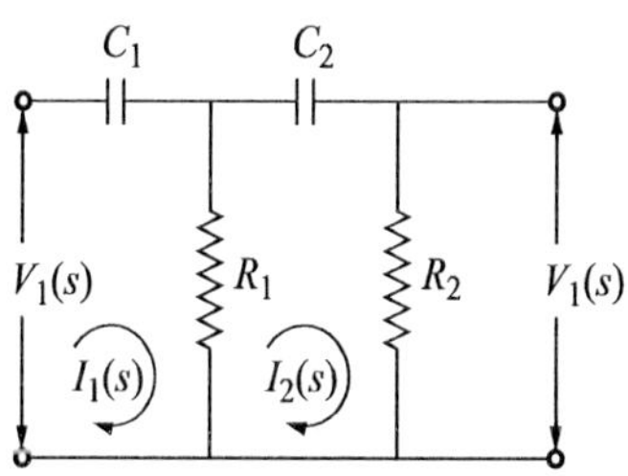

5.6. Determine the open-circuit driving point impedance (Z_{11}), transfer impedance (Z_{21}) and voltage transfer function (G_{21}) for the network shown in figure.

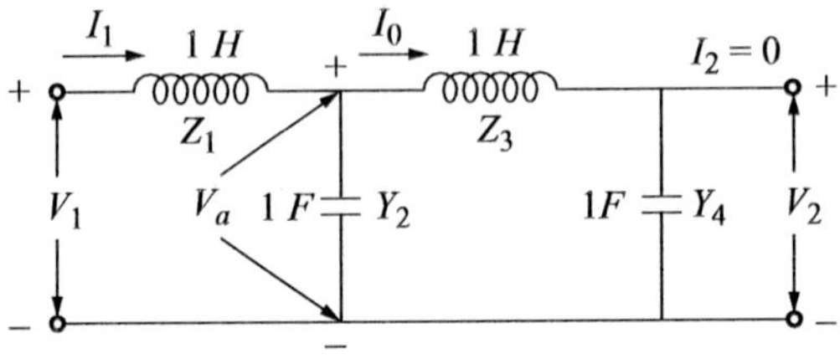

5.7. Obtain the pole zero location for the function

$$T(S) = \frac{(s^2 + 4)(s + 4)}{(s + 1)(s + 3)}$$

5.8. Obtain the pole-zero locations of the following functions:

(*i*) $f_1(1) = e^{-\sigma t}$; a decaying exponential.

(*ii*) $f_2(1) = \cos \omega t$; a sinusoid.

(*iii*) $f_3(1) = e^{\sigma t} \cos \omega t$; a damped sinusoid.

5.9. Find the open circuit driving point impedance at terminals $1 - 1'$ of the ladder work shown in figure.

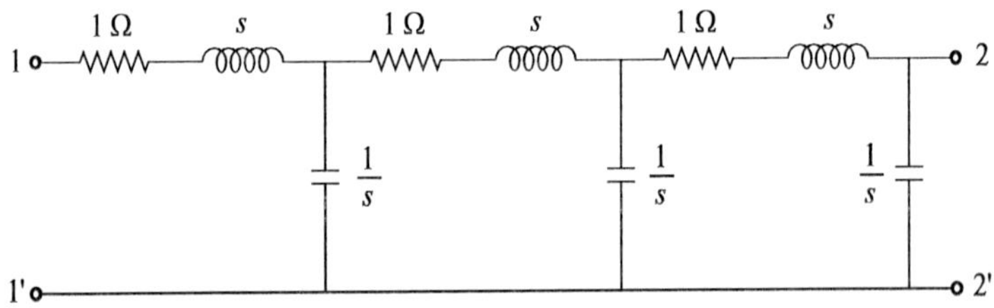

5.10. Check whether given functions are suitable in representing the driving point admittance functions are not.

(*i*) $$Z_{(s)} = \frac{4s^4 + s^2 - 3s + 1}{s^3 + 2s^2 + 2s + 40}$$

(*ii*) $$Y_{(s)} = \frac{15(s^3 + 2s^2 + 3s + 4)}{s^4 + 8s^2 + 6s^2}$$

(*iii*) $$Z_{(s)} = \frac{s^2 + s + 2}{2s^2 + s + 1}$$

5.11. Check whether given functions are suitable in representing the transfer functions are not:

(*i*) $G_{21}(s) = \dfrac{3s+2}{5s^3+4s^2+1}$ (*ii*) $\alpha_{21}(s) = \dfrac{2s^2+5s+1}{5+7}$

(*iii*) $Z_{21}(s) = \dfrac{1}{s^3+2s}$ (*iv*) $G_{21}(s) = \dfrac{2s^2+5}{3s^2 9s+1}$

5.12. For the circuit shown in figure (*a*) obtain driving point admittance $Y_{11} = I_1/V_1$.

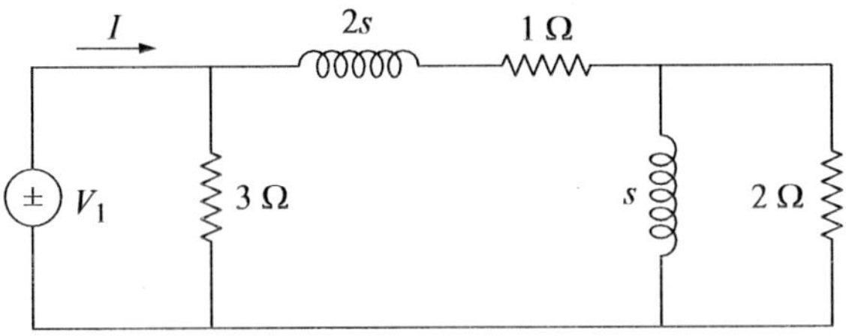

5.13. Find the transfer impedance function $Z_{21}(s) = V_{21}(s)/I_{1(s)}$ of the network shown in figure.

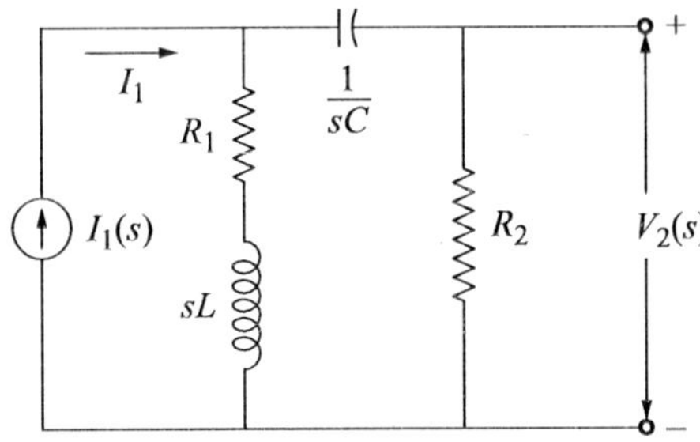

5.14. Find the driving point impedance of the network shown in Fig. (*a*). Find the poles and zeros of the network and locate them in the *s*-plane.

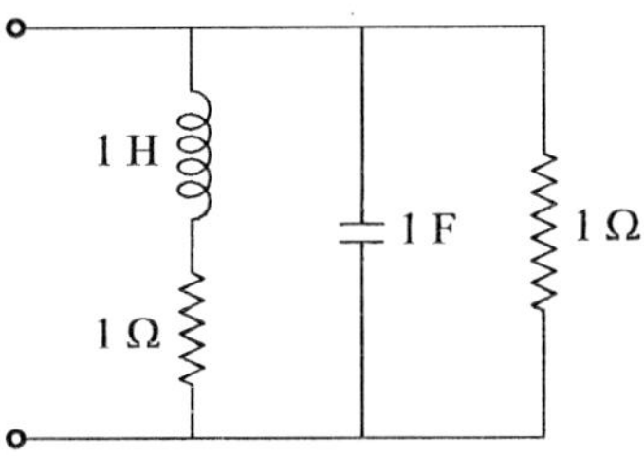

5.15. Find driving point impedance $Z_{11}(s)$, transfer impedance $Z_{21}(s)$ and voltage transfer function $G_{21}(s)$ for the circuit of figure.

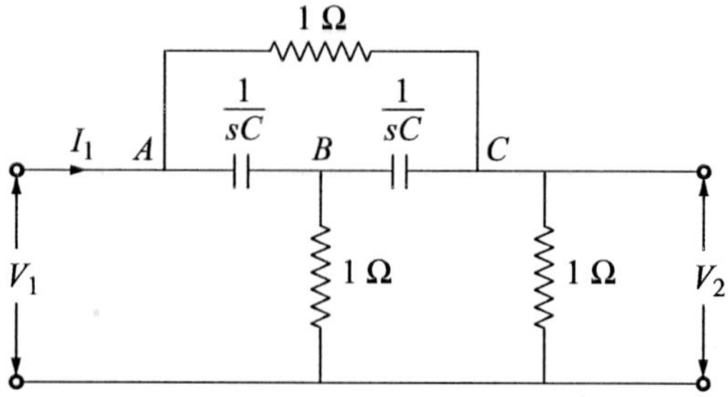

5.16. For the network of figure and the element values specified, find the current ratio transfer function given by $d = (I_2/I_1)$.

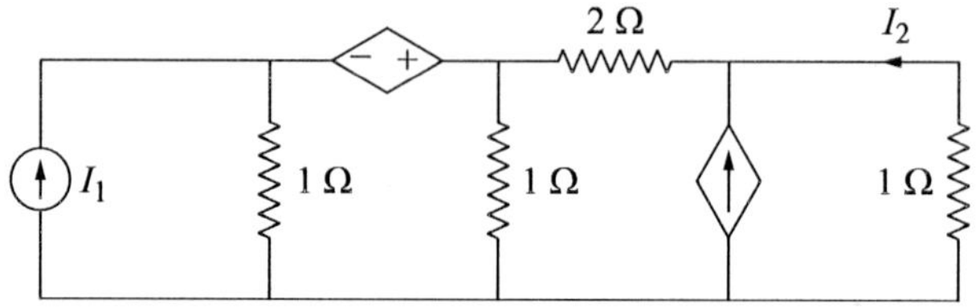

5.17. Find the current transfer ratio (I_2/I_1) and voltage transfer ratio (V_2/V_1) of the network shown in figure.

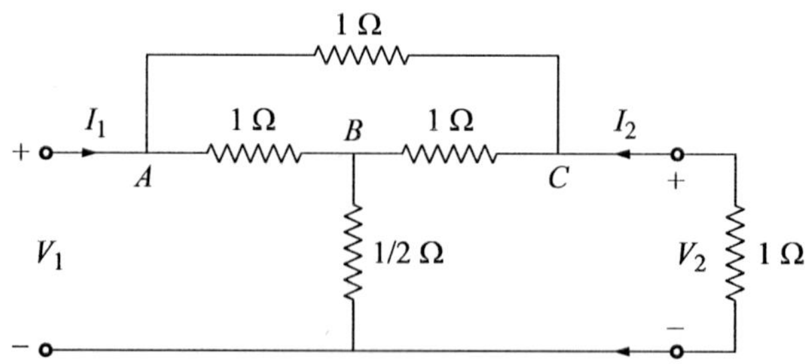

5.18. Find the current transfer ratio (I_2/I_1) and voltage transfer ratio (V_2/V_1) of the network shown in figure:

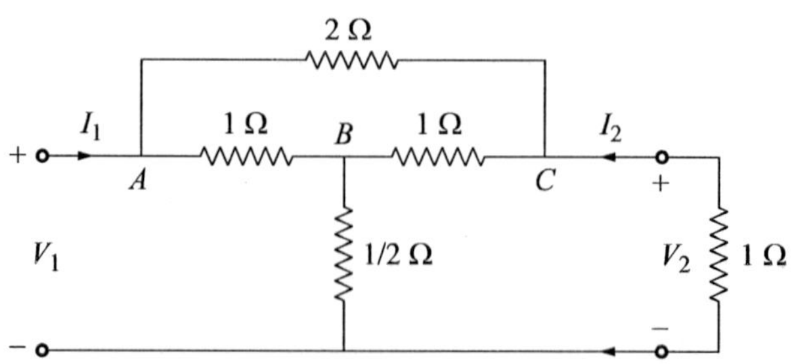

□□□

6

Two Port Networks

6.1 INTRODUCTION

A general network having two pair of terminals, one as input terminals and the other as output terminals is a very important building block in various electrical and electronic networks. A pair of terminals at which signal may enter or leave a network is called a port. Thus, two port network base one input part and one output port as shown in Fig. 6.1 V_1 and I_1 are input voltage and current, whereas V_2 and I_2 are output voltage and current. The directions of currents are defined for analysis purposes as shown in the figure is important to note that current of output port I_2 is also entering the port instead of going out.

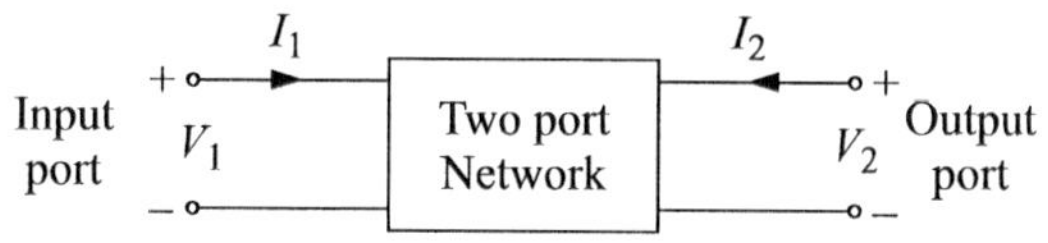

Fig 6.1 *Two port network*

Two-port network representation for analysis is done in automatic control system, transmission and distribution systems, transforms, bridge circuits, filters, etc.

The networks under consideration are only Linear Time Invariant (LTI). *LTI networks have all elements whose parameters are not dependent on time.*

The LTI two-port network characterization is done using following parameters:

(*i*) Open circuit or impedance or *z*-parameters.

(*ii*) Short circuit or admittance or *y*-parameters.

(*iii*) Hybrid or *h*-parameters.

(*iv*) Inverse hybrid or *g*-parameters.

(*v*) Transmission or *A*, *B*, *C*, *D* parameters.

(*vi*) Inverse transmission or *a*, *b*, *c*, *d* parameters.

The analysis using two-port networks presumes that such networks are composed of linear elements and contain no independent sources. Dependent sources are permissible in two-port networks.

These parameters and their definitions are summarized as in Table 6.1:

Table 6.1 Parameter and their definitions

S. No	Name of Parameters	Dependent Variables	Independent Variables	Equations
1	Open circuit or impedance or z-parameters	V_1, V_2	I_1, I_2	$V_1 = z_{11} I_1 + z_{12} I_2$ $V_2 = z_{21} I_1 + z_{22} I_2$
2	Short circuit or admittance or y-parameters	I_1, I_2	V_1, V_2	$I_1 = y_{11} V_1 + y_{12} V_2$ $I_2 = y_{21} V_1 + y_{22} V_2$
3	Transmission or T-parameters	V_1, I_1	V_2, I_2	$V_1 = AV_2 - BI_2$ $I_1 = CV_2 - DI_2$
4	Inverse transmission or inverse T-Parameters	V_2, I_2	V_1, I_1	$V_2 = aV_1 - bI_1$ $I_2 = cV_1 - dI_1$
5	Hybrid or h-parameters	V_1, I_2	I_1, V_2	$V_1 = h_{11} I_1 + h_{12} V_2$ $I_2 = h_{21} I_1 + h_{22} V_2$
6	Inverse hybrid or g-parameters	I_1, V_2	V_1, I_2	$I_1 = g_{11} V_1 + g_{12} I_2$ $V_2 = g_{21} V_1 + g_{22} I_2$

6.2 *z*-PARAMETERS

z-parameters are obtained by expressing voltages at two-ports in terms of currents at two ports. Currents I_1 and I_2 are independent variables and V_1, V_2 are dependent variables,

$$V_1 = f_1(I_1, I_2)$$
$$V_2 = f_2(I_1, I_2)$$

The relationship in equation form are:

$$V_1 = z_{11} I_1 + z_{12} I_2 \quad ...(6.1)$$
$$V_2 = z_{21} I_1 + z_{22} I_2 \quad ...(6.2)$$

It can also be written in matrix form.

$$\begin{bmatrix} V_1 \\ V_2 \end{bmatrix} = \begin{bmatrix} z_{11} & z_{12} \\ z_{21} & z_{22} \end{bmatrix} \begin{bmatrix} I_1 \\ I_2 \end{bmatrix} \quad ...(6.3)$$

or $$[V] = [z]\,[I] \quad ...(6.4)$$

The z-parameters relate to impedance, hence, these are also known as impedance parameters.

z-parameter definitions are:

Port 2 is open	Port 1 is open
$z_{11} = \left.\dfrac{V_1}{I_1}\right\vert_{I_2=0}$	$z_{12} = \left.\dfrac{V_1}{I_2}\right\vert_{I_1=0}$
$z_{21} = \left.\dfrac{V_2}{I_1}\right\vert_{I_2=0}$	$z_{22} = \left.\dfrac{V_2}{I_2}\right\vert_{I_1=0}$

z-parameters are defined, only when the current is one of the ports is zero, thus, these are also known as open circuit parameters.

z-parameters matrix elements are known as:

z_{11} = open circuit to driving point input port impedance

z_{22} = open circuit driving point output port impedance

z_{12} = open circuit reverse transfer impedance.

z_{21} = open circuit forward transfer impedance.

6.2.1 Equivalent Circuit using *z*-parameters

A two-port network can be represented in terms of 3-parameters as shown in Fig. 6.2.

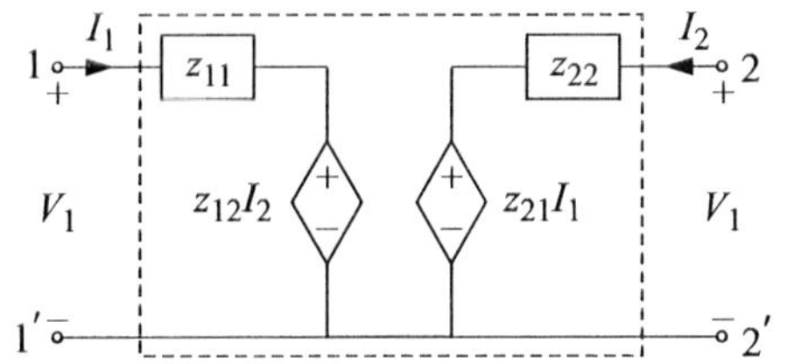

Fig. 6.2 *Equivalent circuit of two-port network using z-parameters.*

The equivalent circuit contains two dependent voltage sources.

6.2.2 Condition for Reciprocity and Symmetry

Consider Fig. 6.3(a) and (b) which use reciprocity theorem. The network will be reciprocal of $I_1' = I_2'$

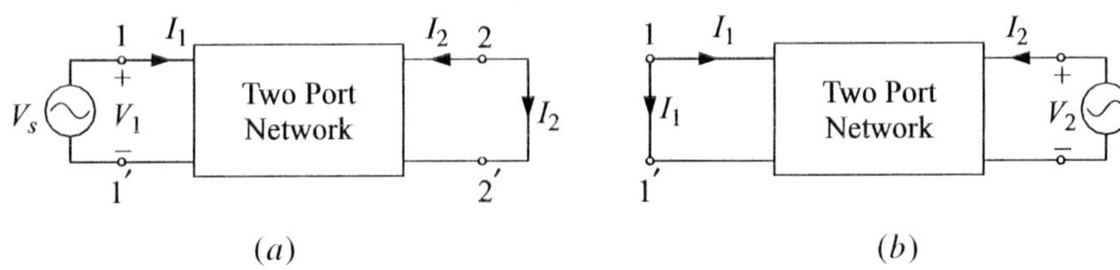

Fig. 6.3 (a) *Voltage source at port* (b) *Voltage source at port 2*

Two port network using reciprocity theorem from Fig. 6.3
From Fig. 6.3 (*a*)
We have: $V_2 = 0$, $I_2 = -I_2'$ and $V_1 = V_s$.
From equation (6.1), we get: $V_s = z_{11} I_1 - z_{12} I_2'$
and $0 = z_{21} I_1 - z_{22} I_2'$
By eliminating I_1 from above equations:

$$I_2' = \frac{z_{21}}{z_{11} z_{22} - z_{12} z_{21}} V_s \qquad ...(6.5)$$

Similarly, from Fig. 6.3(*b*), we get:

$$o = -z_{11} I_1' + z_{12} I_2$$
$$V_s = -z_{21} I_1' + z_{22} I_2$$

By eliminatory I_2 from above equation:

$$I_1' = \frac{z_{12}}{z_{11} z_{22} - z_{12} z_{21}} V_s \qquad ...(6.6)$$

The two-port network shall be reciprocal,
if $z_{12} = z_{21}$ ...(6.7)

Now, if both the ports are interchangeable without changing the port voltages and currents, then it is necessary that:

$$\left.\frac{V_1}{I_1}\right|_{I_2=0} = \left.\frac{V_2}{I_2}\right|_{I_1=0}$$

i.e. $z_{11} = z_{22}$ ...(6.8)

Example 6.1 Obtain the *z*-parameters for the network shown. Draw the *z*-parameter equivalent model and find whether the network is (*a*) reciprocal (*b*) symmetrical.

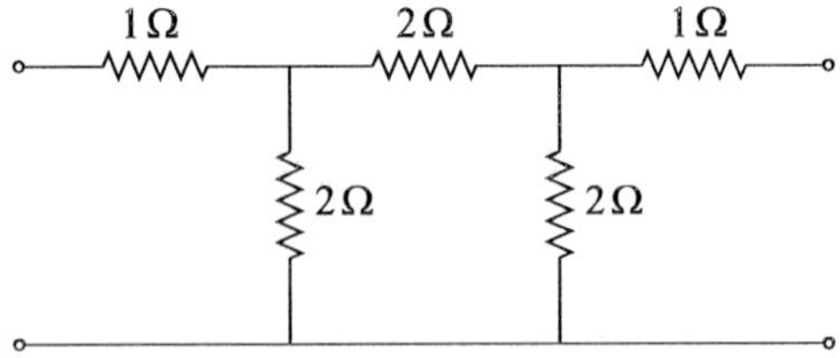

Solution: To find z_{11} and z_{21} consider the Fig. (*a*).

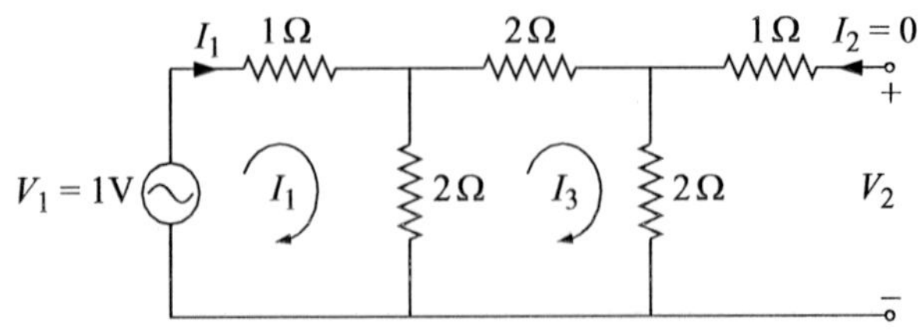

(*a*) *Network port 2 is open*

Using KVL for the mesh a and b, we get:

$$1 = (1 + 2)\, I_1 - 2I_3$$

and $$0 = -2I_1 + (2 + 2 + 2)I_3$$

or $$3I_1 - 2I_3 = 1$$

$$-2I_1 + 6I_3 = 0$$

The values of I_1 and I_3 from above equations are:

$$I_1 = \frac{3}{7}\,A \text{ and } I_3 = \frac{1}{7}\,A$$

Hence, $$V_2 = 2I_2 = \frac{2}{7}\,V \text{ and } I_1 = \frac{3}{7}\,A$$

Thus, $$z_{11} = \left.\frac{V_1}{I_1}\right|_{I_2=0} = \frac{1V}{3/7\,A} = \frac{7}{3}\,\Omega$$

and $$z_{21} = \left.\frac{V_2}{I_1}\right|_{I_2=0} = \frac{\frac{2}{7}V}{\frac{3}{7}A} = \frac{2}{3}\,\Omega$$

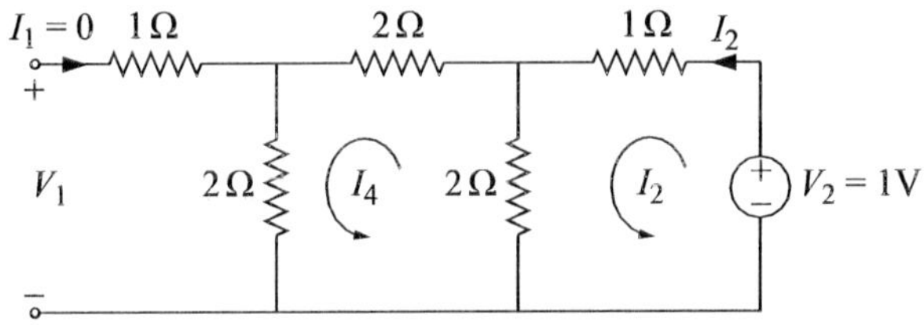

(*b*) *Network port i is open*

By KVL is the loops of Fig. (*b*), we get

$$3I_2 - 2I_4 = 1 \text{ and } -2I_2 + 6I_4 = 0$$

By solving, we get:

$$I_2 = \frac{3}{7}\,A, \quad I_4 = \frac{1}{7}\,A,\ V_1 = 2I_4 = \frac{2}{7}\,V$$

Also $z_{22} = \left.\frac{V_2}{I_1}\right|_{I_1=0} = \frac{1}{3/7} = \frac{7}{3}\,\Omega = A$ and $z_{12} = \left.\frac{V_1}{I_2}\right|_{I_1=0} = \frac{2/7}{3/7} = \frac{2}{3}\,A$

$\therefore$ z-parameters are:

$$\begin{bmatrix} z_{11} & z_{12} \\ z_{21} & z_{22} \end{bmatrix} = \begin{bmatrix} 7/3 & 2/3 \\ 2/3 & 7/3 \end{bmatrix}$$

We find that:

$$z_{11} = z_{22} \text{ } i.e. \text{ network is symmetrical}$$

$z_{12} = z_{21}$ *i.e.* network is reciprocal

z-parameter equivalent model of the network is shown

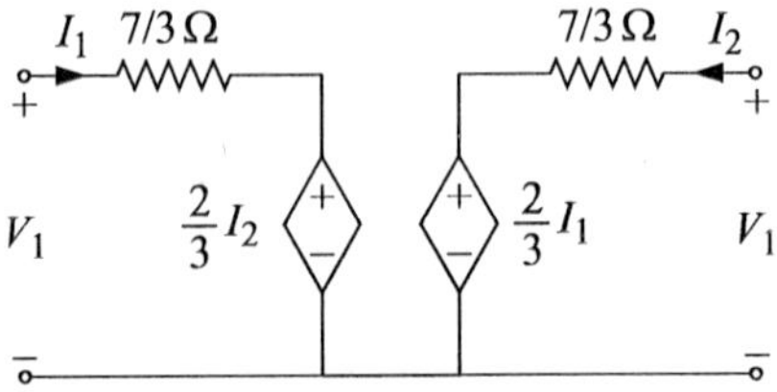

6.3 y-PARAMETERS

There parameters are defined by short circuiting one of the ports, hence, these are also known as short circuit parameters. Further, these parameters are the ratio of current to voltage, hence, these are also known as admittance parameters.

In this case, V_1 and V_2 are independent variables; and I_1 and I_2 are dependent variables, Thus,

$$\begin{aligned} I_1 &= y_{11} V_1 + y_{12} V_2 \\ I_2 &= y_{21} V_1 + y_{22} V_2 \end{aligned} \quad ...(6.9)$$

In matrix form:

$$\begin{bmatrix} I_1 \\ I_2 \end{bmatrix} = \begin{bmatrix} y_{11} & y_{12} \\ y_{21} & y_{22} \end{bmatrix} \begin{bmatrix} V_1 \\ V_2 \end{bmatrix}$$

where

$$[Y] = \begin{bmatrix} y_{11} & y_{12} \\ y_{21} & y_{22} \end{bmatrix} \quad ...(6.10)$$

Is known as admittance matrix.

In short, $[I] = [Y]\,[V]$

The y-parameters are defined as:

Port 2 is shorted	Port 1 is shorted
$y_{11} = \left.\dfrac{I_1}{V_1}\right\|_{V_2=0}$	$y_{12} = \left.\dfrac{I_1}{V_2}\right\|_{V_1=0}$
$y_{21} = \left.\dfrac{I_2}{V_1}\right\|_{V_2=0}$	$y_{22} = \left.\dfrac{I_2}{V_2}\right\|_{V_1=0}$

The y-parameter matrix elements are known as:

y_{11} = Short circuit driving point input admittance

y_{22} = Short circuit driving point output admittance

y_{12} = Short circuit reverse transfer admittance

y_{21} = Short circuit forward transfer admittance.

6.3.1 Equivalent Circuit Using y-Parameters

The two-port networks can be represented by an equivalent circuit using y-parameters shown in Fig. 6.4

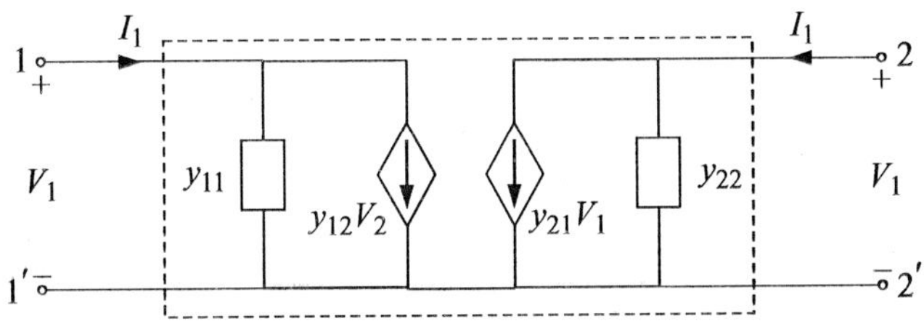

Fig. 6.4 *Equivalent circuit of two-port network using y-parameters.*

The equivalent circuit contains two dependent current sources.

6.3.2 Condition for Reciprocity and Symmetry

Consider Figs. 6.5 (*a*) and (*b*) which house reciprocity theorem. The network will be reciprocal if $I_1' = I_2'$.

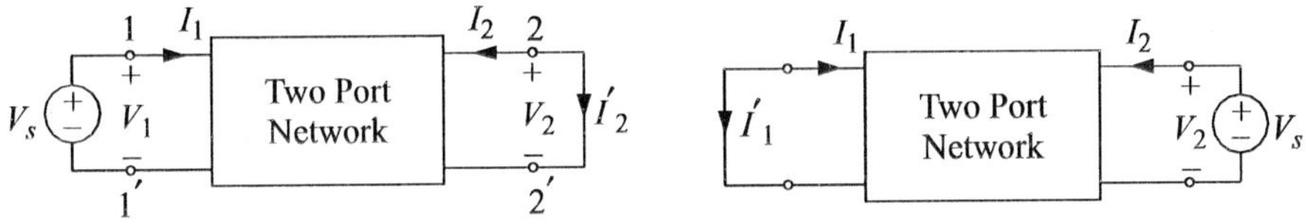

(*a*) *Voltage source of port 1* (*b*) *Voltage source at port 2*

Fig. 6.5 *Two-port network using reciprocity theorem.*

From Fig. 6.5(*a*), $I_2 = -I_2',\ V_2 = 0,\ V_1 = V_s,$

Thus, $I_1 = y_{11}\ V_s$ and $I_2' = -y_{21}\ V_s$

From Fig. 6.5(*b*), $I_1 = -I_1',\ V_1 = 0,\ V_2 = V_s$

$I_1 = y_{12}\ V_s,\ I_1' = -y_{12}\ V_s$

For $I_1' = I_2'$, we need:

$$y_{12} = y_{21} \qquad ...(6.11)$$

The above is a condition for reciprocal network.

The two-port network port can be interchanged without changing the port voltages and currents.

This can be if:

$$\left.\frac{I_1}{V_1}\right|_{V_2=0} = \left.\frac{I_2}{V_2}\right|_{V_1=0}$$

Thus, requirement $y_{11} = y_{22}$ is essential for two port network to be symmetrical.

Example 6.2 Obtain the y-parameters of the network shown in the figure. Find its equivalent circuit using y-parameters and find whether the network is (a) reciprocal, (b) symmetrical.

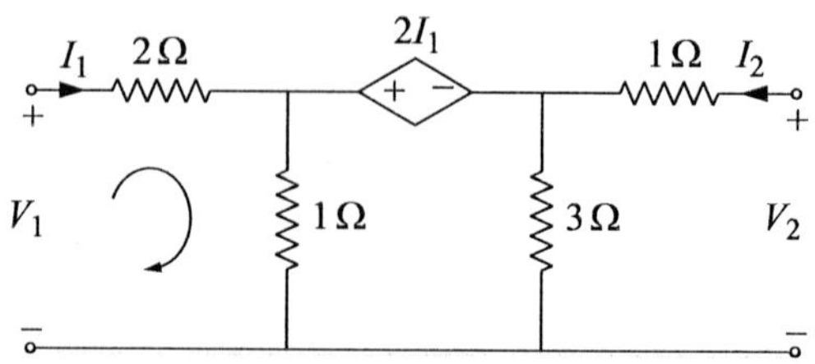

Solution: Let port 2 be shorted for y_{11} and y_{21} parameters as shown in Fig. (a).

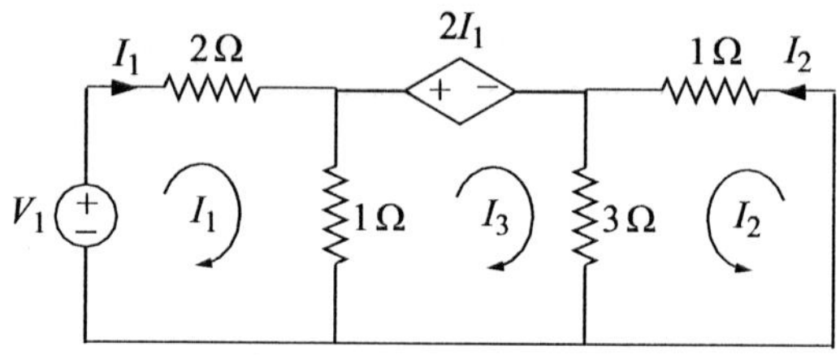

(a) *Port 2 of network is shorted*

By KVL in the loops shown in the Fig. (a), we get:

For loop 1:

$$V_1 = 3I_1 - I_3$$

For loop 2:

$$0 = 4I_2 + 3I_3$$

For loop 3:

$$-2I_1 = -I_1 + 3I_2 + 4I_3$$

or

$$3I_1 - I_3 = V_1$$
$$4I_2 + 3I_3 = 0$$
$$I_1 + 3I_2 + 4I_3 = 0$$

Form above equations, we get:

$$I_1 = \frac{\begin{vmatrix} V_1 & 0 & -1 \\ 0 & 4 & 3 \\ 0 & 3 & 4 \end{vmatrix}}{\begin{vmatrix} 3 & 0 & -1 \\ 0 & 4 & 3 \\ 1 & 3 & 4 \end{vmatrix}} = \frac{7}{25} V_1$$

and

$$I_2 = \frac{\begin{vmatrix} 3 & V_1 & -1 \\ 0 & 0 & 3 \\ 1 & 0 & 4 \end{vmatrix}}{25} = \frac{3}{25} V_1$$

Hence,

$$y_{11} = \left.\frac{I_1}{V_1}\right|_{V_2=0} = \frac{7}{25}\ \mho$$

and $$y_{21} = \left.\frac{I_2}{V_1}\right|_{V_2=0} = \frac{3}{25}\ \mho$$

Similarly, we short port 1 for y_{12} and y_{22} as shown in Fig. (*b*).

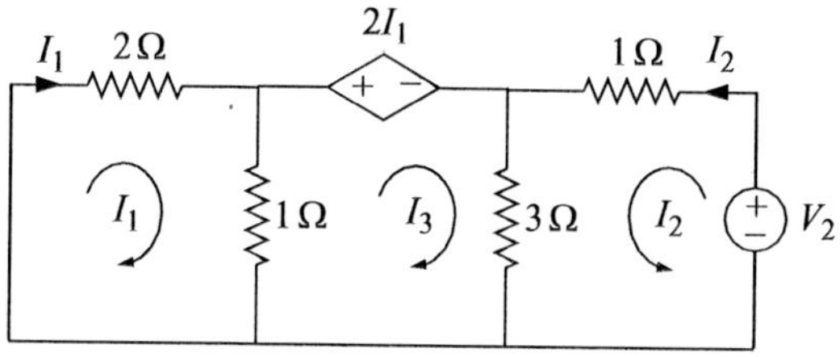

(*b*) *Port 1 of the network is shorted.*

By KVL equations for loops; we get:

$$3I_1 - I_3 = 0$$
$$4I_2 + 3I_3 = V_2$$
$$-I_1 + 3I_2 + 4I_3 = -2I_1$$

or
$$3I_1 - I_3 = 0$$
$$4I_2 + 3I_3 = V_2$$
$$I_1 + 3I_2 + 4I_2 = 0$$

From above equations, we get:

$$I_1 = \frac{\begin{vmatrix} 0 & 0 & -1 \\ V_2 & 4 & 3 \\ 0 & 3 & 4 \end{vmatrix}}{\begin{vmatrix} 3 & 0 & -1 \\ 0 & 4 & 3 \\ 1 & 3 & 4 \end{vmatrix}} = \frac{-3}{25} V_2$$

and $$I_2 = \frac{\begin{vmatrix} 3 & 0 & -1 \\ 0 & V_2 & 3 \\ 1 & 0 & 4 \end{vmatrix}}{25} = \frac{13}{25} V_2$$

Hence, $$y_{22} = \left.\frac{I_2}{V_2}\right|_{V_1=0} = \frac{13}{25}\ \mho$$

and $$y_{12} = \left.\frac{I_1}{V_2}\right|_{V_1=0} = -\frac{3}{25}\ \mho$$

The y-parameter matrix can be written as:

$$\begin{bmatrix} y_{11} & y_{12} \\ y_{21} & y_{22} \end{bmatrix} = \begin{bmatrix} \frac{7}{25} & -\frac{3}{25} \\ \frac{3}{25} & \frac{13}{25} \end{bmatrix}$$

Note that $y_{11} \neq y_{22}$ *i.e.* network is not symmetrical

and $\quad y_{12} \neq y_{21}$ *i.e.* network is not reciprocal.

The equivalent circuit of two-port network given in terms of y-parameter is given in Fig. (c):

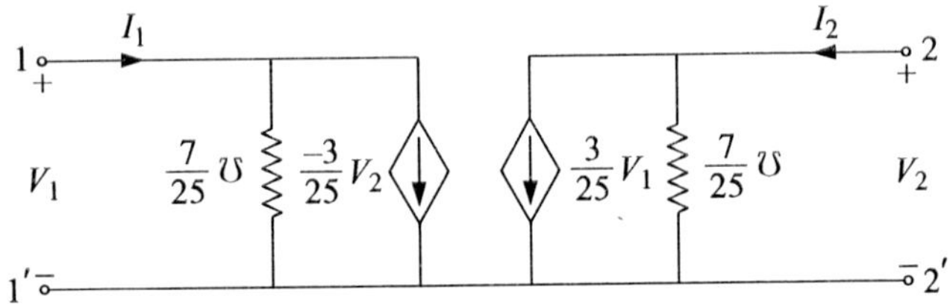

(c) *Equivalent circuit of the network given in terms of y-parameters.*

6.4 ABCD PARAMETERS

ABCD parameters were initially used for transmission lines, therefore, these are also known as transmission (T) or chain parameters. In this case, V_2 and I_2 are independent variable and V_1 and I_1 are dependent variables.

Thus,
$$V_1 = f_1(V_2, I_2)$$
$$I_1 = f_2(V_2, I_2)$$

In this case, I_2 direction is to be taken opposite to conventional two port direction, therefore relationship are:

$$V_1 = AV_2 - BI_2$$
$$I_1 = CV_2 - DI_2$$

or
$$\begin{bmatrix} V_1 \\ I_1 \end{bmatrix} = \begin{bmatrix} A & B \\ C & D \end{bmatrix} \begin{bmatrix} V_2 \\ I_2 \end{bmatrix} \quad \text{...(6.12)}$$

The transmission matrix is defined as:

$$[T] = \begin{bmatrix} A & B \\ C & D \end{bmatrix}$$

The individual transmission parameters are defined below:

Port 2 is open	Port 2 is shorted
$A = \left.\dfrac{V_1}{V_2}\right\vert_{I_2=0}$	$B = \left.\dfrac{V_1}{-I_2}\right\vert_{V_2=0}$
$C = \left.\dfrac{I_1}{V_2}\right\vert_{I_2=0}$	$D = \left.\dfrac{I_1}{-I_2}\right\vert_{V_2=0}$

These parameters are individually named as:

A = Open circuit reverse voltage gain

B = Short circuit transfer impedance.

C = Open circuit transfer admittance

D = Short circuit reverse current gain

ABCD parameters are also known as chain parameters.

There is no equivalent circuit using *ABCD* parameters since input port variables are defined exclusively in terms of output port variables and therefore, it is not possible to satisfy both equations of the *ABCD* parameters simultaneously.

Note: Equivalent circuit is not feasible using *ABCD* parameters.

6.4.1 Condition for Reciprocity and Symmetry

Consider Figs. 6.6 (*a*) and (*b*) to study reciprocity and symmetry of *ABCD* parameters.

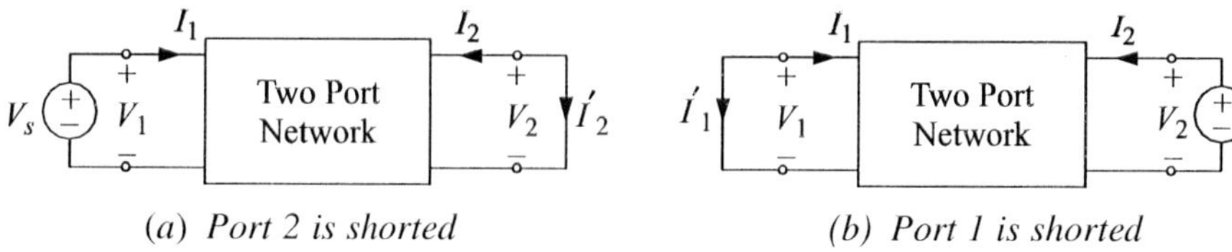

Fig. 6.6 *Networks to study reciprocity and symmetry of ABCD parameters.*

From Fig. 6.6 (*a*), we get:

$V_1 = AV_2 - BI_2$ or $V_s = BI_2'$ as $V_1 = V_s$, $V_2 = 0$, and $I_2 = -I_2'$

and $I_1 = CV_2 - DI_2$ or $I_1 = DI_2'$ or $I_1 = DI_2'$ as $V_2 = 0$ and $I_2 = I_2'$

$$\therefore \qquad I_2' = \frac{1}{B} V_s \qquad \text{...(6.13)}$$

From Fig. 6.6 (*b*), we get:

$V_1 = AV_2 - BI_2$ or $0 = AV_s - BI_2$ as $V_1 = 0$

and $I_1 = CV_2 - DI_2$ or $-I_1' = CV_s - DI_2$ as $I_1 = -I_1'$

By eliminating I_2 from above equations,

we get: $$I_1' = \frac{AD - BC}{B} V_s \quad ...(6.14)$$

For reciprocity condition $I_1' = I_2'$, hence, form equations (6.13) and (6.14), we get:

$$AD - BC = 1 \quad ...(6.15)$$

or $$\begin{vmatrix} A & B \\ C & D \end{vmatrix} = 1$$

For symmetry, $$z_{11} = z_{22}$$

i.e $$\left.\frac{V_1}{I_1}\right|_{I_2=0} = \left.\frac{V_2}{I_2}\right|_{I_1=0}$$

But, $$\left.\frac{V_1}{I_1}\right|_{I_2=0} = \left.\frac{AV_2 - BI_2}{CV_2 - DI_2}\right|_{I_2=0} = \frac{A}{C} = z_{11}$$

We know that: $I_1 = CV_2 - DI_2$

$\therefore$ $$\left.\frac{V_2}{I_2}\right|_{I_1=0} = \frac{D}{C} = z_{22}$$

Thus, for symmetry condition:

$$\frac{A}{C} = \frac{D}{C}$$

or $$A = D \quad ...(6.16)$$

Example 6.3 Obtain the transmission or *ABCD* parameters of the circuit shown in the figure. Find whether the network is (*a*) reciprocal (*b*) symmetrical.

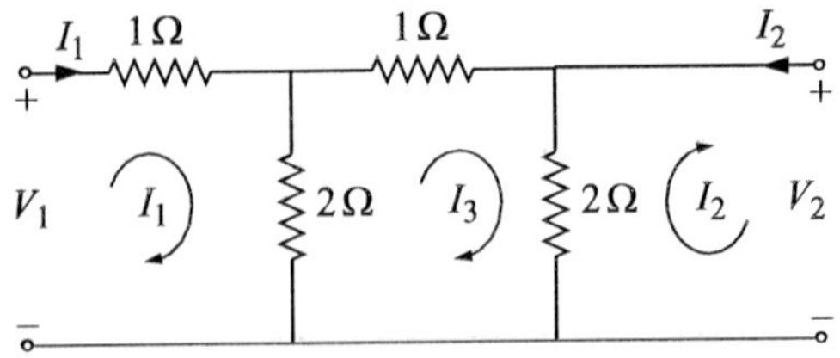

Solution: By keeping port 2 open and writing equations, we get

$$3I_1 - 2I_3 = 1$$

and $$-2I_1 + 5I_3 = 0$$

which give:

$$I_1 = \frac{5}{11} A \text{ and } I_3 = \frac{2}{11} A$$

$$\text{Hence, } V_1 = 1 \times \frac{5}{11} + 2 \times \left(\frac{5}{11} - \frac{2}{11}\right) = \frac{5}{11} + \frac{10}{11} - \frac{4}{11} = 1V$$

and
$$V_2 = 2I_3 = 2 \times \frac{2}{11} = \frac{4}{11}$$

Hence,
$$A = \left.\frac{V_1}{V_2}\right|_{I_2=0} = \frac{1}{(4/11)} = \frac{11}{4}$$

$$C = \left.\frac{I_1}{V_2}\right|_{I_2=0} = \frac{(5/11)}{(4/11)} = \frac{5}{4}\,\mho$$

For parameters B and D, let the port 2 be short; and by writing much equation in that case, we get:

$$3I_1 - 2I_3 = 1$$
$$-2I_1 + 3I_3 = 0$$

and
$$I_2 = -I_3$$

By solving from above equations, we get

$$I_1 = \frac{3}{5}A,\; I_2 = -\frac{2}{5},\; \text{and } I_3 = \frac{2}{5}A$$

$$\text{Hence, } V_1 = 3I_1 - 2I_3 = 3 \times \frac{3}{5} - 2 \times \frac{2}{5} = 1V$$

$$B = \left.\frac{V_1}{-I_2}\right|_{V_2=0} = \frac{1}{-\left(-\frac{2}{5}\right)} = \frac{5}{2}\,\Omega$$

$$D = \left.\frac{I_1}{-I_2}\right|_{V_2=0} = \frac{\left(\frac{3}{5}\right)}{-\left(-\frac{2}{5}\right)} = \frac{3}{2}$$

Now, $ABCD$ parameters can be written as:

$$[T] = \begin{bmatrix} A & B \\ C & D \end{bmatrix} = \begin{bmatrix} \frac{11}{4} & \frac{5}{2} \\ \frac{5}{4} & \frac{3}{2} \end{bmatrix}$$

$$AD - BC = \frac{11}{4} \times \frac{3}{2} - \frac{5}{2} \times \frac{5}{4} = \frac{33}{8} - \frac{25}{8}$$

or $AD - BC = 1$, the network is reciprocal.

As $A \neq D$, the network is not symmetrical.

6.5 *h*-PARAMETERS

h stands for hybrid, hence, these are also known as hybrid parameters. In this case, I_1 and V_2 are independent variables, and V_1 and I_2 are dependent variables. *h*-parameters are defined from the functions given below:

$$V_1 = f_1(I_1, V_2)$$
$$I_2 = f_2(I_1, V_2)$$

The actual relationship with *h*-parameters are:

$$V_1 = h_{11}\, I_1 + h_{12}\, V_2$$
$$I_2 = h_{21}\, I_1 + h_{22}\, V_2 \qquad \text{...(6.17)}$$

or

$$\begin{bmatrix} V_1 \\ I_2 \end{bmatrix} = \begin{bmatrix} h_{11} & h_{12} \\ h_{21} & h_{22} \end{bmatrix} \begin{bmatrix} I_1 \\ V_2 \end{bmatrix}$$

i.e. *h*-matrix is:

$$[h] = \begin{bmatrix} h_{11} & h_{12} \\ h_{21} & h_{22} \end{bmatrix} \qquad \text{...(6.18)}$$

h-parameters are defined as:

Port 2 shorted	Port 1 open
$h_{11} = \left.\dfrac{V_1}{I_1}\right\vert_{V_2=0}$	$h_{12} = \left.\dfrac{V_1}{V_2}\right\vert_{I_1=0}$
$h_{21} = \left.\dfrac{I_2}{I_1}\right\vert_{V_2=0}$	$h_{22} = \left.\dfrac{I_2}{V_2}\right\vert_{I_1=0}$

The individual *h*-parameters are named as:

h_{11} = Short circuit input impedance

h_{21} = Forward short circuit current gain

h_{12} = Reverse open circuit voltage gain

h_{22} = Open circuit output admittance.

6.5.1 Equivalent Circuit Using *h*-parameters

Equivalent circuit using *h*-parameters is shown in Fig. 6.7. It has voltage source in input port and current source in output port.

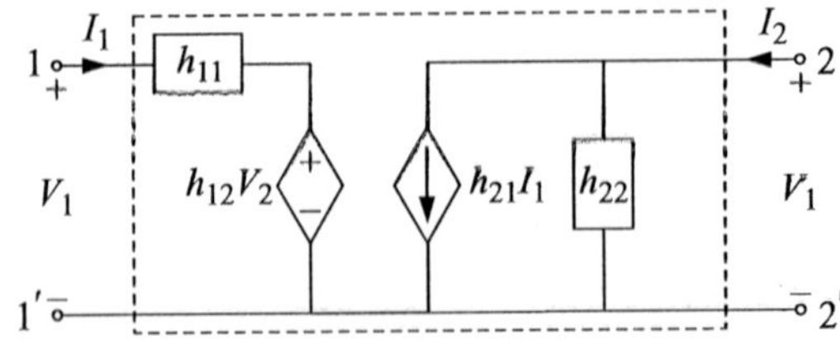

Fig. 6.7 *Equivalent of two-port network using h-parameters*

6.5.2 Condition for Reciprocity and Symmetry

Consider the Fig. 6.8 (*a*) and (*b*) for finding condition for reciprocity and symmetry.

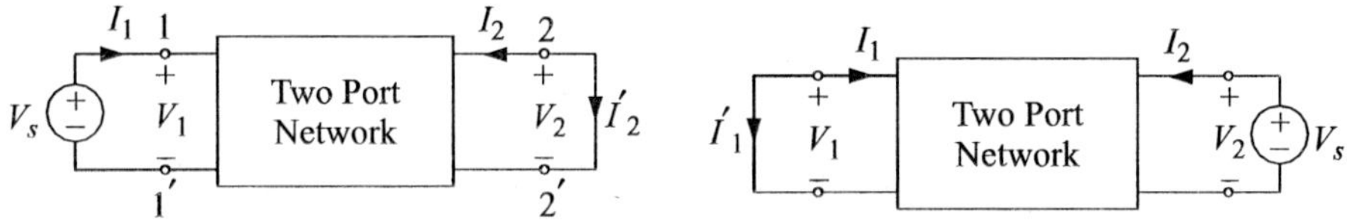

(*a*) *Port 2 shorted* (*b*) *Port 1 shorted*

Fig. 6.8 *Networks for deriving reciprocity and symmetry*

From Fig. 6.8 (*a*), we get:

$V_1 = h_{11} I_1 + h_{12} V_2$ or $V_s = h_{11} I_1$ as $V_2 = 0$

and $I_2 = h_{21} I_1 + h_{22} V_2$ or $-I_2' = h_{21} I_1$ as $V_2 = 0$ and $I_2 = -I_2'$

Hence, $$I_2' = -\frac{h_{21}}{h_{11}} V_s$$

From Fig. 6.8 (*b*), we get

$$V_1 = h_{11} I_1 + h_{12} V_2 \text{ or } 0 = -h_{11} I_1' + h_{12} V_s$$

$$\text{as } I_1 = -I_1',\ V_1 = 0 \text{ and } V_2 = V_s$$

and $$I_2 = h_{21} I_1 + h_{22} V_2 \text{ or } I_2 = -h_{21} I_1 + h_{22} V_s$$

$$\text{as } I_1 = -I_1' \text{ and } V_2 = V_s$$

Hence, $$I_1' = \frac{h_{12}}{h_{11}} V_s$$

For reciprocity condition, $I_1' = I_2'$

$\therefore$ $$-\frac{h_{21}}{h_{11}} V_s = \frac{h_{12}}{h_{11}} V_s$$

or $$h_{12} = -h_{21} \qquad \text{...(6.19)}$$

For symmetry condition, we need: $z_{11} = z_{22}$

or $$\left.\frac{V_1}{I_1}\right|_{I_2=0} = \left.\frac{V_2}{I_2}\right|_{I_1=0}$$

By putting $I_2 = 0$ in equation (6.17), we get:

$$V_1 = h_{11} I_1 + h_{12} V_2$$

and $$0 = h_{21} I_1 + h_{22} V_2$$

By eliminating V_2 from above equations, we get:

$$\left.\frac{V_1}{I_1}\right|_{I_2=0} = \frac{h_{11} h_{22} - h_{12} h_{21}}{h_{22}} = z_{11}$$

By putting $I_1 = 0$ in equation (6.17), we get:

$$V_1 = h_{12} V_2$$

and $$I_2 = h_{22} V_2$$

$$\therefore \quad \left.\frac{V_2}{I_2}\right|_{I_1=0} = \frac{1}{h_{22}} = z_{22}$$

Now, taking $z_{11} = z_{22}$ for symmetry conditions:

we get: $$\frac{h_{11} h_{22} - h_{12} h_{21}}{h_{22}} = \frac{1}{h_{22}}$$

or $$h_{11} h_{22} - h_{12} h_{21} = 1 \qquad \text{...(6.20)}$$

Example 6.4 Obtain h-parameters for the circuit shown in the figure. Find the equivalent circuit asing h-parameters and find whether the network is (a) reciprocal, (b) symmetrical.

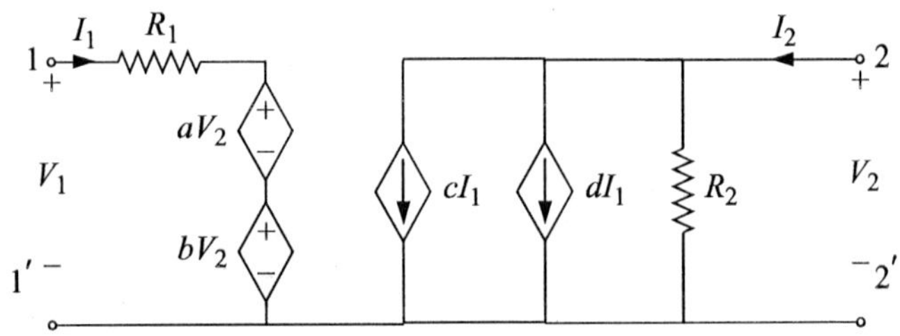

Solution: By KVL equations at the input, we get:

$$R_1 I_1 + (a + b)V_2 = V_1 \qquad \text{...}(i)$$

By KCL equations at the output, we get:

$$(c + d) I_1 + \frac{V_2}{R_2} = I_2 \qquad \text{...}(ii)$$

The h-parameter equations for the two ports are:

$$V_1 = h_{11} I_1 + h_{12} V_2 \qquad \text{...}(iii)$$

and $$I_2 = h_{21} I_1 + h_{22} V_2 \qquad \text{...}(iv)$$

By comparison of equations (i), (ii), (iii) and (iv), we get:

$$h_{11} = R_1, \quad h_{21} = c + d$$

$$h_{12} = a + b, \; h_{22} = \frac{1}{R_2}$$

As $h_{11} \neq -h_{22}$, the given network is not reciprocal, and $h_{11} h_{22} - h_{12} h_{21}$

$$= R_1 \times \frac{1}{R_2} - (a - b)(c + d)$$

or $$h_{11} h_{22} - h_{12} h_{21} \neq \frac{R_1}{R_2} - (a + b)(c + d)$$

hence, network is not symmetrical. The given network in terms of h-parameter is shown in the equivalent circuit the figure given below:

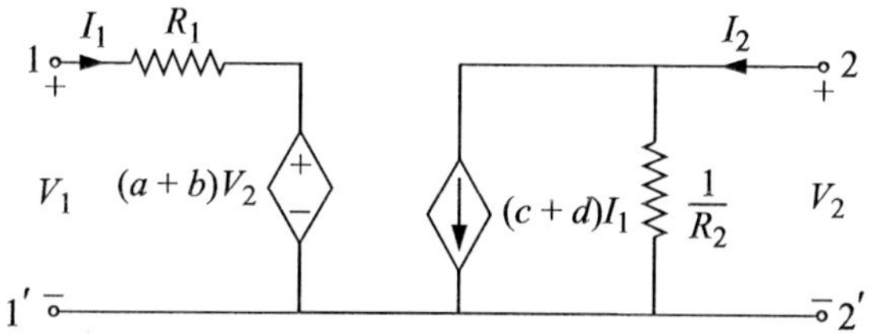

6.6 INTER-RELATIONSHIP BETWEEN THE PARAMETERS

6.6.1 z-parameters in Terms of Other Parameters

z-parameters are given by:

$$V_1 = z_{11} I_1 + z_{12} I_2 \quad ...(6.21)$$

$$V_2 = z_{21} I_1 + z_{22} I_2 \quad ...(6.22)$$

(*i*) In terms of y-parameters

y-parameters are given by:

$$I_1 = y_{11} V_1 + y_{12} V_2 \quad ...(6.23)$$

$$I_2 = y_{21} V_1 + y_{22} V_2 \quad ...(6.24)$$

or
$$\begin{bmatrix} I_1 \\ I_2 \end{bmatrix} = \begin{bmatrix} y_{11} & y_{12} \\ y_{21} & y_{22} \end{bmatrix} \begin{bmatrix} V_1 \\ V_2 \end{bmatrix}$$

V_1 and V_2 from above:

$$V_1 = \frac{\begin{vmatrix} I_1 & y_{12} \\ I_2 & y_{22} \end{vmatrix}}{\begin{vmatrix} y_{11} & y_{12} \\ y_{21} & y_{22} \end{vmatrix}} = \frac{y_{22} I_1 - y_{12} I_2}{y_{11} y_{22} - y_{12} y_{21}}$$

$$V_1 = \frac{y_{22}}{y_{11} y_{22} - y_{12} y_{21}} I_1 - \frac{y_{12}}{y_{11} y_{22} - y_{12} y_{21}} I_2 \quad ...(6.25)$$

Similarly,

$$V_2 = \frac{\begin{vmatrix} y_{11} & I_1 \\ y_{21} & I_2 \end{vmatrix}}{\begin{vmatrix} y_{11} & y_{12} \\ y_{21} & y_{22} \end{vmatrix}} = \frac{y_{11} I_2 - y_{21} I_1}{y_{11} y_{22} - y_{12} y_{21}}$$

or $$V_2 = \frac{y_{11}}{y_{11} y_{22} - y_{12} y_{21}} I_2 - \frac{y_{21}}{y_{11} y_{22} - y_{12} y_{21}} I_1 \quad ...(6.26)$$

Consider, $y_{11} y_{22} - y_{12} y_{21} = \Delta y$

Then, from equations (6.25) and (6.26), we get:

$$V_1 = \left[\frac{y_{22}}{\Delta y}\right] I_1 + \left[\frac{-y_{12}}{\Delta y}\right] I_2 \qquad ...(6.27)$$

and
$$V_2 = \left[-\frac{y_{21}}{\Delta y}\right] I_1 + \left[\frac{y_{11}}{\Delta y}\right] I_2 \qquad ...(6.28)$$

Now by comparing equation (6.21) with equation (6.27) and equation (6.22) with equation (6.28), we get:

$$z_{11} = \frac{y_{22}}{\Delta y}, \qquad y_{21} = \frac{-y_{12}}{\Delta y}$$

$$z_{21} = \frac{-y_{21}}{\Delta y}, \qquad y_{22} = \frac{y_{11}}{\Delta y}$$

i.e. z-parameters in terms of y-parameters in matrix form are:

$$[z] = \begin{bmatrix} \dfrac{y_{22}}{\Delta y} & \dfrac{-y_{12}}{\Delta y} \\ \dfrac{-y_{21}}{\Delta y} & \dfrac{y_{11}}{\Delta y} \end{bmatrix} \qquad ...(6.29)$$

(*ii*) In terms of *ABCD* parameter

We know that:

$$V_1 = AV_2 + B(-I_2) \qquad ...(6.30)$$

$$I_1 = CV_2 + D(-I_2) \qquad ...(6.31)$$

The above can be rewritten as:

$$CV_2 = I_1 + D\,I_2$$

or
$$V_2 = \left(\frac{1}{C}\right) I_1 + \left(\frac{D}{C}\right) I_2 \qquad ...(6.32)$$

By putting value of V_2 into equation (6.30), we get:

$$V_1 = A\left[\left(\frac{1}{C}\right) I_1 + \left(\frac{D}{C}\right) I_2\right] - BI_2$$

or
$$V_1 = \left(\frac{A}{C}\right) I_1 + \left(\frac{AD}{C} - B\right) I_2$$

or
$$V_1 = \left(\frac{A}{C}\right) I_1 + \left(\frac{AD - BC}{C}\right) I_2 \qquad ...(6.33)$$

We know that:

$$V_1 = z_{11}\, I_1 + z_{12}\, I_2 \qquad ...(6.34)$$

$$V_2 = z_{21}\, I_1 + z_{22}\, I_2 \qquad ...(6.35)$$

By comparing equations (6.33) with (6.34), and equation (6.32) with (6.35), we get:

$$z_{11} = \frac{A}{C}, \quad z_{21} = \frac{1}{C}$$

$$z_{12} = \frac{AD - BC}{C}, \quad z_{22} = \frac{D}{C}$$

Hence, z-parameters equivalent of $ABCD$ parameters in matrix form is:

$$[z] = \begin{bmatrix} \frac{A}{C} & \frac{AD - BC}{C} \\ \frac{1}{C} & \frac{D}{C} \end{bmatrix}$$

(*iii*) In terms of h-parameters

Equations of h-parameters are:

$$V_1 = h_{11} I_1 + h_{12} V_2 \quad ...(6.36)$$

$$I_2 = h_{21} I_1 + h_{22} V_2 \quad ...(6.37)$$

Equation (6.37) gives:

$$h_{22} V_2 = -h_{21} I_1 + I_2$$

i.e.

$$V_2 = \left(-\frac{h_{21}}{h_{22}}\right) I_1 + \left(-\frac{1}{h_{22}}\right) I_2 \quad ...(6.38)$$

By substituting value of V_2 from equation (6.38) into equation (6.36), we get:

$$V_1 = h_{11} I_1 + h_{12} \left[\frac{-h_{21}}{h_{22}} I_1 + \frac{1}{h_{22}} I_2\right]$$

or

$$V_1 = \left[h_{11} - \frac{h_{12} h_{21}}{h_{22}}\right] I_1 + \left[\frac{h_{12}}{h_{22}}\right] I_2$$

or

$$V_1 = \left[\frac{h_{11} h_{22} - h_{12} h_{21}}{h_{22}}\right] I_1 + \left[\frac{h_{12}}{h_{22}}\right] I_2 \quad ...(6.39)$$

Now, compare equation (3.39) and (6.38) with the following equations:

$$V_1 = z_{11} I_1 + z_{12} I_2$$

and

$$V_2 = z_{21} I_1 + z_{22} I_2$$

The above gives:

$$z_{11} = \frac{h_{11} h_{22} - h_{12} h_{21}}{h_{22}}, \quad z_{21} = \frac{-h_{21}}{h_{22}}$$

and

$$z_{12} = \frac{h_{12}}{h_{22}}, \quad z_{22} = \frac{1}{h_{22}}$$

The above in matrix form is given as:

$$[z] = \begin{bmatrix} \dfrac{h_{11} h_{22} - h_{12} h_{21}}{h_{22}} & \dfrac{h_{12}}{h_{22}} \\ \dfrac{-h_{21}}{h_{22}} & \dfrac{1}{h_{22}} \end{bmatrix}$$

6.6.2 *y*-parameters in Terms of the Other Parameters

The y-parameters are defined as:

$$I_1 = y_{11} V_1 + y_{12} V_2 \qquad ...(6.40)$$

and

$$I_2 = y_{12} V_1 + y_{22} V_2 \qquad ...(6.41)$$

(*i*) In terms of z-parameters

The z-parameters are defined as:

$$V_1 = z_{11} I_1 + z_{12} I_2 \qquad ...(6.42)$$

and

$$V_2 = z_{21} I_1 + z_{22} I_2 \qquad ...(6.43)$$

or

$$\begin{bmatrix} V_1 \\ V_2 \end{bmatrix} = \begin{bmatrix} z_{11} & z_{12} \\ z_{21} & z_{22} \end{bmatrix} \begin{bmatrix} I_1 \\ I_2 \end{bmatrix}$$

Using Cramer's rule, we get:

$$I_1 = \frac{\begin{vmatrix} V_1 & z_{12} \\ V_2 & z_{22} \end{vmatrix}}{\begin{vmatrix} z_{11} & z_{12} \\ z_{21} & z_{22} \end{vmatrix}} = \frac{z_{22} V_1 - z_{21} V_2}{z_{11} z_{22} - z_{12} z_{21}}$$

or

$$I_1 = \frac{z_{22}}{\Delta z} V_1 + \frac{-z_{12}}{\Delta z} V_2 \qquad ...(6.44)$$

where

$$\Delta z = z_{11} z_{22} - z_{12} z_{21}$$

In similar way,

$$I_2 = \frac{\begin{vmatrix} z_{11} & V_1 \\ z_{21} & V_2 \end{vmatrix}}{\Delta z} = \frac{z_{11} V_2 - z_{21} V_1}{\Delta z}$$

or

$$I_2 = \frac{z_{11}}{\Delta z} V_2 + \frac{-z_{21}}{\Delta z} V_1$$

or

$$I_2 = \left(\frac{-z_{21}}{\Delta z}\right) V_1 + \left(\frac{z_{11}}{\Delta z}\right) V_2 \qquad ...(6.45)$$

From equations (6.40), (6.41), (6.44) and (6.45), we get:

$$y_{11} = \frac{z_{22}}{\Delta z}, \quad y_{21} = \frac{-z_{21}}{\Delta z}$$

and $$y_{12} = \frac{-z_{12}}{\Delta z}, \quad y_{22} = \frac{z_{11}}{\Delta z}$$

y-parameters in matrix are:

$$[y] = \begin{bmatrix} \frac{z_{22}}{\Delta z} & \frac{-z_{12}}{\Delta z} \\ \frac{-z_{21}}{\Delta z} & \frac{z_{11}}{\Delta z} \end{bmatrix}$$

(*ii*) In terms of *ABCD* parameters

ABCD parameters are defined as:

$$V_1 = A\, V_2 + B(-I_2) \qquad ...(6.46)$$

$$I_1 = C\, V_2 + D\,(-I_2) \qquad ...(6.47)$$

Form equation (6.46), we get:

$$V_1 - AV_2 = -BI_2$$

or $$I_2 = \left(-\frac{I}{B}\right)V_1 + \left(\frac{A}{B}\right)V_2 \qquad ...(6.48)$$

By substituting I_2 in equation (6.47), we get:

$$I_1 = CV_2 + D\left[+\frac{V_1}{B} - \frac{A}{B}V_2\right]$$

or $$I_1 = \left(\frac{D}{B}\right)V_1 + \left(\frac{BC - AD}{B}\right)V_2 \qquad ...(6.49)$$

By comparing equations (6.49) and (6.48) with equations (6.40) and (6.41), we get:

$$y_{11} = \frac{D}{B}, y_{21} = \frac{-1}{B}$$

and $$y_{12} = \frac{BC - AD}{B}, y_{22} = \frac{A}{B}$$

In matrix form:

$$[y] = \begin{bmatrix} \frac{D}{B} & \frac{BC - AD}{B} \\ \frac{-1}{B} & \frac{A}{B} \end{bmatrix}$$

(*iii*) In terms of *h*-parameters

h-parameters are defined as:

$$V_1 = h_{11}\, I_1 + h_{12}\, V_2 \qquad ...(6.50)$$

$$I_2 = h_{21}\, I_1 + h_{22}\, V_2 \qquad ...(6.51)$$

Equation (6.50) gives:

$$V_1 - h_{12} V_2 = h_{11} I_1$$

or
$$I_1 = \left(\frac{1}{h_{11}}\right) V_1 + \left(\frac{-h_{12}}{h_{11}}\right) V_2 \qquad \text{...(6.52)}$$

By substituting I_1 from equation (6.52) into equation (6.51), we get:

$$I_2 = h_{21}\left[\frac{1}{h_{11}} V_1 - \frac{h_{21}}{h_{11}} V_2\right] + h_{22} V_2$$

or
$$I_2 = \left(\frac{h_{21}}{h_{11}}\right) V_1 + \left(\frac{h_{11} h_{22} - h_{12} h_{21}}{h_{11}}\right) V_2 \qquad \text{...(6.53)}$$

By comparing equations (6.52) and (6.53) with equations (6.40) and (6.41), we get:

$$y_{11} = \frac{1}{h_{11}}, \quad y_{21} = \frac{h_{21}}{h_{11}}$$

and
$$y_{12} = \frac{-h_{12}}{h_{11}}, \quad y_{22} = \frac{h_{11} h_{22} - h_{12} h_{21}}{h_{11}}$$

In matrix form:

$$[y] = \begin{bmatrix} \dfrac{1}{h_{11}} & \dfrac{-h_{12}}{h_{11}} \\ \dfrac{h_{21}}{h_{11}} & \dfrac{h_{11} h_{22} - h_{12} h_{21}}{h_{11}} \end{bmatrix}$$

6.6.3 *ABCD* Parameters in Terms of Other Parameters

ABCD parameters are defined as:

$$V_1 = AV_2 + B(-I_2) \qquad \text{...(6.54)}$$

$$I_1 = CV_2 + D(-I_2) \qquad \text{...(6.55)}$$

(*i*) In terms of z-parameters

z-parameters are defined as:

$$V_1 = z_{11} I_1 + z_{12} I_2 \qquad \text{...(6.56)}$$

$$V_2 = z_{21} I_1 + z_{22} I_2 \qquad \text{...(6.57)}$$

From equation (6.57), we get:

$$V_2 - z_{22} I_2 = z_{21} I_1$$

or
$$I_1 = \left(\frac{1}{z_{21}}\right) V_2 + \left(\frac{-z_{22}}{z_{21}}\right) I_2 \qquad \text{...(6.58)}$$

By substituting value of I_1 in the equation (6.56), we get:

$$V_1 = z_{11}\left[\frac{V_2}{z_{21}} - \frac{z_{22}}{z_{21}} I_2\right] + z_{12} I_2$$

or $$V_1 = \left(\frac{z_{11}}{z_{21}}\right) V_2 + \left(\frac{z_{11}\, z_{22} - z_{12}\, z_{21}}{z_{21}}\right)(-I_2) \quad ...(6.59)$$

By comparing equations (6.59) and (6.58) with equations (6.54) and (6.55), we get:

$$A = \frac{z_{11}}{z_{21}}, \quad C = \frac{1}{z_{21}}$$

and $$B = \frac{z_{11}\, z_{22} - z_{12}\, z_{21}}{z_{21}}, \quad D = \frac{z_{22}}{z_{21}}$$

In the matrix form:

$$[T] = \begin{bmatrix} \dfrac{z_{11}}{z_{21}} & \dfrac{z_{11}\, z_{22} - z_{12}\, z_{21}}{z_{21}} \\ \dfrac{1}{z_{21}} & \dfrac{z_{22}}{z_{21}} \end{bmatrix}$$

(*ii*) In terms of *y*-parameters

y-parameters are defined as:

$$I_1 = y_{11} V_1 + y_{12} V_2 \quad ...(6.60)$$

$$I_2 = y_{21} V_1 + y_{22} V_2 \quad ...(6.61)$$

Equation (6.61) gives:

$$I_2 - y_{22} V_2 = y_{21} V_1$$

or $$V_1 = \left(\frac{-y_{22}}{y_{21}}\right) V_2 + \left(\frac{-1}{y_{21}}\right)(-I_2) \quad ...(6.62)$$

By substituting value of V_1 in equation (6.60), we get:

$$I_1 = y_{11}\left[\frac{-y_{22}}{y_{21}} V_2 + \frac{1}{y_{21}} I_2\right] + y_{12} V_2$$

or $$I_1 = \left(\frac{y_{12}\, y_{21} - y_{11}\, y_{22}}{y_{21}}\right) V_2 + \left(\frac{-y_{11}}{y_{21}}\right)(-I_2) \quad ...(6.63)$$

By comparing equations (6.62) and (6.63) with equations (6.54) and (6.55), we get:

$$A = \frac{-y_{22}}{y_{21}}, \quad C = \frac{y_{12}\, y_{21} - y_{11}\, y_{22}}{y_{21}}$$

$$B = \frac{-1}{y_{21}}, D = \frac{-y_{11}}{y_{21}}$$

In matrix form:

$$[T] = \begin{bmatrix} \dfrac{-y_{22}}{y_{21}} & \dfrac{-1}{y_{21}} \\ \dfrac{y_{12}\, y_{21} - y_{11}\, y_{22}}{y_{21}} & \dfrac{-y_{11}}{y_{21}} \end{bmatrix}$$

(*iii*) In terms of *h*-parameters

h-parameters are defined as:

$$V_1 = h_{11}\, I_1 + h_{12}\, V_2 \quad ...(6.64)$$

$$I_2 = h_{21}\, I_1 + h_{22}\, V_2 \quad ...(6.65)$$

From equation (6.65), we get:

$$I_2 - h_{22}\, V_2 = h_{21}\, I_1$$

or

$$I_1 = \left(\frac{-h_{22}}{h_{21}}\right) V_2 + \left(\frac{-1}{h_{21}}\right)(-I_2) \quad ...(6.66)$$

By substituting value of I_1 in equations (6.64), we get:

$$V_1 = h_{11}\left[-\frac{h_{22}}{h_{21}} V_2 + \frac{1}{h_{21}} I_2\right] + h_{12}\, V_2$$

or

$$V_1 = \left(\frac{h_{12}\, h_{21} - h_{11}\, h_{22}}{h_{21}}\right) V_2 + \left(\frac{-h_{11}}{h_{21}}\right)(-I_2) \quad ...(6.67)$$

By comparing equations (6.67) and (6.66) with equations (6.54) and (6.55), we get:

$$A = \frac{h_{12}\, h_{21} - h_{11}\, h_{22}}{h_{21}}, \quad B = \frac{-h_{11}}{h_{21}}$$

$$C = \frac{h_{22}}{h_{21}}, \quad D = \frac{-1}{h_{21}}$$

In matrix form:

$$[T] = \begin{bmatrix} \dfrac{h_{12}\, h_{21} - h_{11}\, h_{22}}{h_{21}} & \dfrac{-h_{11}}{h_{21}} \\ \dfrac{-h_{22}}{h_{21}} & \dfrac{-1}{h_{21}} \end{bmatrix}$$

6.6.4 *h*-parameters in Terms of Other Parameters

h-parameters are defined by:

$$V_1 = h_{11} I_1 + h_{12} V_2 \quad ...(6.68)$$
$$I_2 = h_{21} I_1 + h_{22} V_2 \quad ...(6.69)$$

(*i*) In terms of *z*-parameters

z-parameters are defined by:

$$V_1 = z_{11} I_1 + z_{12} I_2 \quad ...(6.70)$$

and

$$V_2 = z_{21} I_1 + z_{22} I_2 \quad ...(6.71)$$

From equation (6.71), we get:

$$I_2 = \left(\frac{-z_{21}}{z_{22}}\right) I_1 + \left(\frac{1}{z_{22}}\right) V_2 \quad ...(6.72)$$

By substituting value of I_2 in equation (6.70), we get:

$$V_1 = z_{11} I_1 + z_{12} \left[\frac{-z_{21}}{z_{22}} I_1 + \frac{1}{z_{22}} V_2\right]$$

or

$$V_1 = \left(\frac{\Delta z}{z_{22}}\right) I_1 + \frac{z_{12}}{z_{22}} V_2 \quad ...(6.73)$$

where

$$\Delta z = z_{11} z_{22} - z_{12} z_{21}$$

By comparing equations (6.73) and (6.72) with equations (6.68) and (6.69), we get:

$$h_{11} = \frac{\Delta z}{z_{22}}, \quad h_{12} = \frac{z_{12}}{z_{22}}$$

and

$$h_{21} = \frac{-z_{21}}{z_{22}}, \quad h_{22} = \frac{1}{z_{22}}$$

In matrix form

$$[h] = \begin{bmatrix} \dfrac{\Delta z}{z_{22}} & \dfrac{z_{12}}{z_{22}} \\ \dfrac{-z_{21}}{z_{22}} & \dfrac{1}{z_{22}} \end{bmatrix}$$

(*ii*) In terms of *y*-parameters

y-parameters are defined as:

$$I_1 = y_{11} V_1 + y_{12} V_2 \quad ...(6.74)$$

and

$$I_2 = y_{21} V_1 + y_{22} V_2 \quad ...(6.75)$$

From equation (6.74), we get:

$$V_1 = \left(\frac{1}{y_{11}}\right) I_1 - \frac{y_{12}}{y_{11}} V_2 \quad \text{...(6.76)}$$

By substituting value of V_1 in equation (6.75), we get:

$$I_2 = y_{21}\left[\frac{1}{y_{11}} I_1 - \frac{y_{12}}{y_{11}} V_2\right] + y_{22} V_2$$

or

$$I_2 = \left(\frac{y_{21}}{y_{11}}\right) I_1 + \frac{\Delta y}{y_{11}} V_2 \quad \text{...(6.77)}$$

where $\Delta y = (y_{11}\, y_{22} - y_{12}\, y_{21})$

By comparing equations (6.76) and (6.77) with equations (6.68) and (6.69), we get:

$$h_{11} = \frac{1}{y_{11}}, \quad h_{12} = \frac{-y_{12}}{y_{11}}$$

and

$$h_{21} = \frac{y_{21}}{y_{11}}, \quad h_{22} = \frac{\Delta y}{y_{11}}$$

In matrix form:

$$[h] = \begin{bmatrix} \dfrac{1}{y_{11}} & \dfrac{-y_{12}}{y_{11}} \\ \dfrac{y_{21}}{y_{11}} & \dfrac{\Delta y}{y_{11}} \end{bmatrix}$$

(*iii*) In terms of *ABCD* parameters

ABCD parameters are defined as:

$$V_1 = A\, V_2 + B\,(-I_2) \quad \text{...(6.78)}$$

and

$$V_1 = CV_2 + D(-I_2) \quad \text{...(6.79)}$$

From equation (6.79), we get:

$$I_2 = \left(-\frac{1}{D}\right) I_1 + \left(\frac{C}{D}\right) V_2 \quad \text{...(6.80)}$$

By substituting value of I_2 in equation (6.78), we get:

$$V_1 = AV_2 + B\left[\frac{1}{D} I_1 - \frac{C}{D} V_2\right]$$

or

$$V_1 = \left(\frac{B}{D}\right) I_1 + \left(\frac{AD - BC}{D}\right) V_2 \quad \text{...(6.81)}$$

By comparing equations (6.81) and (6.80) with equations (6.68) and (6.69), we get:

$$h_{11} = \frac{B}{D}, \quad h_{12} = \frac{AD - BC}{D}$$

and
$$h_{21} = -\frac{1}{D}, \quad h_{22} = \frac{C}{D}$$

In matrix form:

$$[/h] = \begin{bmatrix} \frac{B}{D} & \frac{AD - BC}{D} \\ -\frac{1}{D} & \frac{C}{D} \end{bmatrix}$$

6.6.5 Summary of Inter-relationships between Various Parameters, Reciprocity and Symmetry

(*i*) Inter-relationship between various parameters:

Parameters	*[z]*	*[y]*	*[T]*	*[h]*
[z]	$\begin{bmatrix} z_{11} & z_{12} \\ z_{21} & z_{22} \end{bmatrix}$	$\begin{bmatrix} \frac{y_{22}}{\Delta y} & \frac{-y_{12}}{\Delta y} \\ -\frac{y_{21}}{\Delta y} & \frac{y_{11}}{\Delta y} \end{bmatrix}$		$\begin{bmatrix} \frac{\Delta h}{h_{22}} & \frac{h_{12}}{h_{22}} \\ -\frac{h_{21}}{h_{22}} & \frac{1}{h_{22}} \end{bmatrix}$
[y]	$\begin{bmatrix} \frac{z_{22}}{\Delta z} & -\frac{z_{12}}{\Delta z} \\ -\frac{z_{21}}{\Delta z} & \frac{z_{11}}{\Delta z} \end{bmatrix}$	$\begin{bmatrix} y_{11} & y_{12} \\ y_{21} & y_{22} \end{bmatrix}$	$\begin{bmatrix} \frac{D}{B} & -\frac{\Delta T}{B} \\ -\frac{1}{B} & \frac{A}{B} \end{bmatrix}$	$\begin{bmatrix} \frac{1}{h_{11}} & -\frac{h_{12}}{h_{11}} \\ -\frac{h_{21}}{h_{11}} & \frac{\Delta h}{h_{11}} \end{bmatrix}$
[T]	$\begin{bmatrix} \frac{z_{11}}{z_{21}} & \frac{\Delta z}{z_{21}} \\ \frac{1}{z_{21}} & \frac{z_{22}}{z_{21}} \end{bmatrix}$	$\begin{bmatrix} -\frac{y_{22}}{y_{21}} & -\frac{1}{y_{21}} \\ -\frac{\Delta y}{y_{21}} & \frac{y_{11}}{y_{21}} \end{bmatrix}$	$\begin{bmatrix} A & B \\ C & D \end{bmatrix}$	$\begin{bmatrix} -\frac{\Delta h}{h_{21}} & -\frac{h_{11}}{h_{21}} \\ -\frac{h_{22}}{h_{21}} & -\frac{1}{h_{21}} \end{bmatrix}$
[h]	$\begin{bmatrix} \frac{\Delta z}{z_{22}} & \frac{z_{21}}{z_{22}} \\ -\frac{z_{21}}{z_{22}} & \frac{1}{z_{22}} \end{bmatrix}$	$\begin{bmatrix} \frac{1}{y_{11}} & -\frac{y_{12}}{y_{11}} \\ \frac{y_{21}}{y_{11}} & \frac{\Delta y}{y_{11}} \end{bmatrix}$	$\begin{bmatrix} \frac{B}{D} & \frac{\Delta T}{D} \\ -\frac{1}{D} & \frac{C}{D} \end{bmatrix}$	$\begin{bmatrix} h_{11} & h_{12} \\ h_{21} & h_{22} \end{bmatrix}$

Where $\Delta z = z_{11}\, z_{22} - z_{12}\, z_{21}$; $\Delta y = y_{11}\, y_{22} - y_{12}\, y_{21}$

$\Delta T = AD - BC$; $\Delta h = h_{11}\, h_{12} - h_{12}\, h_{21}$

(*ii*) Condition for reciprocity and symmetry of network

Parameter	*Reciprocity*	*Symmetry*
z	$z_{12} = z_{21}$	$z_{11} = z_{22}$
y	$y_{12} = y_{21}$	$y_{11} = y_{22}$
T	$AD - BC = 1$	$A = D$
h	$h_{12} = h_{21}$	$h_{11}\, h_{22} - h_{12}\, h_{21} = 1$

6.7 INTERCONNECTION OF TWO-PORT NETWORKS

Series, parallel, cascade, series-parallel and parallel-series interconnection configurations of two port networks are in use. These interconnection give rise two a new two-port network. The parameters are chosen on the basis of utility for each configuration.

6.7.1 Series Connection or Series-Series Connection

A series connection of two-port networks N' and N'' is shown in Fig. 6.9.

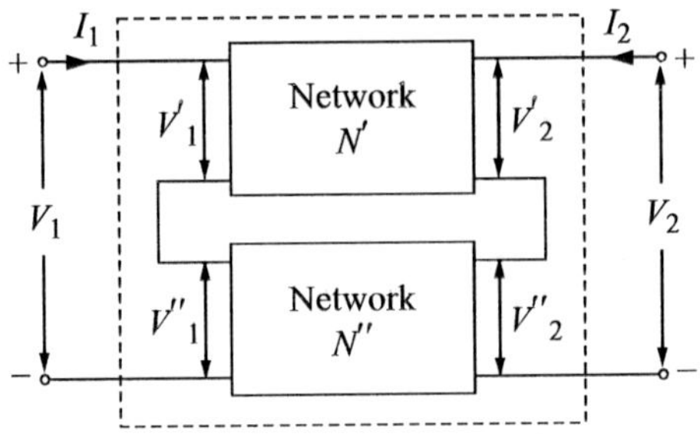

Fig. 6.9 *Series Connection of two-port netwroks*

The open circuit impedances z-parameters are z' and z'' for networks N' and N'' respectively.

For network N':

$$\begin{bmatrix} V_1' \\ V_2' \end{bmatrix} = \begin{bmatrix} z_{11}' & z_{12}' \\ z_{21}' & z_{22}' \end{bmatrix} \begin{bmatrix} I_1 \\ I_2 \end{bmatrix}$$

Similarly, for network N'':

$$\begin{bmatrix} V_1'' \\ V_2'' \end{bmatrix} = \begin{bmatrix} z_{11}'' & z_{12}'' \\ z_{21}'' & z_{22}'' \end{bmatrix} \begin{bmatrix} I_1 \\ I_2 \end{bmatrix}$$

Now, we know: $V_1 = V_1' + V_2'' = (z_{11}' + z_2' I_2) + (z_{11}'' I_1 + z_{12}'' I_2)$

$$= (z_{11}' + z_{11}'')I_1 + (z_{12}' + z_{12}'')\, I_2$$

Similarly, $\quad V_2 = V_2' + V_2'' = (z_{21}' I_1 + z_{22}' I_2) + (z_{21}'' I_1 + z_{22}'' I_2)$

or $\quad V_2 = (z_{21}' + z_{21}'')\, I_1 + (z_{22}' + z_{22}'')\, I_2$

So, the z-parameters of the series-connected combined network is given by:

$$\begin{bmatrix} V_1 \\ V_2 \end{bmatrix} = \begin{bmatrix} z_{11} & z_{12} \\ z_{21} & z_{22} \end{bmatrix} \begin{bmatrix} I_1 \\ I_2 \end{bmatrix}$$

where $z_{11} = z'_{11} + z''_{11}$

$$z_{12} = z'_{12} + z''_{12}$$
$$z_{21} = z'_{21} + z''_{21}$$
$$z_{22} = z'_{22} + z''_{22}$$

or $$[z] = [z'] + [z'']$$

Thus, the overall z-parameter matrix for series connected two two-port networks is simply the sum of z-parameter matrices of each individual two-port network connected in series

6.7.2 Parallel-connected or Parallel-parallel Connection

A parallel-connection of two two-port network N' and N'' is shown in Fig. 6.10.

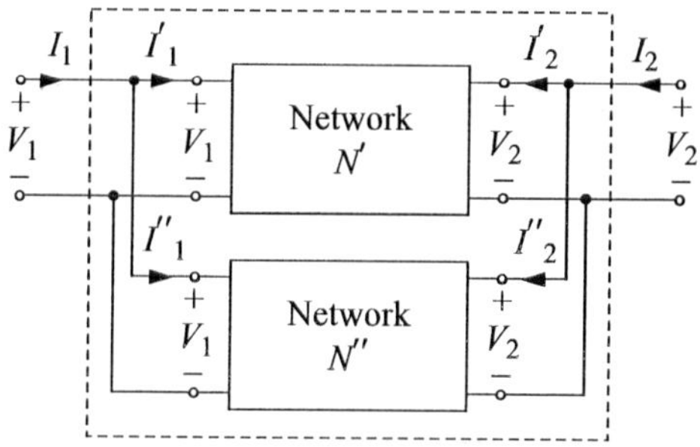

Fig. 6.10 *Parallel-connection of two two port network*

Short-circuit of admittance or y-parameters are Y' N' and N'' networks Y' and Y'' respectively.

For network N',

$$\begin{bmatrix} I_1' \\ I_2' \end{bmatrix} = \begin{bmatrix} Y'_{11} & Y'_{12} \\ Y'_{21} & Y'_{22} \end{bmatrix} \begin{bmatrix} V_1 \\ V_2 \end{bmatrix}$$

Similarly for network N'',

$$\begin{bmatrix} I''_1 \\ I''_2 \end{bmatrix} = \begin{bmatrix} Y''_{11} & Y''_{12} \\ Y''_{21} & Y''_{22} \end{bmatrix} \begin{bmatrix} V_1 \\ V_2 \end{bmatrix}$$

Now, $$I_1 = I'_1 + I''_2 = \left(Y'_{11} V_1 + Y'_{12} V_2\right) + \left(Y''_{11} V_1 + Y''_{12} V_2\right)$$

or $$I_1 = (Y'_{11} + Y''_{11}) V_1 + (Y'_{12} + Y'_{12}) V_2$$

Similarly

$$I_2 = I'_2 + I''_2 = (Y'_{21} + V_1 + Y'_{22} V_2) + (Y'_{21} V_1 + Y''_{22} V_2)$$
$$I_2 = (Y'_{21} + Y''_{21}) V_1 + (Y'_{22} + Y''_{22}) V_2$$

Thus,

$$\begin{bmatrix} I_1 \\ I_2 \end{bmatrix} = \begin{bmatrix} Y_{11} & Y_{12} \\ Y_{21} & Y_{22} \end{bmatrix} \begin{bmatrix} V_1 \\ V_2 \end{bmatrix}$$

Where
$$Y_{11} = Y'_{11} + Y''_{11}$$
$$Y_{12} = Y'_{12} + Y''_{12}$$
$$Y_{21} = Y'_{21} + Y''_{21}$$
$$Y_{22} = Y'_{22} + Y'_{22}$$
or
$$[Y] = [Y'] + [Y'']$$

Hence, the overall y-parameter matrix for parallel connected two-port network is the sum of y-parameter matrices each individual two-port network connected in parallel connection.

7.3 CASCADE-CONNECTION

Cascade connection of two-port networks is shown in Fig. 6.11

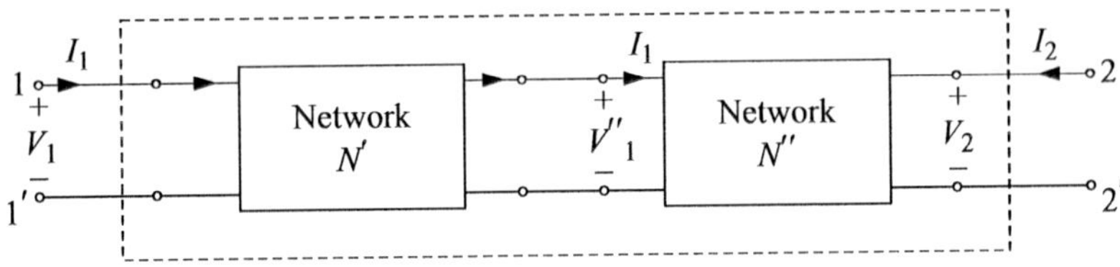

Fig. 6.11 *Cascade connection of two port network cascade connection*

For network N',
$$\begin{bmatrix} V_1 \\ I_1 \end{bmatrix} = \begin{bmatrix} A' & B' \\ C' & D' \end{bmatrix} \begin{bmatrix} V_1'' \\ -I_1'' \end{bmatrix}$$

Similarly for network N'',
$$\begin{bmatrix} V_1'' \\ I_1'' \end{bmatrix} = \begin{bmatrix} A'' & B'' \\ C'' & D'' \end{bmatrix} \begin{bmatrix} V_2 \\ I_2 \end{bmatrix}$$

or
$$\begin{bmatrix} V_1'' \\ -I_1'' \end{bmatrix} = \begin{bmatrix} A'' & B'' \\ C'' & D'' \end{bmatrix} \begin{bmatrix} V_2 \\ I_2 \end{bmatrix}$$

By substituting $\begin{bmatrix} V_1'' \\ I_1'' \end{bmatrix}$ from previous equation, we get:
$$\begin{bmatrix} V_1 \\ I_1 \end{bmatrix} = \begin{bmatrix} A' & B' \\ C' & D' \end{bmatrix} \begin{bmatrix} A'' & B'' \\ C'' & D'' \end{bmatrix} \begin{bmatrix} V_2 \\ -I_2 \end{bmatrix}$$

Take,
$$\begin{bmatrix} A & B \\ C & D \end{bmatrix} = \begin{bmatrix} A' & B' \\ C' & D' \end{bmatrix} \begin{bmatrix} A'' & B'' \\ C'' & D'' \end{bmatrix}$$

or
$$[T] = [T'] [T'']$$

where
$$A = A' A'' + B' C''$$
$$B = A' \mathrm{B}'' + B' D''$$

$$C = C' A'' + D' B''$$
$$D = C' B'' + D' D''$$

Thus, overall T-parameter matrix for cascade connected two-port networks is the matrix product of the T-parameters matrices of each individual two-port network in cascade.

6.8 *T*-REPRESENTATION OF TWO-PORT NETWORK

An equivalent T-network of any two-port network in shown is Fig. 6.12.

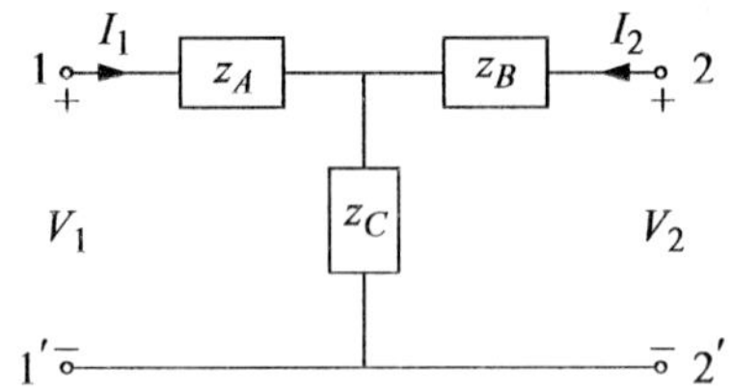

Fig. 6.12 *T-representation of a two-port network*

z-parameters for a two port network are defined as:

$$z_{11} = \left.\frac{V_1}{I_1}\right|_{I_2=0} = z_A + z_C, \qquad z_{21} = \left.\frac{V_2}{I_1}\right|_{I_2=0} = z_C$$

$$z_{12} = \left.\frac{V_1}{I_2}\right|_{I_1=0} = z_C, \qquad z_{22} = \left.\frac{V_2}{I_2}\right|_{I_1=0} = z_B + z_C$$

$$\text{or} \quad [Z] = \begin{bmatrix} z_A + z_C & z_C \\ z_C & z_B + z_C \end{bmatrix}$$

The values of z-network in terms of two-port network, is given by;

$$z_A = z_{11} - z_{12}$$
$$z_B = z_{22} - z_{21}$$
$$z_C = z_{21}$$

6.9 EQUIVALENT π-NETWORK REPRESENTATION OF A TWO-PORT

An equivalent π-network representation of a two-port network is shown in the Fig. 6.13

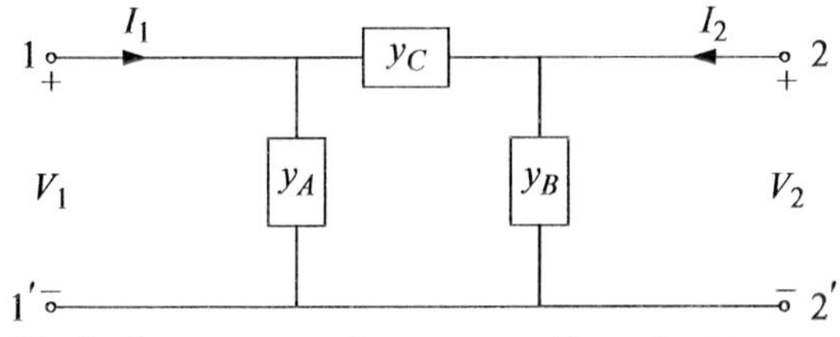

Fig. 6.13 *Equivalent π-network representation of a two port network.*

y-parameters of a port-network are defined as:

$$y_{11} = \left.\frac{I_1}{V_1}\right|_{V_2=0} = y_A + y_C, \qquad y_{12} = \left.\frac{I_1}{V_2}\right|_{V_1=0} = -y_C$$

$$y_{21} = \left.\frac{I_2}{V_1}\right|_{V_2=0} = -y_C, \qquad y_{22} = y_B + y_C$$

$$i.e.\ [Y] = \begin{bmatrix} y_A + y_C & -y_C \\ -y_C & y_B + y_C \end{bmatrix}$$

The y-parameter values of π-network representation of a two-port network is given by:

$$y_A = y_{11} + y_{12} \text{ or } y_{11} + y_{21}$$
$$y_B = y_{22} + y_{12} \text{ or } y_{22} + y_{21}$$
$$y_C = -y_{12} \text{ or } -y_{21}$$

6.10 LATTICE NETWORK

A symmetric lattice network is shown in Fig. 6.14.

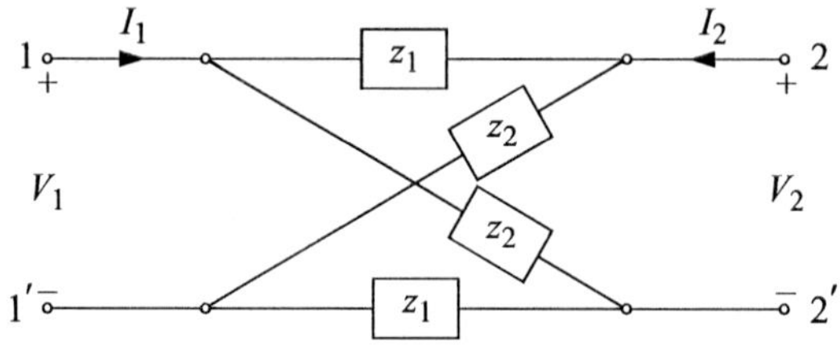

Fig. 6.14 *A symmetric lattice network*

A two-port network can be represented by a symmetric lattice network.

A rearrangement of impudences of Fig. 6.14 can be done as shown in Fig. 6.15.

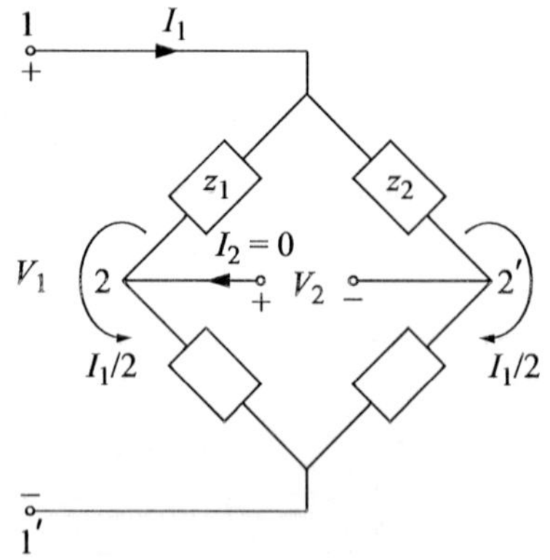

Fig. 6.15 *Rearrangement of lattice network*

To find z-parameter values, consider port 2 is open *i.e.* $I_2 = 0$.
By KVL equations, we get:

000 $$V_1 = \frac{I_1}{2}(z_1 + z_2) \text{ and } V_2 = \frac{I_1}{2}(z_2 - z_1)$$

Hence, $$\left.\frac{V_2}{I_1}\right|_{I_2=0} = z_{21} = \frac{1}{2}(z_2 - z_1) \text{ and } \left.\frac{V_1}{I_1}\right|_{I_2=0} = \frac{1}{2}(z_2 + z_1)$$

As the network symmetrical and reciprocal hence, we get:

$$z_{11} = z_{22} = \frac{1}{2}(z_2 + z_1)$$

$$z_{12} = z_{21} = \frac{1}{2}(z_2 - z_1)$$

i.e. $$z_2 + z_1 = 2z_{11} \text{ and } z_2 - z_1 = 2z_{12}$$

By adding the two equations, we get:

$$z_2 = z_{11} + z_{12}$$

and by subtracting second equation from the first equation, we get:

$$z_1 = z_{11} - z_{12}$$

6.11 LADDER NETWORK

Impedances and admittances are shown in a ladder network as shown in the Fig. 6.16, for the convenience of analysis *i.e.* series arms by impedances and shunt arms by admittances,

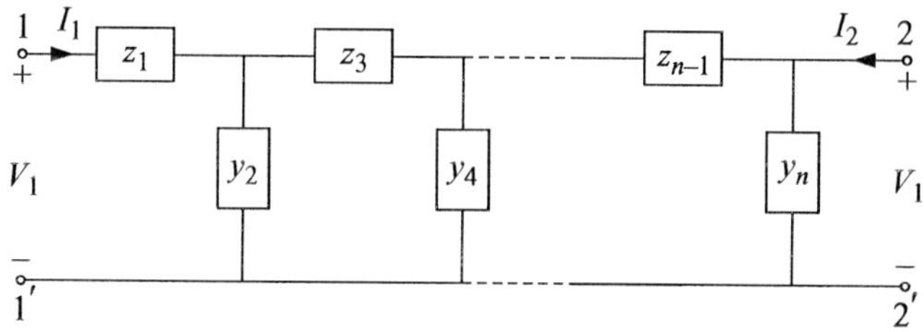

Fig. 6.16 *A ladder network*

In order to find the driving point impedance at port 1 – 1′, we start computation beginning at port 2 – 2′ *i.e.* begin with y_6, *see* Fig. 6.17.

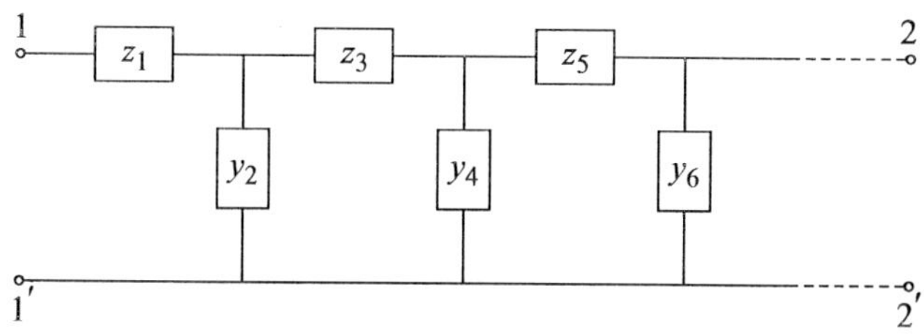

Fig 6.17 *A ladder network for analysis*

$$z_{11} = z_1 + \cfrac{1}{y_2 + \cfrac{1}{z_3 + \cfrac{1}{y_4 + \cfrac{1}{z_5 + \cfrac{1}{y_6}}}}}$$

SUMMARY

1. **Z-parameters:** z-parameters are obtained by expressing voltages at two-ports in terms of currents at two ports. The z-parameters relate to impedance, hence, these are also known as impedance parameters.
2. **Y-parameters:** There parameters are defined by short circuiting one of the ports, hence, these are also known as short circuit parameters. Further, these parameters are the ratio of current to voltage, hence, these are also known as admittance parameters.
3. **ABCD Parameters:** ABCD parameters were initially used for transmission lines, therefore, these are also known as transmission (T) or chain parameters.
4. ***h*-parameters:** h stands for hybrid, hence, these are also known as hybrid parameters.

EXERCISES

6.1. Find the Z, Y and $ABCD$ parameters for the given Network shown in figure.

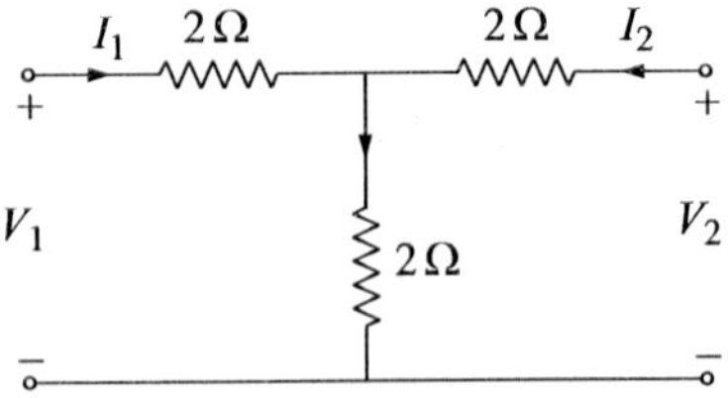

6.2. Determine the Z parameter of Network shown in figure.

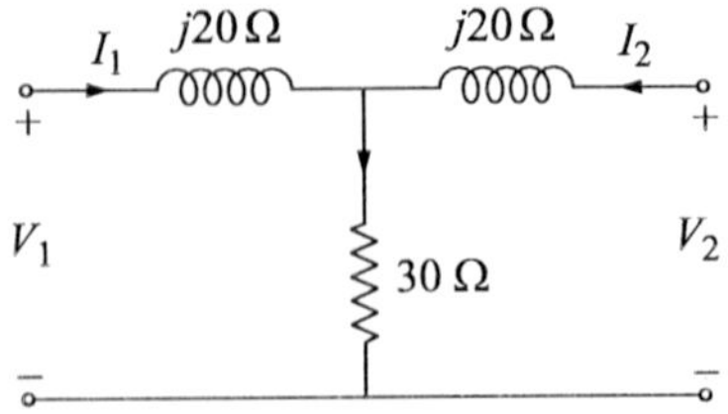

6.3. Find the Y and Z parameters for the Network as shown in figure.

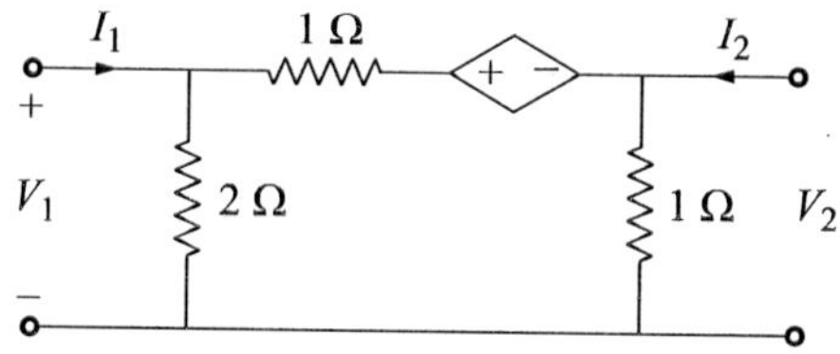

6.4. Obtain the transmission parameters of the cascaded connecting two identical networks as shown in figure.

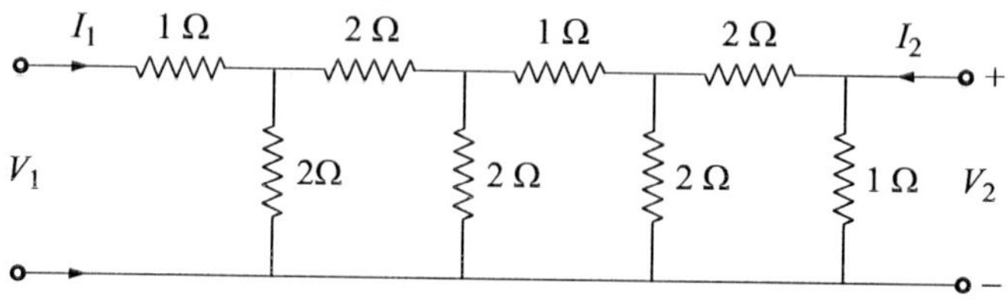

6.5. Find the Z-parameter. For the Network shown in figure.

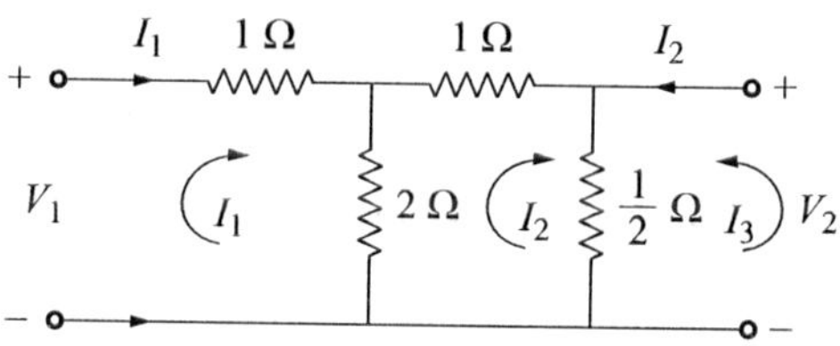

6.6. Determine the image imepedance of the Network shown in figure.

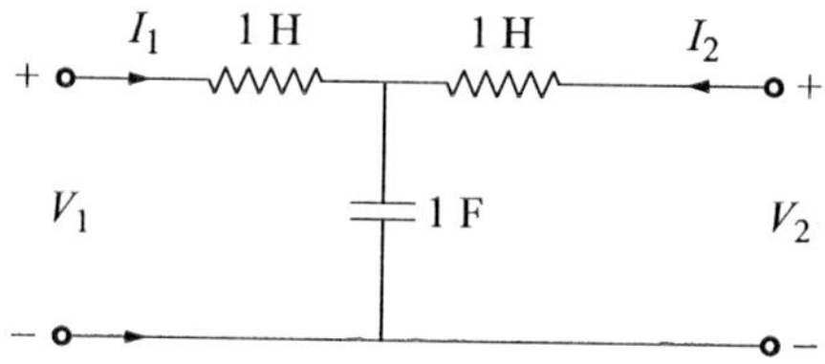

6.7. Find the Y and Z parameters for the circuit as shown in figure.

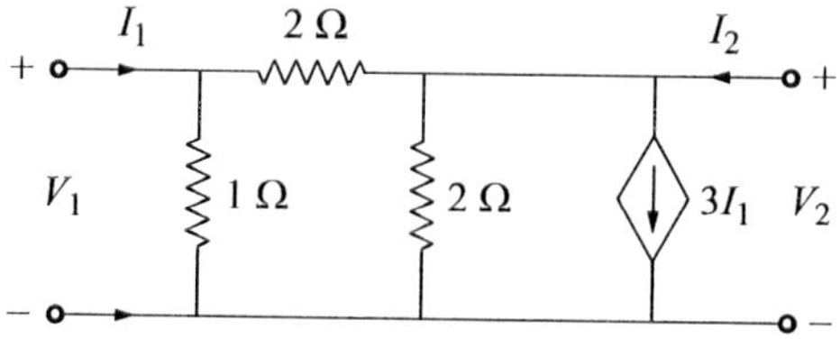

6.8. Determine h-parameter of the network shown in figure.

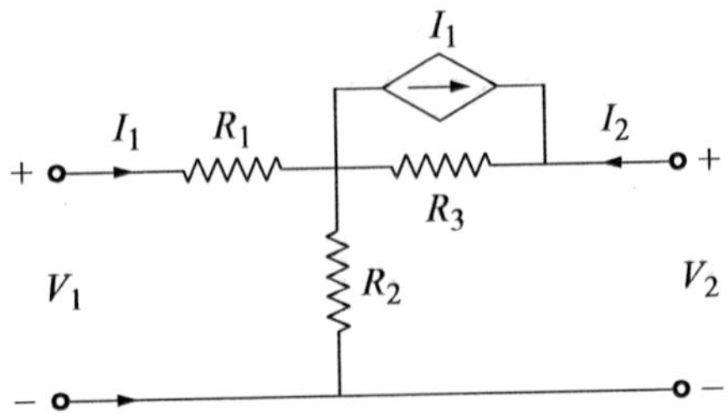

6.9. Determine the transmission parameter of the circuit shown in figure.

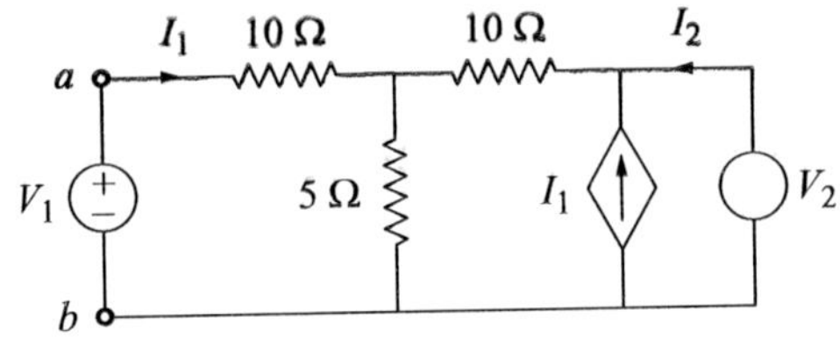

6.10. For the two port Network shown in figure. Find the h-parameter and find $ABCD$ parameter.

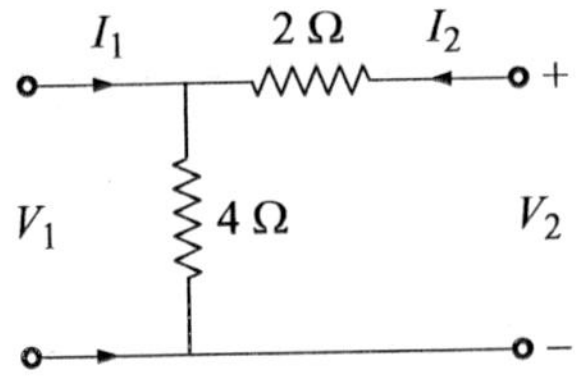

6.11. The Network shown in figure., consist of a resistive T network, connecting in parallel. Determine the h-parameter.

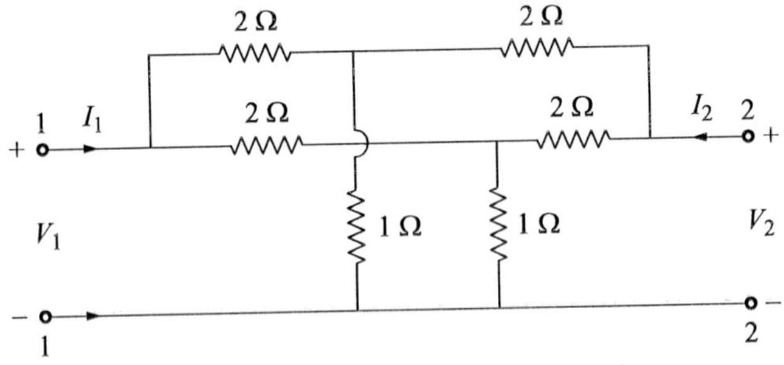

6.12. In the circuit shown in figure. Find the Z and h-parameters.

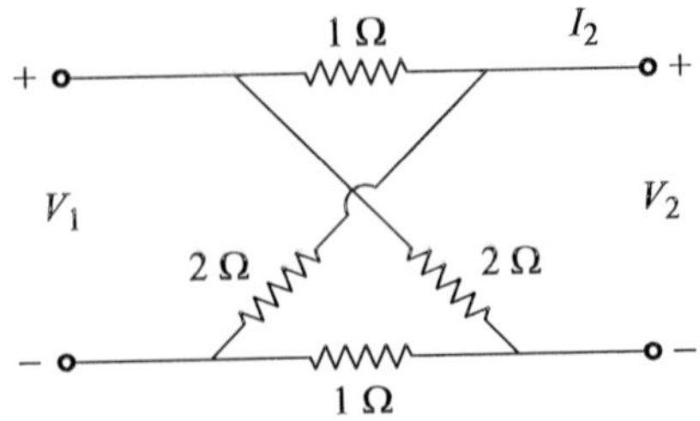

6.13. Find the *z*-parameter, for the Network as shown in figure.

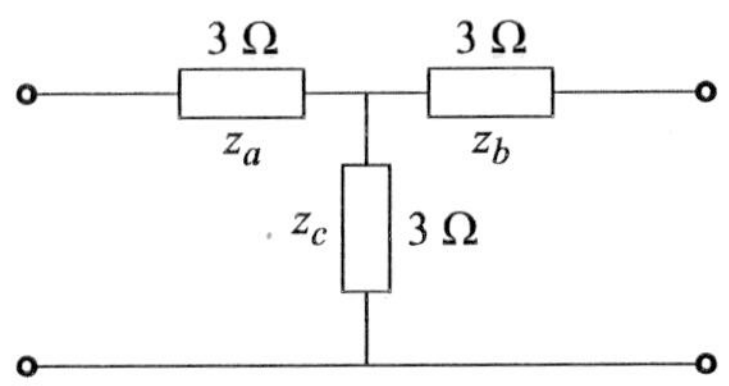

6.14. Determine transmission parameter of a T-Network shown in figure. Considering 3 section as shown in given figure assuming connecting in cascade manner.

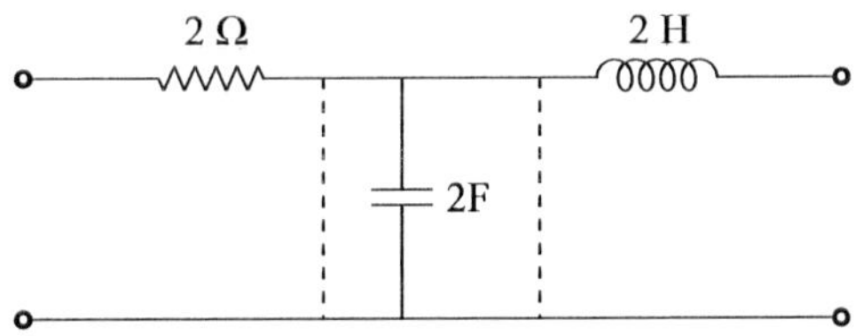

6.15. Find the short circuit and open circuit impedance of the network shown in figure.

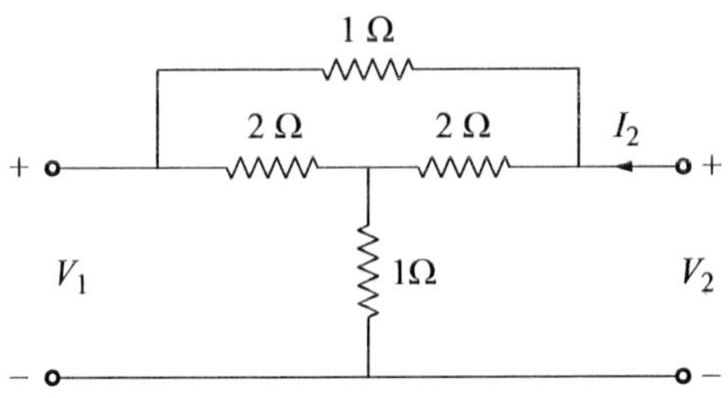

6.16. Find the Y-parameters of network shown in given figure and check whether the circuit is symmetric and reciprocal or net.

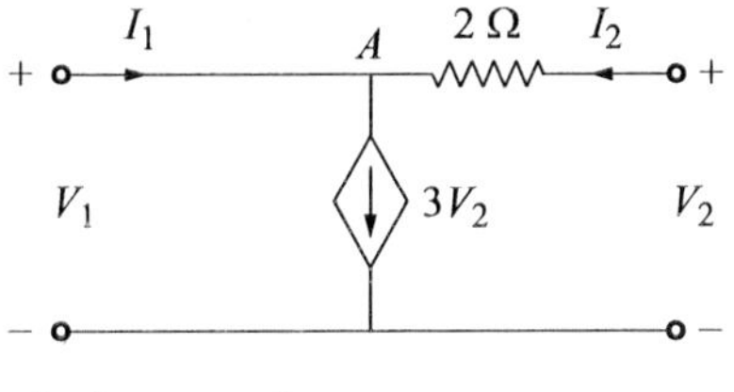

6.17. For the Network the equation are;

$$I_1 = 0.5V_1 - \alpha_2 V_2$$
$$I_2 = -0.2V_1 + V_2$$

Find the *Y* parameters and *ABCD* parameters. Also find its equivalent of Network.

6.18. Obtain the z-parameter of the Network shown in figure.

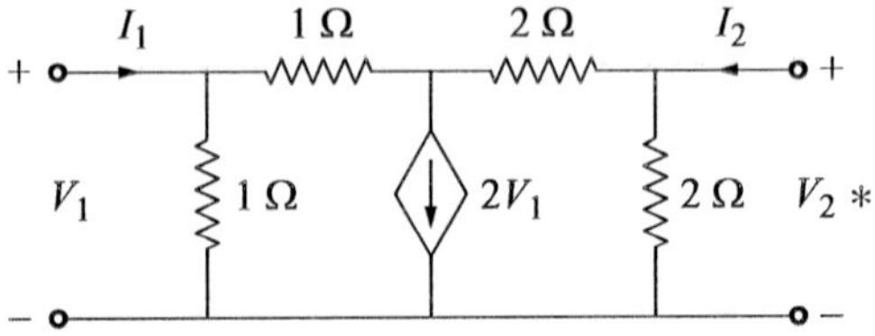

6.19. Find the Z-parameter of the Network as shown in figure.

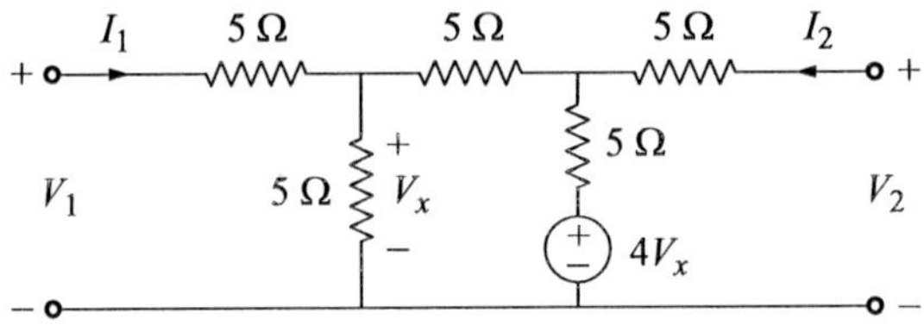

6.20. Show that for the Network in figure. The following equations.

$$\begin{bmatrix} V_2 \\ i_2 \end{bmatrix} = \begin{bmatrix} 1.5 & 6.5 \\ 0.25 & 1.25 \end{bmatrix} \begin{bmatrix} V_1 \\ -i_1 \end{bmatrix}$$

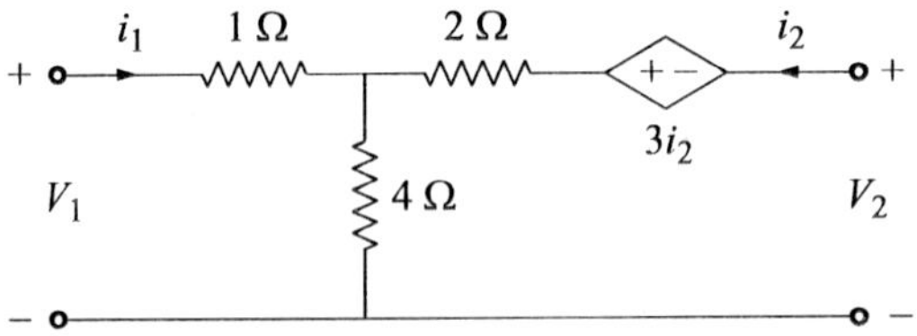

6.21. Determine the y-parameter of the Network as shown in figure.

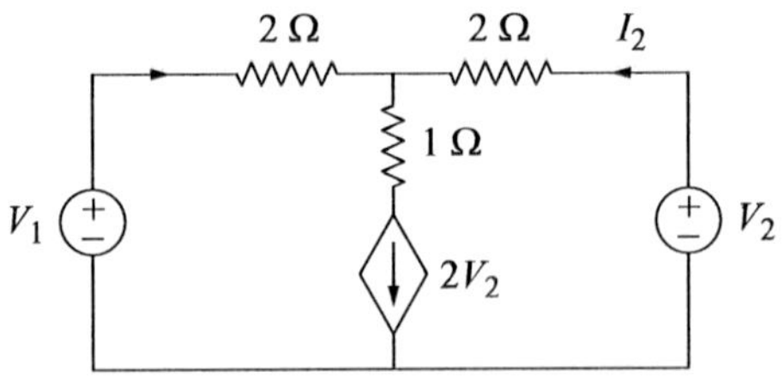

6.22. Find the $ABCD$ parameters of the Network shown in given figure.

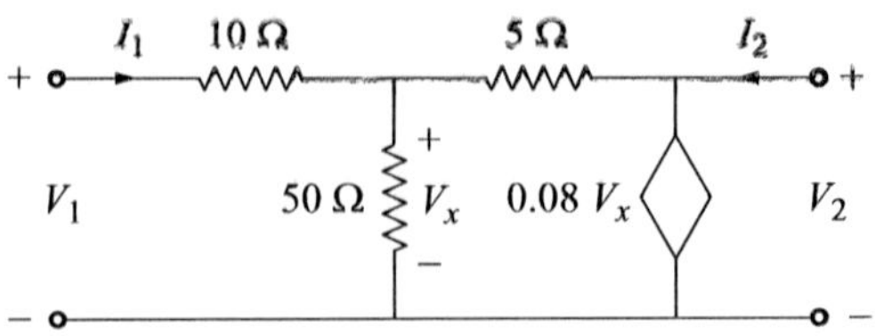

6.23. Find $[T_A]$, $[T_B]$ and $[T_C]$ for the given two port network and also find $[T]$ for. Combined Network as shown in figure.

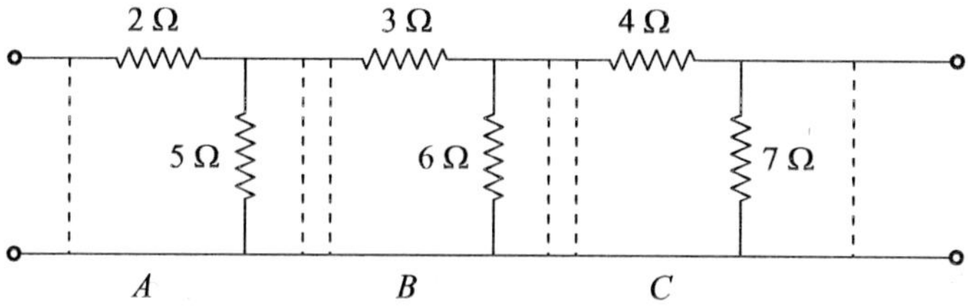

6.24. Find the (*n*) for two port. Network as shown in given figure.

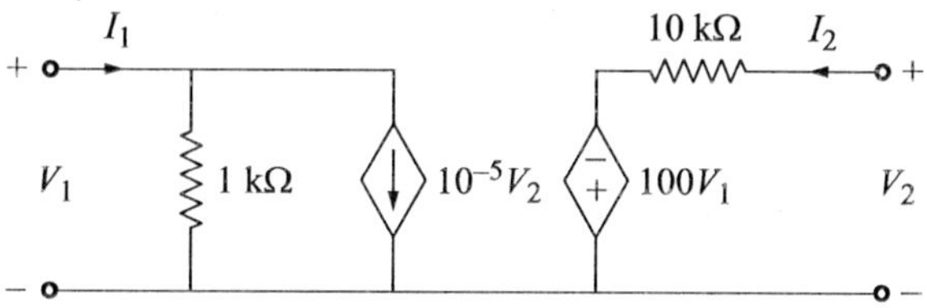

6.25. Find the *ABCD* parameter, for the Network shown in given figure.

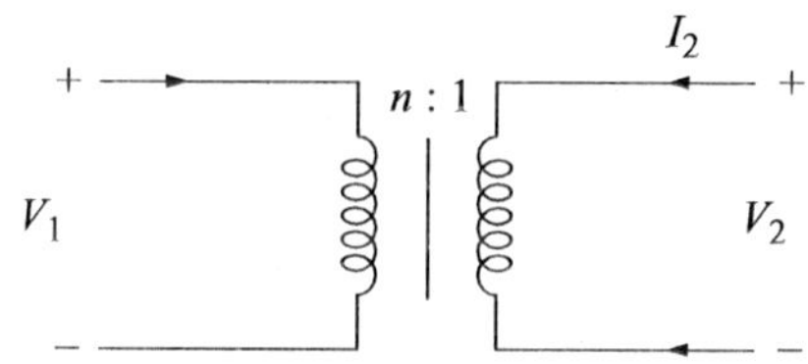

6.26. Find the open circuit impedance parameter of the network shown in given figure.

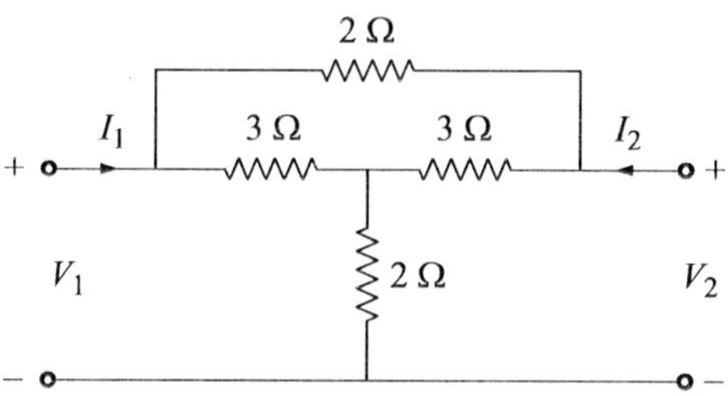

6.27. Find the *Z* and *Y* parameters for the network shown in given figure.

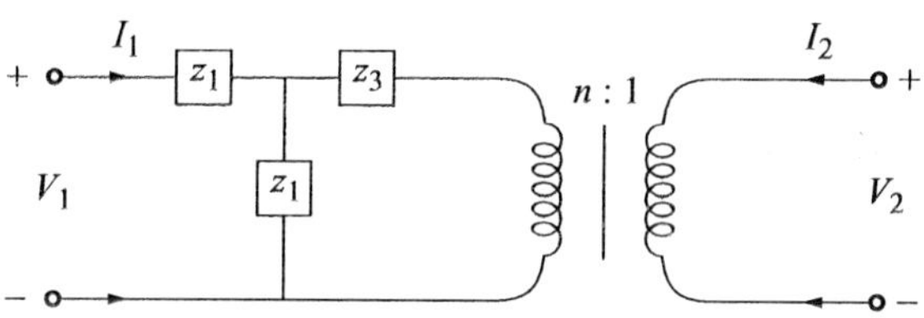

❑❑❑

7

Network Synthesis

7.1 INTRODUCTION

Network elements, input (excitation) to the network and output (response) from the network are the three things associated with any network. In the case of network analysis, network elements and excitation are known. Based on these network response is obtained. The response is unique for a network and the excitation. The network analysis involves several methods and theorems.

Network synthesis is exactly opposite of the network analysis. It is to find the network satisfying a known excitation and known response. It is important to note that analysis given response which is always, unique, whereas, synthesis does not given unique solution. Driving point impedance and admittance functions for all the networks passive elements are rational. These are in the form of ratio of two polynomials in 's', however, it is not possible that every ratio of two polynomials in 's' can be synthesized with passive elements.

In synthesis, it is necessary to know that the given network function is realizable for this, the network functions should satisfy all the properties of Hurwitz polynomials and the positive real functions.

7.2 HURWITZ POLYNOMIALS

A Hurwitz polynomial has to satisfy the following basic properties:

(*i*) The polynomial should be real for real 's'.

(*ii*) The roots of the polynomial should have real parts which are either zero or negative. These polynomials have to satisfy many more properties.

7.2.1 Properties of Hurwitz Polynomial of Order '*n*'

Suppose $P(s)$ is a Hurwitz polynomial of order 'n' as given here:

$$P(s) = a_n s^n + a_{n-1} s^{n-1} + ... + a_1 s + a_0$$

where an, a_{n-1}, ..., a_1, a_0 are the coefficients of the polynomials.

Hurwitz polynomial $p(s)$ should have:

(*i*) All the coefficients of 'a' are real and positive.

(*ii*) All the terms starting from highest power of 's' to lowest power of 's' are present.

The coefficients can be missing, if the polynomial is either even or odd. An even polynomial consists of only even powers of 's' *i.e.*, all odd powers of 's' are missing. However, no even powers of 's' should be missing in between.

An even polynomials. $p(s)$ $s^6 + 2s + s^2 + 5$

Similarly, an odd polynomial consists of only odd powers of 's' and in between no odd powers of 's' are missing. All even powers of 's' must be missing.

An odd polynomial: $p(s) = s^5 + 3s^3 + 2s$

(*iii*) In case, polynomial is odd or even, then all its roots must lie on imaginary $j\omega$ axis.

(*iv*) Hurwitz polynomial can be separated into even and odd parts. Say $m(s)$ is even part and $n(s)$ is the odd part.

Thus, $p(s) = m(s) + n(s)$,

where both $m(s)$ and $n(s)$ parts should have roots on jω axis.

For example: $p(s) = s^4 + s^3 + 5s^2 + 3s + 4 = m(s) + n(s)$.

Even part: $m(s) = s^4 + 5s^2 + 4 = (s^2 + 1)\ (s^2 + 4)$

i.e., roots of $m(s)$ are $s = \pm j1$ and $s = \pm j1$ and $s = \pm j = 2$ which lie on $j\omega$ axis.

odd part: $n(s) = s^3 + 3s = s(s^2 + 3)$

i.e., roots of $n(s)$ are $s = 0$ and $\pm j\sqrt{3}$ which also lie on $j\omega$ axis.

Although, the roots of $m(s)$ and $n(s)$ separately lie on $j\omega$ axis, but it does not mean that all the roots of the polynomial $p(s) = m(s) + n(s)$ will also lie on $j\omega$ axis.

(*v*) The continued fraction expansion of the ratio of even to odd parts or odd to even parts, should yield all positive quotient terms. The highest polynomial out of $m(s)$ and $n(s)$ must be taken as a numerator to start the division. Accordingly, the number of quotients obtained in the continued fraction expansion will be equal to highest power of 's' present in the polynomial of $p(s)$.

Thus, $$\psi(s) = \frac{m(s)}{n(s)} \text{ or } \frac{n(s)}{m(s)}$$

The continued fraction expansion is:

$$\psi(s) = \alpha_1 s + \cfrac{1}{\alpha_2 s + \cfrac{1}{\alpha_3 s + \cfrac{1}{\text{----------} + \alpha_n s}}}$$

All the coefficients α_1, α_2, ..., α_n must be positive for the $p(s)$ to Hurwitz polynomial.

(*vi*) In case, the continued fraction expansion terminates abruptly, then it indicates common factor between $m(s)$ and $n(s)$ in such case, the polynomial $p(s)$ can be expressed as the product of the two facts which are individually Hurwitz polynomial.

If $p(s) = p_1(s)\, p_2(s)$T, then $p(s)$ is Hurwitz polynomial and $p_1(s)$ and $p_2(s)$ are individually Hurwitz polynomial.

(*vii*) In case, $p(s)$ is even or odd, then either $m(s)$ or $n(s)$ is missing. As such, continued expansion is not feasible. In that event, continued fraction expansion of $p(s)$ with its derivative $p'(s)$ is obtained.

$$\frac{p(s)}{p'(s)} = p_1(s)\, p_2(s)$$

Again, if $p_1(s)$ and $p_2(s)$ are individually Hurwitz polynomial, then $p(s)$ is a Hurwitz polynomial.

Example 7.1 Find whether the polynomial given is Hurwitz.

$$p(s) = s^4 + 3s^3 + 4s^2 + 3s + 1 \text{ of 's'}$$

Solution: No coefficients are missing and all the coefficients are real and positive. Now, we can proceed further.

even part: $m(s) = s^4 + 4s^2 + 1$

odd part: $n(s) = 3s^3 + 3s$

$$\psi(s) = \frac{m(s)}{n(s)} = \frac{s^4 + 4s^2 + 1}{3s^3 + 3s}$$

$$\begin{array}{l}
3s^3 + 3s\,\big)\, s^4 + 4s^2 + 1 \,\big(\tfrac{1}{3}s \\
\qquad\quad \underline{-\,s^4 + s^2} \\
\qquad\qquad 3s^2 + 1\,\big)\, 3s^3 + 3s \,\big(s \\
\qquad\qquad\qquad\quad \underline{3s^3 + s} \\
\qquad\qquad\qquad\qquad 25\,\big)\, 3s^2 + 1 \,\big(\tfrac{3}{2}s \\
\qquad\qquad\qquad\qquad\quad \underline{3s^2\sigma} \\
\qquad\qquad\qquad\qquad\qquad 1\,\big)\, 25 \,\big(25 \\
\qquad\qquad\qquad\qquad\qquad\quad \underline{25} \\
\qquad\qquad\qquad\qquad\qquad\quad \times
\end{array}$$

Thus,

$$\psi(s) = \frac{m(s)}{n(s)} = \frac{1}{3}s + \cfrac{1}{s + \cfrac{1}{\frac{3}{2}s + \frac{1}{2s}}}$$

$\therefore\ \alpha_1 = \frac{1}{3},\ \alpha_2 = 1,\ \alpha_3 = \frac{3}{2}$ and $\alpha_4 = 2$ which are all real and positive, hence, the polynomial $p(s)$ is Hurwitz.

Example 7.2 Find whether the given polynomial is Hurwitz.

$$p(s) = s^3 + 2s^2 + 3s + 6$$

Solution: All the coefficients of 's' present, and are real and positive. Now, we can proceed further.

Even part : $m(s) = 2s^2 + 6$

odd part : $n(s) = s^3 + 3s$

$$\therefore \qquad \psi(s) = \frac{n(s)}{m(s)} = \frac{s^3 + 3s}{2s^2 + 6}$$

$$2s^2 + 6)\ s^3 + 3s\ (\tfrac{1}{2}s$$

$$\frac{S^3 + 3S}{\times}$$

It is found, that the division terminates abruptly, by common factor $s^3 + 3s$.

$$\therefore \qquad p(s) = (s^3 + 3s)\left(1 + \frac{2}{s}\right)$$

$$= (s^2 + 3)(s + 2) = p_1(s)\ p_2(s)$$

$p_2(s) = s + 2$ is Hurwitz.

$p_1(s) = s^2 + 3$

has roots $s = \pm\sqrt{3}j$ which are on $j\omega$ axis *i.e.*, $p_1\ (s)$ is Hurwitz.

$p_1(s)\ p_2(s) = p(s)$, hence, $p(s)$ is Hurwitz.

7.2.2 Routh Hurwitz method

An array is constructed using coefficients of given polynomial in a specific way. The polynomial can be decided whether it is Hurwitz by inspection of the array formed.

Consider: $p(s) = a_n\ s^n + a_{n-1}\ s_{n-1} + \ldots + a_1\ s + a_0$

The Routh Hurwitz array is constructed as follows:

s^n	a_n	a_{n-2}	a_{n-4}
s_{n-1}	a_{n-1}	a_{n-3}	a_{n-5}
s_{n-2}	b_n	b_{n-1}	b_{n-2}
s_{n-2}	c_n	c_{n-1}	c_{n-2}
$\vdots$	$\vdots$	$\vdots$	$\vdots$
s_0	a_0		

a_n to a_0 are coefficients of the given polynomial $p(s)$ calculation of other of the array are:

$$b_n = \frac{a_{n-1}\, a_{n-2} - a_n\, a_{n-3}}{a_{n-1}}$$

$$b_{n-1} = \frac{a_{n-1}\, a_{n-4} - a_n\, a_{n-5}}{a_{n-1}}$$

and
$$c_n = \frac{b_n\, a_{n-3} - a_{n-1}\, b_{n-1}}{b_n}$$

$$c_{n-1} = \frac{b_n\, a_{n-5} - a_{n-1}\, b_{n-2}}{b_n}$$

The above process is continued until the n th row has been completed *i.e.*, s^o is obtained. The complete array of coefficients is triangular. The last row always consists of a single coefficient which is equal to the constant term of the given polynomial.

Routh's criterion states that the number of roots of the given polynomial with positive real parts is equal to the number of changes in sign of the coefficients of the first column of the array. Thus, the necessary and sufficient condition for the polynomial $p(s)$ to be a Hurwitz polynomial are:

(*i*) None of the elements of the first row of the array should be zero.

(*ii*) These should not be any sign change in the first column of the array.

These can be a situation that then is a complete row of zeros while generating as array as given below:

s^n	a_1	a_2	a_3	
s_{n-1}	b_2	b_2	b_3	
s_{n-2}	c_1	c_2	c_3	
s_{n-3}	d_1	d_2	d_3	
s_{n-4}	0	0	0	← row containing zeros.

In this situation, an equation is formed using the coefficients of a row which is just above the row of zeros. Such equation is known as an Auxiliary equation, say denoted as $A(s)$.

$A(s)$ is always odd or even polynomial in 's'.

$A(s) = d_1\, s^{n-3} + d_2\, s^{n-4} + d_3\, s^{n-6} + \ldots$

Differentiation of $A(s)$ *w.r.t.* 's' gives:

$$\frac{d\,A(s)}{ds} - d_1(n-3)s^{n-4} + \text{s}\; d_2(n-4)s^{n-5} + \ldots.$$

The array is completed by replacing row of zeros by the coefficients of the equation $\frac{d\,A(s)}{ds}$. Now, if there is no sign change in the first column of the array, then the $p(s)$ is Hurwitz polynomial, except same rare special cases, when the polynomial may not be Hurwitz. These special cases can be found out by solving equation $A(s) = 0$ as it has some roots of $p(s) = 0$, which decide whether polynomial or Hurwitz or not.

7.3 POSITIVE REAL FUNCTIONS

If the driving point immittance (impedence or admittance) is a positive real function, then it is physically realizable using passive R, L and C components. Therefore, immittance functions must be checked for being positive real, before synthesizing. A positive real function satisfies the following conditions:

(*i*) The given function $F(s)$ is real for real 's',

(*ii*) The real part of $F(s)$ is greater than or equal to zero, when the real part of 's' is greater than or equal to zero
i.e., $Re[F(s)] \geq 0$ for $Re\,[s] \geq 0$

(*iii*) The functions $F(s)$ is rational function.

There are some more properties of positive real function, in addition to above basic properties.

7.3.1 Properties of Positive Real Function

$$\text{Consider: } F(s) = \frac{Q(s)}{P(s)} = \frac{b_n\, s^n + b_{n-1}\, s^{n-1} + \ldots + a_1 s + a_0}{a_n\, s^n + a_{n-1}\, s^{n-1} + \ldots + a_1\, s + a_0}$$

$F(s)$ will be positive real function, if the function $F(s)$ has following properties:

(*i*) The coefficients of $Q(s)$ and $P(s)$ are real and positive. Therefore,

(*a*) $F(s)$ is real when s is real.

(*b*) Complex poles and zeros of $F(s)$ are in complex conjugate pairs.

(*c*) $N = \frac{b_n}{a_n}$ known as scale factor, is real and positive.

(*ii*) $F(s)$ poles and zeros have negative or zero real parts.

(*iii*) $F(s)$ poles on the imaginary ($j\omega$) axis are simple. Their residues are real and positive. The same statement is true for $\frac{1}{F(s)}$ poles.

(*iv*) The numerator and denominator degrees in $F(s)$ differ at the most by 1 *i.e.*, degrees of $m(s)$ and $n(s)$ differ by 1. Hence, the number of finite poles and finite zeros of $F(s)$ differ at the most by 1.

(*v*) The terms of lowest degree of in numerator and denominator polynomials of $F(s)$, differ in degree at most by 1. Hence, $F(s)$ has neither multiple poles nor zeros at the origin.

(*vi*) If $F(s)$ polynomial is positive real function, then $\frac{1}{F(s)}$ is also positive and real function. Therefore, for a network, if driving point impedence function $z(s)$ is positive and real, then the driving point admittance function $y(s) = \frac{1}{Z(s)}$ is also positive real.

(*vii*) As the sum of positive real functions is also positive real, hence, if two impedance are in series, the sum of impedances of positive real. Also, if two admittances are in parallel, then their addition is also positive real, However, the difference of two positive real functions is not necessarily a positive real function.

All the above conditions are necessary for the function $F(s)$ to be positive real, but they are not sufficient conditions.

7.3.2 Produce for Testing for Positive Realness

Suppose: $F(s) = \frac{Q(s)}{P(s)}$

Remove all the common factors of $Q(s)$ and $P(s)$. Check wether $F(s)$ meets the necessary conditions for positive realness. If yes, then check for necessary and sufficient conditions of $F(s)$ being positive real.

7.3.2.1 Necessary Conditions

(*i*) All the coefficients of $P(s)$ and $Q(s)$ must be real and positive.

(*ii*) The degrees of $P(s)$ and $Q(s)$ differ at most by one.

(*iii*) Lowest degree in $P(s)$ and $Q(s)$ differ at most by 1.

(*iv*) Imaginary axis ($j\omega$) poles and must be simple and no multiple pole or zero should lie on the imaginary axis.

(*v*) There should be no missing terms in $P(s)$ and $Q(s)$ unless all even or odd terms are missing.

(*vi*) The poles and zeros of $F(s)$ must be in the left half of s-plane.

(*vii*) These should not be multiple poles or zeros of $F(s)$ either at origin ($s = 0$) or at infinity ($s = \infty$).

(*viii*) The simples poles $F(s)$ on the imaginary axis ($j\omega$) should have real and positive residues.

7.3.2.2 Necessary and Sufficient Conditions

(*i*) $F(s)$ must be real when 's' is real.

(*ii*) It $F(s) = \frac{Q(s)}{P(s)}$, then $p(s) + Q(s)$ must be Hurwitz polynomial. For $p(s) + Q(s)$ to be Hurwitz polynomial, all the quotients in the continued fraction expansion must be positive. In addition, the continued fraction expansion must not terminate abruptly.

(*iii*) $Re[F(j\omega)] \geq 0$ for all ω.

For checking this condition, seperate $P(s)$ and $Q(s)$ polynomials into even and odd parts.

$$F(s) = \frac{Q(s)}{P(s)} = \frac{m_1(s) + n_1(s)}{m_2(s) + n_2(s)} = \frac{m_1 + n_1}{m_2 + n_2}$$

where $m_1(s)$ = even function and $n_1(s)$ = odd functions.

or
$$f(s) = \frac{m_1 \times n_1}{m_2 + n_2} \times \frac{m_2 - n_2}{m_2 - n_2}$$

$$= \frac{(m_1 m_2 - n_1 n_2) + (m_2 n + m_1 n_2)}{m_2^2 - n_2^2}$$

$$= \frac{m_1 m_2 - n_1 n_2}{m_2^2 - n_2^2} + \frac{m_2 n_1 - m_1 n_2}{m_2^2 - n_2^2}$$

It may be noted that:

(*i*) Product of two even functions is another even function.

(*ii*) Product of two odd functions is an even function.

(*iii*) Product of an even function and an even function is an odd function.

Hence, $m_1 m_2$ is an even function.

$n_1 n_2$ is an even function.

$m_1 n_2$ is an odd function.

and $m_2 n_1$ is also an odd function.

m_1^2 is an even function.

m_2^2 is an even function.

Thus, the even part of $F(s) = \dfrac{m_1 m_2 - n_1 n_2}{m_2^2 - n_2^2}$

and the odd part of $F(s) = \dfrac{m_2 n_1 - m_1 n_2}{m_2^2 - n_2^2}$

The substitution of $s = j2$ in the even part given the real part of $F(j\omega)$ and the substitution of $s = j\omega$ is odd part given imaginary part of $F(s)$.

Thus, if $\qquad F(j\omega) = Re\ F(j\omega) + jIm\ F(j\omega)$

then $\qquad Re\ F(j\omega) = \text{Even}\ [F(s)]|_{s = j\omega}$

and $\qquad j\ I_m\ F(j\omega) = \text{Odd}\ [F(s)]|_{s = j\omega}$

Thus, to test for the third condition for positive realness, the real part of $f(j\omega)$ is determined by finding the even part of $F(s)$ and patting $s = j\omega$. Subsequently check to see whether $Re[F(j\omega)] \geq 0$ for all values of ω.

The denominator of $Re[F(j\omega)]$ is always a positive quantity *i.e.*,

$[m_2(j\omega)]^2 - [n_2(j\omega)]^2 = m_2(\omega_2) + n_2(\omega_2) \geq 0$

This is because, there is an extra imaginary in $n_2(j\omega)$, which when squared given – 10 Hence, the denominator of $Re\ [F(j\omega)]$ which is the sum of two squared numbers, is always positive. Thus, we just need to determine the numerator.

$$A(\omega_2) = m_1\ (j\omega)\ m_2\ (j\omega) - n_1\ (j\omega)\ n_2\ (j\omega) \geq 0$$

i.e., whether $A\omega_2$ will be positive or zero but never negative. Generally, the condition can be verified by factorizing $A(\omega_2)$. It factorization is not sufficient for the required conclusion, then $A(\omega_2)$ is plotted over sufficently large range of ω and it is ensured that is not negative. In short, the testing procedure for positive realness of a function arc:

(*i*) Ensure that all the necessary conditions are met by inspection.

(*ii*) Ensure that all the poles and zeros of $F(s)$ functions are in the left half of s-plane or on the imaginary ($j\omega$) axis *i.e.*, $p(s)$ and $Q(s)$ polynomials are Hurwitz polynomials.

(*iii*) In case, $F(s)$ function has poles on imaginary axis, the residues at the poles must be real and positive *i.e.*,

find the partial fractions of $F(s)$ where $s = \pm\ j\omega_0$ is the pair of poles on imaginary ($j\omega$) axis:

$$F(s) = ...,\ + \frac{k}{s - j\omega_0} + \frac{k*}{s + j\omega_0}$$

k and $k*$ are residues of complex conjugates and these must be real and positive.

(*iv*) Last but not least test:

$$A(\omega^2) = (m_1\ m_2 - n_1\ n_2)|_{2\,=\,j\omega} \geq 0 \text{ for all } \omega \geq 0.$$

The above can be tested by factorizing $A(\omega^2)$.

7.4 SYNTHESIS OF DRIVING POINT IMMITANCE FUNCTIONS

If the immitance function is positive real function, then it can be synthesized to obtain an electrical network using passive elements *i.e.*, *R*, *L*, *C*.

A network can be synthesized generally in two forms namely Foster form and Cauer form. Further, Foster forms can be Forster I form and Foster II form. Similarly, Cauer forms can be Cauer I form and Cauer II form.

$$Z(s) = \frac{k_0}{s} + \frac{2\,k_2\ s}{s^2 + \omega_2^2} + ...,\ + \frac{2k_i s}{s^2 + \omega_i} + k\ \infty\ s$$

If the given $L - C$ function $F(s)$ in an impedance function $Z(s)$, then it can be realized in the Foster *I* form. The network consists $L - C$ elements as shown in the Fig. 7.1.

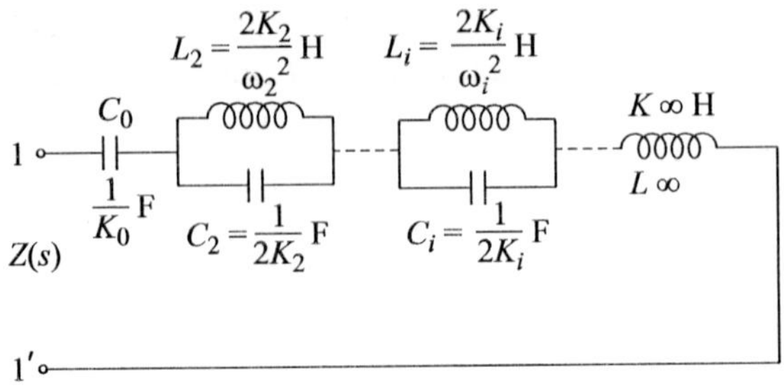

Fig. 7.1 *Foster I form of LC network*

The figure circuit elements are decided as follows:

$\frac{k_0}{s}$ = represents a capacitor of value $c_0 = \frac{1}{k_0} F$.

$K_\infty s$ = represents an inductor of value $L_\infty = K_\infty$

$\frac{2k_i s}{s^2 + \omega_i^2}$ = represents a capacitor of value $C_i = \frac{1}{2k_i} F$

C_0 will be present only of there is a pole at the origin and L_∞ will be present, if there is a zero at infinity.

7.4.1 Foster II Form

Consider the driving point admittance is the function to be realized, then,

$$y(s) = \frac{k_0}{s} + \frac{2k_2 s}{s^2 + \omega_2^2} + \ldots, + \frac{2k_i s}{s^2 + \omega_c^2} + k_\infty s$$

It the given $L - C$ function $y(s)$ is an admittance function $y(s)$, then it can be realized in the Foster II form as shown in the Fig. 7.2.

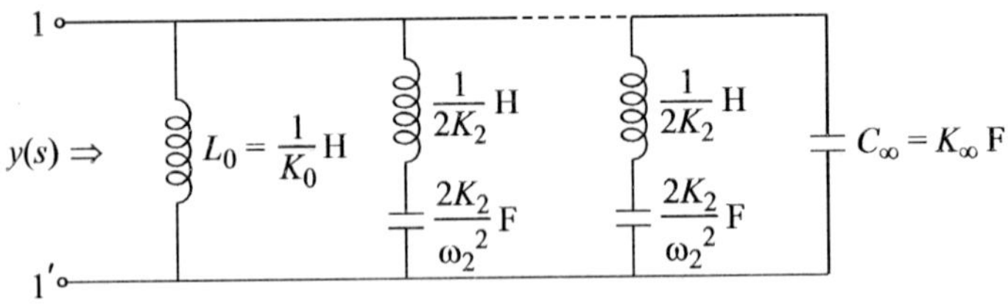

Fig. 7.2 *Foster II form of L C network*

As the function is an admittance function, the partial fraction of $y(s)$ gives a network consisting of parallel branches. L_0 is present, if $y(s)$ has a pole at $s = 0$ and C_∞ is present in $y(s)$ has a pole at ∞.

7.4.2 Cauer I Form

Cauer *I* form realization $z(s)$ is in the form of a ladder network as shown in the Fig. 7.3.

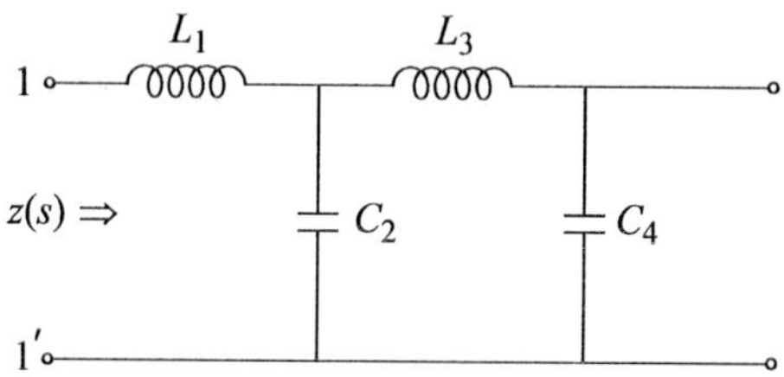

Fig. 7.3 *Cauer I form of L – C network*

The general pattern shall be like a ladder network as shown in Fig. 7.4.

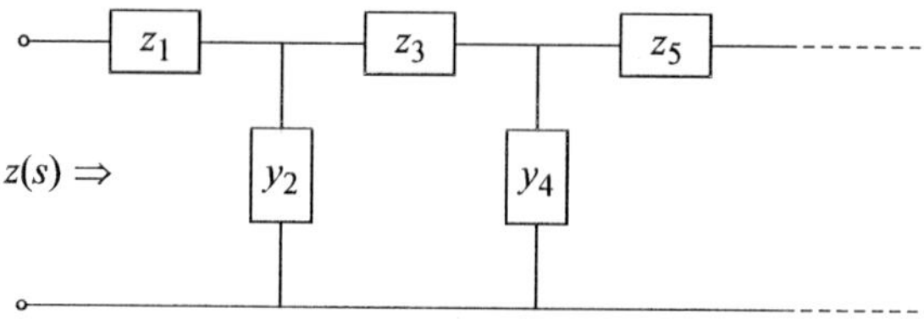

Fig. 7.4 *A general ladder network*

The impedance function is given by:

$$Z(s) = z_1(s) + \cfrac{1}{y_2(s) \cfrac{1}{z_3(s) + \cfrac{1}{z_3(s) + \cfrac{1}{z_4(s) + \cfrac{1}{z_s(s)}}}}}$$

It $z(s)$ function has no pole at infinity, then it has a zero there. In such case, the function is first inverted and then steps are taken as usual. Hence, the nature of first element of the network is fixed by the nature of $z(s)$ at $s = \infty$. If $z(s)$ has a pole at $s = \infty$, the first element is an inductor in series arm, and if it has a zero at $s = \infty$, the first element in a capacitor in short arm.

7.4.3 Cauer II Form

If the first series element in Cauer I form is a capacitor and subsequently a parallel element shall be an inductor, and so on. The $L-C$ network so obtained, is known as Cauer form which is shown in Fig. 7.5.

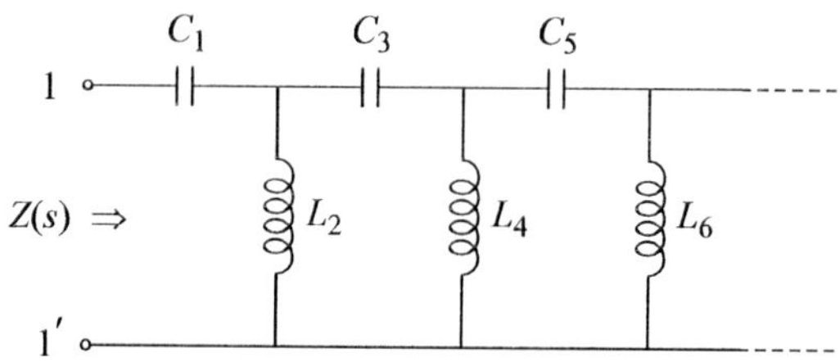

Fig. 7.5 *Cauer II form R C network*

$$Z(s) = \cfrac{1}{c_1 s} + \cfrac{1}{\cfrac{1}{L_2 s} + \cfrac{1}{\cfrac{1}{c_3 s} + \cfrac{1}{\cfrac{1}{c_3 s} + \cfrac{1}{L_4 s + \ldots}}}}$$

The continued fraction expansion decides the values of c_1, c_3, c_5 and L_2, L_4, L_6, etc. In order to obtain Cauer *II* form, the *LC* immittance function is arranged in ascending order of *s*.

$$Z(s) = \frac{b_0 + b_2\, s^2 + \ldots, + b_{n-2}\, s^{n-2} + \phi_n\, s^n}{a_1\, s + a_3\, s^3 + \ldots, + a_{m-2}\, s^{m-2} + q_{m-s}{}^{m}}$$

where $n - m = \pm\, 1$

The lest power of the numerator is divided by the lowest power of the denominator, and then we invert the remainder and divide again.

Foster and Cauer forms of networks give the minimum number of elements for a specified L – C driving point function. These localizations are known as Canonical forms.

7.5 SYNTHESIS OF DRIVING POINT IMMITANCE FUNCTION

LC Network: LC network is also known as reactance or loss less network. The driving point immitance function of LC network can be realized if the function meets the properties of LC driving point immittance function.

Properties of LC Driving Point Immitance Functions

(*i*) The immittance function is a ratio of even to odd or odd to even polynomials.

(*ii*) The highest powers of numerator and denominator must differ by unity.

(*iii*) The lowest powers of numerator and denominator must differ by unity only.

(*iv*) The poles and zeros are simple and are located on $j\omega$ axis only. There are no multiple poles or zeros either at origin or infinity or at any point.

(*v*) The poles and zeros interface (alternate) on the $j\omega$ axis.

(*vi*) The imaginary poles and zeros occur in the form of complex conjugate pairs.

(*vii*) The residues at $j\omega$ axis poles are real and positive.

(*viii*) These must be a either a pole or zero at origin and the infinity.

(*ix*) The number of elements required in any of the four forms of realization is equal to the highest power of '*s*' in the immitance function as a whole.

(*x*) The slope of the reactance versus plot is always positive.

7.5.1 Realization of Immitance Function of LC Networks

The realization of driving immitance functions of *LC* networks are done in any of the four forms *i.e.*, Foster I form, Foster II form, Cauer I form or Cauer II form.

Properties of *RC* driving point Impedance functions

(*i*) The poles and zeros of function $Z_{RC}(s)$ are simple and these are no multiple poles and zeros.

(*ii*) The poles and zeros are located on negative real axis.

(*iii*) The poles and zeros interlace (alternate) on the $j\omega$ axis.

(*iv*) The singularity nearest to or at the origin is a pole.

(*v*) The singularity nearest to or at infinity is a zero.

(*vi*) The partial fraction expansion of the function has real and positive residues.

(*vii*) There is no zero at the origin.

(*viii*) There is no pole at infinity.

(*ix*) The slope of the $Z(\sigma)$ versus *s* plot is negative $d\sigma$ the functions *i.e.*, $dZ_{Rc}(\sigma)$ is negative.

(*x*) The value of at $s = 0$ is greater than the value of functions at $s = \infty$.

7.6 SYNTHESIS OF DRIVING POINT IMPEDANCE FUNCTIONS

RC Networks: The realization of driving point impedance $F(s)$ function requires that meet the properties of driving point functions.

7.6..1 Realization of Impedance Function of RC Network

$Z_{RC}(s)$ functions can be realized in any of the four forms *i.e.*, Foster I form, Foster II form, Cauer I form and Cauer II form. It is important to note that the number of elements (*R* and *C*) are not equal to the highest power of '*s*' in $Z_{RC}(s)$ for *RC* networks.

7.7 SYNTHESIS OF DRIVING POINT IMPEDANCE FUNCTIONS OF RL NETWORKS

$Z_{RL}(s)$ must meet properties of RL driving point impedance functions.

7.7.1 Properties of RL Driving Point Impedance

(*i*) Z_{RL} function poles and zeros are simple and these are no multiple poles and zeros.

(*ii*) The poles and zeros are located on negative real axis.

(*iii*) The poles and zeros interlace (alternate) on the negative real axis.

(*iv*) The singularities nearest to or at the origin is a pole.

(*v*) The singularity nearest to or at infinity is a zero.

(*vi*) The partial function expansion residues of $Z_{RL}(s)$ are negative and real, therefore, $\frac{Z_{RL}(s)}{s}$ residues are obtained which are positive and real.

(*vii*) No pole at the origin.

(*viii*) The value of $Z_{RL}(s)$ at $s = 0$ is always less than the value of $Z_{RL}(s)$ at $s = \infty$.

(*ix*) The slope of Z_{RL} (σ) varsus σ plot is positive *i.e.*, $\frac{d_1 Z_R(\sigma)}{d\sigma}$ is positive.

7.7.2 Realization of Impedance Function of RL Network

$Z_{RL}(s)$ must have the properties $Z_{RL}(s)$, and then it can be realized in all four forms *i.e.*, Foster I, foster II, Cauer I and Cauer II forms. The number of elements (RL) are not equal to highest power of 's' in the realized network.

SUMMARY

1. **Network Synthesis:** Network synthesis is exactly opposite of the network analysis. It is to find the network satisfying a known excitation and known response. It is important to note that analysis given response which is always, unique, whereas, synthesis does not given unique solution.
2. **Hurwitz Polynomials:** A Hurwitz polynomial has to satisfy the following basic properties:
   (*i*) The polynomial should be real for real 's'.
   (*ii*) The roots of the polynomial should have real parts which are either zero or negative. These polynomials have to satisfy many more properties.
3. **Positive Real Functions:** If the driving point immittance (impedence or admittance) is a positive real function, then it is physically realizable using passive R, L and C components. Therefore, immittance functions

must be checked for being positive real, before synthesizing. A positive real function satisfies the following conditions:

(*i*) The given function $F(s)$ is real for real 's',

(*ii*) The real part of $F(s)$ is greater than or equal to zero, when the real part of 's' is greater than or equal to zero

$$i.e., Re[F(s)] \ ^3\ 0 \text{ for } Re\ [s]\ ^3\ 0$$

(*iii*) The functions $F(s)$ is rational function.

There are some more properties of positive real function, in addition to above basic properties.

4. **Synthesis of Driving Point Immitance Functions:** If the immitance function is positive real function, then it can be synthesized to obtain an electrical network using passive elements *i.e.*, R, L, C.

EXERCISES

7.1. Check whether the polynomial

$$s^7 + 7s^6 - 3s^3 + 2s^4 - s^3 + s^2 + s + 1$$

is Hurwitz or not.

7.2. Check whether the polynomials are Hurwitz or not

(*i*) $s^4 + 2s^2 + 4s + 10$

(*ii*) $s^6 + 7s^5 + 9s^4 + s^3 + s^2 + 2s$

7.3. Check whether a polynomial expressed as:

$$q(s) = s^2 + 6s^2 + 11s + 6 \text{ is Hurwitz or not}$$

7.4. Test another the following function a.p.r.f or not.

$$F(s) = \frac{s^2 + 2s^2 + 3s + 5}{s^2 + 6s + 8}$$

7.5. Test whether the following functions are positive real function or nat.

(*i*) $F(S) = \dfrac{s^2 + s + 6}{s^2 + s + 1}$ (*ii*) $F(S) = \dfrac{s^2 + 6s + 5}{s^2 + 9s + 14}$.

7.6. Determine whether the function $z(s) = \dfrac{2s^2 + 5}{s(s^2 + 4)}$ is positive real or not.

7.7. Realize the function

$$Z(s) = \frac{(s^2 + 1)\,(s^2 + 4)}{s(s^2 + 2)}$$

(*a*) Foster-I and IInd form

(*b*) Cauer-I and IInd form.

7.8. An Impedance function is given by

$$Z(s) = \frac{(s+1)(s+3)(s+5)}{s(s+2)(s+4)}$$

Find the R-C representation of

(*a*) Foster-I and II forms.

(*b*) Cauer-I and II forms.

7.9. Given $Z(s) = \dfrac{2(s+1)(s+3)}{(s+2)(s+6)}$

Synthesize it in (*a*) Foster I and II forms

(*b*) Cauer I and II form.

7.10. Synthesis the R–L driving point impedance $Z(s)$ to get Foster I and II forms.

$$Z(s) = \frac{s(s+3)}{(s+2)(s+4)}$$

7.11. Realize that the following RC driving point function in (*i*) First Foster Form (*ii*) First Cauer Form.

$$Z(s) = \frac{(s+1)(s+3)}{s(s+2)}$$

7.12. An impedance function is given by

$$Z(s) = \frac{2(s+1)(s+3)}{(s+2)(s+4)}$$

Find the R-L representation of Foster-first form.

❑❑❑

8

Filters

8.1 INTRODUCTION

Filters are used in the complex communication systems. They are types of networks, the designers generally use. All these networks may be realized in various forms of a four terminal network. A four terminal network is a network having one input port and one output port. In case when electrical properties of the network remain unchanged even after interchanging input and output terminals, the network is known as symmetrical network. When the electrical properties of the network are changed after interchanging input and output terminals the network is called asymmetrical network.

A network which *freely passes definite band of frequencies* which *almost suppresses other band of frequencies* is "*known as filter*". The range of frequencies over which attenuation by filter is zero, is known as pass band. The *range of frequencies* over which *attenuation* is *infinite* is known as *stop band or attenuation band or band elimination filter.*

8.2 IMAGE PARAMETERS

Image parameters are defined for two-port networks. There are two parameters very useful is filters namely image impedance (Z_i) and image transfer constant (e^{θ}).

8.2.1 Image Impedance

A two-port network is shown in Fig. 8.1 and image impedances are denoted by Z_{i_1} and Z_{i_2}

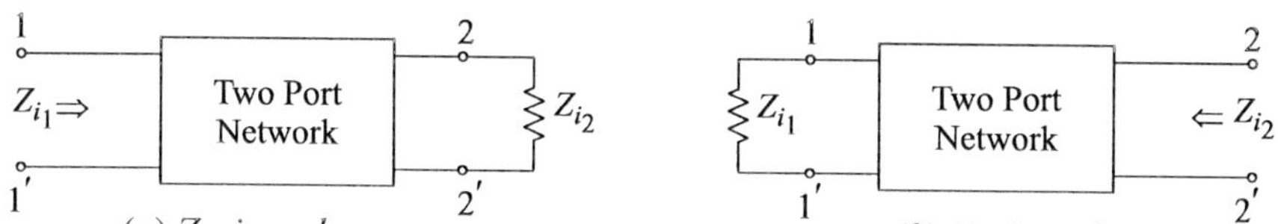

(*a*) Z_{i_1} *impedance* (*b*) Z_{i_2} *impedance*

Fig. 8.1 *Image impedences.*

For port 1 and port 2 respectively. Suppose the network is terminated with image impedance of port 2 *i.e.* Z_{i_2} at output pair of terminals, then the impedance measured at its input pair of terminals will be image impedance of port 1 *i.e.* Z_{i_1}.

Similarly, if port one of the network is terminated with image impedance of port 1 *i.e.* Z_{i_1}, then the impedance measured at port 2 will be image impedance of port 2 *i.e.* Z_{i_2}. Fig. 8.1(*a*) and (*b*) illustrate these conditions.

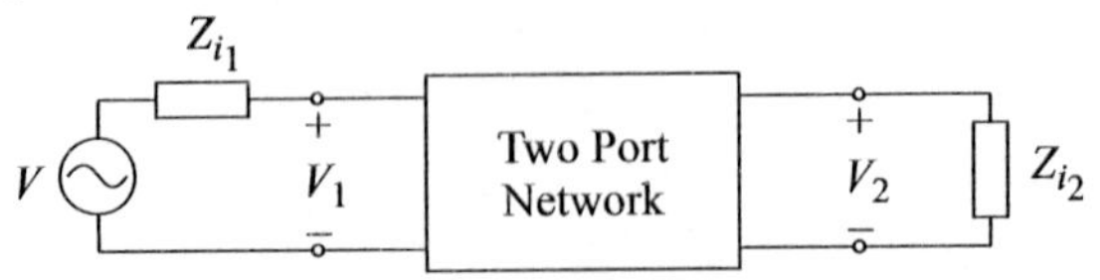

Fig. 8.2 *Properly terminated network.*

8.2.2 Image Transfer Constant

When a network is properly terminated in its image impedances at the both ports as shown in Fig. 8.2, then the *image transfer constant is defined as* follows:

$$e^{\theta} = \sqrt{\frac{V_1}{V_2} \cdot \frac{I_1}{I_2}} \quad \text{...(8.1)}$$

The real port of image transfer constant is known as image attenuation constant, and the imaginary part is known as image phase constant.

8.3 CHARACTERISTIC IMPEDANCE

Characteristic impedance has *relevance where infinite number of identical* networks are connected *in cascade* as shown in Fig. 8.3 (a).

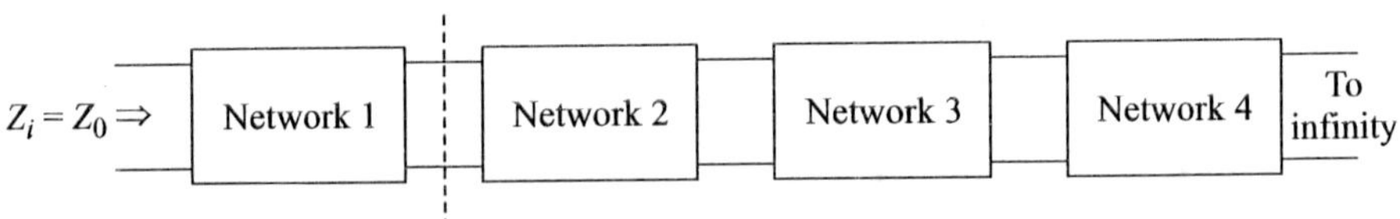

(*a*) *Input impedance at input terminals of the first network of chain of infinite networks*

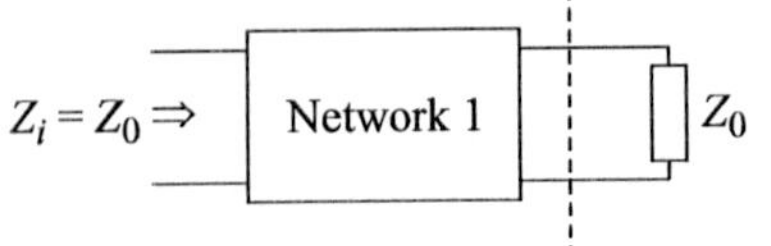

(*b*) *Input impedance of first network terminals impedance*

Fig. 8.3 *Chain of infinite networks and characteristic impedance.*

The *characteristic impedance* (Z_0) of a network is the *impedance measured at the input terminals of the first network* in the chain of infinite networks in cascade. If the chain of infinite networks is reduced by network, the characteristic impedance remains unattached. Thus, the chain of infinite networks can be *replaced by an impedance* Z_0 and connected to the output *terminals* of first network as shown in Fig. 8.3 (*b*). In short, when any network is terminated in its characteristic impedance Z_0, the input impedance will also be Z_0. The network terminated in characteristic impedance at input as well as output terminals known as properly terminated network as shown in the Fig. 8.4.

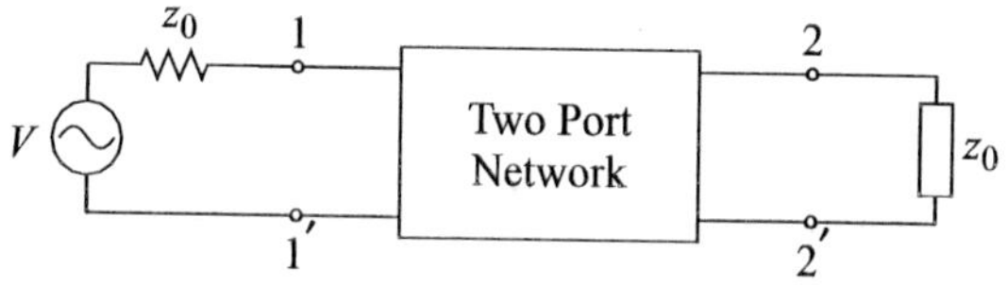

Fig. 8.4 *Properly terminated two-port network.*

8.3.1 Characteristic Impedance of Symmetrical T-network

A symmetrical T-network is shown in Fig. 8.5. It has input and output arm of equal impedance,

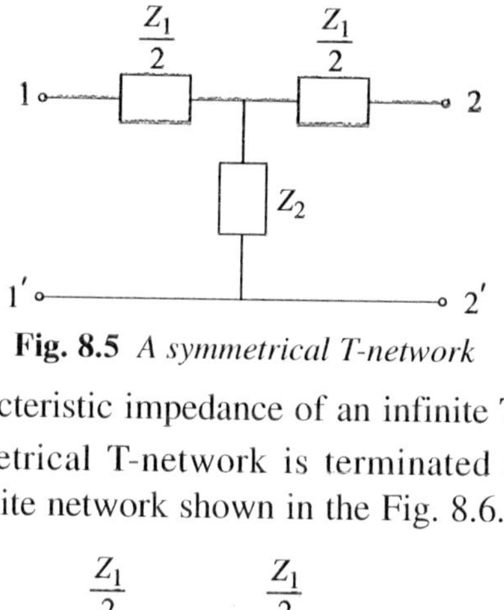

Fig. 8.5 *A symmetrical T-network*

Suppose the characteristic impedance of an infinite T-networks is Z_0.

Now, first symmetrical T-network is terminated in Z_0, characteristic impedance of the infinite network shown in the Fig. 8.6.

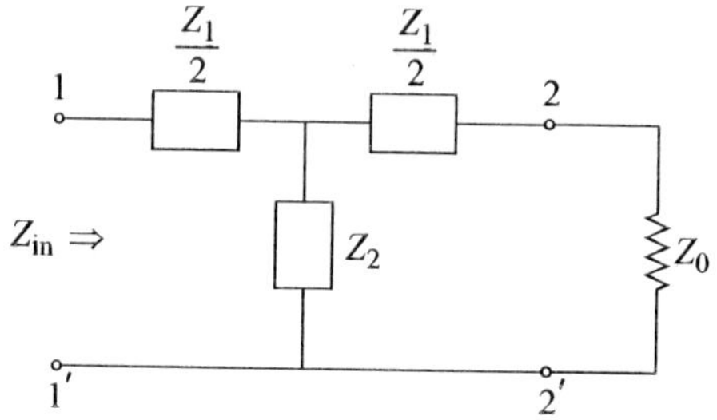

Fig. 8.6 *First T-network terminated in Z_0*

The input impedance of the network is given by:

$$Z_{in} = \frac{Z_1}{2} + \left[Z_2 \parallel \left(\frac{Z_1}{2} + Z_0\right)\right]$$

We know that by definition: $Z_{in} = Z_0$

$$\therefore \qquad Z_0 = \frac{Z_1}{2} + \frac{Z_2\left(\frac{Z_1}{2} + Z_0\right)}{Z_2 + \frac{Z_1}{2} + Z_0}$$

or $$Z_0\left(Z_2 + \frac{Z_1}{2} + Z_0\right) = \frac{Z_1}{2}\left(Z_2 + \frac{Z_1}{2} + Z_0\right) + \frac{Z_1 Z_2}{2} + Z_2 Z_0$$

or $$Z_2 Z_0 + \frac{Z_1 Z_0}{2} + Z_0^2 = \frac{Z_1 Z_2}{2} + \frac{Z_1^2}{4} + \frac{Z_1 Z_0}{2} + \frac{Z_1 Z_2}{2} + Z_2 Z_0$$

or $$Z_0^2 = \frac{Z_1^2}{4} + Z_1 Z_2$$

or $$Z_0 = \sqrt{\frac{Z_1^2}{4} + Z_1 Z_2} \qquad \text{...(8.2)}$$

8.3.2 Characteristic Impedance of Symmetrical π-network

A symmetrical π-network is shown in Fig. 8.7. It has side equal impedance arms in input and output side ports.

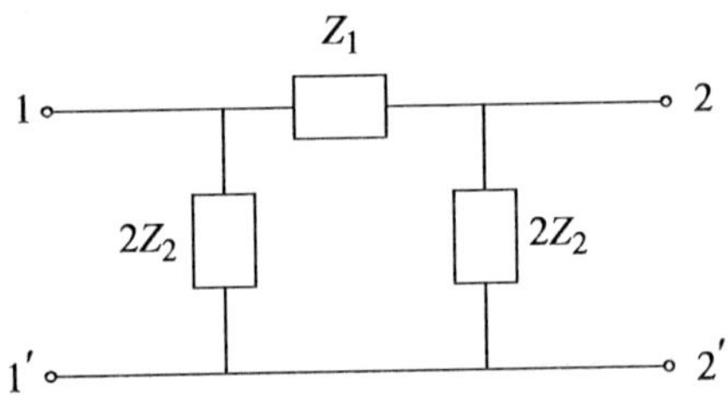

Fig. 8.7 *A symmetrical π-network*

Suppose the characteristic impedance of an infinite π-network is Z_0.

Now, first symmetrical π-network is terminated in Z_0 characteristic impedance of the infinite network, as shown in the Fig. 8.8.

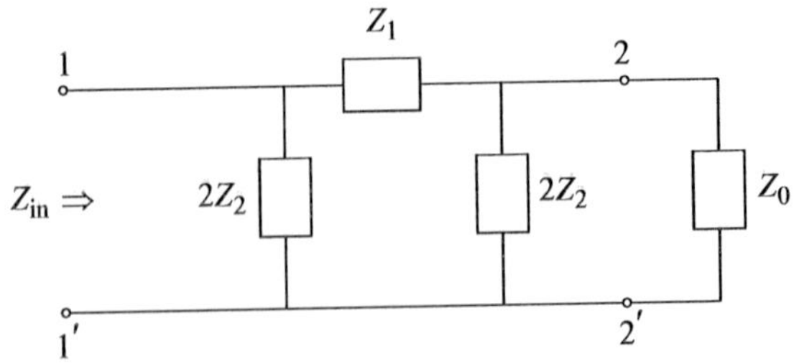

Fig. 8.8 *First network terminated in Z_0*

The input impedance is given by:

$$Z_{in} = \left[\{(2Z_2 \parallel Z_{L0}) + Z_1\} \parallel 2Z_2\right]$$

We know that: $Z_{in} = Z_0$

$$\therefore \quad Z_0 = \left(Z_1 + \frac{2Z_2\, Z_0}{2Z_2 + Z_0}\right) \parallel 2Z_2$$

or
$$Z_0 = \left(\frac{2Z_1Z_2 + Z_1Z_0 + 2Z_2\, Z_0}{2Z_2 + Z_0}\right) \parallel 2Z_2$$

or
$$Z_0 = \frac{(2Z_2)\left(\dfrac{2Z_1Z_2 + Z_1Z_0 + 2Z_2\, Z_0}{2Z_2 + Z_0}\right)}{2Z_2 + Z_0}$$

or
$$Z_0 = \frac{2Z_2\,(2Z_1Z_2 + Z_1Z_0 + 2Z_2\, Z_0)}{2Z_2\,(2Z_2 + Z_0) + 2Z_1\, Z_2 + Z_1\, Z_0 + 2Z_2\, Z_0}$$

or
$$4Z_2^{\,2}\, Z_0 + 2Z_2\, Z_0^{\,2} + 2Z_1\, Z_2\, Z_0 + Z_1\, Z_0^{\,2} + 2Z_2\, Z_0^{\,2} = 4Z_1\, Z_2^{\,2} + 2Z_1\, Z_2\, Z_0 + 4Z_2^{\,2}\, Z_0$$

or
$$4Z_2\, Z_0^{\,2} + Z_1\, Z_0^{\,2} = 4Z_1\, Z_2^{\,2}$$

or
$$Z_0^{\,2} = \frac{4Z_1\, Z_2^{\,2}}{Z_1 + 4Z_2}$$

or
$$Z_0^{\,2} = \frac{Z_1^{\,2}\, Z_2^{\,2}}{\dfrac{Z_1^{\,2}}{4} + Z_1\, Z_2}$$

or
$$Z_0 = \frac{Z_1\, Z_2}{\sqrt{\left(\dfrac{Z_1}{2}\right)^2 + Z_1\, Z_2}}$$

The characteristic impedance of T-network can be written as:

$$Z_{OT} = \sqrt{\left(\frac{Z_1}{2}\right)^2 + Z_1\, Z_2}$$

and similarly, characteristic impedance of π-network can be written as:

$$Z_{o\pi} = \frac{Z_1\, Z_2}{\sqrt{\left(\dfrac{Z_1}{2}\right)^2 + Z_1\, Z_2}}$$

The relationship between T-network characteristic impedance and that of π-network is:

$$Z_{o\pi} = \frac{Z_1 Z_2}{Z_{OT}} \quad ...(8.3)$$

8.4 BASIC TYPES OF FILTERS AND RELATED TERMS ARE AS FOLLOWS

(*i*) *Pass band filters:* It has a band range of frequencies over which attenuation by filter is zero.

(*ii*) *Stop band filters:* It has a band range of frequencies over which attenuation is infinite. It is also known as "attenuation band" filter. The discussion of filters will be confined to filters using symmetrical unbalanced T or π sections as shown in the Fig. 8.7.

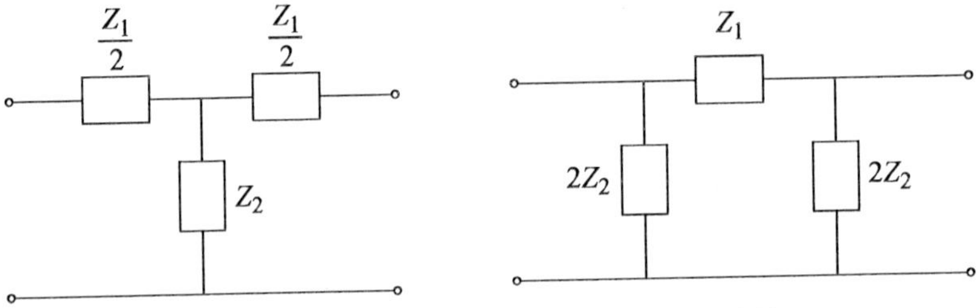

Fig. 8.7 *Symmetrical unbalanced filters.*

Type of pass band filters are low pass filter, high pass filter and band pass filter. Low pass filter allows signals below a frequency known as critical frequency as shown in Fig. 8.8 (*a*). The high pass filter allows signals with

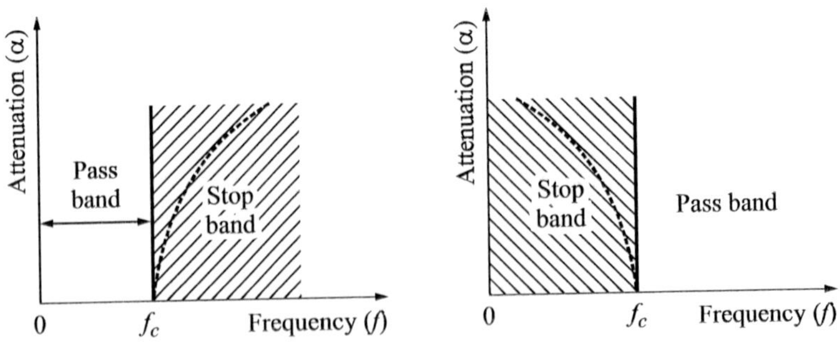

(*a*) Low pass filter (*b*) High pass filters

Fig. 8.8 *Low pass high pass, band filter characteristics*

more than a critical frequency as shown in Fig. 8.8 (*b*). The band pass filter allows signals with frequencies in the range of f_1 to f_2 as shown Fig. 8.9 (*a*).

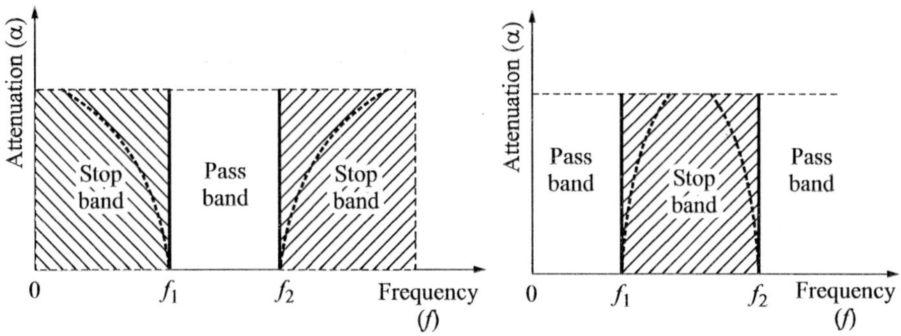

(*c*) Band pass filter (*d*) Band elimination factor

Fig. 8.9 *Pass band and stop band filter characteristics.*

Fig. 8.9 (*b*) shows characteristics of band stop or band elimination filter. It stops signals within frequency range of f_1 to f_2.

8.5 FILTER FUNDAMENTALS

All filters are constructed from *purely reactive elements* so *that attenuation can become* zero. It has been observed from *T* and π sections that the *characteristic impedance depends* on the reactances of Z_1 and Z_2 offered by purely reactive elements used in series and shunt arms of a filter. Thus, the characteristic impedance Z_0 varies with frequency as Z_1 and Z_2 both way with frequency.

Z_0 can be either real or imaginary. Suppose Z_0 is real over the range of frequency. Such filter along with its terminating impedance will absorb power from any power source connected to it. As the filter is purely active, hence, it can absorb power itself. Thus, all the power delivered by the power source is passed to the load. *i.e.* there is no attenuation. This means pass band.

Now, suppose Z_0 is imaginary or purely reactive, the filter and its termination can not absorb any power *i.e.* no power is transferred to the load. Thus, attenuation is very high and ideally attenuation is infinite. This means stop band, cut-off frequency is defined as the frequency at which Z_0 changes form real to being imaginary.

We know for *T*-section:

$$Z_{OT} = \sqrt{\frac{Z_1^{\ 2}}{4} + Z_1 Z_2}$$

It the filter *i.e.* *T*-network is purely reactive, then

We get:

$$Z_{OT} = \sqrt{\frac{(SX_1)^2}{4} + (SX_1)(SX_2)}$$

$$= \sqrt{-\frac{X_1^{\ 2}}{4} - X_1 X_2}$$

or $$Z_{OT} = \sqrt{-X_1\left(\frac{X_1}{4} + X_2\right)} \qquad ...(8.4)$$

i.e. Z_0 is purelyimaginary X_1 and X_2 have the same sign, thus, it gives stop band.

Z_0 is purely real if X_1 and $\frac{X_4}{4} \times X_2$ have the opposite signs, thus it gives pass band.

If we draw the reaction X_1 and $\frac{X_1}{4} \times X_2$ *versus* frequency, cut-off frequency can be calculated. The cut-off frequency is obtained with consideration which follows:

The bond of frequencies for which the curve lie on opposite side of frequency axis is pass band, while the band frequencies for which the curves lie on the same side of the frequency axis is stop band. The change over point from pass band to stop to stop band gives cut-off frequency.

8.6 PASSIVE FILTERS

The filters using passive components inductors (L) and capacitors. The filters discussed so far are low pass filter, high pass filter, band pass filter and stop filters. All these are using inductors (L) and capacitors (C), therefore, are known as passive filters. Now, the passive filters are being taken for detailed discussion.

8.6.1 Low Pass Filters

Low pass filters of T and π types are shown in Fig. 8.10.

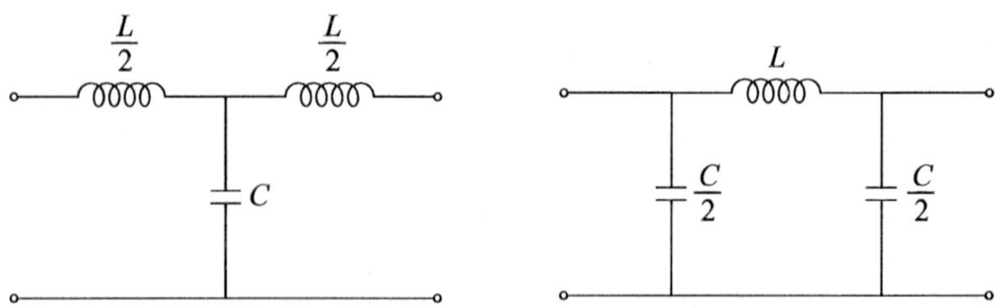

(*a*) *T type low pass filter using inductor and capacitor*

(*b*) *π type low pass filter using inductors and capacitors*

Fig. 8.10 *Low pass filters using inductors and capacitors.*

8.6.1.1 Impedance R_0

In terms of impedances, the low pass filters is shown in Fig. 8.11.

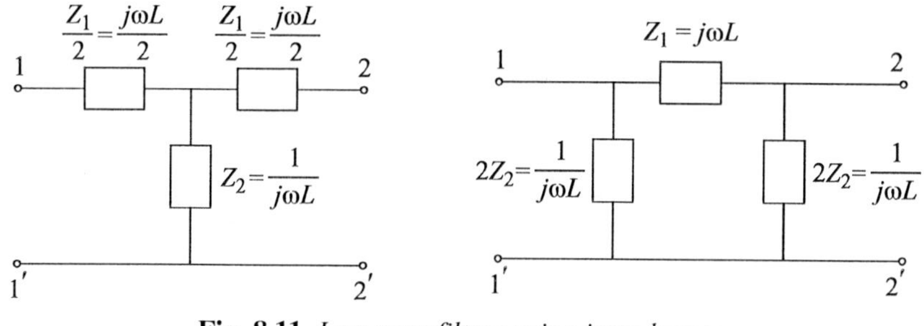

Fig. 8.11 *Low pass filters using impedances.*

The impedances are given by:

$$Z_1 = j\omega L \text{ and } Z_2 = \frac{1}{j\omega c} - \frac{-j}{\omega c}$$

$\therefore$ $$Z_1 Z_2 = (j\omega L)\left(\frac{-j}{\omega c}\right)$$

or $$Z_1 Z_2 = \frac{L}{C} \text{ i.e. a real constant}$$

$\therefore$ $$R_0 = \sqrt{\frac{L}{C}} \text{ i.e. a real constant}$$

$\therefore$ $$R_0 = \sqrt{\frac{L}{C}} \qquad \text{...(8.5)}$$

8.6.1.2 Cut-off Frequency f_c

T and π networks taken here are equivalent, hence, both the filters have same cut-off frequency f_c. Thus, cut-off frequency for T-filter can be easily obtained.

(*i*) Graphical Method

We know that:

$$Z_1 = j\omega L \text{ i.e. } X_1 = \omega L$$

$$Z_2 = \frac{-j}{\omega c} \text{ i.e. } X_2 = -\frac{1}{\omega c}$$

The plot of reactance versus frequency is shown in Fig. 8.12.

It can be noted form the characteristics that all reactance curves have positive slope. The curves upto point '*a*' are on positive as well as negative of reaction (*X*) axis, whereas from point '*a*' onwards all curves are on positive side of reactance axis. This means all the frequencies upto intersection point '*a*' give pass band and beyond point '*a*' give stop band. The intersection point '*a*' gives cut-off frequency f_c.

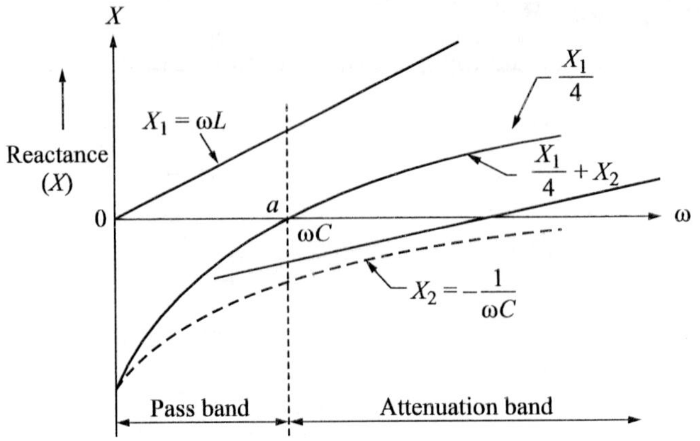

Fig. 8.12 *Plot of reactance versus frequency for low pass filter.*

Hence, at intersection point '*a*' where the curve $\frac{X_1}{4} + X_2$ crosses the frequency axis, we get:

$$\omega = \omega_c \text{ and } \frac{\omega_c L}{4} - \frac{1}{\omega_c C} = 0$$

or $$\frac{\omega_c L}{4} = \frac{1}{\omega_c C}$$

or $$\omega_c^2 = \frac{4}{LC}$$

or $$\omega_c = \frac{2}{\sqrt{LC}}$$

As $\omega_c = 2\pi f_c$, hence, we get:

or $$f_c = \frac{1}{\pi\sqrt{LC}} \qquad \text{...(8.6)}$$

We know that: $$Z_{0T} = \sqrt{\frac{Z_1^2}{4} + Z_1 Z_2}$$

We also know that: $Z_1 = j\omega L$ and $Z_1 Z_2 = \frac{L}{C}$

By substitution, we get

$\therefore$ $$Z_{0T} = \sqrt{\frac{(j\omega L)^2}{4} + \frac{L}{C}}$$

or $$Z_{0T} = \sqrt{-\frac{\omega^2 L^2}{4} + \frac{L}{C}} = \sqrt{\frac{L}{C}}\sqrt{1 - \frac{\omega_2 LC}{4}}$$

or $$Z_{0T} = R_0 \sqrt{1 - \frac{\omega^2 LC}{4}} \text{ as } \sqrt{\frac{1}{C}} = R_0$$

For Z_{0T} to be real $\frac{\omega^2 LC}{4} < 1$ and to be imaginary $\frac{\omega^2 LC}{4} > 1$.

For critical frequency ω_c $Z_{OT} = 0$

or $$\sqrt{\frac{L}{C}} \sqrt{1 - \frac{\omega_c^{\,2} LC}{4}} = 0$$

or $$\omega_c^{\,2} = \frac{4}{LC}$$

or $$\omega_c = \frac{2}{\sqrt{LC}}$$

As $\omega_c = 2\pi f_c$, we get:

$$f_c = \frac{1}{\pi\sqrt{LC}}$$

8.6.1.3 Variation of Z_{OT} and Z_{Op} with Frequency

We know that $Z_{OT} = R_O \sqrt{1 - \frac{\omega_2 LC}{4}}$

By substituting $\frac{1}{\sqrt{LC}} = \omega_c$, we get:

$$Z_{0T} = R_O \sqrt{1 - \left(\frac{\omega}{\omega c}\right)^2}$$

or $$Z_{0T} = R_O \sqrt{1 - \left(\frac{f}{f_c}\right)^2} \quad ...(8.7)$$

Similarly, for $Z_{O\pi}$: or $$Z_{0\pi} = \frac{R_O}{\sqrt{1 - \left(\frac{f}{f_c}\right)^2}} \quad ...(8.8)$$

$$Z_{0\pi} = \frac{Z_1 Z_2}{Z_{0T}}$$

$$Z_{0\pi} = \frac{R_0^2}{R_{0T}\sqrt{1 - \left(\frac{f}{f_c}\right)^2}}$$

Plots of Z_{0T} *versus* f_c and $Z_{0\pi}$ versus f are given in Fig. 8.13.

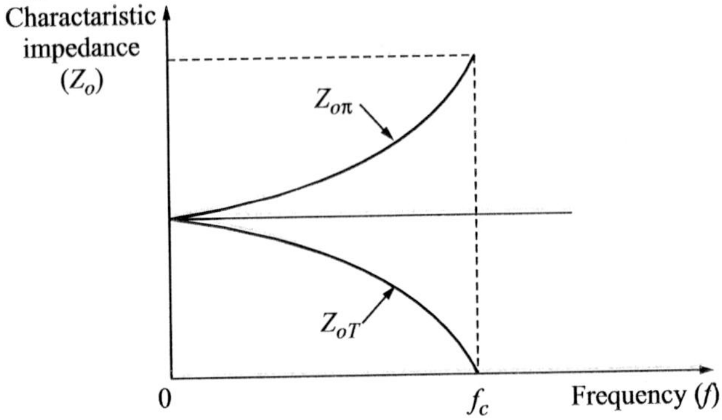

Fig. 8.13 *Z_{OT} and $Z_{O\pi}$ versus f*

8.6.1.4 Inductor and Capacitor Values

We know that:

$$R_0 = \sqrt{\frac{L}{C}} \text{ and } f_c = \frac{1}{\pi\sqrt{LC}}$$

If we divide R_O equations by f_c, then we get:

$$L = \frac{R_O}{\pi f_c} \quad ...(8.9)$$

By multiplying R_0 equation by f_c, we get:

$$C = \frac{1}{(\pi\, f_c)\, R_O} \quad ...(8.10)$$

8.6.2 High Pass Filter

T and π sections of high pass filters are shown in Fig. 8.14.

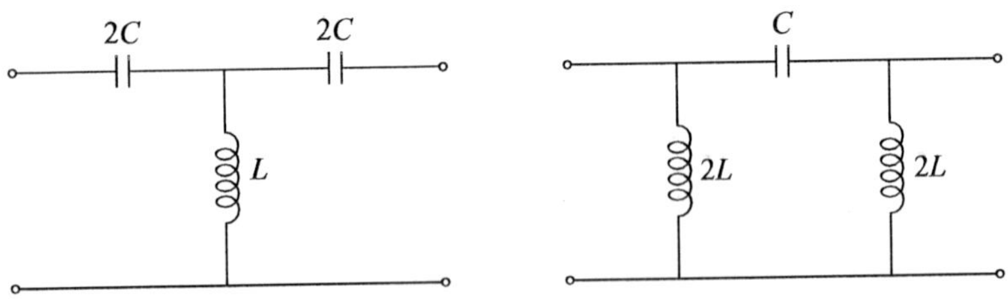

(*a*) *T-section* (*b*) *π-section*

Fig. 8.14 *T and π sections of high pass filter.*

8.6.2.1 Impedance R_0

T and π sections in terms of impedances are shown in the Fig. 8.15.

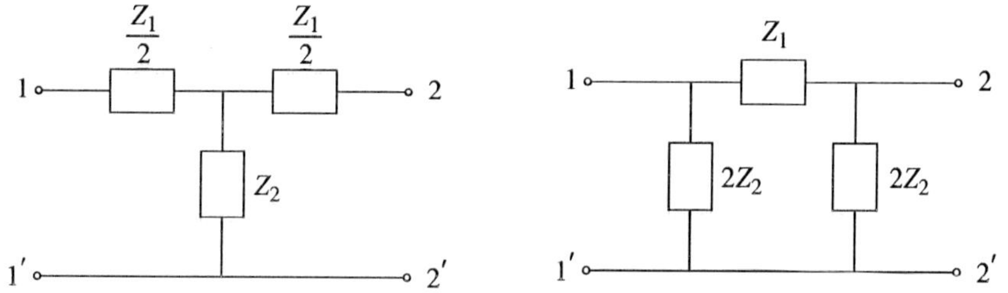

(*a*) *T-section* (*b*) *π-section*

Fig. 8.15 *T and π sections in terms of impedances.*

$$Z_1 = \frac{-j}{\omega c} \text{ and } Z_2 = j\omega L$$

$\therefore$
$$Z_1 Z_2 = \left(\frac{-j}{\omega c}\right)(j\omega L)$$

or
$$Z_1 Z_2 = \frac{L}{C}$$

$\therefore$
$$R_0^2 = \frac{L}{C}$$

or
$$R_0 = \sqrt{\frac{L}{C}} \qquad ...(8.11)$$

8.6.2.2 Cut-off Frequency f_c

(*i*) Graphical Method: T and π-sections of filters have same cut-off frequency, hence, we can take T-section only.

We know:
$$Z_1 = \frac{-j}{\omega c} \text{ i.e. } X_1 = \frac{-1}{\omega c}$$
$$Z_2 = j\omega L \text{ i.e. } X_2 = \omega L$$

$\therefore$
$$\frac{X_1}{4} + X_2 = \frac{-1}{4\omega c} + \omega L$$

or
$$\frac{X_1}{4} + X_2 = \omega L - \frac{1}{4\omega c}$$

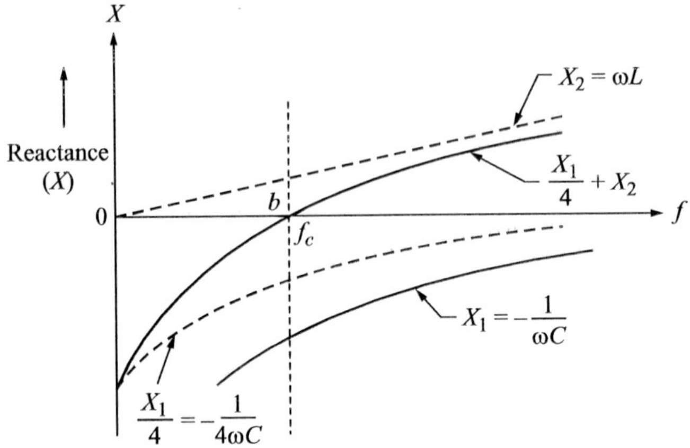

Fig. 8.16 *Reactance versus frequency*

The reactance curves are shown in the Fig. 8.16 for cut-off reactance versus frequency

For cut-off frequency we:

$$\frac{X_1}{4} + X_2 = 0$$

or
$$-\frac{1}{4\omega_c C} + \omega_c L = 0$$

or
$$\omega_c = \frac{1}{2\sqrt{LC}} \quad \text{...(8.12)}$$

(*ii*) Algebraic Method

We know that:
$$Z_{OT} = \sqrt{\frac{Z_1 Z}{4} + Z_1 Z_2}$$

or
$$Z_{OT} = \sqrt{\frac{-1}{4\omega^2 c^2} + \frac{L}{C}} = \sqrt{\frac{L}{C}} \times \sqrt{1 - \frac{1}{4\omega^2 LC}}$$

or
$$Z_{OT} = R_0 \sqrt{1 - \frac{1}{4\omega^2 LC}}$$

At cut-off frequency:

$$1 - \frac{1}{4\omega_c^{\ 2} LC} = 0$$

or
$$\omega_c = \frac{1}{2\sqrt{LC}}$$

or
$$f_c = \frac{1}{4\pi\sqrt{LC}}$$

8.6.2.3 Variation Z_{0T} and Z_{0p} with Frequency

We know that:

$$Z_{0T} = R_0\sqrt{1 - \frac{1}{4\omega^2 LC}}$$

By substituting value of ω_c, we get

$$Z_{0T} = R_0\sqrt{1 - \frac{\omega_c^2}{\omega^2}}$$

or

$$Z_{0T} = R_0\sqrt{1 - \left(\frac{f_c}{f}\right)^2} \qquad ...(8.13)$$

We know that:

$$Z_{0\pi} = \frac{Z_1 Z_2}{Z_{0T}}$$

By substitution, we get:

$$Z_{0\pi} = \frac{R_0^2}{R_0\sqrt{1 - \left(\frac{f_c}{f}\right)^2}}$$

or

$$Z_{0\pi} = \frac{R_0}{\sqrt{1 - \left(\frac{f_c}{f}\right)^2}} \qquad (8.14)$$

The plot of Z_{0T} and Z_{0T} versus f is given in the Fig. 8.17.

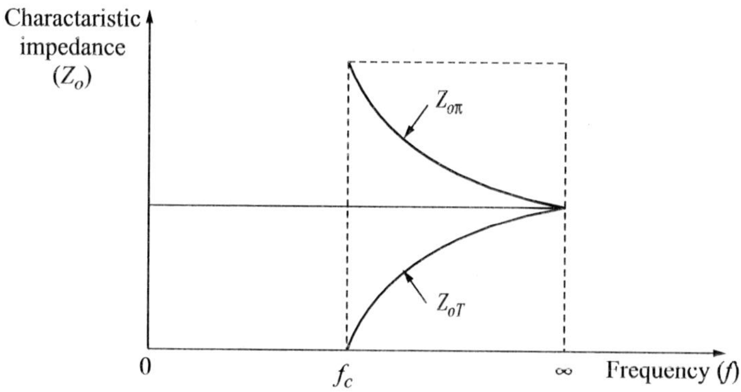

Fig. 8.17 *Z_{0T} and $Z_{0\pi}$ verses f*

8.6.2.4 Inductor and Capacitor Values

We know that: $$R_0 = \sqrt{\frac{L}{C}} \text{ and } f_c = \frac{1}{4\pi\sqrt{LC}}$$

Division of R_0 by f_c gives:

$$L = \frac{R_0}{(4\pi\, f_c)} \qquad \text{...(8.15)}$$

Multiplication of R_0 by f_c gives:

$$C = \frac{1}{(4\pi\, f_c)\, R_0} \qquad \text{...(8.16)}$$

8.6.3 Band Pass Filter

Band pass filters are constructed *by cascading low pass filter* and *high pass filter as shown* in the Fig. 8.18

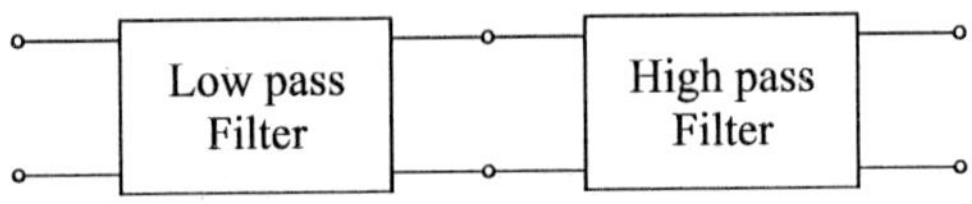

Fig. 8.18 *Construction of band pass filter.*

The cut-off frequency of low-pass filter section should be higher than that of high pass filter section. An economical method is to combine both low-pass filter and high-pass filter in the form of T or π sections as shown in Fig. 8.19.

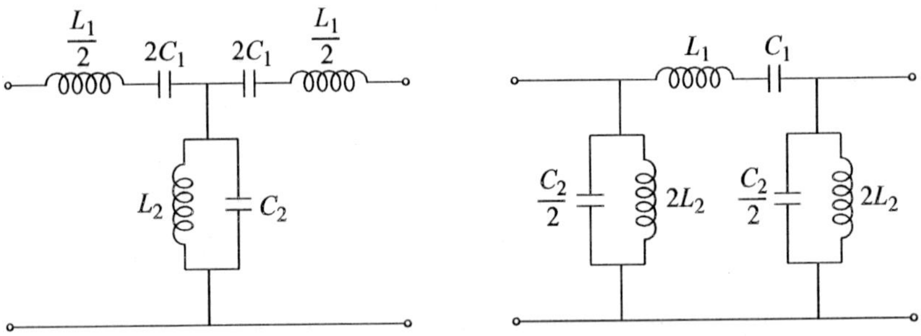

(*a*) *T-type* (*b*) *π-type*

Fig. 8.19 *T and π sections of band pass filter*

8.6.3.1 Impedance R_0

The impedences are given by:

$$Z_1 = j\omega L_1 + \left(\frac{-j}{\omega c_2}\right)$$

or $$Z_1 = j\left[\frac{\omega^2 L_1 C_1 - 1}{\omega c_1}\right]$$

and $$Z_2 = (j\omega L_2) \| \left(\frac{-j}{\omega c_2}\right)$$

or $$Z_2 = \frac{\dfrac{L_2}{C_1}}{\dfrac{j(\omega_2 L_2 C_2 - 1)}{\omega C_2}}$$

or $$Z_2 = \frac{-j\omega L_2}{(\omega^2 L_2 C_2 - 1)}$$

$$\therefore \quad Z_1 Z_2 = \frac{L_2}{C_1} = \frac{L_1}{C_2}$$

Thus, $Z_1 Z_2$ is real and constant.

$$\therefore \quad R_0^2 = Z_1 Z_2 = \frac{L_2}{C_1} = \frac{L_1}{C_2}$$

Hence, $$R_0 = \sqrt{\frac{L_2}{C_1}} = \sqrt{\frac{L_1}{C_2}} \quad \text{...(8.17)}$$

8.6.3.2 Cut-off Frequency

Let $Z_1 = sX_1$ and $Z_2 = pX_2$. We had drawn plot of reactance *versus* frequency curves for low pass and high pass filters. Same way, reactances X_1 and $\left(\frac{X_1}{4} + X_2\right)$ reactances *versus* frequency f plots are done as shown in Fig. 8.20.

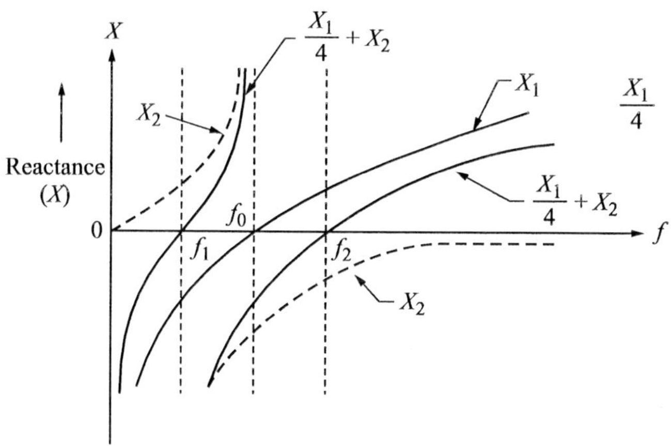

Fig. 8.20 *Reactance, versus frequency for band pass filter.*

The frequencies between f_1 and f_2 constitute pass band, and frequencies below f_1 and frequencies above f_2 give stop band. The frequencies f_1 and f_2 are known as lower and upper cut-off frequencies of pass band filter.

***(ii)* Algebraic Method:** For band pass filter, for Cut-off frequency:

$$\frac{Z_1}{4} + Z_2 = 0$$

or
$$Z_1^2 = -4Z_1 Z_2$$

But
$$Z_1 Z_2 = R_0^2, \text{ hence,}$$

$$Z_1^2 = -4R_0^2$$

i.e
$$Z_1 = \pm j(2R_0)$$

The lower cut-off frequency f_1 corresponds to $Z_1 = -j(R_0)$ and the upper Cut-off frequency f_2 corresponds to $Z_1 = j(R_0)$

Thus,
$$\omega_2 L_1 - \frac{1}{\omega_2 C_1} = -\left(\omega_1 L_1 - \frac{1}{\omega_1 C_1}\right)$$

or
$$\frac{\left(\omega_2^2 L_1 C_1 - 1\right)}{\omega_2 C_1} = -\frac{\left(\omega_1^2 L_1 C_1 - 1\right)}{\omega_1 C_1}$$

or
$$\omega_2^2 L_1 C_1 - 1 = \frac{\omega_2}{\omega_1}\left(1 - \omega_1^2 L_1 C_1\right)$$

Further, $\omega_0^2 = \dfrac{1}{L_1 C_1}$

$$\therefore \quad \frac{\omega_2^2}{\omega_0^2} - 1 = \frac{\omega_2}{\omega_1}\left(1 - \frac{\omega_1^2}{\omega_0^2}\right)$$

or
$$\left(1 - \frac{\omega_1^2}{\omega_0^2}\right) = \frac{\omega_1}{\omega_2}\left(\frac{\omega_2^2}{\omega_0^2} - 1\right)$$

or
$$\frac{\omega_0^2 - \omega_1^2}{\omega_0^2} = \frac{\omega_1}{\omega_2}\left(\frac{\omega_2^2 - w_0^2}{\omega_0^2}\right)$$

or
$$\omega_2 \omega_0^2 - \omega_1^2 \omega_2 = \omega_1 \omega_2^2 - \omega_1 \omega_0^2$$

or
$$\omega_0^2 (\omega_1 + \omega_2) = \omega_1 \omega_2 (\omega_1 + \omega_2)$$

or
$$\omega_0^2 = \omega_1 \omega_2$$

or
$$f_0^2 = f_1 f_2$$

$$\therefore \quad f_0 = \sqrt{f_1 f_2} \qquad \text{...(8.18)}$$

This means that frequency of resonance of individual arms is the geometric mean of two cut-off frequencies,

8.6.3.3 Band Pass Filter Element

If the filter is terminated with impedance R_0 and cut-off frequencies f_1 and f_2, then, at the lower cut-off frequency f_1, we get:

$$j\omega_1 L_1 - \frac{j}{\omega_1 C_1} = -j(2R_0)$$

or $$\frac{1}{\omega_1 C_1} - \omega_1 L_1 = 2R_0$$

or $$1 - \omega_1^2 L_1 C_1 = 2R_0(\omega_1 C_1)$$

or $$1 - \frac{\omega_1^2}{\omega_0^2} = (2\omega_1 C_1) R_0 \text{ as } \omega_0^2 = \frac{1}{L_1 C_1}$$

or $$1 - \frac{f_1^2}{f_0^2} = 4\pi R_0 f_1 C_1$$

we know that: $f_0^2 = f_1 f_2$

Hence, $$1 - \frac{f_1^2}{f_1 f_2} = 4\pi R_0 f_1 C_1$$

or $$\frac{f_2 - f_1}{f_2} = 4\pi R_0 f_1 C_1$$

$\therefore$ $$C_1 = \frac{(f_2 - f_1)}{4\pi R_0 (f_1 f_2)} \quad ...(8.20)$$

We know that: $$f_0 = \frac{1}{2\pi\sqrt{L_1 C_1}}$$

or $$f_0^2 = \frac{1}{4\pi^2 L_1 C_1}$$

or $$L_1 = \frac{1}{4\pi^2 C_1 f_0^2}$$

By substituting value of C_1 from equation (8.20), we get:

$$L_1 = \frac{1}{4\pi^2 f_0^2 \left[\dfrac{f_2 - f_1}{4\pi R_0 f_1 f_2}\right]}$$

We also know that: $f_0^2 = f_1 f_2$

$$\therefore \qquad L_1 = \frac{R_0}{\pi(f_2 - f_1)} \qquad ...(8.21)$$

We know that: $R_0 = \sqrt{\dfrac{L_2}{C_1}} = \sqrt{\dfrac{L_1}{C_2}}$

i.e.,

$$R_0^2 = \frac{L_2}{C_1} = \frac{L_1}{C_2}$$

By substituting value of L_1 in $R_0^2 = \dfrac{L_1}{C_2}$, we get:

$$C_2 = \frac{L_1}{R_0^2} = \frac{R_0}{\pi(f_2 - f_1)} \times \frac{1}{R_0^2}$$

or

$$C_2 = \frac{1}{\pi R_0(f_2 - f_1)} \qquad ...(8.22)$$

We know that: $R_0^2 = \dfrac{L_2}{C_1}$

or

$$L_2 = C_1 R_0^2$$

By substituting value of C_1 from equation (8.20), we get:

$$L_2 = \frac{(f_2 - f_1)}{4\pi R_0} \frac{R_0^2}{(f_1 f_2)}$$

or

$$L_2 = \frac{R_0 (f_2 - f_1)}{4\pi(f_1 f_2)} \qquad ...(8.23)$$

8.6.3.4 Band Elimination Filters

Band elimination filter is also known as stop band filter. It stops a range of frequencies between two cut-off frequencies f_1 and f_2. While pass all the frequencies below f_1 and above f_2. Thus, range of frequencies between f_1 and f_2 constitutes a stop band in which attenuation to the frequencies is $\omega.B_0$. The frequencies below f_1 and above f_2 constitute two separate pass bands in which attenuation to the frequencies is zero.

Band elimination filters can be constructed by connecting a low pass filter section in parallel with a high pass filter, However, this approach is not economical, hence, a combination of the two is cheaper. Based on this, Fig. 8.21 shows T and π type section of a band elimination filter.

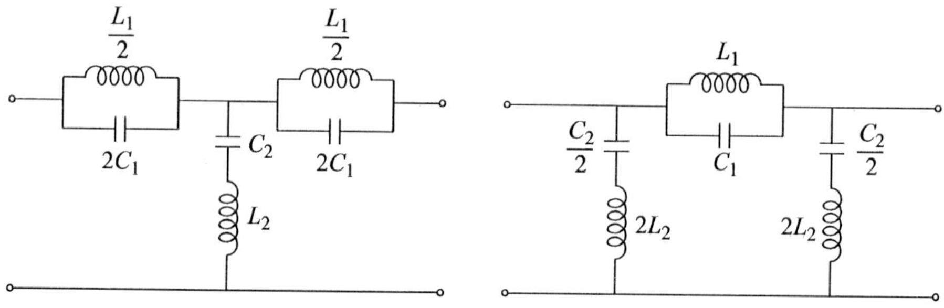

(a) *T-type section* (b) *π-type section*

Fig. 8.21 *T and π type band elimination filters section*

Both *T* and π-type sections are equivalent, hence, solving for *T*-type section serves for π-type as well. The series arm resonant frequency of *T*-type section is given by:

$$\omega_0 = \frac{1}{\sqrt{\left(\frac{L_1}{2}\right)(2C_1)}} = \frac{1}{\sqrt{L_1 C_1}}$$

or $$\omega_0^2 L_1 C_1 = 1 \qquad ...(8.24)$$

Similarly, shunt arm resonant frequency gives:

$$\omega_0 = \frac{1}{\sqrt{(L_2)(C_2)}}$$

or $$\omega_0^2 L_2 C_2 = 1$$

Thus, $$L_1 C_1 = L_2 C_2 \qquad ...(8.25)$$

8.6.3.4.1 Impedance R_0

From series arm,

$$Z_1 = \frac{(j\omega L_1)\left(-\frac{j}{\omega C_1}\right)}{j\omega L_1 - \frac{1}{\omega C_1}} = \frac{\omega L_1}{j(\omega^2 L_1 C_1 - 1)}$$

or $$Z_1 = \frac{-j\omega L_1}{(\omega^2 L_1 C_1 - 1)}$$

Similarly, shunt arm impedance,

$$Z_2 = j\omega L - \frac{j}{\omega C_2} = j\left(\frac{\omega_2 L_2 C_2 - 1}{\omega C_2}\right)$$

∴ $$Z_1 Z_2 = \frac{L_1}{C_2} = \frac{L_2}{C_1}$$

Thus, $$R_0^2 = Z_1 Z_2 = \frac{L_1}{C_2} = \frac{L_2}{C_1}$$

or $$R_0 = \sqrt{\frac{L_1}{C_2}} = \sqrt{\frac{L_2}{C_1}} \quad \text{...(8.26)}$$

8.6.3.4.2 Cut-off Frequencies

$$Z_1 = jX_1 \text{ and } Z_2 = jX_2$$

The reactance versus frequency curves *i.e.,* X_1, X_2, $\frac{X_1}{4} + Y_2$ etc shown in Fig. 8.22.

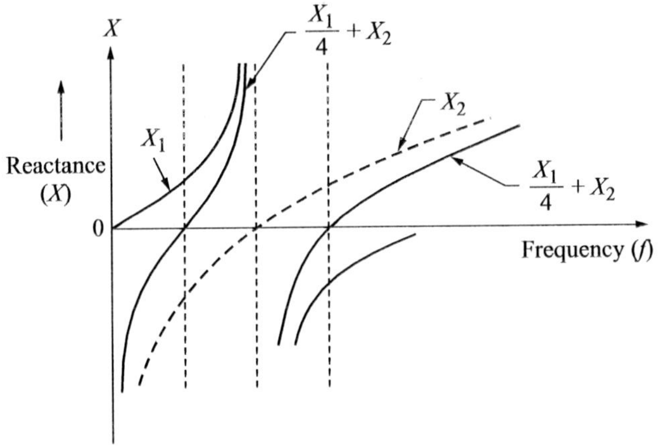

Fig. 8.22 *Reactance versus frequency band for elimination band filter.*

Stop band = between frequencies f_1 and f_2

Pass band = below f_1 and above f_2

f_1 and f_2 = Cut-off frequencies.

For band elimination filter,

$$\frac{Z_1}{4} + Z_2 = 0$$

or $$\frac{Z_1}{4} = -Z_2$$

or $$Z_1^2 = -4Z_1 Z_2$$

But, we know that: $$Z_1 Z_2 = R_0^2$$

∴ $$Z_1 = \pm j(2R_0)$$

From plot, we find that:

$$\frac{\omega_1 L_1}{1-\omega_1^2 L_1 C_1} = -\frac{\omega_2 L_1}{1-\omega_2^2 L_1 C_1}$$

or $$1 - \omega_2^{\,2} L_1 C_1 = \frac{\omega_2}{\omega_1}\left(\omega_1^{\,2} L_1 C_1 - 1\right)$$

We know that: $$\omega_0^{\,2} = \frac{1}{L_1 C_1} \text{ or } L_1 C_1 = \frac{1}{\omega_0^{\,2}}$$

$\therefore$ $$\left[1-\left(\frac{\omega_2^{\,2}}{\omega_0}\right)^2\right] = \frac{\omega_2}{\omega_1}\left[\left(\frac{\omega_1}{\omega_0}\right)^2 - 1\right]$$

or $$1-\frac{\omega_2^{\,2}}{\omega_0^{\,2}} = \left(\frac{\omega_2}{\omega_1}\right)\left(\frac{\omega_1^{\,2}}{\omega_0^{\,2}} - 1\right)$$

or $$\frac{\omega_0^{\,2} - \omega_2^{\,2}}{\omega_0^{\,2}} = \frac{\omega_2}{\omega_1}\left(\frac{\omega_1^{\,2} - \omega_0^{\,2}}{\omega_0^{\,2}}\right)$$

or $$\omega_1\,\omega_0^{\,2} - \omega_1\,\omega_2^{\,2} = \omega_2\,\omega_1^{\,2} - \omega_2\,\omega_0^{\,2}$$

or $$\omega_1\,\omega_0^{\,2} + \omega_2\,\omega_0^{\,2} = \omega_2\,\omega_1^{\,2} + \omega_1\,\omega_2^{\,2}$$

or $$\omega_0^{\,2}\,(\omega_1 + \omega_2) = \omega_1\,\omega_2(\omega_2 + \omega_1)$$

Hence, $$\omega_0^{\,2} = \omega_1\,\omega_2$$

or $$f_0^{\,2} = f_1 f_2$$

or $$f_0 = \sqrt{f_1\,f_2}$$

Elements of the Filter

If cut-off frequencies are f_1 and f_2 with termination impedance of R_0, then

The lower cut-off frequency is derived from:

$$Z_1 = \pm j(Z\,R_0)$$

$$j\left(\frac{\omega_1\,L_1}{1-\omega^2\,L_1\,C_1}\right) = j(2R_0)$$

or $$\omega_1\,L_1 = 2R_0\left(1 - \frac{\omega_1^{\,2}}{1-\omega_0^{\,2}}\right) \text{ as } L_1\,C_1 = \frac{1}{\omega_0^{\,2}}$$

or $$\omega_1\,L_1 = 2R_0\left(1 - \frac{\omega_1^{\,2}}{\omega_1\,\omega_2}\right), \text{ as } \omega_0^{\,2} = \omega_1\,\omega_2$$

or $$L_1 = 2R_0\left(\frac{\omega_2 - \omega_1}{\omega_1\,\omega_2}\right)$$

or $$L_1 = \frac{R_0(f_2 - f_1)}{\pi\,f_1\,f_2}$$

From the resonances of series arm,

$$f_0 = \frac{1}{2\pi\sqrt{L_1 C_1}}$$

or $$f_0^2 = \frac{1}{4\pi^2 L_1 C_1}$$

or $$C_1 = \frac{1}{4\pi^2 f_0^2 L_1} = \frac{1}{4\pi^2 (f_1 f_2) L_1} \quad \text{as } f_0^2 = f_1 f_2$$

By substituting value of L_1, we get:

$$C_1 = \frac{1}{4\pi^2 (f_1 f_2)\left[\dfrac{R_0 (f_2 - f_1)}{\pi f_1 f_2}\right]}$$

or $$C_1 = \frac{1}{4\pi R_0 (f_2 - f_1)}$$

We know that: $$R_0^2 = \frac{L_1}{C_2}$$

i.e., $$C_2 = \frac{L_1}{R_0^2}$$

By substituting value of C_1, we get:

i.e., $$L_2 = R_0^2 C_1$$

By substituting value of C_1, we get:

$$L_2 = R_0^2 \times \frac{1}{4\pi R_0 (f_2 - f_1)}$$

or $$L_2 = \frac{R_0}{4\pi (f_2 - f_1)}$$

8.7 COMPOSITE FILTERS

In composite filters, cut off frequency and design impedance are the two important specifications of designs. The no. of various section in composite filters depends on the attenuation characteristics required.

8.8 LIMITATIONS OF PASSIVE FILTERS

There are several limitations exist in passive filters as given below:

(*i*) Inductor should be in a filter particularly at low frequencies (< 1 kHz). At these frequencies practical inductor having reasonable Q, are large, fully and costly.

(*ii*) There are needs to cascade many sections leading to composite filter. In such cases, isolation employers are required to prevent loading of circuit.

(*iii*) External amplifies are needed to achieve suitable gain.

(*iv*) It is not possible to have high range of Q-factor.

In view of above limitations, the passive filters are rarely used.

8.9 ACTIVE FILTERS

Active filter networks use the *active elements such as op-amps, transistors along with the passive elements* such as resistors, inductors and capacitors. *Inductors* are fully, heavy, *non-linear*, generate stray magnetic fields, dissipate energy, therefore, generally in active filters, inductors are not used. The active filters have several advantages over passive filters and these are:

(*i*) All the active and passive elements can be made in integrated form, hence, there is reduction in size and weight.

(*ii*) Procurements of *ICs* are at very low cost, of quantity is very large.

(*iii*) Modern *ICs* have pushed down prices of op-amps.

(*iv*) Op-amp is easily adjustable in a closed loop way, therefore, active filter inputs signal are not attenuated.

(*v*) Flexibility in gain and frequency adjustments, the tuning of active filter is very easy.

(*vi*) As the op-amp has high input impedance and low output impedance, therefore, active filters using op-amps do not load the sources.

(*vii*) Inductors are not used in modern active filters, therefore, active filters are economical.

(*viii*) Active filters have better response due to usage of high quality components.

(*ix*) Active filters offer voltage gain, which is not feasible in passive filters.

(*x*) The design procedure is easy to passive filter design.

8.10 ACTIVE LOW-PASS FILTER

An active low-pass filter is shown in Fig. 8.23,

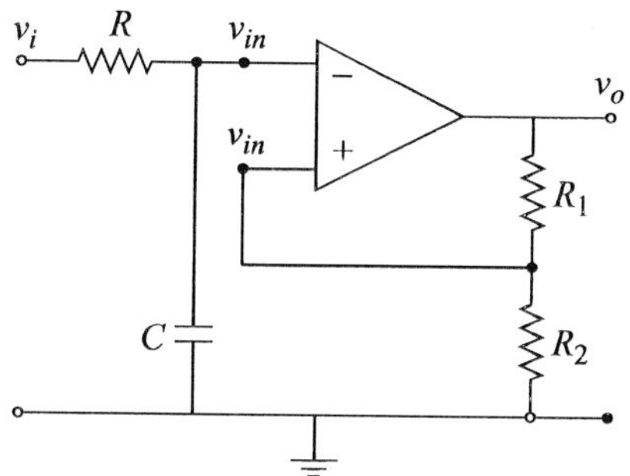

Fig. 8.23 *An active low-pass filter.*

Input impedance of an amplifier is very high, hence, input current is nearly zero. Thus, currents passing through R_1 and R_2 are almost same.

$$\therefore \qquad \frac{v_0 - v_{in}}{R_2} = \frac{v_{in} - o}{R_1}$$

or $$v_{in} = \frac{v_0}{\left(1 + \dfrac{R_2}{R_1}\right)} \qquad ...(i)$$

and $$\frac{v_i - v_{in}}{R} = \frac{v_i}{(1/s_c)}$$

or $$v_{in} = \frac{v_i}{1 + Rcs} \qquad ...(ii)$$

Thus, from equations (*i*) and (*ii*) we get

$$\frac{v_0}{1 + \dfrac{R_2}{R_1}} = \frac{v_1}{1 + Rcs}$$

or $$\frac{v_o}{v_i} = \frac{1 + R_2 / R_1}{1 + Rcs} \qquad ...(iii)$$

Op-amp gain is obtained from equation (*i*), which is:

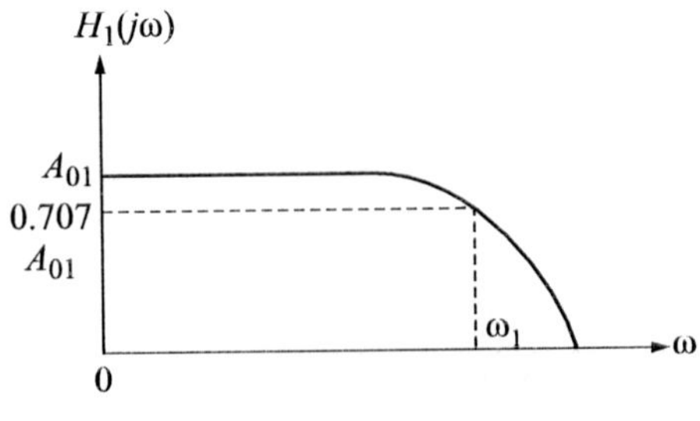

$$A_{01} = \frac{v_0}{v_{in}} = 1 + \frac{R_2}{R_1} \qquad ...(iv)$$

∴ Transfer function of the filter from (*iii*) and (*iv*), we get:

$$A(s) = \frac{v_o}{v_i} = \frac{A_{01}}{1 + (s/\omega_0)}$$

where $\omega_0 = \dfrac{1}{RC}$ = Cut-off frequency.

8.11 ACTIVE HIGH-PASS FILTER

An active high-pass filter is shown in Fig. 8.24.

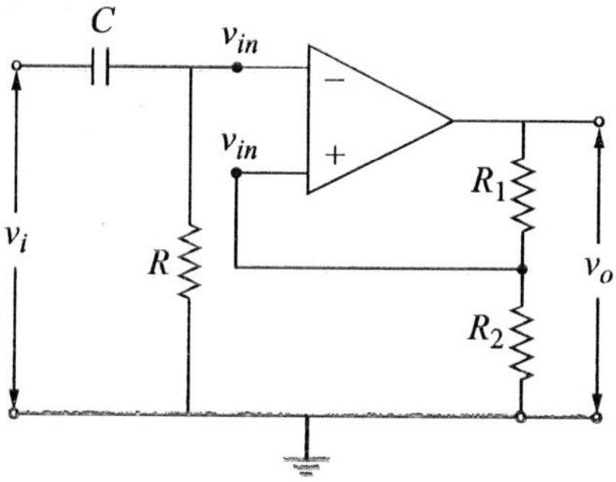

Fig. 8.24 *An active high-pass filter.*

Input impedance of op-amp is very high, hence, input current is almost zero. Thus, currents passing through R_1 and R_2 are same.

$$\therefore \qquad \frac{v_0 - v_{in}}{R_2} = \frac{v_{in} - o}{R_1}$$

or

$$v_{in} = \frac{v_o}{1 + \dfrac{R_2}{R_1}} \qquad ...(i)$$

and

$$\frac{v_i - v_{in}}{\dfrac{1}{sc}} = \frac{v_{in}}{R}$$

or

$$v_{in} = \frac{v_i}{1 + \dfrac{1}{Rcs}} \qquad ...(ii)$$

By equating equations (*i*) and (*ii*), we get:

$$\frac{v_o}{v_i} = \frac{1 + \dfrac{R_2}{R_1}}{1 + \dfrac{1}{Rcs}}$$

or

$$H_2(s) = \frac{v_o}{v_i} = \frac{A_{02}}{1 + \left(\dfrac{1}{Rcs}\right)} = \frac{A_0}{1 + \left(\dfrac{\omega_0}{s}\right)}$$

or

$$H_2(s) = A_{02}\,\frac{(s/\omega_0)}{1 + (s/\omega_0)}$$

where
$$A_{02} = \left(1 + \frac{R_2}{R_1}\right)$$
$$\omega_0 = \frac{1}{RC}$$

It can be noted that formulae for both is same of $\left(\frac{s}{\omega_0}\right)$ low-pass is replaced by $\frac{1}{(s/\omega_0)}$ for high-pass filter or *vice-versa.* The frequency by response of active high-pass filter is shown in figure.

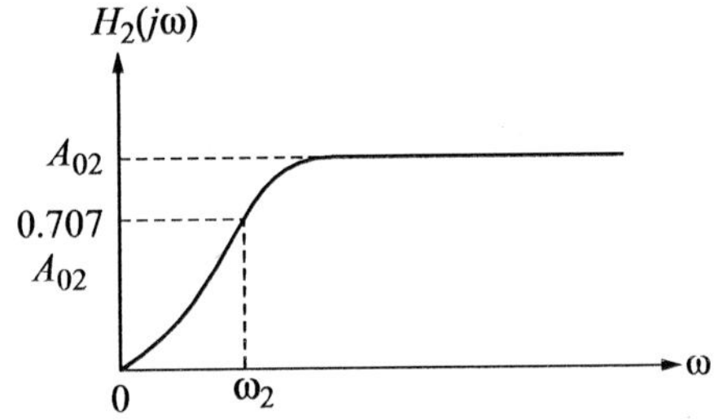

Active Band Pass Filter

Cascading of active low-pass and active high-pass filter are shown in Fig. 8.25 gives the active band pass filter.

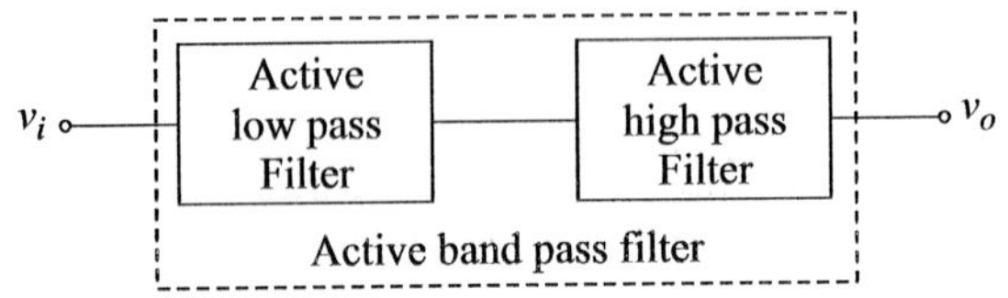

Fig. 8.25 *Active band-pass filter block diagram*

The frequency response of active low-pass figure and active-high filter figure on bond for $\omega_1 > \omega_2$ to give frequency response of active bond pass filter is shown in Fig. 8.26

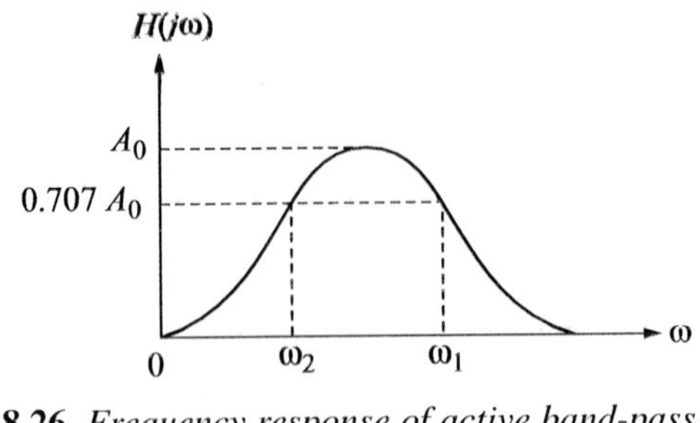

Fig. 8.26 *Frequency response of active band-pass filter.*

Active Band-elimination Filter

Parallel combination of active low-pass and high-pass filters give active band-stop elimination filter as shown in Fig. 8.27

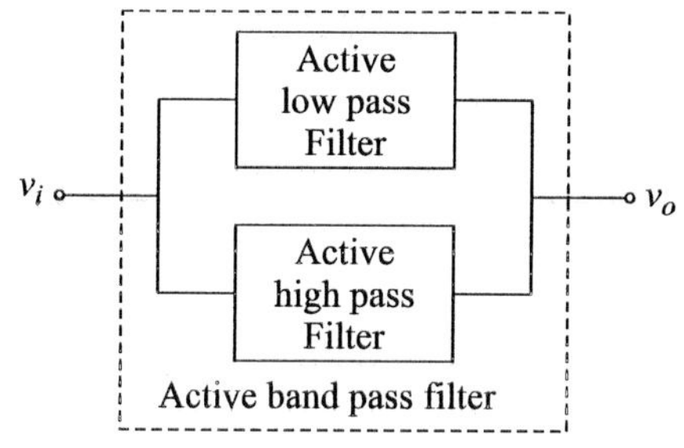

Fig. 8.27 *Active band-elimination block diagram*

The frequency responses of active low band-pass filter figure and active high-band-pass filter figure combine for $\omega_1 < \omega_2$ to give frequency responses of active band-domination filter as shown in Fig. 8.28

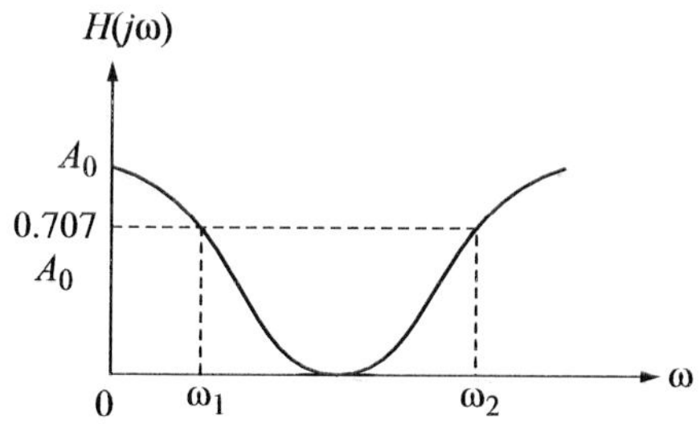

Fig. 8.28 *Frequency response of active band-filter.*

SUMMARY

1. **Filters:** A network which *freely passes definite band of frequencies* which *almost suppresses other band of frequencies* is "*known as filter*". The range of frequencies over which attenuation by filter is zero, is known as pass band. The *range of frequencies* over which *attenuation* is *infinite* is known as *stop band or attenuation band or band elimination filter.*
2. **Image Parameters:** Image parameters are defined for two-port networks. There are two parameters very useful is filters namely image impedance (Z_i) and image transfer constant (e^{θ}).
3. **Characteristic Impedance:** The *characteristic impedance* (Z_0) of a network is the *impedance measured at the input terminals of the first network* in the chain of infinite networks in cascade.

4. **Basic Types of Filters:** Pass band filters; Stop band filters; Symmetrical unbalanced filters; Passive filters; Low Pass filters; High Pass filter; Band elimination filters; Compositive filters; Active filters; Active Low-Pass filter; Active high-pass filter; Active band pass filter; Active band-elimination filter.

EXERCISES

8.1. Design a constant-k low pass filter (both p and T Networks) having a cut off frequency of 1 kHz and terminated load resistance is zero Ω.

8.2. Design a high pass filter (both p and T Networks) having a cut off frequency of 1 kHz with a load resistance of 500 Ω.

8.3. Given figure show a high pass filter. Find (*a*) cut-off frequency, (*b*) z_0 at $f = \infty$.

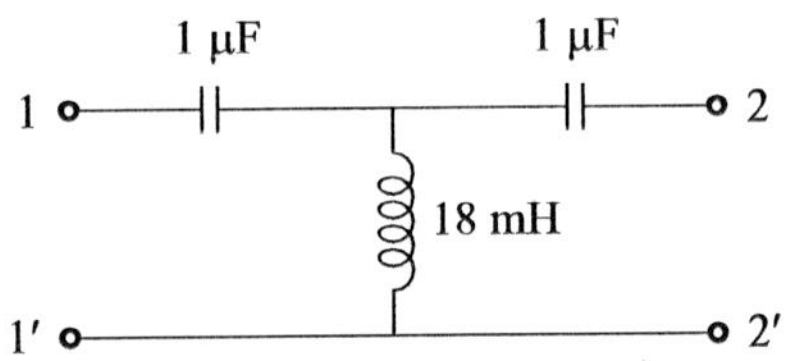

8.4. Design a constant K band pass filter having a passband from 100 Hz to 400 Hz and a characteristics resistance of 400 W. Draw the configuration of the filter, find resonant frequency of series and shunt circuit.

8.5. Design a constant-K band stop filter having a characteristics resistance of 200 Ω and cut off frequencies as 2000 Hz and 5000 Hz. Draw the filter configuration.

8.6. An m derived low pass filter has cut-off frequency of 2000 Hz and frequency of infinite alternation of 2500 Hz. Find the value of constant m.

8.7. An m derived high pass filter has a cut-off frequency of 2000 Hz. If $m = 0.5$. Find the frequency of infinite attenuation.

8.8. Design the T and π section of m derived high pass filter having $f_c = 3000$ Hz. $f_\infty = 2700$ Hz and $R_0 = 600\ \Omega$.

8.9. Design an m derived high pass filter having a design impedance of 600 Ω, cutt off frequency of 5 kHz and $m = 0.35$. Also determine the frequency of infinite attenuation.

❑❑❑

Appendix A: Laplace Transform

I. LAPLACE TRANSFORM

When we talk about the Laplace transform, we are actually talking about the version of the Laplace transform known as the **unilinear Laplace Transform.** The other version, the **Bilinear Laplace Transform** (not related to the Bilinear Transform, below) is not used in this book.

The Laplace Transform is defined as:

$$F(s) = \pounds[f(t)] = \int_0^{\infty} f(t)e^{-st}\,dt \qquad \textbf{[Laplace Transform]}$$

And the Inverse Laplace Transform is defined as:

$$f(t) = \pounds^{-1}\{F(S)\} = \frac{1}{2\pi i}\int_{c-i\infty}^{c+\infty} e^{ft} F(s)\,ds \qquad \textbf{[Inverse Laplace Transform]}$$

Table of Common Laplace Transforms

S.No.	Time Domain $(t) = \pounds^{-1}\{X(s)\}$	Laplace Domain $X(s) = \pounds\{x(t)\}$
1	$\frac{1}{2\pi n}\int_{c-j\infty}^{c+j\infty} X(s)\,e^{st}\,ds$	$\int_{-\infty}^{\infty} x(t)\,e^{-st}\,dt$
2	$\delta(t)$	1
3	$\delta(t-a)$	e^{-as}
4	$u(t)$	$\frac{1}{s}$
5	$u(t-a)$	$\frac{e^{-as}}{s}$
6	$tu(t)$	$\frac{1}{s^2}$

Contd ...

S.No.	**Time Domain** $(t) = £^{-1}\{X(s)\}$	**Laplace Domain** $X(s) = £\{x(t)\}$
7	$t^n u(t)$	$\frac{n!}{s^{n+1}}$
8	$\frac{1}{\sqrt{\pi t}} u(t)$	$\frac{1}{\sqrt{s}}$
9	$e^{at} u(t)$	$\frac{1}{s-a}$
10	$t^n e^{at} u(t)$	$\frac{n!}{(s-a)^{n+1}}$
11	$\cos(\omega t) u(t)$	$\frac{s}{s^2+\omega^2}$
12	$\sin(\omega t)\ u(t)$	$\frac{\omega}{s^2+\omega^2}$
13	$\cosh(\omega t) u(t)$	$\frac{s}{s^2-\omega^2}$
14	$\sinh(\omega t) u(t)$	$\frac{\omega}{s^2-\omega^2}$
15	$e^{at}\ \cos(\omega t) u(t)$	$\frac{s-a}{(s-a)^2+\omega^2}$
16	$e^{at}\ \sin(\omega t) u(t)$	$\frac{\omega}{(s-a)^2+\omega^2}$
17	$\frac{1}{2\omega^3}(\sin \omega t - \omega t \cos \omega t)$	$\frac{1}{(s^2+\omega^2)^2}$
18	$\frac{t}{2\omega} \sin \omega t$	$\frac{s}{(s^2+\omega^2)^2}$
19	$\frac{1}{2\omega}(\sin \omega t + \omega t \cos \omega t)$	$\frac{s^2}{(s^2+\omega^2)^2}$

Table of the most important properties of the Laplace transform:

Property	Definition
Linearity	$£\{af(t) + bg(t)\} = aF(s) + bG(s)$
Differentiation	$£\{f'\} = s£\{f\} - f(0^-)$ $£\{f''\} = s^2£\{f\} - sf(0^-) - f'(0^-)$ $£\{f^{(n)}\} = s^n£\{f\} - s^{n-1}f(0^-) - \ldots - f(n-1)(0^-)$
Frequency Division	$£\{t(f(t)\} = -F'(s)$ $£\{t^nf(t)\} = (-1)^nF^{(n)}(s)$
Frequency Integration	$£\left\{\frac{f(t)}{t}\right\} = \int_s^{\infty} F(\sigma)\, d\sigma$
Time Integration	$£\left\{\int_0^t f(\tau)\, d\tau\right\} = £\{u(t) * f(t)\} = \frac{1}{s}F(s)$
Scaling	$£\{f(at)\} = \frac{1}{a}F\left(\frac{s}{a}\right)$
Initial value theorem	$f(0^+) = \lim_{x \to \infty} sF(s)$
Final value theorem	$f(\infty) = \lim_{s \to 0} sF(s)$
Frequency Shifts	$£\{e^{at} f(t)\} = F(s - a)$ $£^{-1}\{f(s - a)\} = e^{at} f(t)$
Time Shifts	$£\{f(t - a)yu(t - a)\} = e^{-as} F(s)$ $£^{-1}\{e^{-as} F(s)\} = f(t - a)u(t - a)$
Convolution Theorem	$£(f(t) * g(t)\} = F(s)G(s)$ **Where:** $f(t) = £^{-1}\{F(s)\}$ $g(t) = £^{-1}\{G(s)\}$ $s = \sigma + j\omega$

II. TABLE OF INVERSE LAPLACE TRANSFORMS

One Pole		
$F(s)$	s plane	$f(t)$
$\frac{1}{s+a}$	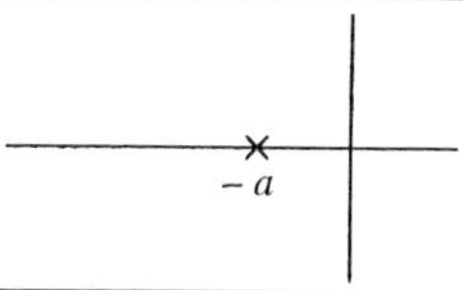	e^{-at}
Two Poles		
$\frac{1}{(s+a_1)(s+a_2)}$	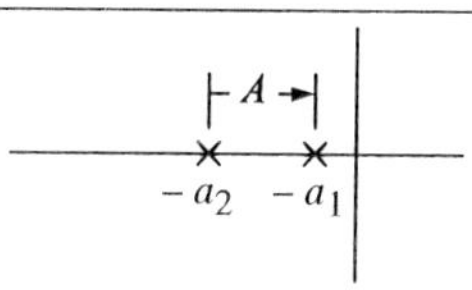	$\frac{1}{A}(e^{-a_1t} - e^{-a_2t})$

Two Poles (*continued*)		
$F(s)$	s plane	$f(t)$
$\dfrac{(s+b)}{(s+a_1)(s+a_2)}$	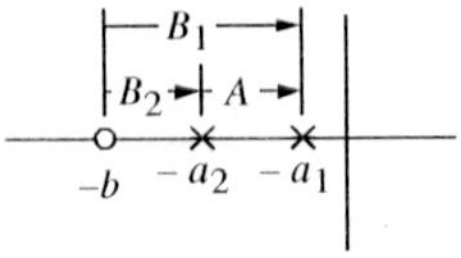	$\dfrac{1}{A}(B_1 e^{-a_1 t} - B_2 e^{-a_2 t})$
$\dfrac{1}{(s+a)^2}$	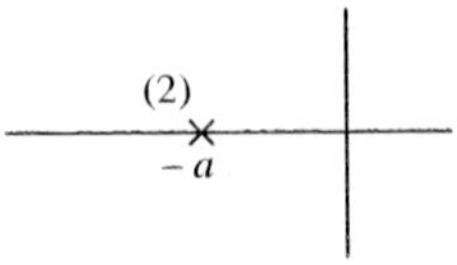	te^{-at}
$\dfrac{(s+b)}{(s+a)^2}$	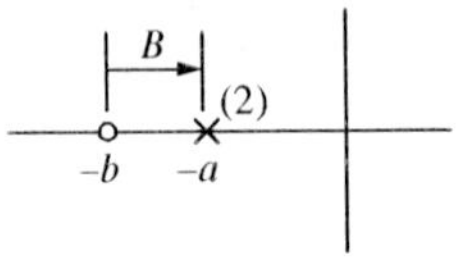	$(Bt + 1)e^{-at}$
$\dfrac{1}{(s+a)^2+\beta^2}$	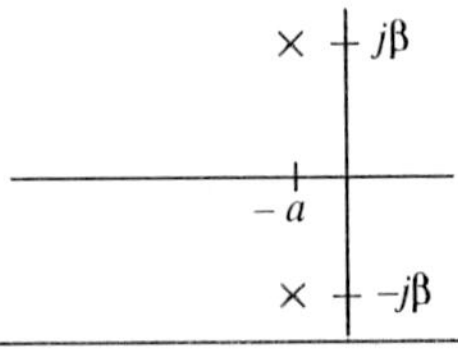	$\dfrac{1}{\beta}e^{-at}\sin\beta t$
$\dfrac{s+b}{(s+a)^2+\beta^2}$	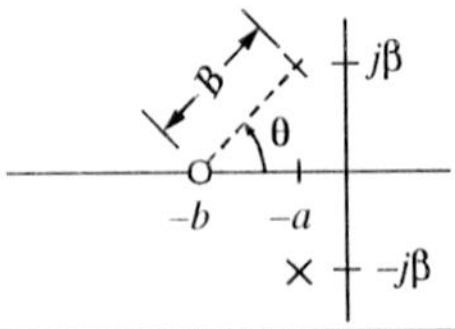	$\dfrac{1}{\beta}e^{-at}\sin(\beta t+\theta)$
Three Poles		
$\dfrac{1}{(s+a_1)(s+a_2)(s+a_3)}$	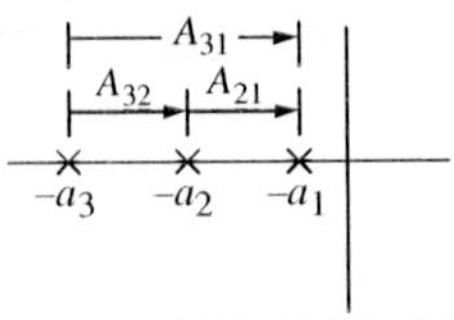	$\dfrac{1}{A_{21}A_{31}}e^{-a_1 t} - \dfrac{1}{A_{21}A_{32}}e^{-a_2 t} + \dfrac{1}{A_{31}A_{32}}e^{-a_3 t}$
$\dfrac{s+b}{(s+a_1)(s+a_2)(s+a_3)}$	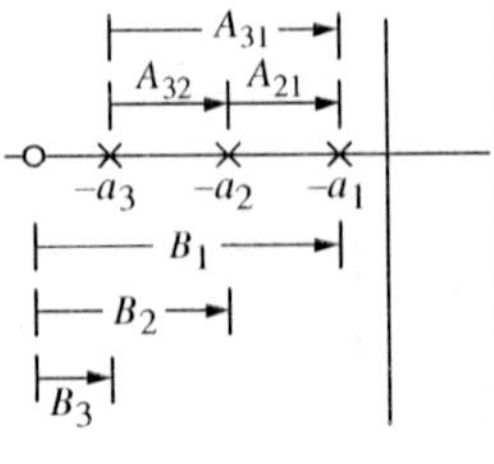	$\dfrac{B_a}{A_{21}A_{31}}e^{-a_1 t} - \dfrac{B_2}{A_{21}A_{32}}e^{-a_2 t} + \dfrac{B_3}{A_{31}A_{32}}e^{-a_3 t}$

$\dfrac{s^2}{(s+a_1)(s+a_2)(s+a_3)}$	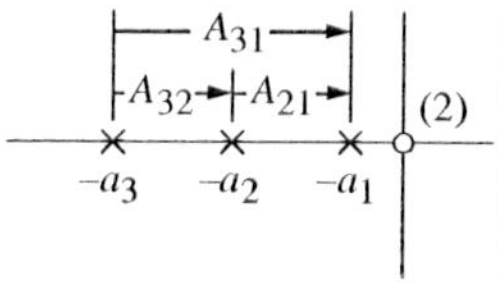	$\dfrac{a_1^2}{A_{21}A_{31}}e^{-a_1t} - \dfrac{a_2^2}{A_{21}A_{32}}e^{-a_2t} + \dfrac{a_3^2}{A_{31}A_{32}}e^{-a_3t}$

Three Poles (*Continued*)

$F(s)$	s plane	$f(t)$
$\dfrac{s^2}{(s+a_1)^2(s+a_2)}$	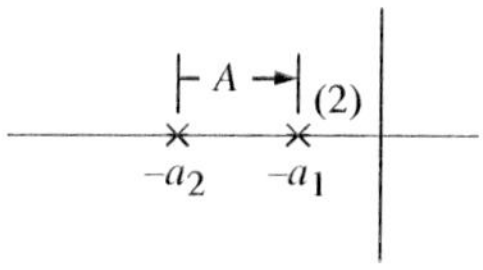	$\dfrac{1}{A^2}[e^{-a_1t} + (At-1)e^{-a_1t}]$
$\dfrac{s+b}{(s+a_1)^2(s+a_2)}$	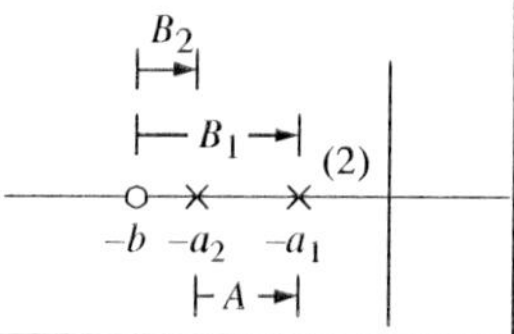	$\dfrac{1}{A^2}[B_2e^{-a_1t} + (AB_1t - B_2)e^{-a_1t}]$
$\dfrac{s^2}{(s+a_1)^2(s+a_2)}$	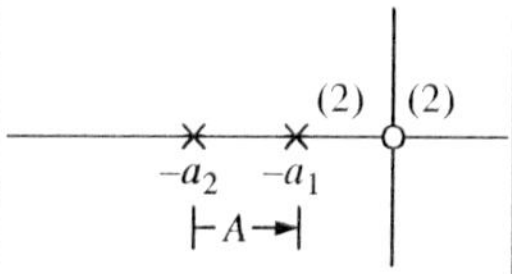	$\dfrac{1}{A^2}[a_2^2e^{-a_2t} + (a_1^2At - a_1A - a_1a_2)e^{-a_1t}]$
$\dfrac{1}{(s+a_2)[(s+\alpha)^2+\beta^2]}$	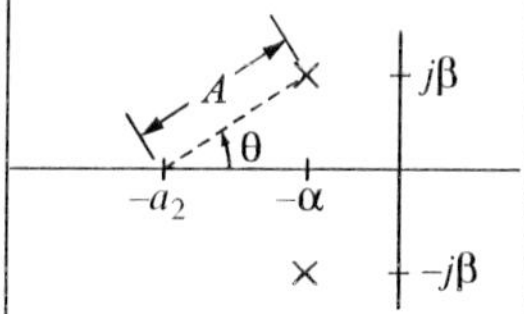	$\dfrac{1}{A^2}e^{-a_2t} + \dfrac{1}{A\beta}e^{-\alpha t}\sin(\beta t - \theta)$
$\dfrac{s+b}{(s+a_1)[(s+\alpha)^2+\beta^2]}$	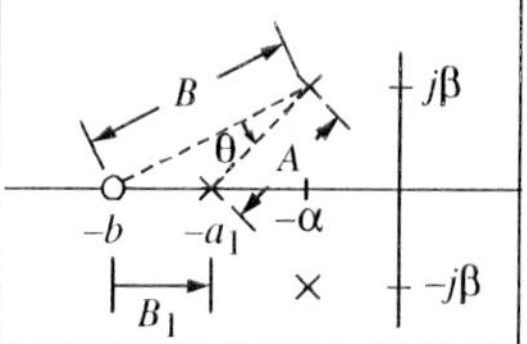	$\dfrac{B_1}{A_2}e^{-a_1t} + \dfrac{B}{\beta A}e^{-\alpha t}\sin(\beta t - \theta)$
$\dfrac{s^2}{(s+a_1)[(s+\alpha)^2+\beta^2]}$	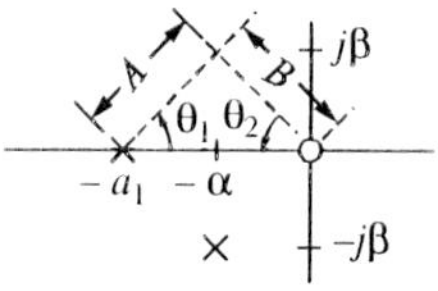	$\dfrac{a_1^2}{A^2}e^{-a_1t} + \dfrac{B^2}{\beta A}e^{-\alpha t}\sin(\beta t - \theta_1 - 2\theta_2)$

Appendix B: Complex Number

1. DEFINITIONS

The complex number A = $a + jb$ is represented in the complex plane as shown in Fig. 1.1(a). If we indicate the real part by Re and the imaginary part by Im, then

$$\text{Re } \mathbf{A} = a, \qquad \text{Im } \mathbf{A} = b \tag{1.1}$$

Observe that if $\mathbf{B} = c + jd$ and $B = \mathbf{A}$, then $a = c$ and $b = d$. In polar or exponential form, the same complex number is written

$$\mathbf{A} = Ae^{j\theta} \tag{1.2}$$

and represented as in Fig. 1.1(b). The magnitude and phase A are written as

$$|\mathbf{A}| = A, \ \arg \mathbf{A} = \theta \tag{1.3}$$

Equations useful for converting from one form to the other are derived starting from Euler's identity,

$$e^{\pm j\theta} = \cos\theta \pm j\sin\theta \tag{1.4}$$

Then equation (1.2) becomes

$$\mathbf{A} = A\cos\theta + jA\sin\theta \tag{1.5}$$

Comparing this equation with equation (1.1), we have

$$a = A\cos\theta, \quad b = A\sin\theta \tag{1.6}$$

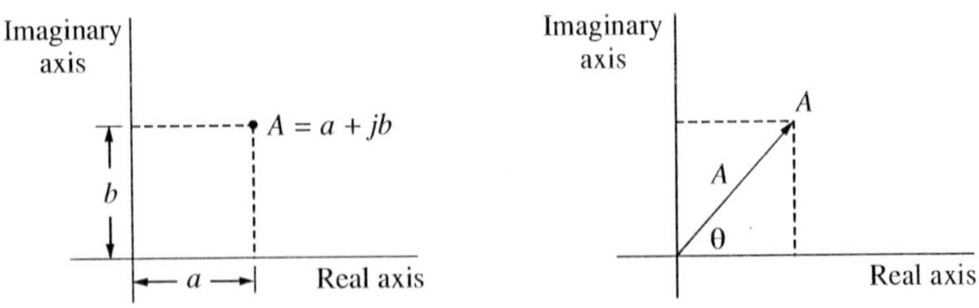

(a) Rectilinear (b) Polar

Fig. 1.1 *Complex Number*

Squaring the two equations and adding, we find that

$$A^2 = a^2 + b^2 \tag{1.7}$$

Dividing b by a, we have

$$\theta = \tan^{-1}\left(\frac{b}{a}\right) \tag{1.8}$$

1.2 ADDITION AND SUBSTRACTION

If $A = a + jb$ and $B = c + jd$, then

$$A + B = B + A = C \tag{1.9}$$

and

$$C = (a + c) + j(b + d) \tag{1.10}$$

In other words, the sum of two complex numbers is another complex number which is found by summing real parts and imaginary parts. The addition and subtraction operations are shown in the complex plane in Fig. 1.2.

Example 1. If $A = 1 + j2$ and $B = 1 - j1$, then

$$A + B = C = 2 + j1 \tag{1.11}$$

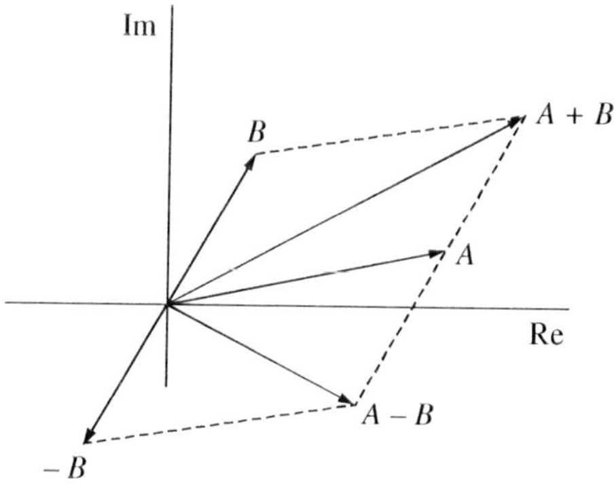

Fig. 1.2 *Addition and subtraction of complex numbers*

and

$$A - B = D = 0 + j3 \tag{1.12}$$

1.3 MULTIPLICATION

Using the complex numbers A and B of the last section, their production is

$$AB = BA = (a + jb)(c + jd) = (ac - bd) + j(ad + bc) \tag{1.13}$$

showing that the product of two complex numbers is another complex number. In the polar or exponential form, we have

$$A = Ae^{i\theta_e}, \qquad B = Be^{i\theta_v} \tag{1.14}$$

and

$$C = AB = ABe^{i(\theta_a + \theta_v)} \tag{1.15}$$

so that

$$|C| = AB \text{ and } \arg C = \theta_a + \theta_b \tag{1.16}$$

The product of two complex numbers is shown in the complex plane in Fig. 1.3. Observe that the multiplication of complex number is more easily accomplished in the polar or expoenential form than in the complex form.

We are frequently required to multiply a complex number A by its conjugate A^*. By definition, if $A^- = A + jb$, then $A^* = a - jb$. Observe that

$$AA^* = (a + jb)(a - jb) = a^2 + b^2 \tag{1.17}$$

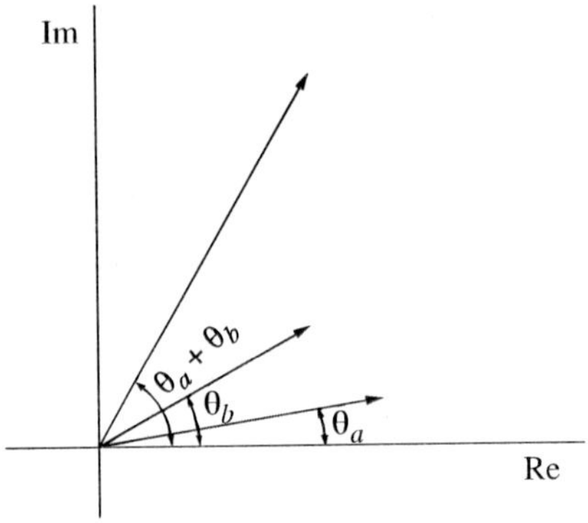

Fig. 1.3 *Product of complex numbers*

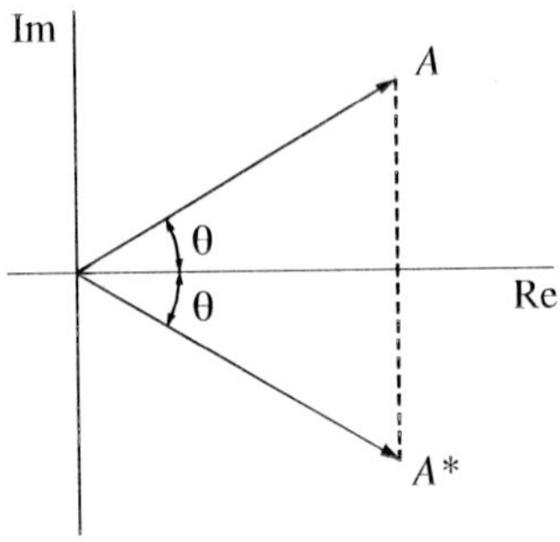

Fig. 1.4 *Exponential form*

showing that the product is a real number. In exponential form,

$$A = Ae^{j\theta}, \text{ then } A^* = Ae^{-j\theta} \tag{1.18}$$

as shown in Fig. 1.4. Then

$$AA^* = A^2 = a^2 + b^2 \tag{1.19}$$

in agreement with equation (1.7).

1.4 DIVISION

If $CB = A$, then for complex numbers

$$C = \frac{A}{B} = \frac{A}{B}\frac{B^*}{B^*} = \frac{AB^*}{[B]^2} \tag{1.20}$$

in which multiplication by B^*/B^* is known as *rationalization*. Then

$$\frac{A}{B} = \frac{(a+jb)(c-jd)}{c^2+d^2} = \frac{ac+bd}{c^2+d^2} + j\frac{bc-ad}{c^2-d^2} \tag{1.21}$$

In the polar or exponential form

$$\frac{A}{B} = \frac{Ae^{i\theta_v}}{Be^{i\theta_v}} = \frac{A}{B}e^{j(\theta_a-\theta_b)} \tag{1.22}$$

so that

$$\left|\frac{A}{B}\right| = \frac{A}{B} \text{ and } \arg\frac{A}{B} = \theta_a - \theta_b \tag{1.23}$$

These relationships are illustrated in the complex plane in Fig. 1.5.

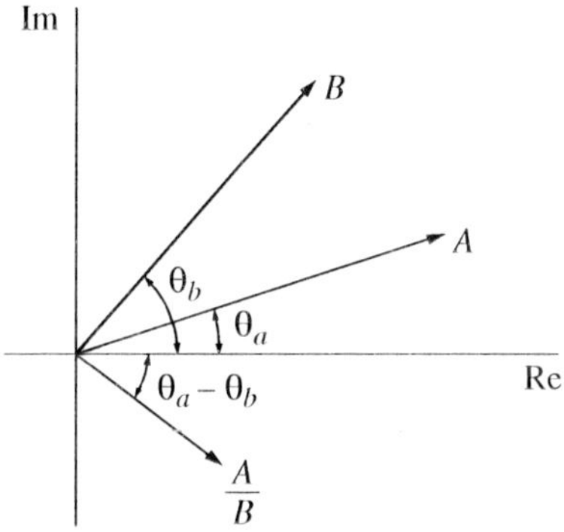

Fig. 1.5 *Relationships in complex plane*

Example 2. Suppose that it is required to divide $1 + j\sqrt{3}$ by $j1$ and to express the result in the form $a + jb$. Then

$$\frac{1+j\sqrt{3}}{j1}\frac{-j1}{-j1} = \sqrt{3} - j1 = a + jb \tag{1.24}$$

In exponential form

$$\frac{2e^{j60°}}{1e^{j90°}} = 2e^{-j30°} = 2\left(\frac{\sqrt{3}}{2} - j\frac{1}{2}\right) = \sqrt{3} - j1 \tag{1.25}$$

In obtaining this result, use was made of Eq. (A-5).

1.5 LOGARITHM OF A COMPLEX NUMBER

The logarithm of a complex number is found by expressing that number in exponential form and also observing that $\theta = \theta + k2\pi$ for any integer value of k with θ in radians. Then

$$\text{In } A = \ln \;=\ln Ae^{i(\theta+k2\pi)} = \ln A + j\theta(+k2\pi) \qquad (1.26)$$

The value with $k = 0$ is known as the *principal value.*

Example 3. We are required to find ln (3 + j4) in the form $a + jb$. Then

$$\ln(3 + j4) = \ln 5 + j\left(\tan^{-1}\frac{4}{3} + k2\pi\right), \quad k = 0, 1, \ldots \qquad (1.27)$$

1.6 ROOTS AND POWERS OF COMPLEX NUMBERS

Roots and powers of complex numbers are found by making use of the law of exponents. Thus

$$A^n = \left(Ae^{j\theta}\right)^n = A^n e^{jn\theta} \qquad (1.28)$$

If we substitute $1/m$ for n and also add k2p to q, we have

$$A^{1/m} = +\sqrt[m]{A}\; e^{(j\theta + k2n)/m} \qquad (1.29)$$

where k is an integer, and $k = 0$ yields the *principal value.* This result is known as *De Moivre's theorem.*

Example 4. It is required to determine the three roots of +1 + j0. Using Eq. (1.29), these are:

$$1e^{j\theta^\circ}, \quad 1e^{j120^\circ}, \quad \text{and} \quad 1e^{j240^\circ} \qquad (1.30)$$

as shown in Fig. 1.6.

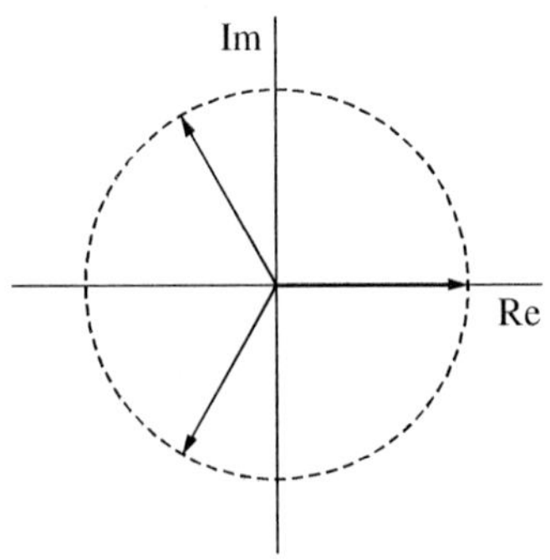

Fig. 1.6 *Roots in polar*

Appendix C: Matrices

1.1 DEFINITIONS

The rectangular array of numbers of functions

$$A = \begin{bmatrix} a_{11} & a_{12} & \dots & a_{1n} \\ a_{21} & a_{22} & \dots & a_{2n} \\ \hdashline a_{m1} & a_{m2} & \dots & a_{mn} \end{bmatrix} \tag{1.1}$$

is known as a *matrix* of order (m, n) or an $m \times n$ matrix. Double bars || || and parentheses ( ) are also used in place of the square brackets. The numbers or functions a_{ij} are the *elements* of the matrix, with the first subscript indicating row position and the second subscript column position. We also use the notation $[a_{ij}]_{m \times n}$ to identify a matrix of order $m \times n$ whose elements are a_{ij}.

A number of matrices are given special names. A matrix of one column but any number of rows is known as a *column* matrix or a *vector.* A matrix of order (n, n) is a *square matrix of order n.* The main or principal diagonal of a square matrix consists of the elements a_{11}, a_{22}, ..., a_{nn}. A square matrix in which all elements except those on the principal diagonal are zero is known as a *diagonal matrix.* Further, if all elements of a diagonal matrix have the value 1, the matrix is known as a *unit or identity matrix, U.* If all elements of a matrix are zero, $a_{ij} = 0$, the matrix is known as a *zero* matrix 0. If $a_{ij} = a_{jt}$ in a matrix, it is known as a *symmetrical* matrix.

The three matrices

$$A = \begin{bmatrix} 1 & -1 & 2 \\ -1 & 0 & 3 \end{bmatrix}, \quad B\begin{bmatrix} 1 & 2 \\ -1 & 1 \\ 0 & 2 \end{bmatrix}, \quad C = \begin{bmatrix} 1 \\ 2 \\ -1 \end{bmatrix} \tag{1.2}$$

illustrate matrices of orders 2 × 3 for A, 3 × 2 for B and 3 × 1 for C. Matrix C is a column matrix or a vector. The matrix

$$U = \begin{bmatrix} 1 & 0 & 0 \\ 0 & 1 & 0 \\ 0 & 0 & 1 \end{bmatrix} \tag{1.3}$$

is square, diagonal, unit, and also symmetrical.

In the next section, we shall be concerned with the *equality of matrices.* If two matrices are equal, then they have the same order and have equal corresponding elements.

1.2 ADDITION AND SUBTRACTION OF MATRICES

The *sum* of two matrices of the same order is found by adding the corresponding elements. Thus, if the elements of A are a_{ij} and of B are b_{ij} and if $C = A + B$, then

$$c_{ij} = a_{ij} + b_{ij} \tag{1.4}$$

Clearly, $A + B = B + A$ for matrices.

If a matrix A is multiplied by a constant α, then every element of A is multiplied by α:

$$\alpha A = [\alpha a_{ij}] \tag{1.5}$$

In particular, if $a = -1$, then

$$-A = [-a_{ij}] \tag{1.6}$$

From this, we see that to subtract B from A, we multiply all elements of B by -1 and add. For example,

$$A - B = \begin{bmatrix} 1 & 2 & 1 \\ -1 & 0 & 1 \end{bmatrix} - \begin{bmatrix} 2 & 1 & 1 \\ 2 & 3 & -1 \end{bmatrix} = \begin{bmatrix} -1 & 1 & 0 \\ -3 & -3 & 2 \end{bmatrix} \tag{1.7}$$

1.3 MULTIPLICATION OF MATRICES

The multiplication of matrices A and B is defined only if the number of columns of A is equal to the number of rows of B. Thus, if A is of order $m \times n$ and B is of order $n \times p$, then the product AB is a matrix C of order $m \times p$; then

$$A_{m\times n} B_{n\times p} = C_{m\times p} \tag{1.8}$$

The elements of C are found from the elements of A and B by multiplying the ith row of A and the jth column of B and summing these products to give c_{ij}. In equation form

$$c_{ij} = a_{i1}b_{1j} + a_{i2}b_{2j} + \dots + a_{ip}b_{pj} = \sum_{k=1}^{p} a_{ik}b_{kj} \tag{1.9}$$

Matrix multiplication is not commutative in general, *i.e.*,

$$AB \neq BA, \text{ usually} \tag{1.10}$$

even when BA is defined [meaning that the orders specified by equation. (1.8) are satisfied].

Example 1. The equations

$$\begin{aligned} z_{11}I_1 + z_{12}I_2 &= V_1 \\ z_{21}I_1 + z_{22}I_2 &= V_2 \end{aligned} \tag{1.11}$$

are written in the following matrix form:

$$\begin{bmatrix} z_{11} & z_{12} \\ z_{21} & z_{22} \end{bmatrix} \begin{bmatrix} I_1 \\ I_2 \end{bmatrix} = \begin{bmatrix} V_1 \\ V_2 \end{bmatrix} \tag{1.12}$$

as may be verified by applying Equation. (1.9).

Example 2. Several matrix multiplications will illustrate the use of Equation. (1.9):

$$\begin{bmatrix} 1 & 1 \\ 1 & 1 \end{bmatrix} \begin{bmatrix} 1 & -2 \\ -1 & 2 \end{bmatrix} = \begin{bmatrix} 0 & 0 \\ 0 & 0 \end{bmatrix} \tag{1.13}$$

$$\begin{bmatrix} 1 & 0 & 2 \\ 1 & -1 & 3 \end{bmatrix} \begin{bmatrix} 1 & 0 & 2 \\ -1 & 2 & -1 \\ 1 & 2 & 0 \end{bmatrix} = \begin{bmatrix} 3 & 4 & 2 \\ 5 & 4 & 3 \end{bmatrix} \tag{1.14}$$

Finally, we note that

$$UA = AU = A \tag{1.14}$$

if U *and* A are both square and of order n.

1.4 OTHER DEFINITIONS

A number of additional definitions will be used in discussing the solution of a matrix equation.

The *transpose* of a matrix A is A^T and is formed by interchanging the rows and columns of A. For example, if

$$A = \begin{bmatrix} 1 & 2 & -1 \\ 0 & -1 & 1 \end{bmatrix}, \text{ then } A^T = \begin{bmatrix} 1 & 0 \\ 2 & -1 \\ -1 & 1 \end{bmatrix} \tag{1.16}$$

The *determinant* of a square matrix has elements which are the elements of the matrix. In other words,

$$\det A = \det [a_{ij}] = [a_{ij}] \tag{1.17}$$

From example,

$$\det\begin{bmatrix}1 & -1\\ 3 & 2\end{bmatrix} = \begin{bmatrix}1 & -1\\ 3 & 2\end{bmatrix} = 5 \tag{1.18}$$

The cofactor A_{ij} of the element of a matrix a_{ij} is defined for a square matrix A. It has the value $(-1)^{i+j}$ times the determinant formed by deleting the ith row and the jth column in det A.

The *adjoint matrix* of a square matrix A is formed by replacing each element a_{ij} by the cofactor A_{ij} and transposing. Thus

$$\text{adjoint of } A = [a_{ij}]^T \tag{1.19}$$

Finally, we define the *inverse matrix* of A, A^{-1}, as the adjoint matrix divided by the determinant of A:

$$A^{-1} = \frac{\text{adjoint of } A}{\det A}, \qquad \det A \neq 0 \tag{1.20}$$

Example 3. Let it be required to determine A^{-1} given

$$A = \begin{bmatrix}1 & 2\\ -1 & 1\end{bmatrix} \tag{1.21}$$

Note that

$$\det A = 3, \quad [A_{ij}] = \begin{bmatrix}1 & 2\\ -1 & 1\end{bmatrix}, \quad [A_{ij}]^T = \begin{bmatrix}1 & -2\\ 1 & 1\end{bmatrix} \tag{1.22}$$

Hence $$A^{-1} = \frac{1}{3}\begin{bmatrix}1 & -2\\ 1 & 1\end{bmatrix} = \begin{bmatrix}\frac{1}{3} & -\frac{1}{3}\\ \frac{1}{3} & \frac{1}{3}\end{bmatrix} \tag{B-23}$$

1.5 MATRIX SOLUTION OF SIMULTANEOUS LINEAR EQUATIONS

In application to network analysis, we are concerned with

$$ZI = V \tag{1.24}$$

where Z is a square matrix of order n, I is a column matrix with elements $I_1, I_2, ..., I_n$, and V is a column matrix of elements $V_1, V_2, ..., V_n$. It was shown in Chapter 6 that I_j may be found using Cramer's rule as

$$I_j = \frac{1}{\det Z}\sum_{i=1}^{n} V_i Z_{ij} \tag{1.25}$$

where Z_{ij} is the cofactor of z_{ij} in det Z.

To obtain this same result using matrices, we multiply equation. (1.24) by Z^{-1} and make use of the relationship $Z^{-1}Z = U$. Thus

$$Z^{-1}ZI = UI = I \tag{1.26}$$

and the solution of equation (1.24) is

$$I = Z\text{–}1V \tag{1.27}$$

This may also be written in the form

$$I = \frac{\text{Adjoint of } Z}{\det Z} V \tag{1.28}$$

where it seems to be in the same form as equation (1.25) but generalized for all I.

Example 4. A network of resistors and batteries is described by the Kirchhoff voltage equations written in matrix form

$$\begin{bmatrix} 2 & -1 \\ -1 & 3 \end{bmatrix} \begin{bmatrix} I_1 \\ I_2 \end{bmatrix} = \begin{bmatrix} 2 \\ 1 \end{bmatrix} \tag{1.29}$$

As in Example 3, we find that

$$Z^{-1} = \frac{1}{5} \begin{bmatrix} 3 & 1 \\ 1 & 2 \end{bmatrix} \tag{1.30}$$

so that by equation. (1.27),

$$\begin{bmatrix} I_1 \\ I_2 \end{bmatrix} = \frac{1}{5} \begin{bmatrix} 3 & 1 \\ 1 & 2 \end{bmatrix} \begin{bmatrix} 2 \\ 1 \end{bmatrix} = \begin{bmatrix} \frac{7}{5} \\ \frac{4}{5} \end{bmatrix} \tag{1.31}$$

In the analysis of networks, especially by the computer, other methods are employed rather than the evaluation of Z^{-1}. In 1.7, we studied the Gauss elimination method, and showed how to obtain the solution of a matrix equation by successive steps of traingularization and back substitution. The result is that the matrix equation

$$ZI = V \tag{1.32}$$

is reduced to the form

$$\begin{bmatrix} c_{11} & & & 0 \\ & c_{22} & & \\ & & c_{33} & \\ 0 & & & c_{nn} \end{bmatrix} \begin{bmatrix} I_1 \\ I_2 \\ I_3 \\ \vdots \\ I_n \end{bmatrix} = \begin{bmatrix} k_1V_1 \\ k_2V_2 \\ k_3V_3 \\ \vdots \\ k_nV_n \end{bmatrix} \tag{1.33}$$

where the square matrix has non-zero terms only on the principal diagonal. The solution is, directly,

$$I_1 = \frac{k_1}{c_{11}} V_1 \tag{1.34}$$

or, in general,

$$I_1 = \frac{k_1}{c_{11}} V_i, \quad i = 1, \ldots, n \tag{1.35}$$

A basic operation in the computer solution of matrix equations like equation. (1.32) is *Lu factorization* where L is a square matrix with non-zero elements on the below the principal diagonal, and U is a square matrix with non-zero elements on the above the principal diagonal, each matrix being in $n \times n$, and

$$LU = Z \tag{1.36}$$

Substituting this equation into equation. (1.32) we have

$$LUI = V \tag{1.37}$$

If we identify $UI = y$, then we may solve

$$Ly = V \tag{1.38}$$

by forward Gauss elimination for y. Once y is known, we solve

$$UI = y \tag{1.39}$$

for I by the same straightforward Gauss elimination procedure. The advantage of this method is that if V changes in equation (1.32), then Z does not have to be refactored.

Finally, the computation of L and U is straightforward. If we let the elements of Z be z_{ij}, then

$$\begin{bmatrix} z_{11} & z_{12} & z_{13} \\ z_{21} & z_{22} & z_{23} \\ z_{33} & z_{32} & z_{33} \end{bmatrix} = \begin{bmatrix} l_{11} & 0 & 0 \\ l_{21} & l_{22} & 0 \\ l_{31} & l_{32} & l_{33} \end{bmatrix} \begin{bmatrix} 1 & u_{12} & u_{13} \\ 0 & 1 & u_{23} \\ 0 & 0 & 1 \end{bmatrix} \tag{1.40}$$

$$= \begin{bmatrix} l_{11} & l_{11}u_{12} & l_{11}u_{13} \\ l_{21} & l_{21}u_{12} + l_{22} & l_{21}u_{13} + l_{22}u_{23} \\ l_{31} & l_{31}u_{12} + l_{32} & l_{31}u_{13} + l_{32}u_{23} + l_{33} \end{bmatrix} \tag{1.41}$$

The solution of these equations for the l's and u's is routine.s

For further reading on this subject, including a consideration of special techniques that may be employed when Z is *sparse,* see Donald A. Calahan, *Computer-Aided Network Design*, McGraw-Hill Book Co., New York, 1972, or George Forsythe and Cleve B. Moler, *Computer Solutions of Linear Algebraic Systems,* Prentice-Hall, Inc., Englewood Cliffs, N.J., 1967.

Index

□□□